Emerging Technologies for Sustainable Desalination Handbook

Emerging Technologies for Sustainable Desalination Handbook

Veera Gnaneswar Gude

Butterworth-Heinemann is an imprint of Elsevier
The Boulevard, Langford Lane, Kidlington, Oxford OX5 1GB, United Kingdom
50 Hampshire Street, 5th Floor, Cambridge, MA 02139, United States

Library of Congress Cataloging-in-Publication Data
A catalog record for this book is available from the Library of Congress

British Library Cataloguing-in-Publication Data
A catalogue record for this book is available from the British Library

ISBN: 978-0-12-815818-0

For information on all Butterworth-Heinemann publications
visit our website at https://www.elsevier.com/books-and-journals

Publisher: Mathew Deans
Acquisition Editor: Ken McCombs
Editorial Project Manager: Peter Jardim
Production Project Manager: Surya Narayanan Jayachandran
Cover Designer: Miles Hitchen

Typeset by SPi Global, India

Dedication

To my late father Gude Bhaskar Rao, to my mother Gude Fathima,
to my sister Kalyani, to my brothers Kiran and Karunakar,
and to Preeti Muire.

Contents

Contributors

Abimola Adebesi Nanotechnology and Catalysis Research Center (NANOCAT), Kuala Lumpur, Malaysia

Aliakbar Akbarzadeh RMIT University, Melbourne, VIC, Australia

Mohammad Amin Alaei Shahmirzadi Amirkabir University of Technology (Tehran Polytechnic), Tehran, Iran

Adnan A. Alanezi The Public Authority for Applied Education and Training (PAAET), Sabah Alsalem, Kuwait

Md. E. Ali Nanotechnology and Catalysis Research Center (NANOCAT), Kuala Lumpur, Malaysia

Abdullah Alkhudhiri King Abdulaziz City for Science and Technology (KACST), Riyadh, Saudi Arabia

Ali Altaee University of Technology Sydney, Sydney, NSW, Australia

Li Ang King Abdullah University of Science & Technology (KAUST), Thuwal, Saudi Arabia

Brian Bolto CSIRO Manufacturing, Clayton South, VIC, Australia

Muhammad Burhan King Abdullah University of Science & Technology (KAUST), Thuwal, Saudi Arabia

José A. Caballero University of Alicante, Alicante, Spain

Qian Chen National University of Singapore, Singapore, Singapore

Zaira Z. Chowdhury Nanotechnology and Catalysis Research Center (NANOCAT), Kuala Lumpur, Malaysia

Kian J. Chua National University of Singapore, Singapore, Singapore

Abhijit Date RMIT University, Melbourne, VIC, Australia

Saeed Dehghani RMIT University, Melbourne, VIC, Australia

Eric S. Fraga University College London, London, United Kingdom

Manuel J. González-Ortega Technical University of Cartagena, Cartagena, Spain

Alaa H. Hawari Qatar University, Doha, Qatar

Nidal Hilal Swansea University, Swansea, United Kingdom

Md. S. Islam Royal Melbourne Institute of Technology (RMIT), Melbourne, VIC, Australia

Rafie B. Johan Nanotechnology and Catalysis Research Center (NANOCAT), Kuala Lumpur, Malaysia

Ali Kargari Amirkabir University of Technology (Tehran Polytechnic), Tehran, Iran

Kum J. M National University of Singapore, Singapore, Singapore

Na Li Xi'an Jiaotong University, Xi'an, China

Jose F. Maestre-Valero Technical University of Cartagena, Cartagena, Spain

Farzaneh Mahmoudi RMIT University, Melbourne, VIC, Australia

Victoriano Martínez-Alvarez Technical University of Cartagena, Cartagena, Spain

Bernardo Martin-Gorriz Technical University of Cartagena, Cartagena, Spain

Khaled Moustafa Conservatoire National des Arts et Métiers, Paris, France

Peter Nasr American University in Cairo, Cairo, Egypt

Kim C. Ng King Abdullah University of Science & Technology (KAUST), Thuwal, Saudi Arabia

Viviani C. Onishi University of Coimbra, Coimbra, Portugal

Kaushik Pal Wuhan University, Wuchang, PR China

Rahman F. Rafique The State University of New Jersey, New Brunswick, NJ, United States

Juan A. Reyes-Labarta University of Alicante, Alicante, Spain

Suresh Sagadevan AMET University, Chennai, India

Hani Sewilam RWTH Aachen University, Aachen, Germany

Syed T. Shah Nanotechnology and Catalysis Research Center (NANOCAT), Kuala Lumpur, Malaysia

Muhammad W. Shahzad King Abdullah University of Science & Technology (KAUST), Thuwal, Saudi Arabia

Mariano Soto-García Technical University of Cartagena, Cartagena, Spain

Qinzhuo Wang CSIRO Manufacturing, Clayton South, VIC, Australia; Xi'an Jiaotong University, Xi'an, China

Zongli Xie CSIRO Manufacturing, Clayton South, VIC, Australia

Wageeh A. Yehye Nanotechnology and Catalysis Research Center (NANOCAT), Kuala Lumpur, Malaysia

Sahar Zare Amirkabir University of Technology (Tehran Polytechnic), Tehran, Iran

Domingo Zarzo Valoriza Agua, Madrid, Spain

Biography

Veera Gnaneswar Gude is an associate professor of Civil and Environmental Engineering at Mississippi State University (MSU). He has more than 15 years of academic, research and industrial experience in desalination, water-wastewater treatment, biofuel scientific, and technological areas. He received a BS degree in chemical engineering technology from Osmania University in 2000 and worked for Du Pont Singapore after his graduation from 2000 to 2004. He received an MS degree in environmental engineering from the National University of Singapore in 2004 and a PhD degree in environmental engineering from the New Mexico State University in 2007, under the direction of Prof. Nagamany Nirmalakhandan, for his research in low-temperature thermal desalination. He expanded his research interests into water and wastewater treatment, microbial desalination, biofuel synthesis using sustainable chemistry principles and process intensification topics during his postdoctoral research, industrial and academic appointments. He is a licensed professional engineer and a board certified environmental engineer by the American Academy of Environmental Engineers and Scientists (AAEES).

Dr. Gude has published more than 70 scientific research articles on desalination (thermal, membrane, hybrid, and microbial), water-wastewater treatment, and on biofuels research in well-regarded journals. He has published 3 books on desalination research (Elsevier) and 2 books on biofuels research (CRC Press), 15 invited book chapters, 50 conference proceedings papers, 15 technical reports, several popular press articles and media releases, and 2 patents on low-temperature desalination and microalgae biofuels technologies. He has delivered 35 invited lectures including 6 plenary/keynote lectures and more than 140 scientific research and educational presentations. He has organized workshops on water-energy-environment nexus topics at national and international conferences. He was the chair and board representative for clean energy and water division of ASES between 2011 and 2016. He serves on numerous scientific advisory boards and task committees across the world including ASCE-EWRI, ASEE, ASES, and AWWA. He is a member of several editorial boards and editor for many scientific journals including *ASCE Journal of Environmental Engineering*, *Heliyon*, *Nature npj Clean Water*, *Renewable Energy*, and *Water Environment Research Journal*. His research is supported by NSF, USEPA, USGS, USDA, and by other industrial and international agencies. He has received much recognition for his research, education, and service activities at regional, national, and international conferences and from professional societies

Preface

Desalination industry has advanced in recent years due to the increasing global demands for freshwater associated with population growth. Membrane and thermal desalination processes and technologies have been widely implemented in many parts of the world. Despite more than 50 years of operational and process management experiences, the energy costs still prohibit desalination applications in many areas of critical need for freshwater. On the other hand, due to global water scarcity, non-conventional water sources such as reclaimed water, produced waters, and other industrial wastewaters are being considered for desalination technology applications. Often these water sources contain pollutants that are problematic to the performance of desalination processes. At the same time, the freshwater needs for irrigation continue to grow in arid regions for food production. This also requires careful consideration of water production costs from desalination plants. In addition, environmental pollution related to desalination technology implementation should be minimized. In view of all these concurrent issues, there has been a growing interest among the researchers and industrialists to develop energy-efficient and cost-effective hybrid or novel desalination processes in recent decades. These processes can be developed by combining the principles of physical separation and/or thermal evaporation to serve as standalone or pretreatment processes for desalination and zero-liquid discharge technologies.

Many novel processes were developed in recent years and notable advances have been made in these technologies. These include adsorption desalination, forward osmosis, humidification and dehumidification, membrane distillation, pervaporation, and spray-type thermal processes. In addition, novel membrane materials such as nanocomposite and carbon nanotube membranes were also explored. Each of these processes or membrane materials carries its own set of benefits and drawbacks. Equally important are the desalination technology applications in agriculture, crop irrigation, concentrate or brine valorization, and shale gas-produced water treatment. This book covers the recent advances in each of these technological areas and provides directions for their future development and implementation for various beneficial uses. This book is intended to serve as a resource and guidance manual for understanding the principles, mechanisms, design, and successful implementation of desalination technologies for undergraduate and graduate researchers, science and engineering professionals, and industrial practitioners. This book is a critical resource for researchers from chemical engineering and science, environmental science and engineering, mechanical and electrical engineering, and industrial and process engineering, and management disciplines with a desire to understand the science, engineering, and process management related to sustainable desalination.

This book is presented in two parts and 15 chapters. The first part presents the principles, design, process performance, and recent advances in emerging desalination technologies while the second part discusses the implementation of emerging and conventional desalination processes in various beneficial applications. Details of the chapters are presented next. These chapters are contributed by leading authors and researchers from the relevant fields. The complete list of chapters is shown below.

Chapter No.	Title
Part I: Emerging Technologies	
Chapter 1	Adsorption Desalination
Chapter 2	Forward Osmosis
Chapter 3	Membrane Distillation—Principles and Recent Advances
Chapter 4	Membrane Properties in Membrane Distillation
Chapter 5	Permeate Gap MD and Solar Applications
Chapter 6	Desalination by Pervaporation
Chapter 7	Humidification and Dehumidification
Chapter 8	Spray Low-temperature Desalination
Chapter 9	Electrochemically Active Carbon Nanotube Membranes
Part II: Recent Trends and Applications	
Chapter 10	Carbon Nanotube Membranes
Chapter 11	Beneficial Uses and Valorization of RO Brines
Chapter 12	Desalination of Shale Gas-produced Water
Chapter 13	Forward Osmosis for Agriculture
Chapter 14	Desalination for Crop Irrigation
Chapter 15	Seawater for Sustainable Agriculture

Chapter 1

Considering the issues at water-energy-environment nexus, innovative solutions are required to lower specific energy consumption by utilizing low-grade heat of renewable and nonrenewable sources. This chapter presents the adsorption desalination (AD) process as an innovative solution to overcome some of the performance limitations faced by conventional thermal desalination processes such as MED (multieffect distillation) and MSF (multistage flash evaporation). In addition, hybridization of MED and AD processes called MEDAD to improve the process thermodynamic efficiency and product output is discussed. The basics of adsorption phenomenon, AD cycle, and its hybrid configurations and desalination processes and their economics are presented in detail.

Chapter 2

This chapter elaborates the principles and applications of the forward osmosis (FO) process. The use of the FO process as a pretreatment technique for seawater reverse osmosis (RO) process is discussed. The effect of water recovery rate on the specific energy consumption is presented. Membrane fouling and scaling and other limiting factors for FO performance are discussed. The critical element of FO, that is, draw solution and its selection is also discussed. Membrane types and modeling studies are presented along with an analysis of specific energy consumption in FO-RO hybrid systems at different fouling factors and water recovery rates. The effect of FO pretreatment on RO desalinated water costs and practical implications for implementation are discussed in detail.

Chapter 3

This chapter provides an overview of the membrane distillation (MD) process which combines both evaporation and physical separation principles of thermal and membrane desalination processes.

A comprehensive review of MD process principles, history of development, design and process configurations and modules, membrane materials and properties, and membrane characterization are presented. The transport phenomena related to mass and heat transfer for different MD types are also presented with different models used in the literature. Correlations to estimate the heat and mass transfer coefficients are presented. Concentration polarization and fouling issues are discussed. Operating parameters, process performance in terms of thermal energy efficiency, specific energy consumption, water production costs, and gained output ratios are discussed in detail. MD applications as standalone or hybrid processes are also covered in detail with process performance-related information.

Chapter 4

One of the main reasons that has limited industrialization of MD for water desalination application is the lack of a proper membrane that could exhibit long-term acceptable permeation flux and salt rejection. This chapter explains the required specifications of a membrane for its application in the MD process. Material selection, fabrication, and characterization methods (such as scanning electron microscopy, atomic force microscopy, contact angle, and other methods) are presented. Membrane fabrication techniques, characteristic evaluation (liquid entry pressure, porosity, tortuosity, pore size distribution, and conductivity), and modification methods (surface coating, surface grafting, plasma polymerization, and surface-modified molecules) are discussed. The range of experimental design parameters that affect the process performance are provided in each section. The effect of each process parameter on MD performance is described. Then, the modification methods for improving the performance of the current available membranes will be presented. Next, the economics of MD will be discussed, and finally a conclusion containing the future trend and perspective of MD will be given.

Chapter 5

This chapter presents the permeate gap membrane distillation (PGMD) process as a novel and sustainable MD design having internal heat recovery characteristics. Transport phenomena including numerical modeling of the heat and mass transfer and hydraulics are discussed. Experimental projects at laboratory- and pilot-scale studies are discussed to elaborate the performance of the process in terms of energy efficiency and water production. A detailed account of the literature survey based on recent case studies involving MD integrated with different sources of solar energy and those driven by waste heat is included. A general techno-economic feasibility framework of PGMD systems is also presented.

Chapter 6

This chapter elaborates on desalination by pervaporation principle, its transport mechanism, process design and operation, and techno-economic analysis, with main focus on membrane development including polymers, inorganic materials, and their hybrids. Most common pervaporation membranes such as hybrid organic-inorganic membranes, ZSM-5, cellulose membranes, silica, ionic polyethylene, and various polyether membranes and GO/PAN membranes are discussed. The critical parameters affecting the process performance such as feed temperature and permeate vapor pressure are discussed. Pervaporation technology and the urgent problems to be resolved are emphasized and future trends are discussed. The need for the development of novel membranes and membrane materials is discussed. The possibility of integrating pervaporation with other existing processes such as RO and suggestions for improvement of pervaporation membranes and module design are also proposed.

Chapter 7

This chapter discusses the humidification-dehumidification desalination (HDH) working principle with different process configurations (closed air open water-water heated cycles, multieffect closed air open water-water heated cycles, closed air open water-air heated cycles, closed water open air-water heated cycles, closed water open air-air heated cycles). Theoretical analysis (mass and energy balances), humidifier and dehumidifier and their effectiveness relationships are presented in detail. Heat and mass transfer relationships and efficiency factors are discussed. The effect of process parameters on its performance such as heat capacity ratio and gained output ratio and process control techniques are discussed. Various heat sources, materials, applications, and water desalination costs are provided.

Chapter 8

This chapter deals with spray-assisted low-temperature desalination technology. It applies direct contact heat and mass transfer mechanism for both evaporation and condensation. The working principle of the spray evaporator is first described, and a heat and mass transfer model based on detailed droplet analysis is developed to enable precise performance prediction. A simple spray desalination system is then presented, and the major sources of energy consumption are judiciously quantified. Next, a multistage spray-assisted low-temperature desalination system is introduced. A detailed thermodynamic model is also developed to highlight the transport phenomena within the multistage system. Using this model, the production rate and thermal efficiency of the multistage spray-assisted low-temperature desalination system is analyzed and optimized under different operating parameters.

Chapter 9

Recent breakthroughs in nanostructured materials such as carbon nanotubes (CNTs) and porous graphene membranes for water purification and desalination applications are covered in this chapter. More specifically, the potential of the electrochemically active carbon nanotube (CNTs) membranes is discussed in detail. The unique characteristics of the CNT membrane to adsorb chemical and biological contaminants as well as ion separation from seawater due to their high stability, great flexibility, and large specific surface area has been disclosed. Electrochemically active CNT filters for electro-oxidation of the adsorbed contaminants are presented. The advantages of CNT-based filters for salt separation and bacteria removal are discussed. This chapter provides an explicit and systematic overview of the recent progress of electrochemically active CNT membranes addressing the current prevalent problems associated with water treatment and desalination. The physiochemical aspects including the working principles of this type of membrane have been discussed. The prevailing challenges and future perceptions are also discussed.

Chapter 10

This chapter focuses on the historical and current development of the nanocomposite membrane, which has received growing attention in the field of desalination methods. The incorporation of nanomaterials into the polymer matrix to overcome current drawbacks, including trade-off between flux and salt rejection, and low resistance to fouling and scaling has been discussed. Various types of nanocomposite membranes based on the membrane structure and location of the nanomaterial such as conventional nanocomposites, thin-film nanocomposites, thin-film composites with the nanocomposite substrate, and surface-located nanocomposites are presented.

This chapter also attempts to give the readers insights into the fabrication methods and challenges associated with new functionalities introduced by nanomaterials and their influence on membrane properties. It covers the feasibility and future prospective for nanocomposite membranes with desirable performance.

Chapter 11

Brine or concentrate discharge and management have become a challenge for the implementation of desalination plants. However, brines or concentrate streams can provide numerous opportunities for energy and resource recovery along with beneficial environmental applications.

In this chapter, brine characteristics including chemical and heavy metal composition in different sources are presented. Management strategies such as evaporation ponds, land application, deep well injection, and discharge to sewer networks are discussed. Various valorization methods are presented. The use of desalination brine within the desalination plants, valuable mineral recovery methods, evaporation crystallization, solar ponds, and salt production options are discussed. Zero-liquid discharge options are also covered. The possibility of energy recovery and energy production using brine streams is presented. Environmental applications of brine streams for aquaculture and fish farming, microalgae cultivation, and agriculture have been presented.

Chapter 12

Natural gas exploration from unconventional shale formations, known as "shale gas," is produced recently by the development of horizontal drilling and hydraulic fracturing ("fracking") technologies. In this chapter, the capability of desalination technologies to allow water recycling and/or water reuse in shale gas industry for their sustainable operations is discussed. Water consumption and wastewater generation in shale gas operations and its management options are presented. The challenges of shale gas wastewater desalination and management options are discussed. Advances in zero-liquid discharge (ZLD) desalination processes for treating hypersaline shale gas wastewater to mitigate public health and environmental impacts, and to improve the overall process sustainability are discussed. This chapter outlines the most promising thermal- and membrane-based alternatives for ZLD desalination of shale gas wastewater.

Chapter 13

In this chapter, the forward osmosis process is presented as an energy-efficient desalination technique for providing water for irrigation.. The basic scheme of fertilizer-drawn forward osmosis (FDFO) process as an innovative application of forward osmosis (FO) desalination process is presented.

The FO principles, draw solution characteristics, and selection criteria are discussed. Advantages (energy efficiency and fertilized irrigation) and disadvantages (need for membranes, choice of fertilizer drawn solution, lower water flux and salt rejection issues, fouling and biofouling) are discussed. Pretreatment and posttreatment options for meeting the irrigation water quality standards are discussed. Several options are proposed to improve the process performance to supply irrigation water, minimize soil salinity, control fertilizer application, and to close the irrigation-brackish water-drainage vicious cycle.

Chapter 14

Desalinated seawater (DSW) is considered a stable water source which effectively removes the hydrological constraints for crop production under arid and semiarid conditions. This chapter reviews current irrigation experiences with DSW worldwide, analyzing the key issues for its successful implementation, including the main agronomic concerns, such as low-nutrient concentration, crop toxicity risk due to high boron and chloride concentration, or the sodicity risk affecting the physical properties of soil; the energy requirements for DSW production and allocation, the associated greenhouse gas emissions and the derived cost, as the current limiting factors for its agricultural application; and future research for promoting its sustainability and its development perspectives. In addition, two case studies involving planning and development of DSW supply in semiarid southeastern Spain and blending strategies for optimizing the DSW use together with other available water resources at farm scale, respectively, are presented.

Chapter 15

Water scarcity, desertification, and endangered biodiversity in barren lands are strong incentives to develop genuine solutions and resilient policies toward greening drylands and turning them into more viable ecosystems for people, native animals, and plant species. This chapter presents a few, possibly sustainable methods based on seawater and solar energy to increase agriculture surfaces and to enhance food production systems in dry lands that suffer from water scarcity and food insecurity issues. Solar desert houses built on the land surface or floating farms built on the seawater surface are proposed. The prospective outcomes from both approaches (on land and on the sea surface) to help provide new sources for food production and to restore endangered biodiversity in dry regions are discussed in detail with potential challenges and implementation issues.

Acknowledgments

I would like to express my appreciation to the colleagues in the Civil and Environmental Engineering Department, the Bagley College of Engineering, and the Office of Research and Economic Development at Mississippi State University (MSU) for their support. My sincere appreciation and gratitude are extended to my doctoral and postdoctoral advisors Dr. Nagamany Nirmalakhandan, Dr. Ricardo Jacquez, Dr. Shuguang Deng, and Dr. Adrian Hanson of New Mexico State University (NMSU) for their continuous support in my professional career. Prof. Khandan is gratefully acknowledged for introducing me to desalination research during my doctoral research studies at NMSU. His encouragement and guidance had tremendous impact on my professional development. Colleagues at the Southwest Technology Development Institute of NMSU, especially Corey Asbill and Andrew Rosenthal and Robert Keith Payne of Engineering Technology Department are thanked for their support in my research while at NMSU. Dr. Craig Ricketts is thanked for sharing his workplace at the engineering research facility. Industrial collaborations with Thomas Leggiore, Charles Oborne, and George Forbes are also acknowledged. My graduate research students at Mississippi State University and New Mexico State University are appreciated for their diligence and energy in the work environment and for their research contributions, especially Dr. Bahareh Kokabian and Dr. Edith Martinez-Guerra of MSU for their work on microbial desalination and solar desalination research projects, respectively. Finally, I thank Ms. Preeti Muire for her support and love.

Veera Gnaneswar Gude

Part One

Emerging technologies

Adsorption desalination—Principles, process design, and its hybrids for future sustainable desalination

1

Muhammad W. Shahzad, Muhammad Burhan, Li Ang, Kim C. Ng
King Abdullah University of Science & Technology (KAUST), Thuwal, Saudi Arabia

Nomenclature

Symbols	
P	pressure (kPa)
P_0	saturation pressure (kPa)
R	universal gas constant (kJ/kmol K)
T	temperature (K)
T	surface heterogeneity for Tóth
Greek symbols	
Γ	Dubinin-Astakhow power factor
E	adsorption site energy (kJ/kmol or kJ/kg)
ε_c	equilibrium adsorption site energy (kJ/kg)
ε_1 to ε_3	reference energy (kJ/kmol)
θ	adsorption uptake (kg/kg of adsorbent)
$\tilde{\theta}$	local adsorption uptake (kg/kg of adsorbent)
M	chemical potential (kJ/kmol)
$\chi(\varepsilon)$	energy distribution function
Subscript	
A	adsorbed phase
G	gaseous phase

1.1 Introduction

Water and energy are two major needs for life continuation. Presently, about 36% of global population (2.5 billion people) live in water-scarce regions, yet contributing to 22% of the world's GDP (9.4 trillion USD) [1]. The global water demand (in terms of water withdrawals) is projected to increase by 55% by 2050, mainly because of growing demands from manufacturing (400%), thermal electricity generation (140%), and domestic use (130%). As a result, freshwater availability will be increasingly strained over this time period, and more than 40% of the global population is projected to be living in areas of severe water stress through 2050 [2].

Emerging Technologies for Sustainable Desalination Handbook. https://doi.org/10.1016/B978-0-12-815818-0.00001-1

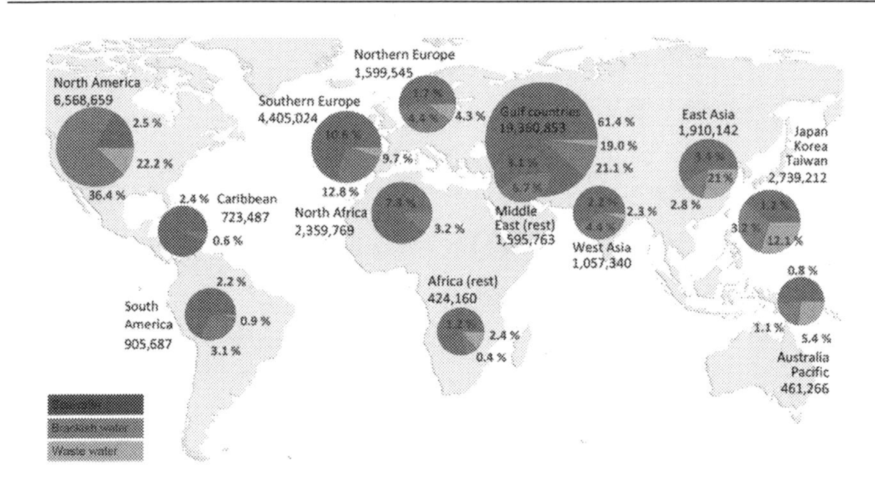

Fig. 1.1 Desalination capacities installed in the world and percent share on the basis of feed water type [3].

Presently, more than 18,000 desalination plants in 150 countries produce ~38 billion m^3 per year as shown in Fig. 1.1 [4–6]. It is projected to increase to 54 billion m^3 per year by 2030, 40% more compared with 2016 [7–9]. Desalination is the most energy-intensive water treatment process that consumes 75.2 TWh per year, about 0.4% of global electricity [10]. Fossil-fuel-operated energy-intensive desalination processes are the major source of CO_2 emission. Presently, globally installed desalination capacities are contributing 76 million tons (Mt) of CO_2 per year and it is expected to grow to 218 million tons of CO_2 per year by 2040 [11,12]. In 2019, the global CO_2 emission is estimated to grow to 43.2 Gt per year, 20% higher than the 2013 value of 36.1 Gt per year [13]. Two-thirds share of the CO_2 emission for the COP21 goal, maintaining the environment temperature increase to below 2°C, has already been used and the remaining will be exhausted by 2050 [13–17].

Fig. 1.2 shows the percentage increase of world water withdrawals and consumption, population and CO_2 emission from 1900. It can be seen that the CO_2 emission is more than 1500% and it is expected to grow to 2200% by 2040 [3]. Similarly, water withdrawals and consumption also increased to more than 1000%. The increase in primary energy consumption is also plotted (1970 baseline at 100%) and it is expected to grow to 500% by 2040 [3,18,19].

In the Gulf Cooperation Council (GCC) countries, a similar water scarcity situation is predicted with the water stress level deemed as severe primarily due to the low rainfall, high evaporation rates, depletion of nonrenewable subsurface water, high industrial and population growth rates. As shown in Fig. 1.3, the annual per capita (APC) renewable water availability in the Gulf countries has decreased rapidly between 1962 and 2010. The reported APC of renewable water varied from a low value of 7.3 m^3 in Kuwait, 20 m^3 in UAE, 33 m^3 in Qatar, 92 m^3 in Bahrain to a higher value of 503 m^3 of Oman [20,21]. The rapid shrinkage of renewable water availability can be attributed to three factors, namely, an increase in industrial and agricultural activities, as well as the

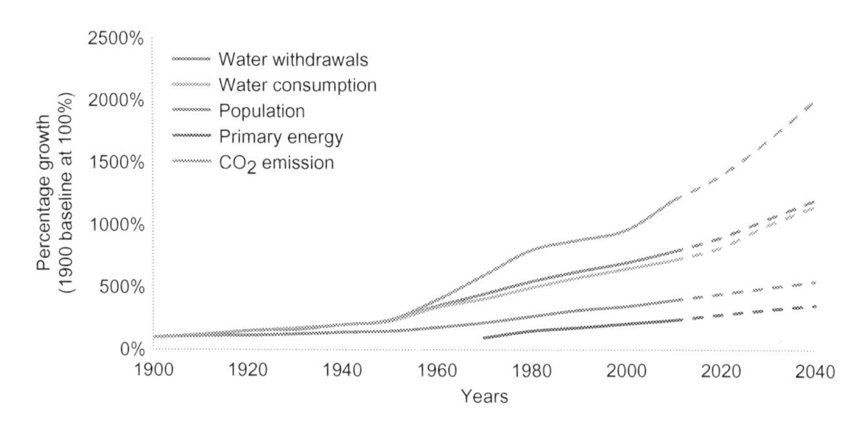

Fig. 1.2 Percentage growth rate of life necessity from 1900 to 2040. Water, population and CO_2 emission: 1900 baseline at 100%, primary energy: 1970 baseline at 100%.

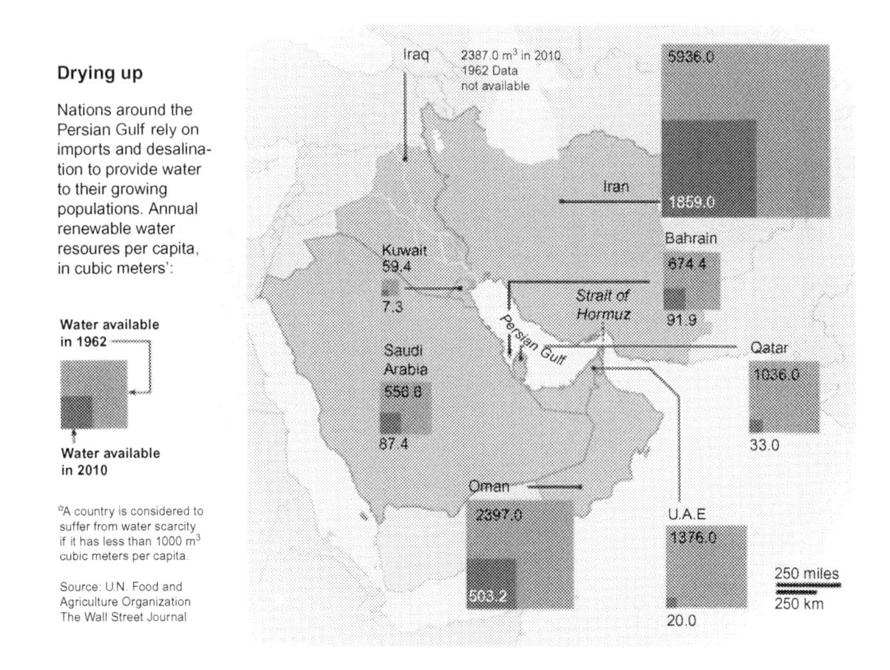

Fig. 1.3 Annual per capita renewable water resource in GCC countries between 1962 and 2010.

exponential growth rate of local population. It is estimated that the water demand in most of the GCC countries will be tripled by 2020 from the 2000 level as shown in Table 1.1 [22].

The currently available water sources are unable to fill the gap between the water supply-demand of most economies in the GCC. To maintain the same economic growth rates in the region, the only source for freshwater supply in the future is by

Table 1.1 Projected water demand in selected GCC countries from 2000 to 2020 [22] (millions of imperial gallons)

Country	2000	2010	2020
Saudi Arabia	170,476	246,065	373,444
Bahrain	27,930	43,181	43,181
Qatar	32,303	56,222	104,780
Dubai	41,354	98,178	155,109

seawater desalination. In the last four decades, the GCC countries have implemented different desalination technologies including: (1) thermal technologies such as multistage flash (MSF) and multiple-effect distillation (MED); and (2) membrane technologies such as Sea Water Reverse Osmosis (SWRO).

Thermally driven MSF operates at a top-brine temperature (TBT) of 120°C and consumes 191–290 MJ/m^3 thermal energy and 2.5–5.0 kWh_{elec}/m^3 electric energy. Typically, MSF can have 40 number of flashing stages and GOR in the range of 8–10. Their CO_2 emission varies from 20 to 25 kg/m^3 as a standalone operation to 14–16 kg/m^3 as a cogeneration operation with the steam power plant. On the other hand, the MED operating temperature varies from 65°C to 70°C and they consume 140–230 MJ/m^3 thermal energy and 2.0–2.5 kWh_{elec}/m^3 electric energy. The MED effects varies from 8 to 10 and GOR from 12 to 15 with TVC operation. Their CO_2 emission varies from 12 to 19 kg/m^3 as a standalone operation to 8–9 kg/m^3 as a cogeneration operation with steam power plants [23–32]. The RO processes only required electricity for desalination and the energy consumption is dependent on the recovery ratio and total dissolved solids (TDS) in the feed since the osmotic pressure is related to TDS. Today's SWRO processes required from 3 to 8 kWh_{elec}/m^3 (55–82 bar pump pressure) for seawater and 1.5–2.5 kWh_{elec}/m^3 for brackish water for large to medium sized plants. Since the RO-specific energy consumption is the lowest among all desalination technologies, it causes the lowest CO_2 emission, 2.79 kg/m^3 in steam cycle operation, and 1.75 kg/m^3 in combined CCGT power plants [33–39].

Despite the fact that the reverse osmosis (RO) technology is the world leading technology due to its low energy intensity, the thermal desalination (MSF and MED) are still dominating in the GCC and account for 68% of the total desalination units in 2010 leaving the rest 32% for RO [40,41]. All these technologies are not only energy intensive but also not environment friendly. Fig. 1.4 shows that the annual electricity demand in the GCC will increase from 467 TWh in 2011 to about 1400 TWh in 2030 and reaching about 2000 TWh in 2040. At the same time, the demand for freshwater in the GCC will grow significantly from 6470 GL in 2011 to 16,000 GL in 2030 and 21,000 GL in 2040 as shown in Fig. 1.4 [42].

From the viewpoint of water-energy-environment nexus, innovative solutions are required to address three major issues; (i) lowering specific energy consumption in terms of kWh/m^3 (ii) utilization of low-temperature heat of renewable and nonrenewable sources, and (iii) lowering of pollutants in brine discharge as well as carbon dioxide emission of burned fossil fuels.

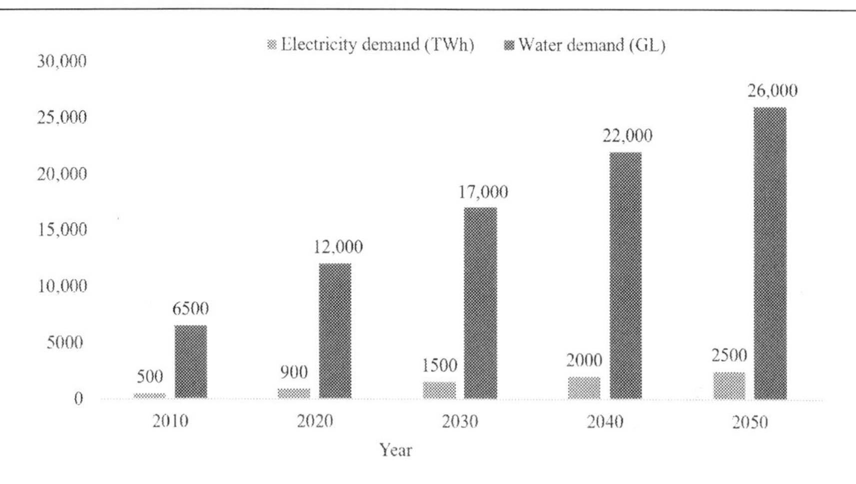

Fig. 1.4 GCC countries electricity demand trend from 2011 to 2050 (TWh).

The adsorption (AD) cycle is an innovative solution that can operate using low-temperature waste heat and can produce water. Extensive studies have been conducted to investigate adsorption isotherms of various pairs [43–52]. The hybridization of both cycles, MED and AD called MED-AD, extend the range of downstream (last stage) temperature of conventional MED system typically from 40°C to 5°C. The additional number of stages enhances the water production of the MED cycle by two- to threefold at the same top-brine temperatures (TBTs). In addition, AD integration cycle has the following advantages: (i) it increases interstage temperature differential of each MED stage due to the lowering of the bottom-brine temperature; (ii) the AD cycle utilizes only low-temperature waste heat; (iii) it has almost no major moving parts; (iv) it reduces the chances of corrosion and fouling due to high concentration exposed to a low temperature (5°C) in the last stages; (v) it produces additional cooling effect from the last stages of MED operating below ambient temperature; and (vi) significant increase in system performance. The basics of the adsorption phenomenon, AD cycle, and its hybrids and desalination processes and economics are presented in detail in this chapter.

1.2 Basic adsorption theory

In designing an adsorption-based system, the basic equilibria-vapor uptake information at assorted pressures and temperatures of the vapor must be available. These are commonly represented in the form of isotherms of an adsorbent-adsorbate pair which, hitherto, have been captured by many empirical and semiempirical models. The simplest adsorption isotherm model is the classical Langmuir model [53], where it assumes a homogeneous surface with a monolayer vapor uptake with all surfaces containing a uniform charged energy (a flat distribution). Each pore vacant site is filled by a single vapor molecule to form a single sorption event. Invoking the fundamental rate of gas molecules filling the adsorption sites, as given by Ward and coworkers [54–57], the expression of the Langmuir isotherm model can be derived as follows:

$$\frac{d\theta}{dt} = K'\left[\exp\left(\frac{\mu_g - \mu_\alpha}{RT}\right) - \exp\left(\frac{\mu_\alpha - \mu_g}{RT}\right)\right] \tag{1.1}$$

$$\theta = \frac{K\exp\left(\frac{\varepsilon}{RT}\right)P}{1 + K\exp\left(\frac{\varepsilon}{RT}\right)P} \tag{1.2}$$

However, on a real solid surface where geometrical roughness is inevitably introduced during formation, the energetic heterogeneity cannot be ignored. The gas molecules experience different potential as adsorption sites of uneven energy are reached. As such, the surface is divided into infinitesimal pieces in which the local adsorption energy sites can be treated equally. Hence, the total adsorption of a heterogeneous surface is expressed:

$$\theta = \sum_i \chi(\varepsilon_i)\cdot\widetilde{\theta}(\varepsilon_i) = \int_0^\infty \chi(\varepsilon)\cdot\widetilde{\theta}(\varepsilon)d\varepsilon \tag{1.3}$$

where $\widetilde{\theta}(\varepsilon_i)$ is the local surface coverage on which the Langmuir model can be applied. It corresponds to the probability of the adsorption site energy distribution and it inherently satisfies the following condition:

$$\int_0^\infty \chi(\varepsilon)\cdot d\varepsilon = 1 \tag{1.4}$$

Applying the condensation approximation (CA) for moderate temperatures, the local surface coverage can be simplified to a step-like Dirac-delta function as shown in Fig. 1.5, and expressed below,

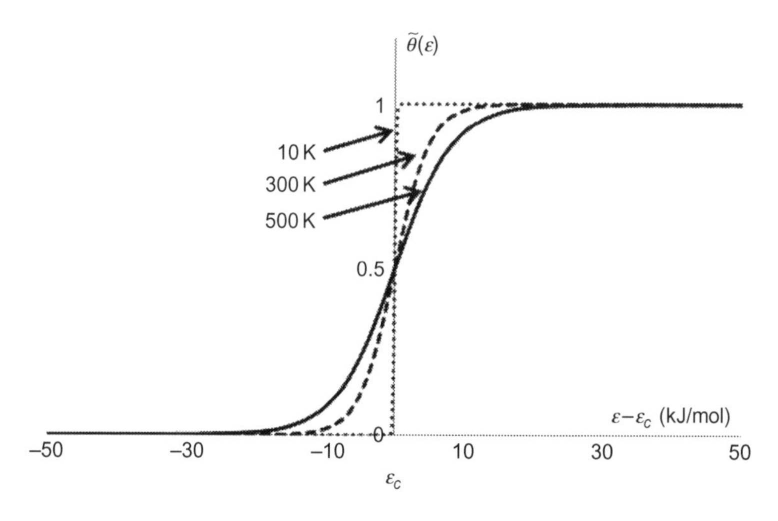

Fig. 1.5 Step-like profile yielded from Eq. (1.5) at moderate temperatures.

$$\lim_{T\to 0}\widetilde{\theta}(\varepsilon)=\begin{cases}0 \text{ for } \varepsilon<\varepsilon_c\\ 1 \text{ for } \varepsilon\geq\varepsilon_c\end{cases} \tag{1.5}$$

where ε_c relates to the adsorption energy in equilibrium conditions. Hence, Eq. (1.3) can be simplified to

$$\begin{aligned}\theta &= \int_0^{\varepsilon_c} 0\cdot\chi(\varepsilon)d\varepsilon+\int_{\varepsilon_c}^{\infty} 1\cdot\chi(\varepsilon)d\varepsilon\\ &= \int_{\varepsilon_c}^{\infty}\chi(\varepsilon)d\varepsilon\end{aligned} \tag{1.6}$$

Obtaining a solution of $\chi(\varepsilon)$ to represent the site energy allocation for a real adsorbent surface is difficult. However, it is possible to simplify the solution by approximating $\chi(\varepsilon)$ to a continuous probability distribution function. Depending on the adsorbent surface characteristics during adsorption interaction, symmetrical or asymmetrical Gaussian functions can be assumed, as illustrated in Fig. 1.6.

Table 1.2 summarizes the functions of $\chi(\varepsilon)$ for the classic Langmuir-Freundlich [58], Dubinin-Astakhov [59], Dubinin-Raduskevich [60], and Tóth isotherm models. By integrating distribution functions from ε_c to ∞, the exact form of these isotherm models can be yielded.

In the recent development of adsorption isotherm theory, Li [61] proposed a universal model that was able to fit all types of isotherms, as specified in Eq. (1.7). Type I—V patterns at various temperatures could be directly captured by the equation with four regression parameters:

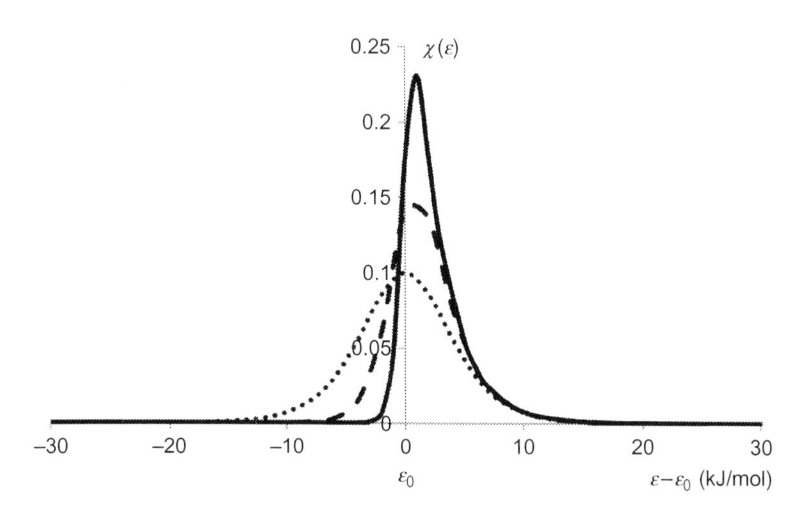

Fig. 1.6 An illustration of symmetrical or asymmetrical Gaussian function to represent the adsorption site energy distribution for classical isotherm models.

Table 1.2 Forms of adsorption site energy distribution functions for the classical Langmuir-Freundlich, Dubinin-Astakhov, Dubinin-Radushkevich, and Tóth isotherm models

Model name	Adsorption site energy distribution $\chi(\varepsilon)$	Isotherm equation
Langmuir-Freundlich	$\chi(\varepsilon)=\dfrac{\frac{1}{c}\exp\left(\frac{\varepsilon-\varepsilon_0}{c}\right)}{\left[1+\exp\left(\frac{\varepsilon-\varepsilon_0}{c}\right)\right]^2}$	$\theta=\dfrac{\left[KP\exp\left(\frac{\varepsilon_0}{RT}\right)\right]^{\frac{RT}{c}}}{1+\left[KP\exp\left(\frac{\varepsilon_0}{RT}\right)\right]^{\frac{RT}{c}}}$
Dubinin-Astakhov	$\chi(\varepsilon)=\dfrac{r(\varepsilon-\varepsilon_1)^{r-1}}{E^r}\exp\left[-\left(\frac{\varepsilon-\varepsilon_1}{E}\right)^r\right]$	$\theta=\exp\left[-\left(\frac{RT}{E}\ln\frac{P_0}{P}\right)^r\right]$
Dubinin-Radushkevich	$\chi(\varepsilon)=\dfrac{2(\varepsilon-\varepsilon_1)}{E^2}\exp\left[-\left(\frac{\varepsilon-\varepsilon_2}{E}\right)^2\right]$	$\theta=\exp\left[-\left(\frac{RT}{E}\ln\frac{P_0}{P}\right)^2\right]$
Tóth	$\chi(\varepsilon)=\dfrac{\frac{1}{RT}\left[\exp\left(\frac{\varepsilon-\varepsilon_3}{RT}\right)\right]^t}{\left\{1+\left[\exp\left(\frac{\varepsilon-\varepsilon_3}{RT}\right)\right]^t\right\}^{\frac{t+1}{t}}}$	$\theta=\dfrac{KP\exp\left(\frac{\varepsilon_3}{RT}\right)}{\left\{1+\left[KP\exp\left(\frac{\varepsilon_3}{RT}\right)\right]^t\right\}^{\frac{1}{t}}}$

$$\frac{q}{q_0}=\frac{A\phi\exp\left(\beta\frac{P}{P_s}\right)\frac{P}{P_s}+C\frac{P}{P_s}}{\left\{1+\phi\exp\left(\beta\frac{P}{P_s}\right)\frac{P}{P_s}\right\}^t} \tag{1.7}$$

and

$$\beta=\exp\left(\frac{E_c}{RT}\right) \tag{1.8}$$

$$A=\frac{[1+\phi\exp(\beta)]^t-C}{\phi\exp(\beta)} \tag{1.9}$$

where the alphabet ϕ and C are constants; t is the surface heterogeneity factor; and E_c denotes the characteristic energy of the adsorbent-adsorbate pair. These four parameters are required to calculate in the regression process. The rest of the letters have their usual means.

Using the unified adsorption isotherm framework, a universal adsorption site energy distribution function (EDF) was proposed that relates directly to their isotherm types, and the proposed EDF fitted well with the statistical rate theory of adsorption. The EDF yielded a single peak asymmetrical distribution for type I which was similar to that for the classical LF, DA, DR, and Tóth isotherm models. The EDF is given as below:

$$\chi(\varepsilon) = \frac{\beta^*}{RT}\left[\phi\exp(\beta^*)\frac{\beta^*}{\beta}+1\right]^{-t-1}\left\{\begin{array}{l}\left[A\phi\exp(\beta^*)\dfrac{(\beta^*+1)}{\beta}+\dfrac{C}{\beta}\right]\\ \left[1-\phi\exp(\beta^*)\dfrac{\beta^*}{\beta}(t-1)\right]\\ -Ct\phi\exp(\beta^*)\left(\dfrac{\beta^*}{\beta}\right)^2\end{array}\right\} \tag{1.10}$$

where the variable β^* is a function of the adsorption site energy ε and expressed as

$$\beta^* = \exp\left(\frac{2E_c - \varepsilon}{RT}\right) \tag{1.11}$$

From the data available from the literature and Eqs. (1.9), (1.10), the assorted isotherms categorized by the IUPAC can be successfully captured succinctly by these equations using only four coefficients of regression, and these EDFs and isotherms are depicted in Table 1.3. For each type of isotherm, the corresponding EDFs have been developed:

The adsorbent-adsorbate interactions is the key in the design of adsorption cycle, which can be functionalized to adopt to warmer ambient temperatures, particularly for operation in the summer period of semiarid or desert regions. Fig. 1.7A and B depict two major types of useful adsorbents in water uptake: The former is the silica gel type 3A, suitable for an AD cycle operating below 33°C ambient such as the tropical weather conditions. The latter figure depicts the isotherms of zeolite (Z0-alumina phosphate oxide, $Al_XPh_YO{\cdot}nH_2O$). It has properties that can be tailored for adsorption/desorption at ambient temperatures of up to 50°C (corresponding to desorption at 3.8–4.5 kPa (60°C isotherm) and adsorption at 2–2.8 kPa (40°C isotherm). The thermophysical properties of both adsorbents are tabulated in Table 1.4 and Fig. 1.8 shows the field emission microscopic view of type 3A silica gel.

1.3 Basic adsorption cycle

The main components of AD cycle are: (i) evaporator, (ii) adsorption and desorption reactor beds, (iii) condenser, (iv) pumps, and (v) pretreatment facility. The detailed process diagram is shown in Fig. 1.9. The AD operation involves two main processes: (i) adsorption-assisted evaporation and (ii) desorption-activated condensation.

(i) **Adsorption-assisted-evaporation**: in which the vapors generated in the AD evaporator are adsorbed onto the pore surface area of the adsorbent. The heat source is circulated through the tubes of the evaporator and seawater is sprayed on the tube bundle. It is observed that the evaporation is initiated by the heat source, but during the adsorption process the high affinity of water vapor of the adsorbent drops the evaporator pressure and contributes in evaporation. The AD evaporator operation temperature can be controlled by the heat source temperature that is normally circulated in terms of chilled water. The AD evaporator can

Table 1.3 A summary of the energy distribution functions and isotherms as categorized by the IUPAC

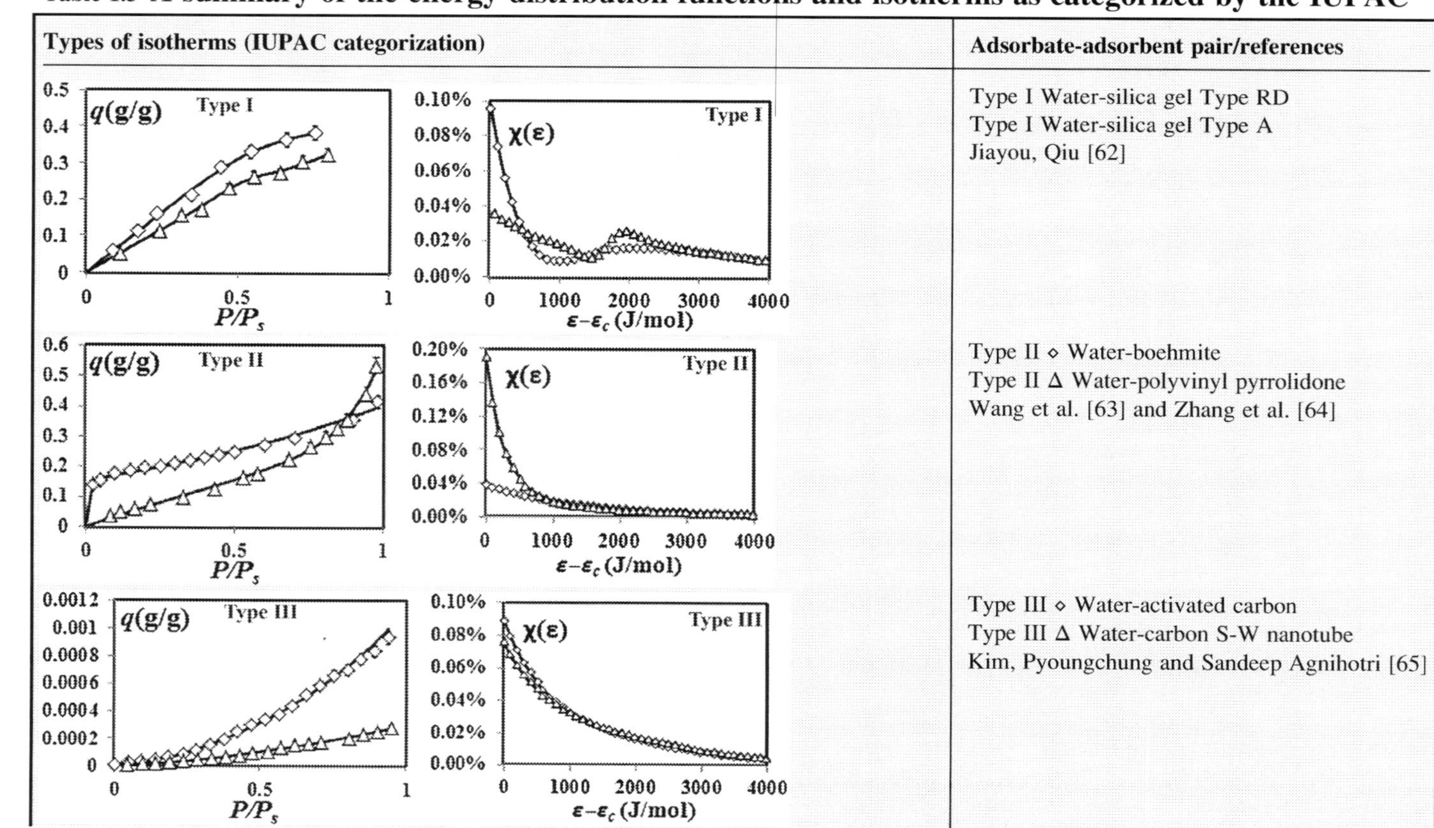

Types of isotherms (IUPAC categorization)	Adsorbate-adsorbent pair/references
Type I	Type I Water-silica gel Type RD Type I Water-silica gel Type A Jiayou, Qiu [62]
Type II	Type II ◇ Water-boehmite Type II Δ Water-polyvinyl pyrrolidone Wang et al. [63] and Zhang et al. [64]
Type III	Type III ◇ Water-activated carbon Type III Δ Water-carbon S-W nanotube Kim, Pyoungchung and Sandeep Agnihotri [65]

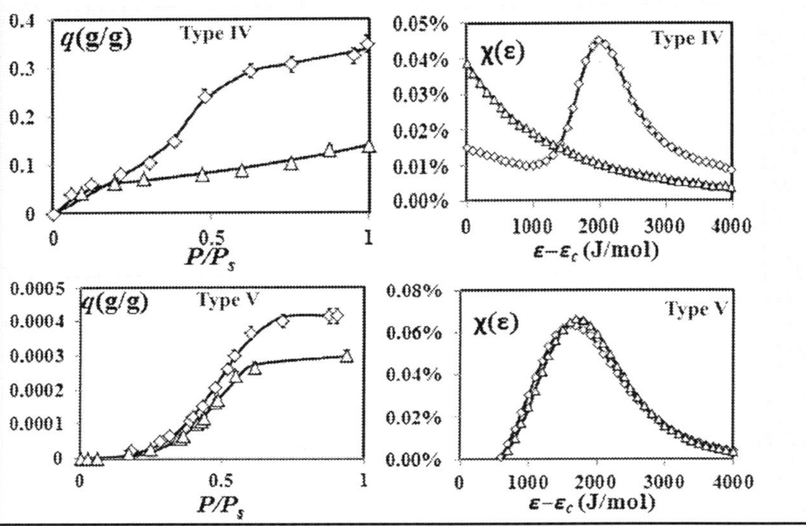	Type IV ◇ Water-boehmite Type IV Δ Water-polyvinyl pyrrolidone Bansal and Dhami [66]
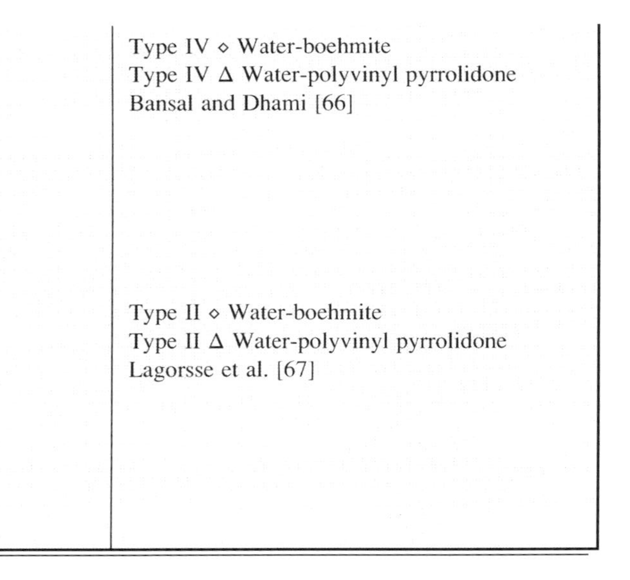	Type II ◇ Water-boehmite Type II Δ Water-polyvinyl pyrrolidone Lagorsse et al. [67]

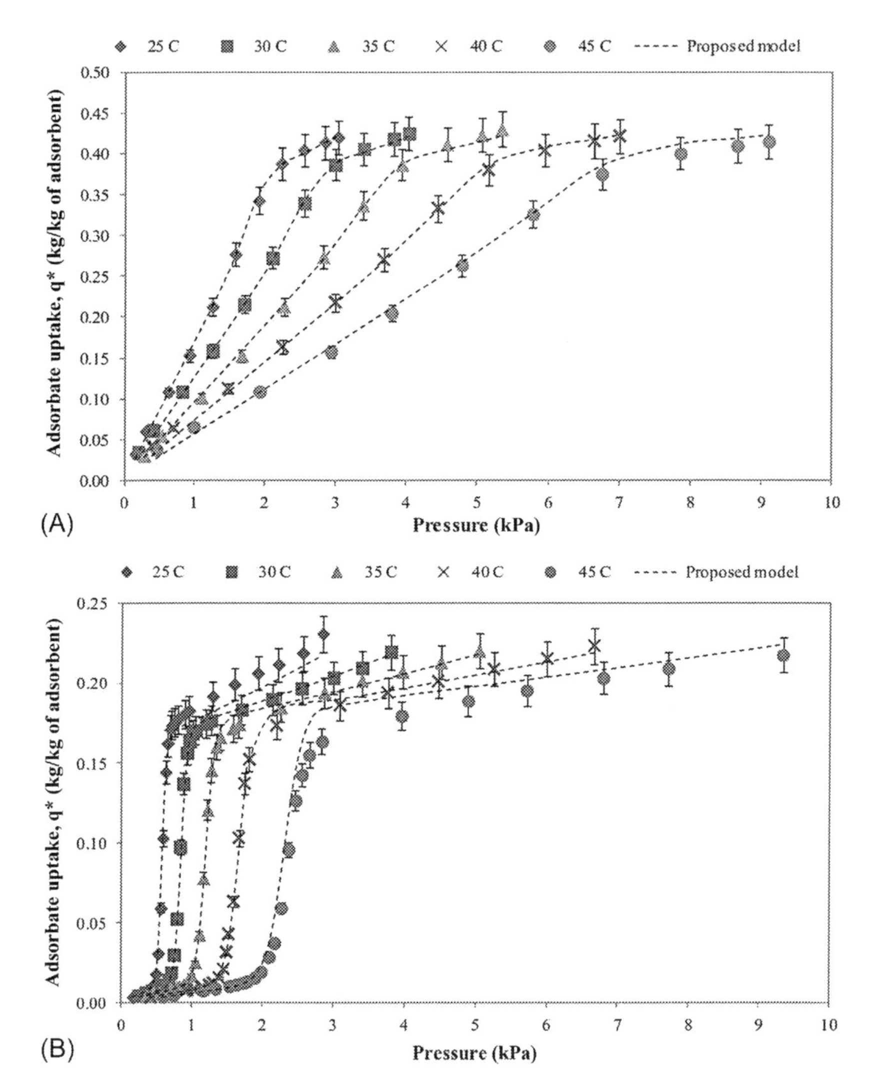

Fig. 1.7 (A) Shows the isotherms of water-silica gel type 3A at four temperatures: 30–45°C at increasing pressure up to 10 kPa, while (B) depicts the isotherms of zeolite (Z01, alumina phosphate oxide) from 30°C to 60°C.

Table 1.4 Thermophysical properties of the silica gel type 3A and Zeolite FAM Z01

Properties	Silica gel type 3A	Zeolite FAM Z01
BET surface area (m^2/g)	680	147.3
Porous volume (mL/g)	0.47	0.071
Apparent density (kg/m^3)	770	600–700
Thermal conductivity (W/mK)	0.174	0.113 (30°C)
Heat of adsorption (H_2O) (kJ/kg of H_2O)	2800	3110 (25°C)
Specific heat capacity (kJ/kgK)	0.921	0.805 (30°C)

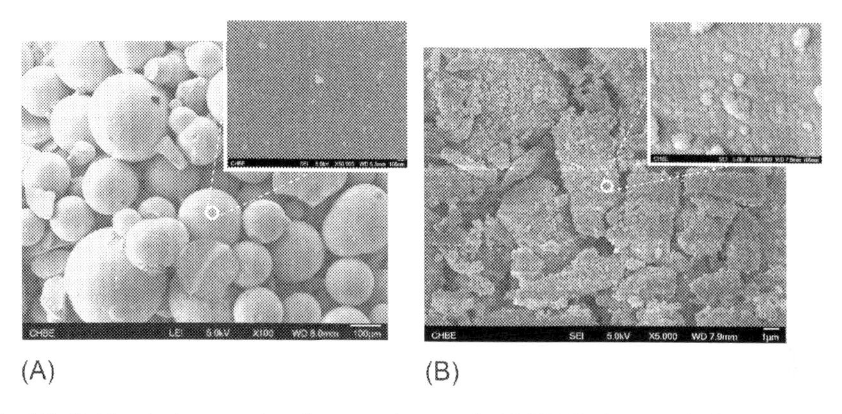

Fig. 1.8 Field emission scanning electron microscopic (FESEM) photographs for (A) silica gel type 3A and (B) zeolite FAM Z01.

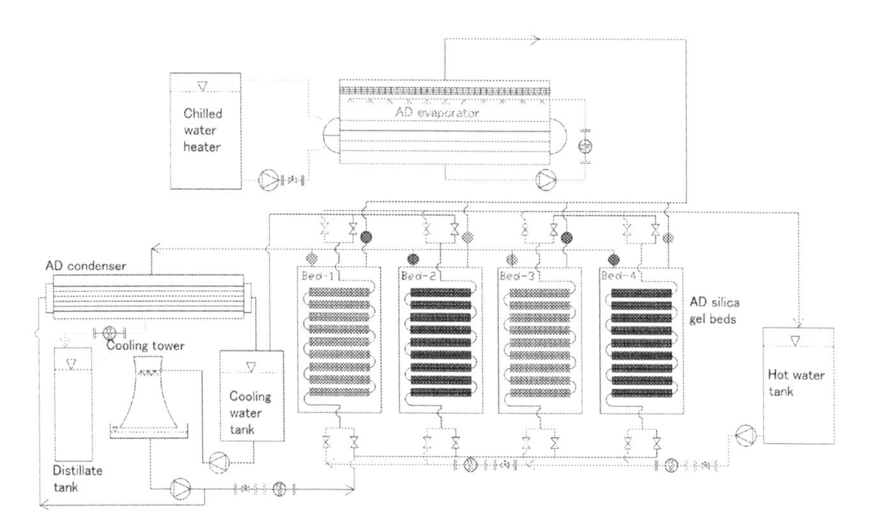

Fig. 1.9 Detailed flow schematic of the adsorption cycle.

operate at a wide range of chilled water temperature varying from 5°C to 30°C to produce the cooling effect as well at low-temperature operation. The vapor adsorption process continues until the adsorbent bed reaches a saturation state.

(ii) Desorption-activated-condensation: in which saturated adsorbent is regenerated using the low-grade industrial waste heat or renewable energy (desorption temperature varies from 55°C to 85°C) and desorbed vapors are condensed in a water-cooled condenser and collected as a distillate water.

In a multibed AD system, the operation and switching technique is used. During operation, one or a pair of adsorbent reactor beds undergo the adsorption process and at same time one or a pair of adsorbent reactor beds execute the desorption process. The time for adsorbent reactor bed operation, either adsorption or desorption, is dependent on the heat source temperature and silica gel quantity packed in a bed. Prior to changing the reactor duties, there is a short time interval called switching in which the adsorber bed is preheated while the desorber bed is precooled to enhance the performance of the cycle. In AD cycles, the operation (adsorption and desorption) and switching processes are controlled by an automated control scheme that can open and close the respective bed hot/cold water valves. During the switching operation, all vapor valves are closed so that no adsorption or desorption takes place. It can be seen that the AD cycle can produce two useful effects: cooling effect during adsorption-assisted evaporations and freshwater during desorption-activated condensation.

The AD cycle can produce two useful effects, cooling by adsorption-assisted evaporation and freshwater by desorption-activated condensation simultaneously with only one heat input. To investigate the performance of the basic adsorption cycle, a four-bed adsorption cycle was installed in KAUST, Saudi Arabia, as shown in Fig. 1.10A. Experiments were conducted to calculate specific daily water production and specific cooling as presented in Fig. 1.10B and C. It can be seen that the AD cycle can be operated from cooling to desalination mode just by adjusting the chilled water temperature with the same hardware [68].

1.4 The MED–AD hybrid cycle

In a conventional MED system, the vapor produced in a steam generator is condensed in successive stages to recover the heat of condensation and this improves the energy utilization for desalination. The number of stages in a MED is controlled by TBT and the lower-brine temperature (LBT). The TBT of typically 70°C is restricted by soft scaling components in the feedwater and the LBT is controlled by ambient conditions due to the water-cooled condenser to condense the last stage vapors. The thermal desalination system can be more efficient if these two limitations can be removed to increase the number of recoveries. Researchers have found that TBT can be increased to 125°C by inducing nanofiltration (NF) before introducing the feed into the system. This NF process helps to remove the soft scaling components and prevent scaling and fouling on the tubes of evaporators. The interstage temperature and the last stage operating temperature limitations can be overcome by hybridization with the adsorption cycle. The AD cycle can operate below ambient conditions typically as low as 5°C due the high affinity of water vapors of the adsorbent (silica gel). MED last stage temperature can be lowered down to 5°C by introducing the AD at the downstream.

MED-AD is the hybrid of the multieffect desalination (MED) system and the adsorption desalination (AD) cycle. The main components of this novel thermal hybrid system are: (1) the multieffect distillation (MED) system, (2) adsorption desalination (AD) cycle, and (3) auxiliary equipment. In this hybridization system, the last stage of the MED is connected to adsorption beds of the AD cycle for the direct vapor communication to adsorption beds. Fig. 1.11 shows the detailed process flow schematic of the MED-AD hybrid system.

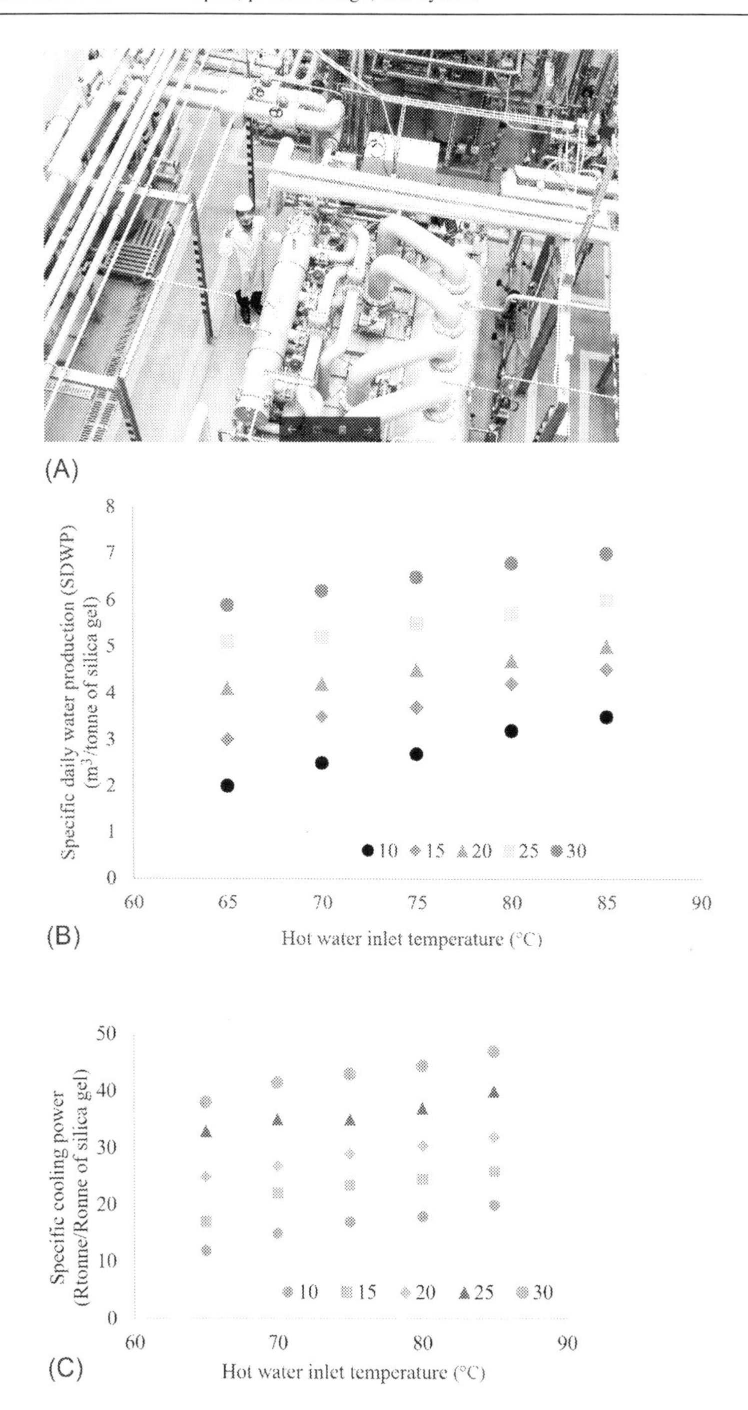

Fig. 1.10 (A) Adsorption cycle pilot at KAUST, (B) specific daily water production, and (C) specific cooling power.

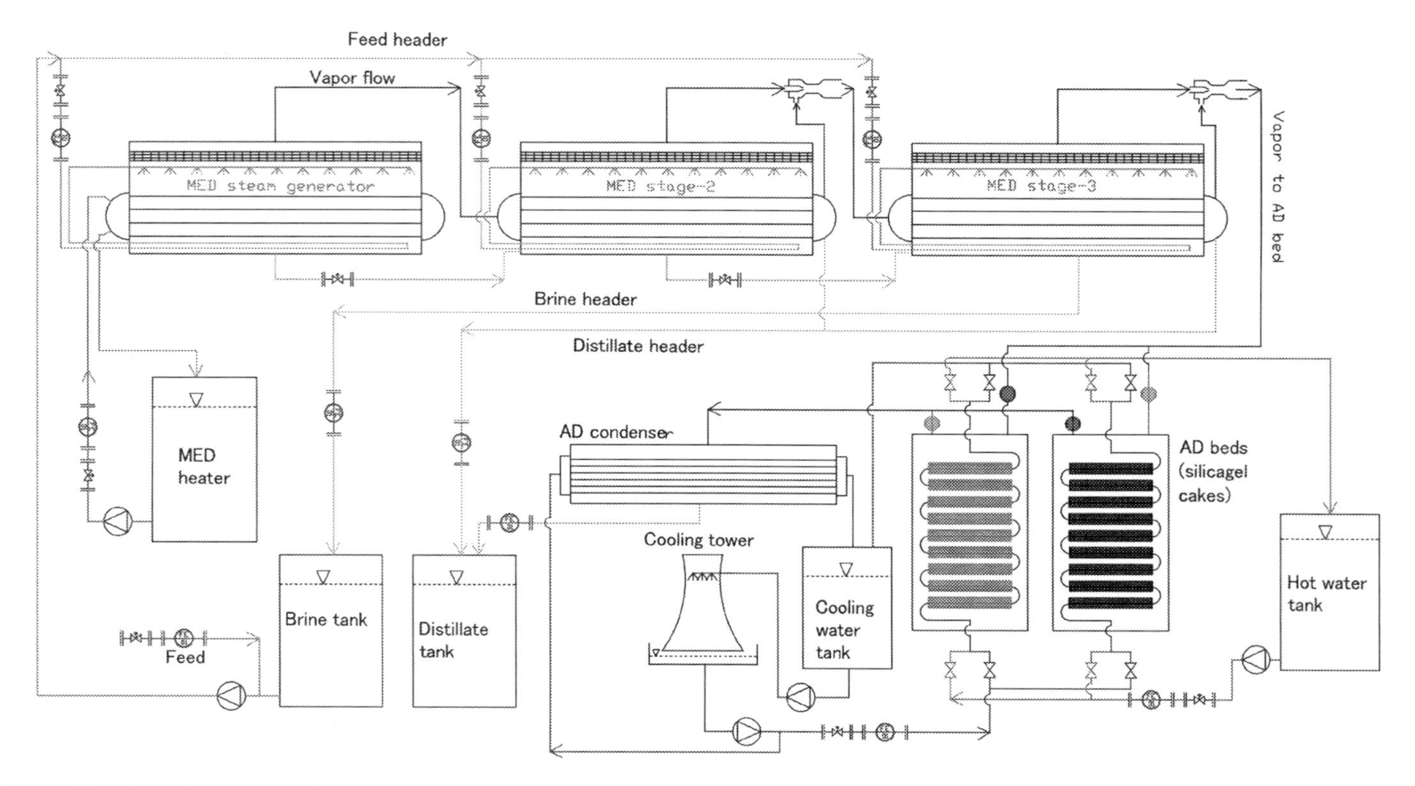

Fig. 1.11 Schematic representation of MED-AD hybrid cycle process flow.

The detailed process is discussed in different forms of streams involved in the operation as discussed below.

Feed stream: The MED system is designed for parallel feed and spray nozzle headers are installed in each MED stage to atomize the feed water for better film evaporation. Each stage is equipped with a built-in feed preheater loop to preheat the feed by extracting the energy from the hot brine sump. A special magnetic coupling pump is used to pressurize the feed water for nozzle operation. Feed flow is controlled via a regulating valve and measured with turbine magnetic pickup flow meter to ensure the required amount of feed to all stages.

Vapor stream (in MED operation): Hot media or heat source is circulated through the tubes during the first stage of MED also called a steam generator to produce the primary vapors. The sprayed brine generates saturated vapors at the state of saturation pressure (P_{TBT}) and temperature (T_{TBT}). These vapors are cascaded to the tube side of the second stage where the vaporization process is repeated by extracting the heat of condensation to evaporate the sprayed brine. The vapors condensed inside the tubes produce the desalted water. These vapors energy reutilization process continues until the last stage where the vapors are condensed in a water-cooled condenser. Low-pressure steam jet ejectors are provided in between the MED stages which extract any noncondensed vapor that may be accumulated at the distillate boxes of MED stages and the removal of noncondensable is a key aspect of MED smooth operation.

Vapor stream (in MED-AD operation): In the case of MED-AD operation, the last stage is connected with AD silica gel beds to adsorb the last-stage vapors. This combination allows removing the condenser with a conventional MED system and hence the last stage can operate as low as 5°C. In the hybrid system, in the last stage the evaporation is observed by the two drivers namely: (1) heat of condensation of tube side vapors and (2) pulling effect of the adsorbent. The AD cycle is operated in a batch manner and the desorption temperature (55–85°C) can be obtained from low-grade waste heat or from renewable energy. The desorbed vapors are than condensed in a AD condenser.

The system investigated here is only having three stages, but in real plants it can be up to 15–20 stages with a combination of NF-MED-AD and it is expected that it can be better than the RO system energetically and efficiently in terms of kWh_{pe}/m^3.

Distillate stream: To maintain the pressure difference between the MED stages, the distillate box of each stage is connected to a distillate header via a U-tube to collect each stage distillate production. This distillate header is then connected to a distillate tank. Aichi-Tokei high accuracy turbine flow meter is used to measure and for logging the total distillate production from MED and AD systems.

Brine stream: The excess brine from each stage is collected to a brine header via a U-tube. This brine header is connected to the brine collection tank to collect brine. This brine can be recirculated for more recovery depending on the concentration and for that it has a communication line with the feed tank to mix with seawater.

Vacuum system: Since the system is under vacuum, so to maintain the required vacuum level a vane-type vacuum pump is installed. All the equipment are directly connected to the vacuum pump to pull any air ingress into the system at any point.

System vacuum is tested for 48 h before experimentation and is found to have very good vacuum holding capacity.

This novel desalination cycle can mitigate the limitations of conventional MED systems to increase the system performance. This combination allows the MED last stage to operate below the ambient temperature typically at 5°C as compared with the traditional MED at 40°C. This not only reduces the corrosion chances but also increases the distillate production to almost twofold as compared with traditional MED systems. The details of the MED-AD hybrid cycle can be found in the literature [69–78].

1.5 MED-AD pilot experimentation

MED-AD hybrid experiments were conducted at different heat source temperature ranges from 70°C to 15°C. Fig. 1.12 shows the instantaneous temperatures of MED-AD components at a heat source temperature of 50°C. It can be seen that steady-state events (minimum temperature fluctuations) occur after 1 h from start-up and experiments for distillate collection are continued for 4–5 h. It can be seen clearly that the interstage temperature difference (ΔT) is more than twice per stage as compared with the conventional MED stages. This is attributed to the vapor uptake by the adsorbent of AD cycle, resulting in the increase of vapor production. The MED + AD cycle yields a stage ΔT from 3°C to 4°C as compared with 1°C or less in the case of MED alone.

Figs. 1.13 and 1.14 show the steady-state temperature and pressure values of MED-AD components for heat source temperatures varying from 15°C to 70°C. It can be observed that in all cases, interstage temperature varies from 3°C to 4°C. The ambient energy scavenging effect can also be seen at 15°C and 20°C heat source temperature

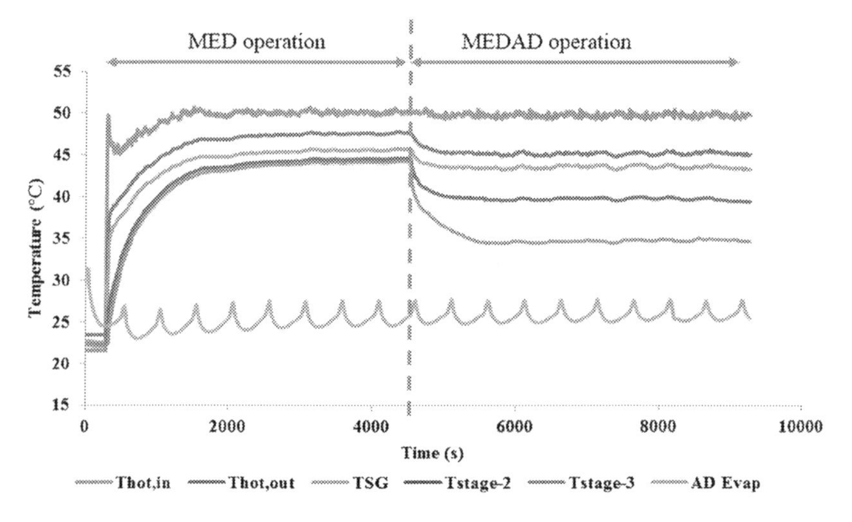

Fig. 1.12 Temperature profiles of MED and MED-AD components at a 50°C heat source.

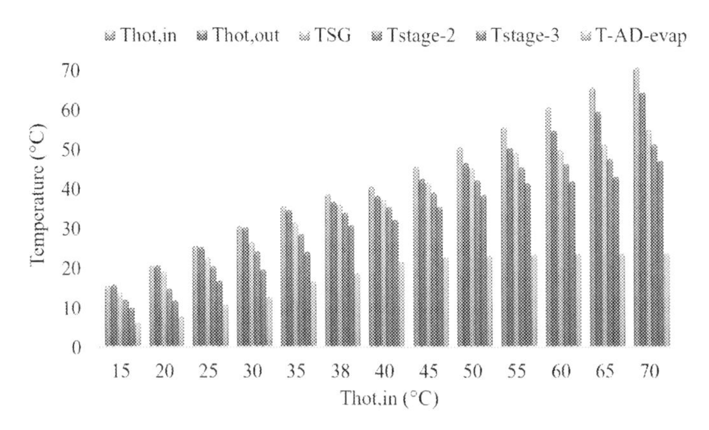

Fig. 1.13 Steady-state temperatures of MED-AD components at different heat sources.

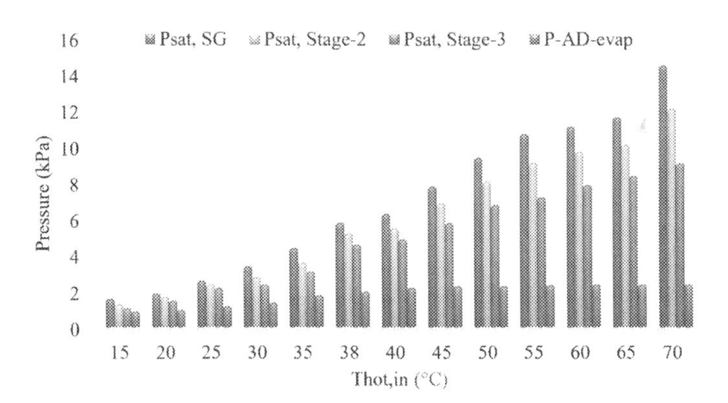

Fig. 1.14 Steady-state pressures of MED-AD components at different heat sources.

where the heat source outlet temperature is higher. It is predicted that in a full-scale 25–30-stage plant, the ambient energy can be scavenged in the last few stages working below ambient temperatures. An additional cooling effect can also be extracted from the stages working below ambient temperatures. Even with only three stages of MED, the hybridization effect can be seen clearly in terms of interstage temperature distribution.

MED stage distillate production is collected in a MED collection tank and AD condenser production in another tank. Both side productions are measured by a high-accuracy turbine flow meter. Fig. 1.15 shows the distillate production trace at a 50°C heat source from MED stages, AD condenser, and combined. The batch-operated AD production can be seen clearly. At the start of desorption, the production is higher and it drops with time to zero during the switching period, while the MED stage production is quite stable. Small fluctuations in MED water production may be due to the fluctuations in the spray of the feed that affect the condensation rate.

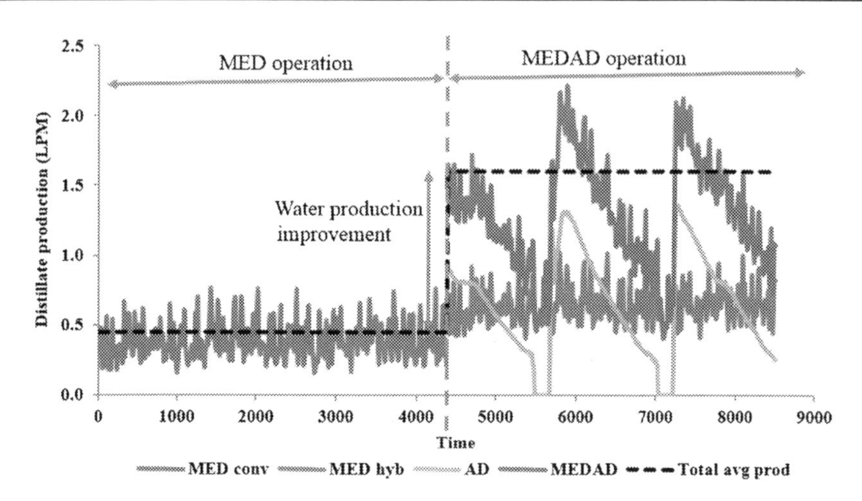

Fig. 1.15 Water production profiles of MED and MED-AD cycles at a 50°C heat source.

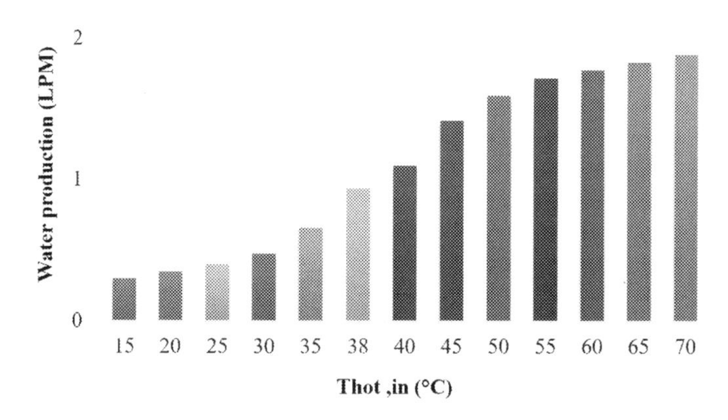

Fig. 1.16 Water production MED-AD cycles at different heat source temperatures.

Experiments are conducted for 4–5 h for stable distillate production measurement. Fig. 1.16 shows the average water production at heat source temperature varies from 15°C to 70°C. The drop in production with lowering the heat source temperature is due to the drop in saturation temperature and hence the saturation pressure. Even though the last stage production is lower, it is still feasible due to energy harvesting from ambient conditions. Another useful effect from the last stage operating below ambient conditions is cooling that can be produced simultaneously with water production. This additional benefit cannot be obtained from conventional MED systems.

Although the experimental facility had only three stages, experiments were conducted in a piecewise manner where the saturation temperature of the last stage was reproduced as the input temperature to the steam generator. In this succession manner, it is possible to collate the performances of the MED-AD hybrid as though

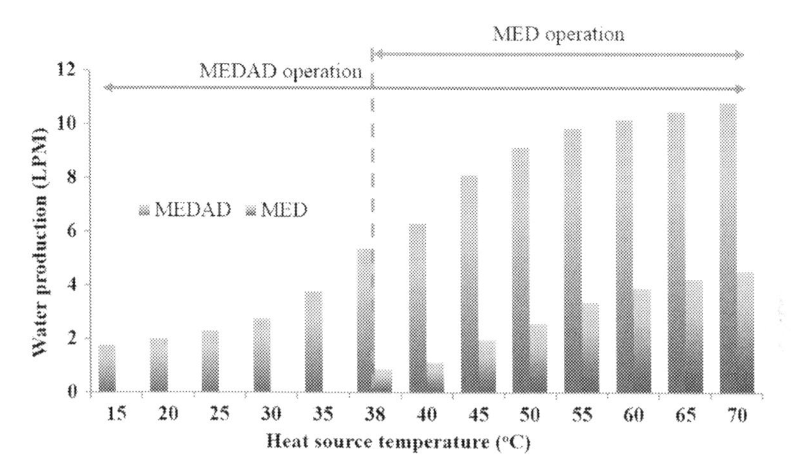

Fig. 1.17 Water production comparison of MED and hybrid MED-AD cycles at assorted heat source temperatures.

the AD cycle is integrated to many stages of MED, a design concept akin to what can be permitted by the available ΔT between the top-brine to the bottom-brine stages. Fig. 1.17 shows the water production measured from both the three-stage MED and from MED-AD hybrid (same three-stage) for an assorted range of stage temperatures. At each steam generator temperature, a marked improvement to the water production is observed with a two- to threefold quantum jump.

1.6 Desalination costing: An exergy approach

In desalination, the cost of fuel or input energy is a major contribution to the overall production cost. Since, the desalination processes are energy intensive, they are normally associated with power generation systems as a dual purpose plant.

In dual purpose plants, primary energy apportionment is very important for fair costing of power production and desalination. In the literature, many researchers [79–82] provided a detailed economic analysis but all studies use energy-based cost apportionment. Energy-based cost apportionment is not the true cost distribution because desalination plants utilized low-grade bleed steam. This low-grade bleed steam has very less ability to perform work in turbines but it can produce a substantial work in desalination because of low-pressure operation. So in dual-purpose power and desalination plants, consideration of same level energies for cost apportionment is irrespective of the water production systems.

For dual-purpose plants, the cost apportioning on the basis of "quality" of energy utilized by the process is the true cost distribution. Cost apportionment utilizing exergy analysis shows the real cost of primary fuel energy for power plants and desalination processes since the quality of the energy is differentiated.

Exergy is an extensive thermodynamic function defined as the maximum theoretically achievable work from an energy carrier under an environmentally imposed condition (T_o, P_o) at a given amount of chemical element. At environmental state also called as dead state, system exergy is considered as zero. Therefore, the general exergy term includes thermomechanical and chemical exergy. The thermomechanical exergy is the maximum available work when system conditions (T and P) approach environmental conditions (T_o, P_o) without changes in the chemical composition of the process stream. The chemical exergy is defined as the maximum work obtained when the concentration (C_o) of each substance in process streams changes to that of environmental conditions (T_o, P_o).

The main difference between energy and exergy is that energy always remains conserved in a process according to the first law of thermodynamics, while exergy is destroyed due to irreversibilities in the process. Exergy destruction or annihilation is caused by entropy generation in a process within each component of the system. In addition, there are some exergy losses to the ambient due to the temperature difference and it is called effluent exergy losses. The main point of interest is to find the potential of work of the flow stream also called specific exergy across each component of the system.

Many researchers [83–103] conducted the detailed exergy analysis for different thermal and membrane desalination processes as a standalone and with dual-purpose plant configuration but for costing they used energy-apportionment techniques. The method of exergy-based cost apportionment is discussed in following sections in detail.

1.6.1 System description

The exergy destruction analysis accurately apportions the amount of primary energy consumed by the major components of the plant. This is demonstrated by considering a cogeneration of power (594 MW) and desalination (2813 m^3/h), as shown schematically in Fig. 1.18 [104]. The major components are arranged synergistically where the gas turbine (GT) generators is operated with high-exergy gases while both the steam turbines and the thermally driven desalination processes are powered by the recovered exergy from the GT exhaust gases.

The distribution of exergy destruction of combined-cycle gas turbines (CCGT) is 75% and the heat recovery steam generator (HRSG) has the remaining 25%. The latter exergy is converted into steam which operates the multistage steam turbines (ST), the steam condenser, and the multi-effect distillation (MED). Approximately $23 \pm 1.5\%$ of the available exergy is converted into steam of high pressure and temperature which is then supplied to the steam turbines for further power production and the thermally driven MED desalination processes. Only less than $3 \pm 0.5\%$ of the available exergy is purged to the ambient as exhaust of combustion products.

In view of the imbalanced exergy destruction in cogeneration processes, it is proposed to convert all derived energies to primary energy using the appropriate conversion factors to calculate the proposed universal performance ratio (UPR) as presented

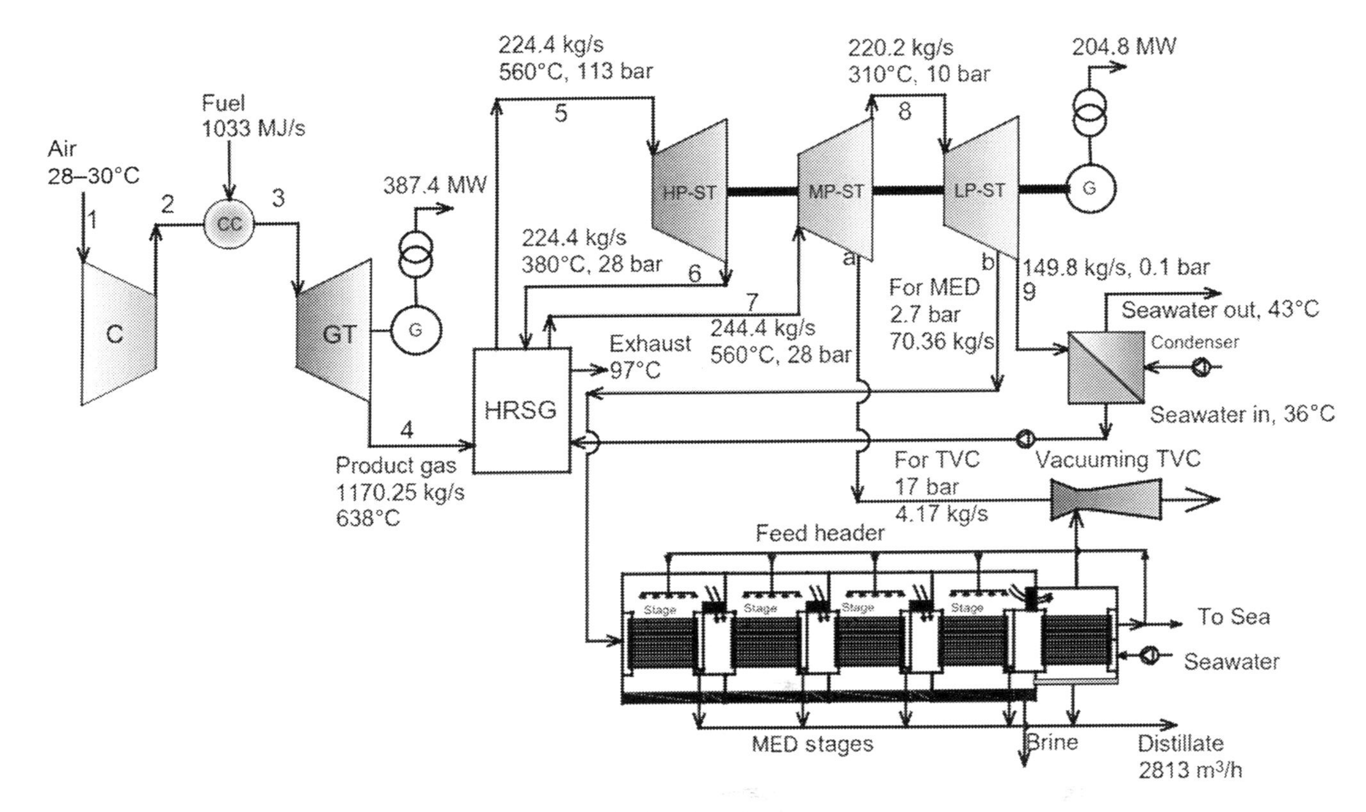

Fig. 1.18 Schematic diagram of the combined power and water desalination plant.

in Eq. (1.12). The proposed UPR is based on the consumption of the primary energy for desalination.

$$UPR = \frac{h_{fg}}{3.6\sum_{i=1}^{2}\{\varphi_1(\mathrm{kWh_{elec}/m^3}) + \varphi_2^{(\mathrm{TBT})}(\mathrm{kWh_{ther}/m^3})\}} \tag{1.12}$$

where φ_1 and $\varphi_2^{(\mathrm{TBT})}$ are the conversion factors needed for converting the derived electricity and thermal input at a given TBT to the respective primary energy. The h_{fg} is the latent heat of potable water produced, that is, 2326 kJ/kg. Based on the primary energy consumption, the detailed UPR values of desalination processes are shown in Table 1.5.

On the basis of the above exergy analysis, a life cycle costing analysis is conducted for a dual purpose plant (power+desalination) and it is found that MED-AD has the lowest water production cost of less than US\$0.5/m^3—the lowest specific cost ever reported in the literature as shown in Fig. 1.19. We also won the GE-ARAMCO global water challenge award for desalination cost of less than US\$ 0.5/m^3.

1.7 Conclusions

Recent developments in adsorption theory, adsorption desalination (AD), and conventional MED desalination cycles have been reviewed in this chapter. We highlight the key role of AD cycles which can be hybridized with the proven cycles such as the MED cycle, exploiting the thermodynamic synergy between the thermally driven cycles that significantly improve the water production yields. Experiments were conducted on a laboratory-scale pilot MED-AD and confirmed the superior synergetic effects that boosted the water production by up to two- to threefold over the conventional MED. We also presented UPR to compare all kinds of desalination processes on the basis of primary energy consumption. It shows that the MED-AD hybrid cycle UPR is the highest among all technologies. The projected LCC of water production can be lowered to as low as US\$0.485/m^3.

1.8 Future roadmap

Thermally driven processes MSF/MED have lower performance because of their process limitations. In MED processes, top-brine temperature (TBT) is limited at 70°C due to the soft scaling components such as magnesium (Mg^{2+}), calcium (Ca^{2+}), and sulfate (SO_4^{2-}) ions in the feed that contribute in system degradation at high TBT typically more than 70°C. As a solution, the researchers have found that these scaling agents can be suppressed by pretreating the feed through nanofiltration (NF) or antiscalant dosing and TBT can be raised to 130°C. The last-stage operating temperature limitations, 40°C, can be overcome by adsorption cycle hybridization that can operate below ambient conditions typically as low as 10°C. This tri-hybrid

Table 1.5 **Universal performance ratio of desalination processes**

Desalination process	Derived energy		Conversion factors		Primary energy	UPR
	kWh_{elec}/m^3	kWh_{ther}/m^3	Electrical energy conversion factor	Thermal energy conversion factor		
SWRO	3.5	NA	47.0%	3.4%	7.45	86
MSF	3.0	80.6			9.1	71
MED	2.3	71.7			7.36	87
MEDAD	2.5	36			6.56	98

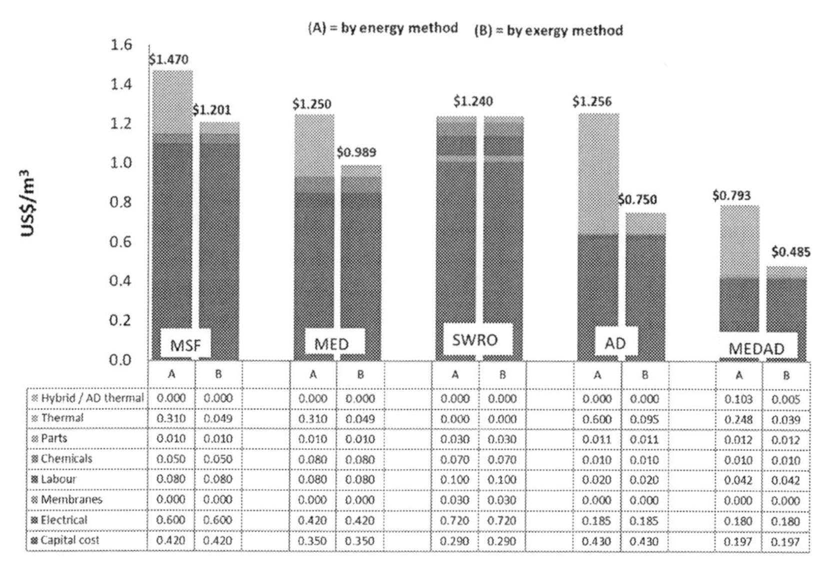

	MSF A	MSF B	MED A	MED B	SWRO A	SWRO B	AD A	AD B	MEDAD A	MEDAD B
Hybrid / AD thermal	0.000	0.000	0.000	0.000	0.000	0.000	0.000	0.000	0.103	0.005
Thermal	0.310	0.049	0.310	0.049	0.000	0.000	0.600	0.095	0.248	0.039
Parts	0.010	0.010	0.010	0.010	0.030	0.030	0.011	0.011	0.012	0.012
Chemicals	0.050	0.050	0.080	0.080	0.070	0.070	0.010	0.010	0.010	0.010
Labour	0.080	0.080	0.080	0.080	0.100	0.100	0.020	0.020	0.042	0.042
Membranes	0.000	0.000	0.000	0.000	0.030	0.030	0.000	0.000	0.000	0.000
Electrical	0.600	0.600	0.420	0.420	0.720	0.720	0.185	0.185	0.180	0.180
Capital cost	0.420	0.420	0.350	0.350	0.290	0.290	0.430	0.430	0.197	0.197

Fig. 1.19 Desalination methods costing by energetic and exergetic approach.

desalination cycle, NF+MED+AD, can operate from a heat source temperature of 130°C to a last-stage temperature of 10°C with more than 20 number of effects and hence the UPR = 250, over 20% of TL. The other hybrid combinations such as NF+RO+MSF, NF+MSF+MED were also proposed for higher performance and maximum thermodynamic synergy. In terms of robustness and commercialization, all individual technologies (NF, MED, MSF, and AD) are well proven and are readily available in the market. Today, thermally driven technologies are available on the shelf to achieve a COP21 goal for sustainable water supplies [3].

Acknowledgment

The authors acknowledge the King Abdullah University of Science & Technology (KAUST) (Project no. 7000000411) for the financial support for MED-AD pilot project.

References

[1] Finding the blue path for a sustainable economy, a white paper by Veolia water. http://www.veolianorthamerica.com/sites/g/files/dvc596/f/assets/documents/2014/10/19979IFPRI-White-Paper.pdf.

[2] Water and energy. The United Nations world water development report; 2014.

[3] Shahzad MW, Burhan M, Li A, Ng KC. Energy-water-environment nexus underpinning future desalination sustainability. Desalination 2017;413:52–64.

[4] Desalination by the numbers, International Desalination Association; 2016. http://idadesal.org/desalination-101/desalination-by-the-numbers/.

[5] Ghaffour N, Missimer TM, Amy GL. Technical review and evaluation of the economics of water desalination: current and future challenges for better water supply sustainability. Desalination 2013;309:197–207.
[6] Elimelech M, Phillip WA. The future of seawater desalination: energy, technology, and the environment. Science 2011;333:712–7.
[7] World Water Development Report (WWDR). U.N.'s World Water Assessment Program; 2014.
[8] UN Stresses Water and Energy Issues, World Water Day 2014 Report; 2014.
[9] Water and energy: facts and figures, The United Nations World Water Development (WWDR) Report; 2014.
[10] UN Water Report, The United Nations inter-agency mechanism on all related issues, including sanitation; 2014. http://www.unwater.org/statistics/statistics-detail/es/c/211827/.
[11] Concept chapter, global clean water desalination alliance "H20 minus CO2". http://www.diplomatie.gouv.fr/fr/IMG/pdf/global_water_desalination_alliance_1dec2015_cle8d61cb.pdf.
[12] Redrawing the energy-climate map, World energy outlook special Report, International Energy Agency (IEA); 2013. www.worldenergyoutlook.org/aboutweo/workshops.
[13] Friedlingstein P, et al. Persistent growth of CO2 emissions and implications for reaching climate targets. Nat Geosci 2014;7:709–15.
[14] COP 21: UN climate change conference, Paris; 2015. http://www.cop21.gouv.fr/en/why-2c/; http://www.c2es.org/facts-figures/international-emissions/historical.
[15] Francey RJ, et al. Atmospheric verification of anthropogenic CO2 emission trends. Nat Clim Chang 2013;3:520–4.
[16] Raupach MR, Le Quéré C, Peters GP, Canadell JG. Anthropogenic CO2 emissions. Nat Clim Chang 2013;3:603–4.
[17] Carbon Dioxide Information Analysis Center, Oak Ridge National Laboratory, Center for climate and Energy Solutions; 2016.
[18] Sustaining growth via water productivity: 2030/2050 scenarios. http://growingblue.com/wp-content/uploads/2011/05/IFPRI_VEOLIA_STUDY_2011.pdf.
[19] Desalination for water supplies, a review of current technologies. A report by Foundation for Water Research, UK. www.fwr.uk.
[20] Yousef Almulla, Gulf cooperation council (GCC) countries 2040 energy scenario for electricity generation and water desalination, [Master of Science thesis], KTH, EGI2014.
[21] Aidrous IAZ. How to overcome the fresh water crisis in the Gulf, http://russiancouncil.ru/en/inner/?id_4=4190#top-content; 2014.
[22] The GCC in 2020: resources for the future, a report from the Economist Intelligence Unit Sponsored by the Qatar Financial Centre Authority; 2010. http://graphics.eiu.com/upload/eb/GCC_in_2020_Resources_WEB.pdf.
[23] Al-Fulaij HF. Dynamic modeling of multi stage flash (MSF) desalination plant, [PhD thesis]. Department of Chemical Engineering University College London; 2011, http://discovery.ucl.ac.uk/1324506/1/1324506.pdf.
[24] Al-Karaghouli A, Kazmerski LL. Energy consumption and water production cost of conventional and renewable-energy-powered desalination processes. Renew Sust Energ Rev 2013;24:343–56.
[25] ESCWA (Economic and Social Commission for Western Asia). Role of desalination in addressing water scarcity, http://www.escwa.un.org/information/publications/edit/upload/sdpd-09-4.pdf; 2009.
[26] Cooley H, Gleick PH, Wolff G. Desalination, with a grain of salt: a California perspective, http://pacinst.org/app/uploads/2015/01/desalination-grain-of-salt.pdf; 2006.

[27] Mabrouk A-NA, Nafey AS, Fath HES. Steam, electricity and water costs evaluation of power-desalination co-generation plants. Desalin Water Treat 2010;22:56–64.

[28] Sommariva C, Hogg H, Callister K. Environmental impact of seawater desalination: relations between improvement in efficiency and environmental impact. Desalination 2004;167:439–44.

[29] Raluy RG, Serra L, Uche J, Valero A. Life-cycle assessment of desalination technologies integrated with energy production systems. Desalination 2004;167:445–58.

[30] Reddy KV, Ghaffour N. Overview of the cost of desalinated water and costing methodologies. Desalination 2007;205:340–53.

[31] Leon Awerbuch, Corrado Sommariva, MSF distillate driven desalination process and apparatus, WO 2006021796 A1 (2006).

[32] Al-Rawajfeh AE, Ihm S, Varshney H, Mabrouk AN. Scale formation model for high top brine temperature multi-stage flash (MSF) desalination plants. Desalination 2014;350:53–60.

[33] Fritzmann C, Löwenberg J, Wintgens T, Melin T. State-of-the-art of reverse osmosis desalination. Desalination 2007;216:1–76.

[34] Dundorf S, MacHarg J, Sessions B, Seacord TF. In: Optimizing lower energy seawater desalination, the affordable desalination collaboration. IDA World Congress-Maspalomas, Gran Canaria, Spain; 2007.

[35] Peñate B, de la Fuente JA, Barreto M. Operation of the RO kinetic® energy recovery system: description and real experiences. Desalination 2010;252:179–85.

[36] Dundorf S, MacHarg J, Sessions B, Seacord TF. In: Optimizing lower energy seawater desalination: the affordable desalination collaboration. IDA World Congress, Dubai UAE; 2009.

[37] Ludwig H. In: Energy consumption of seawater reverse osmosis: expectations and reality for state-of-the-art Technology. IDA World Congress, Dubai UAE; 2009.

[38] Farooque AM, Jamaluddin ATM, Al-Reweli AR. Comparative study of various energy recovery devices used in SWRO process. SWCC Technical Report No. TR.3807/EVP 02005, 2004.

[39] Sauvet-Goichon B. Ashkelon desalination plant-a successful challenge. Desalination 2007;203:75–81.

[40] Eltawil MA, Zhengming Z, Yuan L. In: Renewable energy powered desalination systems: technologies and economics-state of the art. Twelfth International Water Technology Conference, IWTC12 2008 Alexandria, Egypt; 2008.

[41] Mabrouk AN, Fath HES. Technoeconomic study of a novel integrated thermal MSF–MED desalination technology. Desalination 2015;371:115–25.

[42] Almulla Y. Gulf cooperation council (GCC) countries 2040 energy scenario for electricity generation and water desalination. [Master of Science thesis], KTH, 2014.

[43] Thu K, Ng KC, Saha BB, Chakraborty A, Koyama S. Operational strategy of adsorption desalination system. Int J Heat Mass Transfer 2009;52(7–8):1811–6. https://doi.org/10.1016/j.ijheatmasstransfer.2008.10.012.

[44] B.B Saha, K.C. Ng, A. Chakraborty, K. Thu (2009), Most energy efficient approach of desalination and cooling, Cooling India, May–June, pp 72–78, (2009).

[45] Saha BB, Ng KC, Chakraborty A, Thu K. Energy efficient environment friendly adsorption cooling cum desalination system. Cooling India, July–August, 2010. p. 22–6.

[46] Chakraborty A, Leong KC, Thu K, Saha BB, Ng KC. Theoretical insight of adsorption cooling. Appl Phys Lett 2011;98:1. https://doi.org/10.1063/1.3592260.

[47] Ng KC, Thu K, Saha BB, Chakraborty A, Chun WG. Study on a waste heat-driven adsorption cooling cum desalination cycle. Int J Refrig 2012;35(3):685–93. https://doi.org/10.1016/j.ijrefrig.2011.01.008.
[48] Thu K, Saha BB, Chakraborty A, Chun WG, Ng KC. Thermo-physical properties of silica gel for adsorption desalination cycles. Appl Therm Eng 2013;50:1596–602. https://doi.org/10.1016/j.applthermaleng.2011.09.038.
[49] Myat A, Ng KC, Thu K, Kim Y-D. Experimental investigation on the optimal performance of Zeolite–water adsorption chiller. Appl Energy 2013;102:582–90. https://doi.org/10.1016/j.apenergy.2012.08.005.
[50] Thu K, Chakraborty A, Kim Y-D, Myat A, Saha BB, Ng KC. Numerical simulation and performance investigation of an advanced adsorption desalination cycle. Desalination 2013;308:209–18. https://doi.org/10.1016/j.desal.2012.04.021.
[51] Ng KC, Thu K, Kim Y-D, Chakraborty A, Amy G. Adsorption desalination: an emerging low-cost thermal desalination method. Desalination 2013;308:161–79. https://doi.org/10.1016/j.desal.2012.07.030.
[52] Thu K, Yanagi H, Saha BB, Ng KC. Performance analysis of a low-temperature waste heat-driven adsorption desalination prototype. Int J Heat and Mass Transfer 2013;65:662–9. https://doi.org/10.1016/j.ijheatmasstransfer.2013.06.053.
[53] Langmuir I. The adsorption of gases on plane surfaces of glass, mica and platinum. J Am Chem Soc 1918;40:1361–403.
[54] Ward C, Findlay R, Rizk M. Statistical rate theory of interfacial transport. I. Theoretical development. J Chem Phys 1982;76:5599–605.
[55] Elliott JAW, Ward CA. Statistical rate theory description of beam-dosing adsorption kinetics. J Chem Phys 1997;106:5667–76.
[56] Elliott JAW, Ward CA. Statistical rate theory and the material properties controlling adsorption kinetics, on well defined surfaces. Stud Surf Sci Catal 1997;104:285–333.
[57] Elliott JAW, Ward CA. Temperature programmed desorption: a statistical rate theory approach. J Chem Phys 1997;106:5677–84.
[58] Rudzinski W, Borowiecki T, Dominko A, Panczyk T. A new quantitative interpretation of temperature-programmed desorption spectra from heterogeneous solid surfaces, based on statistical rate theory of interfacial transport: the effects of simultaneous readsorption. Langmuir 1999;15:6386–94.
[59] Rudzinski W, Lee S-L, Panczyk T, Yan C-CS. A fractal approach to adsorption on heterogeneous solids surfaces. 2. Thermodynamic analysis of experimental adsorption data. J Phys Chem B 2001;105:10857–66.
[60] Rudzinski W, Lee S-L, Yan C-CS, Panczyk T. A fractal approach to adsorption on heterogeneous solid surfaces. 1. The relationship between geometric and energetic surface heterogeneities. J Phys Chem B 2001;105:10847–56.
[61] Li A. Experimental and theoretical studies on the heat transfer enhancement of adsorbent coated heat exchangers [Doctor of Philosophy]. Department of Mechanical Engineering, National University of Singapore; 2014.
[62] Jiayou Q. Characterization of silica gel-water vapor adsorption [MEng thesis]. NUS; 2004.
[63] Wang S-L, Johnston CT, Bish DL, White JL, Hem SL. Water-vapor adsorption and surface area measurement of poorlycrystalline boehmite. J Colloid Interface Sci 2003;260(1):26–35.

[64] Zhang J, Zografi G. The relationship between "BET" and "free volume"-derived parameters for water vapor absorption into amorphous solids. J Pharm Sci 2000;89(8):1063–72.
[65] Kim P, Agnihotri S. Application of water-activated carbon isotherm models to water adsorption isotherms of single-walled carbon nanotubes. J Colloid Interface Sci 2008;325(1):64–73.
[66] Bansal RC, Dhami TL. Surface characteristics and surface behaviour of polymer carbons-II: adsorption of water vapor. Carbon 1978;16(5):389–95.
[67] Lagorsse S, Campo MC, Magalhaes FD, Mendes A. Water adsorption on carbon molecular sieve membranes: experimental data and isotherm model. Carbon 2005;43 (13):2769–79.
[68] Al-kharabsheh S, Goswami DY. Theoretical analysis of a water desalination system using low grade solar heat. J Sol Energy Eng Trans ASME 2004;126:774–80.
[69] Shahzad MW, Ng KC. On the road to water sustainability in the Gulf. Nature Middle East 2016, https://doi.org/10.1038/nmiddleeast.2016.50.
[70] Shahzad MW, Thu K, Ng KC. A waste heat driven hybrid ME+AD cycle for desalination, water technology. R Soc Chem Environ Sci Water Res Technol 2016, https://doi.org/10.1039/C5EW00217F.
[71] Thu K, Kim Y-D, Shahzad MW, Saththasivam J, Ng KC. Performance investigation of an advanced multi-effect adsorption desalination (MEAD) cycle. Appl Energy 2015;159:469–77.
[72] Bidyut Baran Saha, Ibrahim I. El-Sharkawy, Muhammad Wakil Shahzad, Kyaw Thu, Li Ang and Kim Choon Ng, Fundamental and application aspects of adsorption cooling and desalination, Appl Therm Eng, https://doi.org/10.1016/j.applthermaleng.2015.09.113
[73] Shahzad MW, Thu K, Kim Y-d, Ng KC. An experimental investigation on MEDAD hybrid desalination cycle. Appl Energy 2015;148:273–81.
[74] Shahzad MW, Thu K, Ng KC, WonGee C. Recent development in thermally-activated desalination methods: achieving an energy efficiency less than 2.5 kWh_{elec}/m^3. Desalin Water Treat 2015;1–10. https://doi.org/10.1080/19443994.2015.1035499.
[75] Ng KC, Thu K, Oh SJ, Ang L, Shahzad MW, Ismail AB. Recent developments in thermally-driven seawater desalination: energy efficiency improvement by hybridization of the MED and AD cycles. Desalination 2015;356:255–70.
[76] Shahzad MW, Ng KC, Thu K, Saha BB, Chun WG. Multi effect desalination and adsorption desalination (MEDAD): a hybrid desalination method. Appl Therm Eng 2014;72:289–97.
[77] Kim Choon NG, Thu K, Shahzad MW, Chun WG. Progress of adsorption cycle and its hybrid with conventional MSF/MED processes in the field of desalination. Int Desalin Assoc J Water Desalin Reuse 2014;6(1):44–56. https://doi.org/10.1177/2051645214Y.0000000020.
[78] Shahzad MW, Myat A, WonGee C, Ng KC. Bubble-assisted film evaporation correlation for saline water at sub-atmospheric pressures in horizontal-tube evaporator. Appl Therm Eng 2013;50:670–6. https://doi.org/10.1016/j.applthermaleng.2012.07.003.
[79] Moran MJ. Availability analysis: a guide to efficient energy use. Englewood Cliffs, NJ: Prentice Hall; 1981.
[80] Kotas TJ. The exergy method of thermal plant analysis. Malabar, FL: Krieger; 1995.
[81] Tribus M, Evans RB, Crellin GL. In: Spiegler KW, editor. Principles of desalination. New York: Academic Press; 1966.
[82] Gaggioli RA. Second law analysis for process and energy engineering. ACS symposium series, Washington, DC: American Chemical Society; 1983.

[83] Al-Sulaiman FA, Ismail B. Exergy analysis of major re-circulating multi-stage flash desalting plants in Saudi Arabia. Desalination 1995;103:265–70.
[84] O.A. Hamed, A.M. Zamamir, S. Aly, N. Lior, Thermal performance and exergy analysis of a thermal vapor compression desalination system, Energy Convers Manag 37 (4) (1996) 379-387.
[85] Spliegler KS, El-Sayed YM. The energetics of desalination processes. Desalination 2001;134:109–28.
[86] Sharqawy MH, Lienhard V JH, Zubair SM. In: Formulation of seawater flow exergy using accurate thermodynamic data. Proceeding of the International Mechanical Engineering Congress and Exposition IMECE-2010, November 12–18, Vancouver, British Columbia, Canada; 2010.
[87] Cerci Y. Improving the thermodynamic and economic efficiencies of desalination plants [PhD dissertation]. Mechanical Engineering, University of Nevada; 1999.
[88] Cengel YA, Cerci Y, Wood B. In: Second law analysis of separation processes of mixtures. ASME international mechanical engineering congress and exposition, Nashville, Tennessee, November 14e19; 1999.
[89] Cerci Y. Exergy analysis of a reverse osmosis desalination plant in California. Desalination 2002;142:257–66.
[90] N. Kahraman, Y.A. Cengel, B. Wood, Y. Cerci, Exergy analysis of a combined RO, NF, and EDR desalination plant, Desalination 171 (2004) 217e232.
[91] Kahraman N, Cengel YA. Exergy analysis of a MSF distillation plant. Energy Convers Manag 2005;46:2625–36.
[92] Banat F, Jwaied N. Exergy analysis of desalination by solar-powered membrane distillation units. Desalination 2008;230:27–40.
[93] Nafeya AS, Fath HE, Mabrouka AA. Exergy and thermo-economic evaluation of MSF process using a new visual package. Desalination 2006;201:224–40.
[94] Mabrouka AA, Nafeya AS, Fath HE. Analysis of a new design of a multi-stage flash-mechanical vapor compression desalination process. Desalination 2007;204:482–500.
[95] Nafeya AS, Fath HE, Mabrouka AA. Thermo-economic design of a multi effect evaporation mechanical vapor compression (MEEeMVC) desalination process. Desalination 2008;230:1–15.
[96] Kamali RK, Mohebinia S. Experience of design and optimization of multi-effects desalination systems in Iran. Desalination 2008;222:639–45.
[97] Kamali RK, Abbassi A, Sadough Vanini SA, Saffar Avval M. Thermodynamic design and parametric study of MED-TVC. Desalination 2008;222:596–604.
[98] Ameri M, Mohammadi SS, Hosseini M, Seifi M. Effect of design parameters on multi-effect desalination system specifications. Desalination 2009;245:266–83.
[99] Wang Y, Lior N. Performance analysis of combined humidified gas turbine power generation and multi-effect thermal vapor compression desalination systems-part 1: the desalination unit and its combination with a steam-injected gas turbine power system. Desalination 2006;196:84–104.
[100] Chacartegui R, Sánchez D, di Gregorio N, Jiménez-Espadafor FJ, Muñoz A, Sánchez T. Feasibility analysis of a MED desalination plant in a combined cycle based cogeneration facility. Appl Therm Eng 2009;29:412–7.
[101] Darwish MA, Al Otaibi S, Al Shayji K. Suggested modifications of power-desalting plants in Kuwait. Desalination 2007;216:222–31.
[102] Darwish MA, Alotaibi S, Alfahad S. On the reduction of desalting energy and its cost in Kuwait. Desalination 2008;220:483–95.

[103] Deng R, Xie L, Lin H, Liu J, Han W. Integration of thermal energy and seawater desalination. Energy 2010;35:4368–74.
[104] Palenzuela P, Zaragoza G, Alarcón D, Blanco J. Simulation and evaluation of the coupling of desalination units to parabolic-trough solar power plants in the Mediterranean region. Desalination 2011;281:379–87.

Further reading

[1] Li A, Thu K, Ismail AB, Shahzad MW, Ng KC. Performance of adsorbent-embedded heat exchangers using binder-coating method. Int J Heat Mass Transfer 2016;92:149–57.
[2] Ng KC, Chua HT, Chung CY, Loke CH, Kashiwagi T, Akisawa A, et al. Experimental inventogation of the silica gel-water adsorption isotherm characteristics. Appl Therm Eng 2001;21:1631–42.

Forward osmosis feasibility and potential future application for desalination

2

Ali Altaee, Adnan A. Alanezi†, Alaa H. Hawari‡*
*University of Technology Sydney, Sydney, NSW, Australia, †The Public Authority for Applied Education and Training (PAAET), Sabah Alsalem, Kuwait, ‡Qatar University, Doha, Qatar

2.1 Introduction

There is an urgent need for securing water supplies to large cities in the near future to overcome the water scarcity problem. Several environmental and socioeconomic factors, such as climate change, ground contamination, population increase, extensive agriculture activities, and industrial growth, have compromised freshwater availability not only in arid areas but also affected other regions elsewhere [1–5]. Traditional water reservoirs are not able to meet the high water demands; hence, new supply sources are needed. Several measures are being explored to alleviate the intensity of the water shortage problem, such as wastewater recycling/reuse and seawater desalination. Water reuse is more economical than seawater desalination, but there are many ethical, psychological, and health issues preventing its widespread application for human use [6,7]. Therefore, seawater desalination appears to be the most effective solution for freshwater supply in terms of the product water quality and technology reliability [8]. The main technical options for desalination are thermal processes, multistage flashing (MSF) and multieffect distillation (MED), and membrane processes, such as reverse osmosis (RO) and nanofiltration (NF) [3,9].

Desalination technologies are energy intensive processes and require significant operating experiences to ensure successful implementation. Currently, thermal technologies are almost strictly used in the Middle East because of the extreme water shortage and the cheap energy sources. However, RO remains the most popular desalination technology in the rest of the world because of its significantly lower power consumption compared to thermal technologies [10]. The rapid advance in membrane technologies has resulted in the development of high-performance RO membranes with a high water permeability and a high salt rejection rate [3,11]. Current RO membranes are capable of treating a wide range of seawater salinities and producing high-quality product water. Nevertheless, the many advantages of RO technology for seawater desalination are compromised by a number of drawbacks [3] including intensive pretreatment requirements and performance deterioration over time due to membrane fouling and repeated cleaning [12]. Furthermore, RO power consumption is

Emerging Technologies for Sustainable Desalination Handbook. https://doi.org/10.1016/B978-0-12-815818-0.00002-3

significantly affected by the salt concentration in the feed with a greater feed salinity requiring greater power consumption. Unlike thermal processes, RO performance decreases with an increase in the feed salinity and hence is more efficient at lower feed salinities [13]. Extensive research efforts have been focused on reducing the cost of seawater desalination. For example, dual stage NF and dual stage NF-brackish water RO (BWRO) processes have been proposed as alternatives to the conventional RO process [14]. The application of high permeability and high rejection rate NF and BWRO membranes in seawater desalination was demonstrated to be more energy efficient than RO alone. Unfortunately, the complexity of dual stage NF processes requires significant hands-on experience for successful operation [14].

More recently, there has been growing interest in forward osmosis (FO) for its potential to reduce the desalination cost [15–19]. The FO process operates close to atmospheric pressure with the driving force for water transport provided by the osmotic pressure gradient across the semipermeable membrane. Two solutions of different concentrations and osmotic pressure are pumped into the FO membrane; the high concentration solution is called the draw solution, whereas the low concentration solution is called the feed solution. Freshwater transports from the feed to the draw solution due to the osmotic pressure gradient across the semipermeable FO membrane. Ionic species in the feed solution are retained by the FO membrane because of its high selectivity and rejection rate which may exceed 90% to the NaCl solution [20]. Water transport across the FO membrane dilutes the draw solution; the diluted draw solution requires further treatment for freshwater extraction and draw solution regeneration. For the membrane regeneration processes, membrane processes have been proposed to regenerate of the draw solution depending upon the nature of the osmotic agent [15]. Practically, technical limitations of regeneration processes, such as fouling propensity, energy efficiency, and viability, should be considered in the design stage of the desalination system.

2.2 FO pretreatment of seawater to RO

FO has been proposed as a pretreatment process for the RO membrane in seawater desalination and wastewater treatment [15]. The idea was initially proposed to reduce the energy consumption and to increase the water recovery rate of the desalination process. Seawater enters the FO membrane module as a feed or a low concentration solution and a high concentration or a draw solution is pumped on the other side of the membrane (Fig. 2.1). The osmotic pressure difference between the draw and feed solution induces freshwater transport across the FO membrane separating the feed and draw solution. Several compounds have been suggested for the draw solution including inorganic metal salts, organic solutions, electromagnetic particles, and thermolytic solutions. By all accounts, no standardized draw solution offers the option to select a suitable draw solution formula for different applications. The high purity draw solution can be engineered to reduce fouling of the RO membrane; this will also reduce chemical and antiscalant use for fouling control in the RO system.

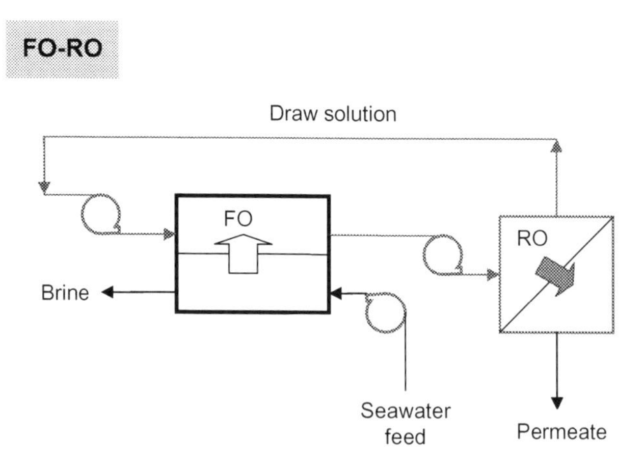

Fig. 2.1 Schematic diagram of the FO pretreatment of seawater to RO.

Furthermore, high RO recovery rates can be achieved using the FO-RO process, which is problematic in conventional RO systems. Increasing the recovery rate of the RO regeneration process to an optimum amount would reduce the cost of desalination. Fig. 2.2 illustrates the impact of the RO recovery rate on the specific power consumption.

The optimal recovery rate in the RO desalination process, which is defined as the recovery rate that minimizes the specific energy consumption, depends on the feed salinity of the feed solution (Fig. 2.2). The specific energy consumption of the RO

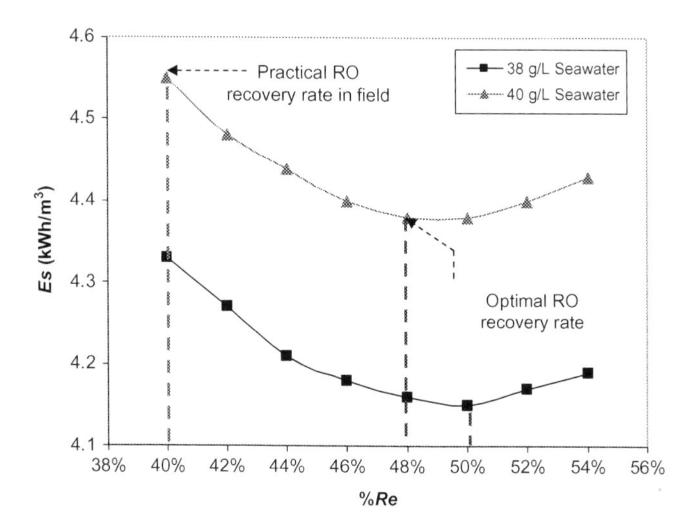

Fig. 2.2 Impact of the recovery rate on the specific power consumption at different feed salinities. Calculations performed using the Reverse Osmosis System Analysis software (ROSA); the feed flow rate was 7 m^3/h using the SW30HRLE-400i module.

membrane decreases gradually with an increase in the membrane recovery rate. This is due to the increase in the permeate flow rate according to the following equation:

$$Es = \frac{P_f}{\eta \times \mathrm{Re}} \tag{2.1}$$

where Es is the specific power consumption of the RO process (kWh/m^3), P_f is the feed pressure (bar), Re is the recovery rate of the RO system (%), and η is the pump efficiency (assumed $\eta = 0.8$). As the recovery rate increases over a certain level, Es increases as well due to the greater feed pressure required to obtain the greater permeate flux and overcome the osmotic pressure of the concentrated brine. The optimum specific power consumption is achieved at 50% and 48% recovery rates for 38 and 40 g/L seawater salinities, respectively. Practically, the maximum recovery rates of 36 and 40 g/L feed salinities are maintained below 40%; increasing the recovery rate over 40% has a negative impact on the performance of the RO system in the long term [16,21]. Fouling and scaling of the RO membrane have been reported at increased recovery rates [22]. The results suggest that the difference between the operating and optimal recovery rate depends on the concentration of the feed solution. Previous studies by McGovern and Lienhard [16] have found that RO is more energy efficient than FO-RO systems at 35 g/L seawater salinity; hence, FO application for seawater pretreatment should be focused on high salinities.

Membrane fouling and scaling are the limiting factors to increase the recovery rate of the RO system. Therefore, the field recovery rate is typically less than the optimal recovery rate to avoid membrane fouling (Fig. 2.3). Such effect is more obvious at

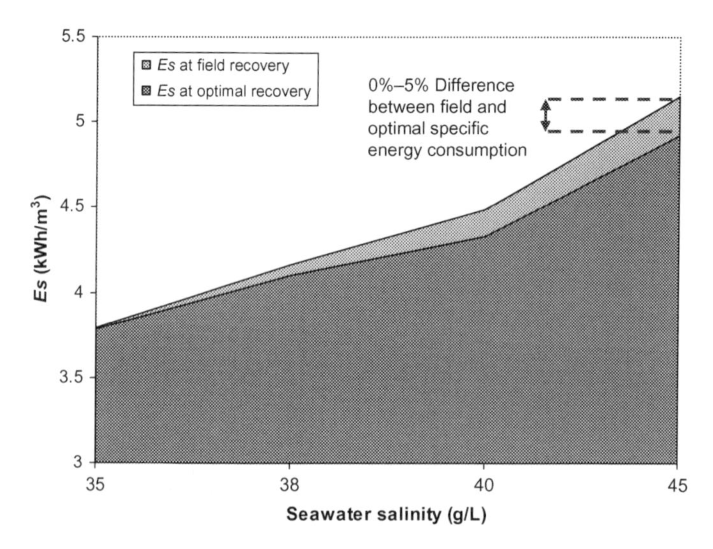

Fig. 2.3 Specific power consumption of the RO system at field and optimal recovery rates. Calculations performed using Reverse Osmosis System Analysis software (ROSA); the feed flow rate was 7 m^3/h using the SW30HRLE-400i module.

high seawater salinity, which is a challenging environment for an RO system. Initially, the high feed salinity requires more pressure and energy for treatment, which cannot practically be achieved because it accelerates the membrane fouling problem. Commercial RO plants treating seawater having a salinity of >45 g/L operate at a recovery rate as low as 38% [23–25], whereas the optimal recovery at such feed salinity is ~46% based on the model calculations in Fig. 2.3. In such environmental conditions, FO pretreatment would, however, have a niche application because the lower fouling propensity of FO compared to the RO process. RO is more adaptable for feedwater at a standard seawater salinity, that is, 35 g/L. At such a feed salinity, a recovery rate of 48%–50% can be achieved by commercial RO systems, which is close or equal to the optimal recovery rate of 50%. Previous studies have found that FO pretreatment of seawater with 35% salinity would increase the energy consumption of RO desalination [13]. The calculations for a 35 g/L seawater feed solution assuming a 50% recovery rate and an energy recovery device (ERD) were provided. Elevated salinity has hence been suggested for FO pretreatment due to its high fouling propensity.

2.3 Potential use of FO pretreatment for RO

FO pretreatment would increase the capital cost of the RO desalination process due to the additional membrane stage, chemicals, and pumping energy required to feed and draw solutions to the FO process. However, the justification for using FO is to reduce RO membrane fouling, which is the most energy intensive stage of an FO-RO system [16,21]. The RO process is responsible for freshwater separation from the diluted draw solution and the rejuvenation of draw solution for reuse. Elevated hydraulic pressures in excess to the osmotic pressure of seawater are often applied in the RO stage for the regeneration of draw solution. This assumption applies to an ideal FO process in which the concentration of the draw solution at the membrane outlet (C_{Do}) is equal to the concentration of the seawater at the membrane entrance (C_{Fi}), that is, $C_{Do} = C_{Fi}$. Conventional RO systems suffer from flux decline due to membrane fouling and scaling; annual flux decline may reach 10% of the original flux [23]. Flux decline caused by lost membrane permeability has consequences on the power consumption of the RO system. As such, the feed pressure is typically increased to maintain the membrane flux and the RO recovery rate; this increases the energy consumption of the desalination process. RO membrane flux decline is reduced when FO pretreatment is applied because of the high purity of the draw solution.

The membrane flux in the RO membrane can be estimated from the solution diffusion equation:

$$J_w = A_w(\Delta\pi - \Delta P) \tag{2.2}$$

where $\Delta\pi$ is the osmotic pressure gradient (bar), A_w is the water permeability coefficient (L/m^2 h bar), and ΔP is the pressure difference across the RO membrane (bar). Eq. (2.2) estimates the initial water flux in a new membrane but this flux declines over time due to fouling effects. The water flux in a fouled RO membrane can be roughly estimated from the following expression assuming a 7% flux decline per annum [21]:

$$J_n = J_o - (0.07n \cdot J_o) \tag{2.3}$$

In Eq. (2.3), J_n is the permeate flux in year n (in L/m^2/h), J_o is the permeate flux of the new RO membrane, and n is number of years. Typically, the RO membrane lasts for 5 years in a commercial desalination plant but may be replaced earlier due to membrane fouling or bad operating conditions. For a 20% flux loss, the term $(0.07n\ J_o)$ in Eq. (2.3) will be $(0.2\ J_o)$. The specific power consumption, Es (kWh/m^3), of the new and fouled membrane can be calculated from the following equation:

$$Es_n = \frac{P_f \times Q_f}{\eta \times J_n \times A} \tag{2.4}$$

where P_f is the feed pressure (bar), Q_f is the feed flow rate (m^3/h), η is the pump efficiency ($\eta = 0.8$), and A is the RO membrane area (m^2). The specific power consumption of the new membrane, $n = 0$, is expected to be less than the specific power consumption of the old membrane, $n > 0$. Realistically, fouling affects the membrane water permeability coefficient, A_w, which appears in Eq. (2.2). This, in turn, reduces the water flux in the RO membrane and causes energy loss over time.

Typically, flux decline depends on the feedwater quality and operating conditions of the RO membrane. This is inevitable in a membrane filtration process but increases due to bad feedwater quality and high operating pressures. The silt density index (SDI) is the parameter often used to describe the feedwater quality to an RO membrane. SDI values vary between 1 and 5 with an SDI value equal to 1 indicative of high-quality feedwater, such as RO permeate, and an SDI value equal to 5 indicates low-quality feedwater, such as surface seawater after conventional sand filtration pretreatment. Most commercial RO plants use multimedia sand filtration for the pretreatment of seawater to reduce the capital and operation costs [26,27]. The SDI of the feedwater after conventional sand filtration pretreatment is approximately SDI 5, which is used when calculating the specific power consumption for the new and fouled RO membrane. The impact of membrane fouling can be evaluated based on the expected flux reduction assuming that 80% of the nominal flux could be maintained after 3 years of RO operation. This is primarily due to membrane fouling due to scaling, fouling, and compaction. Model calculations based on 20% loss of water flux over 3 years reveal the impact of the flux decline on the specific power consumption (Fig. 2.4). There is a 25% increase in the specific power consumption due to the flux decline over 3 years.

2.4 Selection of draw solution

There is a wide range of chemicals and synthetic compounds that have been proposed for the draw solution [28]. The list is long and continues to increase to find an ideal draw solution for different filtration purposes. The ideal draw solution should be

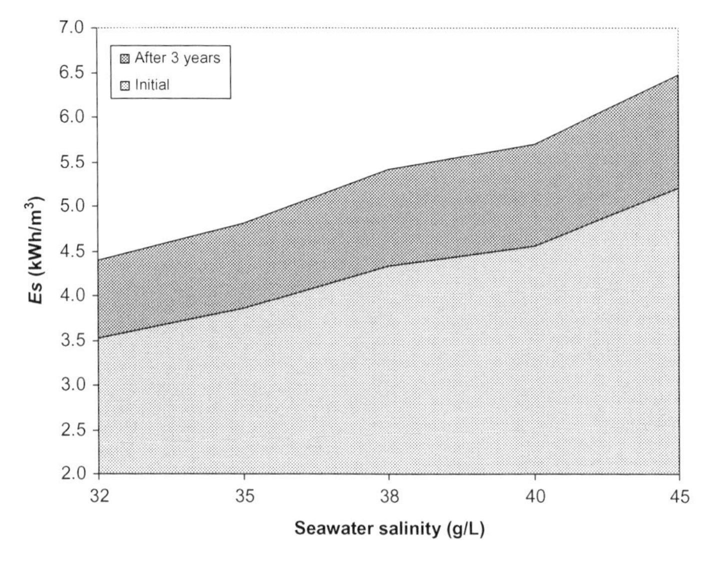

Fig. 2.4 Specific power consumption of an RO membrane after 3 years of operation. Simulation was performed using the ROSA software and the SW30HRLE-400i module assuming 20% flux loss over 3 years; the feed flow rate was 7 m^3/h.

inexpensive, commercially available, easy to use and regenerate, and environmentally acceptable [21]. The long list of draw solutions has made the process of selecting a suitable draw solution slightly laborious and not straightforward. When choosing a draw solution, a researcher should consider the amount of draw solution required for the process. Desalination plants typically have a large capacity that may be in excess of 100,000 m^3/day; hence, the amount of draw solution and ease of regeneration should be considered in the design stage. One ideal draw solution for desalination is an NaCl solution because of its high osmotic pressure and ease of regeneration. NaCl forms 85% of the seawater's composition, which minimizes cross contamination of the draw and feed solution due to solute diffusion across the FO membrane. $MgCl_2$ and $MgSO_4$ are among other ionic solutions which have been suggested for the FO draw solution, but they are more expensive than NaCl and cross contamination due to the diffusion of NaCl and other ionic species in seawater is highly likely during the FO process. Furthermore, $MgSO_4$ should be highly concentrated to generate a desirable osmotic pressure. More information about draw solutions and osmotic pressures can be found in the literature [29].

2.5 Seawater salinity

Seawater salinities vary depending on location with a high salinity of 48,000 mg/L in the gulf region of the Middle East. High salinity means that more osmotic pressure is required for freshwater extraction from seawater and greater energy requirements for

Table 2.1 **Seawater composition**

SW TDS (mg/L)	Ions concentration (mg/L)							
	K^+	Na^+	Mg^{2+}	Ca^{2+}	HCO_3^-	Cl^-	SO_4^{2-}	SiO_2
35,000	387	10,778	1293	421	142	19,406	2710	1.0
40,000	441	12,278	1473	480	162	22,105	3086	1.1
45,000	496	13,812	1657	539	182	24,868	3472	1.2

the regeneration of the draw solution. For conventional RO, the system recovery rate decreases with an increase in the salinity to reduce membrane fouling. Recovery rates of 46%, 40%, and 38% were suggested for seawater salinities in this study [30]. Seawater with standard salinity (35 g/L) and 40 and 45 g/L were evaluated for RO and FO-RO treatment. These concentrations represent the low, medium, and high range of seawater feed salinities that may be encountered in a real situation. Table 2.1 shows the composition of seawater for the selected salinities.

2.6 FO membrane and model

The water flux, J_w (L/m^2/h), in the RO membrane can be estimated from the solution-diffusion expression below:

$$J_w = A_w(\Delta\pi - \Delta P) \tag{2.5}$$

where A_w is the membrane permeability coefficient (L/m^2/h/bar), ΔP is the pressure gradient (bar), and $\Delta\pi$ is the osmotic pressure gradient (bar). Eq. (2.5) is strictly valid for a fully retentive membrane with a reflection coefficient equal to unity. Eq. (2.5) does not account for the concentration polarization effects in the FO membrane. The FO concentration polarization develops on both sides of the membrane, that is, internal and external, which makes it more complicated and severe than concentration polarization in the conventional RO filtration process. The current FO model has been derived from the classical expression of water flux in an RO membrane and has been experimentally validated on a plate and frame FO unit using a flat sheet membrane coupon [20]:

$$J_w = A_w\left(\frac{\pi_{Db}e^{\left(\frac{-J_w}{k}\right)} - \pi_{Fb}e^{(J_wK)}}{1+\frac{B}{J_w}\left(e^{(J_wK)} - e^{\left(\frac{-J_w}{k}\right)}\right)} - \Delta P\right) \tag{2.6}$$

Table 2.2 **FO model parameters**

Parameter	k (m/h)	K (s/m)	A_w (L/m^2 h bar)	B (m/h)
Value	0.306	1.4×10^5	0.67	4×10^{-4}

where K is the solute resistivity for the diffusion within the porous support layer (s/m), k is the bulk mass transfer coefficient (m/s), and π_{Db} and π_{Fb} are the osmotic pressures of the bulk draw and feed solutions (bar), respectively. The Van't Hoff equation was used to estimate the osmotic pressure of the draw and feed solutions. The equation provides a rough estimation of a solution's osmotic pressure but the accuracy decreases at a high concentration: >1 mol/L for NaCl, for instance. However, this equation has been widely accepted because of its simplicity and relative accuracy [31]. The specific power consumption of the FO process, E_{s-FO}, was estimated from the following expression [15]:

$$E_{s-FO} = \frac{(P_f Q_f + P_d Q_d)}{\eta \times Q_p} \tag{2.7}$$

where P_f and P_d are the feed pressures of the feed and draw solutions (bar), respectively, Q_f and Q_d are the flow rates of the feed and draw solutions (m^3/h), respectively, Q_p is the permeate flow rate (m^3/h), and $\eta = 0.8$ is the pump efficiency. Finally, the model parameters of the FO process shown in Table 2.2 are based on a cellulose triacetate (CTA) flat sheet membrane provided by the Hydration Technology Innovation Company (HTI), United States. More information about the experimental parameters and model validation can be found in the literature [20].

2.7 RO energy recovery

Despite the relatively low-energy consumption of RO desalination, conventional RO is still considered an energy intensive process. Typically, the power consumption of conventional RO desalination for standard seawater salinity is ~4 kWh/m^3 [16]. Of the power consumed, 40% is attributed to the high-pressure pump for seawater filtration. RO with an ERD system is often designed for large capacity desalination plants; the ERD function is to recover the energy from the brine concentrate before discharging to the sea.

There are different types of ERD systems with operating efficiencies that vary from 60% to 98% [27,30]. Pelton wheels, pressure exchangers [PX], and turbochargers are types of ERDs that have been used for energy recovery in the RO plant. The efficiency of a turbocharger is between 55% and 60%, whereas it is between 80% and 90% for a Pelton wheel [32]. PX with >95% efficiency are also available but the cost is rather high compared to a Pelton wheel and turbocharger. An ERD is a very common option in large capacity RO plants but may be overlooked in small capacity installations because of the increased cost [21]. The formula used to estimate the specific power consumption of the RO process with the ERD is [32]

$$Es = \frac{\frac{Q_{HPP}(P_{HPP} - P_F)}{\eta_{HPP}} + \frac{Q_{BP}(P_{HPP} - P_{BPin})}{\eta_{BP}}}{Q_p} \tag{2.8}$$

where Q_{HPP} and Q_{BP} are the feed flow rates to the high-pressure pump and the booster pump, respectively (m^3/h), η_{HPP} *and* η_{BP} *are* the high-pressure pump and booster pump efficiency (assumed to be 80%), P_{HPP} and P_{BPin} are the high-pressure pump outlet pressure and the booster pump inlet pressure (bar), and P_F is the pressure of the feed flow (bar). In this study, three ERD efficiencies were evaluated for RO desalination. These efficiencies were 60%, 80%, and 98%, which represent the performance of the turbocharger, Pelton wheel, and PX ERD systems, respectively.

2.8 RO membrane fouling and FO pretreatment

Fouling is inevitable during the membrane filtration processes and results in RO flux decline. The percentage of flux decline depends on the pretreatment method of seawater, which is represented by the SDI value. In seawater with conventional pretreatment, the SDI is ~5, and water flux declines between 7.3% and 9.9% per annum [23]. For MF pretreatment with an $SDI < 3$, the annual flux decline is between 4.4% and 7.3%. For an RO permeate with an $SDI < 1$, the water flux declines between 2.3% and 4.4% per annum; this case applies to the FO pretreatment of seawater for RO desalination. The following assumptions were considered in the evaluation of FO-RO and RO desalination of seawater:

1. Seawater salinities of 35, 40, and 45 g/L were evaluated for the FO-RO and RO systems.
2. The SDI of the feedwater to the RO system in the FO-RO system was equal to 1, whereas an SDI between 3 and 5 was used for the RO system using MF and conventional pretreatment, respectively.
3. Flux decline in the RO membrane of the FO-RO system was fixed at 3% per year. For the RO system, a flux decline between 5% and 10% per year was evaluated.
4. Three ERD efficiencies of 60%, 80%, and 98% were evaluated
5. The specific power consumption of the FO-RO system was the total power consumption of the FO process, as shown in Eq. (2.7), and the RO process, as shown in Eq. (2.8).
6. The feed flow rate was 4 and 7 m^3/h for the RO in the FO-RO process and the conventional RO process, respectively. Typically, the feed flow rate of the RO process increases with the SDI of the feedwater.
7. NaCl concentrations of 1, 1.2, and 1.4 M were used for the draw solution for the 35, 40, and 45 g/L seawater salinities, respectively.

The RO membrane life was estimated to be ~5 years for the FO-RO and RO systems. The water flux of the RO membrane was calculated using the ROSA software, and the specific power consumption of the FO-RO and RO systems was calculated over 5 years of membrane life. Furthermore, we assumed that FO fouling was reversible, which is in agreement with the pilot plant results carried out by Modern Water and showed insufficient flux decline over a period of 12 months [33,34]. Therefore, the flux decline in the FO-RO was mainly due to the RO membrane (3% per annum).

The results in Fig. 2.5A show the impact of the flux decline on the average power consumption of the FO-RO and conventional RO process provided with a 60% efficiency ERD system. The average specific power consumption was obtained by calculating the specific power consumption of the desalination system over 5 years. The results show a stable FO-RO performance over a period of 5 years due to the constant flux decline of the RO system: 3% per year. The FO pretreatment provides high-quality feed to the RO membrane and forms a physical barrier to organic and inorganic fouling. Conversely, the specific power consumption of the RO process increased with an increase in the flux decline, which reflects the direct impact of the feedwater pretreatment and quality on the RO membrane. Apparently, the FO-RO process was more energy efficient than the RO process for all seawater salinities and water flux decline percentages except at a 5% flux decline. This suggests that, for seawater of high quality and proper pretreatment, using RO process desalination in FO pretreatment would have no advantage but would increase the seawater desalination power consumption. Using MF pretreatment or well water would help achieve the targeted feed quality for the RO membrane. This, however, may increase the cost of pretreatment, which should be added to the final product cost. It is still debatable whether replacing conventional pretreatment with MF would reduce the cost of desalination. The capital cost of an MF pretreatment system is greater than the cost of conventional sand filtration pretreatment [3]. Practically, MF pretreatment reduces membrane fouling and the cleaning frequency of the RO membrane. This, in turn, will reduce the plant shutdown period and chemical use. For new installations, a pilot plant test should be performed to determine the proper pretreatment based on the feed quality and the advantages of using expensive pretreatment options in the long term.

In the FO-RO process, the FO pretreatment reduced the power consumption of the desalination, but there is a capital cost incurred due to the extra FO membranes. The cost of the FO membranes is rather high but the lifetime could be extended over 5 years because of their low deterioration factor. Nevertheless, the FO-RO system could reduce the RO energy consumption by up to 18% at a flux decline value of 10%. This represents significant power consumption savings, which is particularly important for large capacity desalination plants. The results suggest that FO should be considered as an alternative to RO when the quality of the feedwater is low or in the case of an insufficient pretreatment process. Expensive pretreatment options, such as using a dissolved air floatation (DAF) system for the removal of colloidal particles and organic matters, have been applied for RO desalination [35]. The Fujairah, UAE RO plant has recently applied a DAF system for the treatment of seawater feed to an RO plant. The system was initially installed to protect the RO in the event of red tide incidents. However, the DAF system has little or no effect on the removal of inorganic matters from the seawater, which are responsible for membrane scale fouling.

For 80% ERD, the impact of the seawater salinity and flux decline on the specific power consumption of the RO system is illustrated in Fig. 2.5B. In general, the

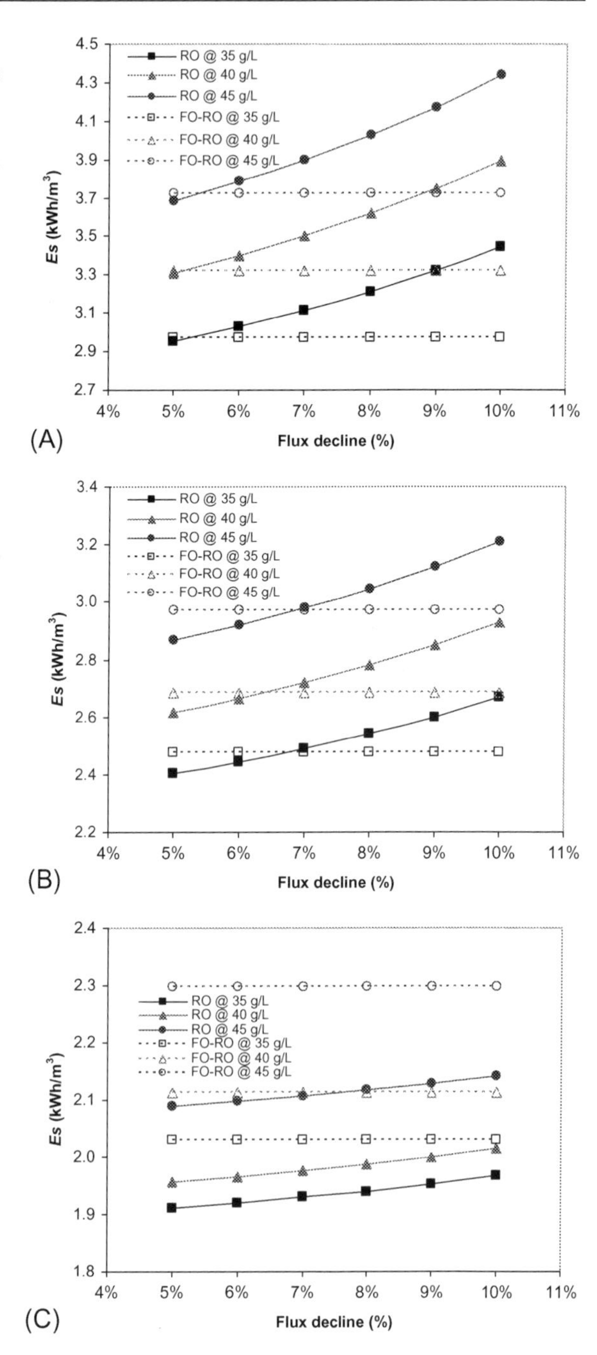

Fig. 2.5 Average specific power consumption of FO-RO and RO systems at different fouling factors. Seawater salinities and compositions are shown in Table 2.1. (A) ERD efficiency is 60%, (B) ERD efficiency is 80%, and (C) ERD efficiency is 98%.

average specific power consumption of the seawater desalination was less than that of the RO and FO-RO systems with a 60% efficient ERD. The performance of the FO-RO system remained constant regardless of the percentage of flux decline due to fouling. These results are based on the assumption that FO fouling was reversible and that there was 3% RO fouling in the FO-RO system. Using an 80% efficient ERD results in 16%–20% savings in the average power consumption of the FO-RO system compared to the FO-RO system with 60% ERD. This is significant particularly for large desalination capacity plants. For the conventional RO process, the average specific power consumption for desalination increased with the percentage of flux decline and was greater for 45 g/L seawater salinity. Similar to the FO-RO process, for the RO process the average power consumption was less with a 60% efficient ERD system. The simulation results reveal that RO performance was very competitive to the FO-RO performance, especially at low percentage flux decline values, emphasizing the significance of feedwater quality and pretreatment technologies. Flux decline values of 5% to 7% indicate a feed SDI of between 1 and 3, which can only be achieved using MF pretreatment for normal quality seawater feed. The average specific power consumption of the RO system was 3% less than for the FO-RO system at 5%–7% flux decline. However, the gap in the average power consumption between the RO and FO-RO systems increased at high flux decline values. Energy savings of ~8% were achieved when FO-RO was used for the desalination of seawater at a 10% flux decline value.

The results show that the RO process was the better desalination option when the membrane fouling and flux decline percentage were low. Although the RO desalination energy was 3% less than that for the FO-RO process, the capital cost of RO is less than that for the FO-RO process. However, RO requires a strict pretreatment method to be competitive with the FO-RO process. The seawater SDI should be <3 to reduce the fouling propensity of the RO membrane. MF pretreatment would achieve that but at the expense of capital cost. As seawater quality becomes more challenging or the pretreatment is insufficiently efficient, the FO-RO process becomes more energy efficient than the conventional RO process. It should be mentioned that MF pretreatment is not enough for the RO process to compete with the FO-RO process unless high-efficiency ERD is applied. Bad water quality could be handled by upgrading the pretreatment process but that would only be possible with a proper ERD system.

The significance of ERD efficiency on the desalination energy demands is more obvious in Fig. 2.5C. Using a 98% ERD efficiency resulted in a sharp reduction in the RO energy demands, and the RO method became more competitive and cost-effective for desalination than for FO-RO. While the average specific power consumption of the FO-RO system remained unaffected by the change in the flux decline, there was a slight increase in the average RO specific power consumption due to the change in the flux decline from 5% to 10%. A maximum 3% increase in the RO average power consumption occurred due to the increase in the flux decline from 5% to 10%. Low feed quality increases the percentage of RO flux decline and the desalination energy requirements, but this effect is much lower for the RO membrane provided with a high-efficiency ERD system. Compared to other ERD systems, a 12% and 18% increase in the average power consumption could be incurred due to an increase in the

RO flux decline from 5% to 10%, respectively, for ERD efficiencies of 80% and 60%, respectively. ERD systems, such as PX, which have a performance ratio >95%, are good candidates for reducing the RO energy consumption. However, it should be mentioned that, although high-efficiency ERD systems would reduce the desalination energy cost, membrane fouling at a high RO flux decline may require membrane replacement. This could increase the maintenance cost of the RO desalination plant.

In general, for an RO plant with a 98% efficient ERD system, the average specific power consumption of the RO was 6%–10% lower than that of the FO-RO system. Furthermore, RO requires less membrane area and footprint than the FO-RO system. The capital investment for a high-performance ERD will pay off the low energy consumption during desalination.

2.9 Pretreatment cost and impact on desalinated water cost

To date, we have discussed the impact of RO flux decline of the energy consumption of RO and FO-RO coupled with ERD systems with 60%–98% performance efficiencies. The impact of pretreatment on the SDI of the feedwater was briefly discussed, and its effect on the RO flux decline was succinctly explained. However, the operating requirement for the RO membrane in the FO-RO system is different from the operating requirement for a conventional RO membrane. FO pretreatment provides a high-quality feed solution to the RO in the FO-RO system; the SDI of the feedwater would be comparable to that of the RO permeate (SDI < 5). This not only reduces membrane fouling but also reduces the operating cost of the desalination plant. For SDI < 1, the RO in the FO-RO system would require a lower feed flow rate than for the RO system. Conventional RO has a feed SDI between 5 and 2; a high concentrate flow rate is required to prevent membrane fouling. As such, RO systems require a higher feed flow rate than FO-RO systems to meet the recommended operating conditions. In this study, we assumed that 4 and 7 m^3/h were the feed flow rates for the RO and FO-RO systems, respectively. The power consumption of the seawater pretreatment, therefore, should be added to the energy cost of desalination.

We performed model calculations to estimate the overall desalination cost, which includes pretreatment, the RO and FO process, if applicable, for the two scenarios. The first scenario was an RO flux decline of 10% using ERD efficiencies of 60% and 98%, and in the second scenario, the RO flux decline was 5% and the ERD efficiencies were 60% and 98%. These scenarios compare the energy efficiency of the RO with the energy efficiency of the FO-RO system for a wide range of membrane fouling and ERD performance efficiencies. The impact of pretreatment on the total average power consumption is illustrated in Table 2.3. The pretreatment processes include the power consumption for pumping seawater from the intake system and the conventional sand filtration sand [36]. The FO power consumption was added to the RO and pretreatment power consumptions in the FO-RO system. For an RO with an ERD efficiency of 98% and a 10% RO flux decline, the total specific power consumption of the FO-RO system was slightly less than that of the RO system. A total power consumption savings of

Table 2.3 **Average specific power consumption for the RO and FO-RO systems at 10% RO flux decline and different seawater salinities**

System	SW (mg/L)	Process	*P* (bar)	*Qf* (m^3/h)	Pump (eff%)	Power (kW)	*Qp – ave* (m^3/h)	*Es* (kWh/m^3)	*Es-FO* (kWh/m^3)	*Es – ave* RO	Est. (kWh/m^3)
RO	35	SW intake	4	7	0.8	0.97	2.44	0.40		3.44[a]	4.04
		Pretreat	2	7	0.8	0.49	2.44	0.20		1.97[b]	2.56
	40	SW intake	4	7	0.8	0.97	2.12	0.46		3.89[a]	4.58
		Pretreat	2	7	0.8	0.49	2.12	0.23		2.01[b]	2.70
	45	SW intake	4	7	0.8	0.97	2.02	0.48		4.34[a]	5.06
		Pretreat	2	7	0.8	0.49	2.02	0.24		2.14[b]	2.86
FO-RO	35	SW intake	4	4	0.8	0.56	1.67	0.33	0.14	2.83[a]	3.47
		Pretreat	2	4	0.8	0.28	1.67	0.17		1.89[b]	2.53
	40	SW intake	4	4	0.8	0.56	1.45	0.38	0.16	3.16[a]	3.90
		Pretreat	2	4	0.8	0.28	1.45	0.19		1.95[b]	2.69
	45	SW intake	4	4	0.8	0.56	1.38	0.40	0.17	3.55[a]	4.32
		Pretreat	2	4	0.8	0.28	1.38	0.20		2.13[b]	2.90

The average power consumption was taken over 5 years.
[a]60% ERD efficiency.
[b]98% ERD efficiency.

between 4% and 0.5% was achieved in favor of the FO-RO system with the upper range occurring at 35 g/L seawater salinity. This insignificant energy saving was not enough for the FO-RO system to compete with the state-of-the-art RO and would be difficult to encourage future investment in FO pretreatment for RO desalination. Conversely, the FO-RO system was 14% more energy efficient than the RO system at 60% ERD efficiency. The results emphasize the importance of ERD performance efficiency on the total power consumption of the desalination process. Despite the low water quality and high RO flux decline, the RO system with a high-efficiency ERD would still be a competitive option for desalination compared to FO-RO, which demonstrated no significant energy saving. However, low performance ERD systems are still in use; for such an RO plant, FO-RO has some advantage over the RO system. The option of installing an extra FO pretreatment process would be weighed against using a high-efficiency ERD system. ERD is a proven technology that has an advantage over the new FO process. More FO pilot plants and small commercial plants should be tested to demonstrate the benefits of the FO process. As mentioned above, when seawater quality is very low, the RO membrane may require more cleaning and maintenance, which increases the operation cost. The RO membrane may require earlier replacement in severe fouling events.

Table 2.4 shows the average specific power consumption if the FO-RO and RO systems are operating at a 5% RO flux decline. The 5% RO flux decline reflects the high feed quality and/or proper pretreatment of seawater and poses a challenging operating condition for the FO-RO system, which is designed for the pretreatment of low-quality water. Apparently, the FO-RO system performed slightly better than the RO system when coupled with a 60% efficient ERD: an ~2% reduction in power consumption was achieved with the FO-RO system. The results hold for the range of seawater salinities evaluated here. Such a small power saving is insufficient for the FO-RO to replace the RO as a competitive technology. For a 98% efficient ERD, the situation is different in which the RO outperformed the FO-RO system. Less than 2% energy saving was achieved with the RO system; however, the RO system requires less membrane area and maintenance than the FO-RO system. This suggests that RO remains the option for seawater desalination when the water quality and pretreatment is good enough to provide a high-quality feed solution. The cost of the FO membranes and replacement should be considered, although there has not yet been any experimental work to provide such data.

2.10 Practical implications

The FO process is relatively a new technology for the pretreatment of seawater for thermal and RO desalinations. The process by itself does not provide desalinated water after pretreatment but a purified ionic saline solution with an osmotic pressure at least equal to that of the feed solution. Previous studies gave misleading inferences about the FO capabilities as an alternative option to the conventional desalination process. A plethora of studies and funds were used investigating the potential of the FO process for seawater desalination with the hope to reduce the energy consumption of

Table 2.4 **Average specific power consumption for the RO and FO-RO systems at 5% RO flux decline and different seawater salinities**

System	SW (mg/L)	Process	*P* (bar)	*Qf* (m^3/h)	Pump (eff%)	Power (kW)	*Qp – ave* (m^3/h)	*Es* (kWh/m^3)	*Es – FO* (kWh/m^3)	*Es – ave* RO	Est. (kWh/m^3)
RO	35	SW intake	4	7	0.8	0.97	2.44	0.40		2.95[a]	3.55
		Pretreat	2	7	0.8	0.49	2.44	0.20		1.91[b]	2.51
	40	SW intake	4	7	0.8	0.97	2.12	0.46		3.31[a]	4.00
		Pretreat	2	7	0.8	0.49	2.12	0.23		1.96[b]	2.64
	45	SW intake	4	7	0.8	0.97	2.02	0.48		3.68[a]	4.40
		Pretreat	2	7	0.8	0.49	2.02	0.24		2.09[b]	2.81
FO-RO	35	SW intake	4	4	0.8	0.56	1.67	0.33	0.14	2.83[a]	3.47
		Pretreat	2	4	0.8	0.28	1.67	0.17		1.89[b]	2.53
	40	SW intake	4	4	0.8	0.56	1.45	0.38	0.16	3.16[a]	3.90
		Pretreat	2	4	0.8	0.28	1.45	0.19		1.95[b]	2.69
	45	SW intake	4	4	0.8	0.56	1.38	0.40	0.17	3.55[a]	4.32
		Pretreat	2	4	0.8	0.28	1.38	0.20		2.09[b]	2.86

The average power consumption was taken over 5 years.
[a]60% ERD efficiency.
[b]98% ERD efficiency.

current technologies. This includes coupling FO with membrane technologies or thermal evaporators for the regeneration and reuse of the draw solution [28].

However, there is no large-scale FO plant for seawater desalination despite the large amount of funds spent in this field. Conversely, recent studies are in favor of the RO technology because of its higher performance and energy efficiency compared to the FO-RO system [15,16,21]. Unfortunately, thus far, there have been no comprehensive FO-RO studies providing experimental data about the power consumption for desalination. Most of the data were theoretical and most of the experimental data were about the FO process. Large-scale and pilot plant studies were just a few and were focused on the performance of the FO process, but no data about RO or thermal plants were provided [33,34]. Incomprehensive and unobjective studies led to the impression that FO would be more energy efficient, overlooking the intensive energy stage of draw solution regeneration. Future studies should be performed on the experimental side and provide data about the entire system energy requirements including the recovery rate.

In general, FO would have niche application in desalinating low-quality seawater, which is a challenging environment for conventional RO systems. In such an environment, more sophisticated pretreatment is required, including MF filtration coupled with DAF to protect the RO membrane; this would increase the cost of pretreatment and the cost of the desalinated water. FO pretreatment could be more competitive to MF and DAF systems, which are currently installed in some large commercial RO plants [30], but unfortunately FO is a rather new technology, whereas MF and DAF are proven technologies. More medium and large pilot plant tests are required to underline the benefit and advantages of FO pretreatment. For low-quality feeds, such as high concentrations of organic matters and colloidal particles, FO pretreatment provides high-quality feed to the RO. In such a case, FO application can be justified for reducing the RO cleaning frequency and membrane replacement due to a premature fouling effect. In this light, future experimental works should focus on the application of FO for pretreatment of high fouling feed solutions to address the issue of extreme RO flux decline over time. This is particularly important for large declination plants or those equipped with low efficiency ERD systems.

2.11 Conclusions

Desalination by FO has been addressed frequently in the literature, but no commercial application has been installed to date. FO may have a niche market for the pretreatment of low-quality feed solutions where conventional pretreatment processes are not sufficient for RO protection. A sharp flux decline in the RO has detrimental effects on the RO performance and results in operating problems. In such a case, FO pretreatment may be recommended to alleviate RO fouling and ensure continuous RO performance. A pilot plant test is a perquisite to support the FO application; power consumption is a key parameter since there are no experimental data about the energy demands of an FO-RO system. The advantage of FO pretreatment is more obvious in the case of a low efficiency ERD system. Many RO plants use a turbocharger or Pelton wheel because of their low to moderate investment in the capital cost compared to the expensive option of a PX. Future work should focus on large FO-RO pilot tests for the pretreatment of low-quality feedwater. Without experimental data, FO application will be doubtful in the near future for seawater desalination and most applications will be limited to small and average installations, such as industrial wastewater.

References

[1] Amy G, Ghaffour N, Li Z, Francis L, Linares RV, Missimer T, et al. Membrane-based seawater desalination: present and future prospects. Desalination 2017;401:16–21.
[2] Sharqawy MH, Lienhard V JH, Zubair SM. Incomplete, on exergy calculations of seawater with application in desalination systems. Int J Therm Sci 2011;50:187–96.
[3] Peñate B, García-Rodríguez L. Current trends and future prospects in the design of seawater reverse osmosis desalination technology. Desalination 2012;284:1–8.
[4] Hamed OA, Kosaka H, Bamardouf KH, Al-Shail K, Al-Ghamdi AS. Concentrating solar power for seawater thermal desalination. Desalination 2016;396:70–8.
[5] Gordon JM, Hui TC. Thermodynamic perspective for the specific energy consumption of seawater desalination. Desalination 2016;386:13–8.
[6] Wester J, Timpano KR, Çek D, Lieberman D, Fieldstone SC, Broad K. Psychological and social factors associated with wastewater reuse emotional discomfort. J Environ Psychol 2015;42:16–23.
[7] Salgot M. Water reclamation, recycling and reuse: implementation issues. Desalination 2008;218:190–7.
[8] Rastogi NK, Cassano A, Basile A. Water treatment by reverse and forward osmosis. In: Advances in membrane technologies for water treatment. Woodhead Publishing Series in Energy; 2015. p. 129–54 [chapter 4].
[9] Bandi CS, Uppaluri R, Kumar A. Global optimization of MSF seawater desalination processes. Desalination 2016;394:30–43.
[10] Shrivastava A, Rosenberg S, Peery M. Energy efficiency breakdown of reverse osmosis and its implications on future innovation roadmap for desalination. Desalination 2015;368:181–92.
[11] Ihm S, Al-Najdi OY, Hamed OA, Jun G, Chung H. Energy cost comparison between MSF, MED and SWRO: case studies for dual purpose plants. Desalination 2016;397:116–25, http://www.sciencedirect.com/science/article/pii/S0011916416307263-item1.
[12] Löwenberg J, Baum JA, Zimmermann Y-S, Groot C, van den Broek W, Wintgens T. Comparison of pre-treatment technologies towards improving reverse osmosis desalination of cooling tower blow down. Desalination 2015;357:140–9.
[13] Wilf M, Klinko K. Optimization of seawater RO systems design. Desalination 2001;138:299–306.
[14] AlTaee A, Sharif AO. Alternative design to dual stage NF seawater desalination using high rejection brackish water membranes. Desalination 2011;273:391–7.
[15] Altaee A, Zaragoza G, Rost van Tonningen H. Comparison between forward osmosis-reverse osmosis and reverse osmosis processes for seawater desalination. Desalination 2014;336:50–7.
[16] McGovern RK, Lienhard JH. On the potential of forward osmosis to energetically outperform reverse osmosis desalination. J Membr Sci 2014;469:245–50.
[17] Thompson NA, Nicoll PG. In: Forward osmosis desalination: a commercial reality. Proceedings IDA World Congress, Perth, Western Australia, September; 2011.

[18] Nicoll PG, Thompson NA, Bedford MR. In: Manipulated osmosis applied to evaporative cooling make-up water—revolutionary technology. Proceedings IDA World Congress, Perth, Western Australia, 4–9 September; 2011.
[19] Blandin G, Verliefde ARD, Tang CY, Le-Clech P. Opportunities to reach economic sustainability in forward osmosis–reverse osmosis hybrids for seawater desalination. Desalination 2015;363:26–36.
[20] Achilli A, Cath TY, Childress AE. Power generation with pressure retarded osmosis: an experimental and theoretical investigation. J Membr Sci 2009;343:42–52.
[21] Altaee A, Millar G, Zaragoza G, Sharif A. Energy efficiency of RO and FO-RO system for high salinity seawater treatment. Clean Techn Environ Policy 2017;19:77–91.
[22] Altaee A, Sharif A. A conceptual NF/RO arrangement design in the pressure vessel for seawater desalination. Desalin Water Treat 2015;54:624–36.
[23] www.membranes.com/docs/trc/Dsgn_Lmt.pdf [Accessed 15 July 2015].
[24] Iwahori H, Ando M, Nakahara R, Furuichi M, Tawata S, Yamazato T. In: Seven years operation and environmental aspects of 40,000 m^3/d seawater RO plant at Okinawa, Japan. Proceedings of IDA Congress, Bahamas; 2003.
[25] Mancha E, DeMichele D, Shane Walker W, Seacord TF, Sutherland J, Cano A. Part II. Performance evaluation of reverse osmosis membrane computer models. Carallo technical report, 2014.
[26] Bates W., Bartels C., Polonio L., Improvements in RO technology for difficult feed waters, hydranautics Nito Dinko, www.membranes.com/docs/papers/New%20Folder/Improvements%20in%20RO%20Technology%20for%20Difficult%20Feed%20Waters%20final%20020311.pdf [Accessed 1 July 2015].
[27] Rachman RM, Ghaffour N, Wali F, Amy GL. Assessment of silt density index (SDI) as fouling propensity parameter in reverse osmosis (RO) desalination systems. Desalin Water Treat 2013;51:1091–103.
[28] Altaee A, Osmosis F. Potential use in desalination and water reuse. J Memb Separ Tech 2012;1:79–93.
[29] Cath TY, Childress AE, Elimelech M. Forward osmosis: principles, applications, and recent developments. J Membr Sci 2006;281:70–87.
[30] The Fujairah 2 reverse osmosis desalination plant, Veolia water, http://www.veolia.com/sites/g/files/dvc171/f/assets/documents/2014/05/fujairah_contract.pdf [Accessed 20 April 2015].
[31] Altaee A, Millar G, Zaragoza G. Integration and optimization of pressure retarded osmosis with reverse osmosis for power generation and high efficiency desalination. Energy 2016;103:110–8.
[32] Stover RL. Seawater reverse osmosis with isobaric energy recovery devices. Desalination 2007;203:168–75.
[33] Tan CH, Ng HY. A novel hybrid forward osmosis-nanofiltration (FO-NF) process for seawater desalination: draw solution selection and system configuration. Desalin Water Treat 2010;13(1–3):356–61.
[34] Bamaga OA, Yokochi A, Zabara B, Babaqi AS. Hybrid FO/RO desalination system: preliminary assessment of osmotic energy recovery and designs of new FO membrane module configurations. Desalination 2011;268:163–9.
[35] Aliff Radzuan MR, Abia-Biteo Belope MA, Thorpe RB. Removal of fine oil droplets from oil-in-water mixtures by dissolved air flotation. Chem Eng Res Des 2016;115:19–33.
[36] Gude VG. Energy consumption and recovery in reverse osmosis. Desalin Water Treat 2011;36:239–60.

Membrane distillation—Principles, applications, configurations, design, and implementation

Abdullah Alkhudhiri, Nidal Hilal*†
*King Abdulaziz City for Science and Technology (KACST), Riyadh, Saudi Arabia,
†Swansea University, Swansea, United Kingdom

3.1 Introduction

3.1.1 Water problem

More than 75% of the earth's surface is covered by water. Saline water represents >97% of it; on the other hand, only 2.5% is fresh water, which is utilized for different uses such as potable water, agriculture, and industry. The average global fresh water consumption is about 300 L/day per person, which equals 100,000 L of fresh water per person annually [1]. Industrial water demands represent a quarter of all water used throughout the world. Today, 700 million people do not have access to clean water. There is a potential water shortage problem in the future. Half of the world's population will face water shortage in 2025 and the situation will become worse in 2050, to reach 75%. This problem can be attributed to the high rate of the human population, limited clean water source, poor water management and programs, and the increasing number of people that live in urban areas (urbanization). In the Middle East, desalinated water is essential and no longer a supplemental water resource. For example, Qatar and Kuwait depend completely on desalination technology for domestic and industrial supplies [2].

The lack of fresh water or potable water in several countries around the world is a result of the shortage of natural water resources. Therefore, it has been necessary to plan and create new methods to provide fresh water that is suitable for human and animal consumption and irrigation.

3.1.2 Desalination technology

According to the International Desalination Association (IDA) [3], the number of people around the world who depend on desalinated water for some or their entire daily use in 2013 is more than 300 million. In the same year, more than 17,000 desalination plants were distributed over 150 countries with total capacity of about 80 million m^3/day. Ghaffour et al. [4] report that 63.7% of the total capacity is produced by membrane processes and 34.2% by thermal processes. The desalination source water is split with about 58.9% from seawater and 21.2% from brackish groundwater sources,

Emerging Technologies for Sustainable Desalination Handbook. https://doi.org/10.1016/B978-0-12-815818-0.00003-5

and the remaining percentage from surface water and saline wastewater. Therefore, there is an urgent need to discover new alternatives and effective sources of fresh water.

3.1.3 Introduction to MD

Membrane distillation (MD) is a promising technology for desalting saline waters and for water treatment. MD is a thermally driven separation (microfiltration) process, in which only vapor molecules are able to pass through a porous hydrophobic membrane. This separation process is driven by the vapor pressure difference existing between the porous hydrophobic membrane surfaces. MD has many attractive features, such as low operating temperatures in comparison to those encountered in conventional processes and the solution (i.e., feed water) is not necessarily heated up to the boiling point. Moreover, the hydrostatic pressure in MD is lower than that used in pressure-driven membrane processes like reverse osmosis (RO). Therefore, MD can be a cost-effective process, which demands less from membrane characteristics too [5]. In this respect, less expensive materials can be used in it such as plastic, for example, to alleviate corrosion problems. According to the principle of vapor-liquid equilibrium, the MD process has a high rejection factor. As a matter of fact, theoretically, complete separation takes place. MD is able to treat high concentrated saline solution near to saturation.

In addition, the membrane pore size required for MD is relatively larger than those for other membrane separation processes, such as RO. The MD process, therefore, suffers less from fouling. The MD system has the feasibility to be combined with other separation processes to create an integrated separation system, such as ultrafiltration (UF) [6] or with an RO unit [7]. Furthermore, MD has the ability to utilize alternative energy sources, such as solar energy [8,9]. The MD process is competitive for the desalination of brackish water and seawater [10,11]. It is also an effective process for removing organic and heavy metals from aqueous solution [12], and from wastewater [13,14]. MD has also been used to treat radioactive waste, where the product could be safely discharged to the environment [15].

However, MD is also attended by some drawbacks such as low permeate flux (compared to other separation processes, like RO), high susceptibility permeate flux to the concentration, and temperature of the feed conditions due to the concentration and temperature polarization phenomenon. Also, trapped air within the membrane introduces a further mass transfer resistance, which also limits the MD permeate flux. Moreover, the heat lost by conduction is quite large and there is no appropriate membrane for MD too.

3.1.4 Historical review for MD

The history of MD can be summarized as the following:

- The first MD patent was filed in 1963 by Bodell [16].
- In 1967, Weyl suggested locating the hot and cold streams in direct contact. PTFE was used with 3.2 mm thickness, 9 μm pore size, and 42% porosity [17].

- The first scientific paper was published Findley in 1967. The basic theory and result were discussed [18].
- In 1968, a second patent was filed by Bodell. The aqueous solution was treated by a tubular silicone membrane [19,20].
- In 1969, a second scientific paper was published by Findley. Mass and heat transfer were studied [19].
- In 1982, a spiral membrane (Gore-Tex membrane) was used for MD application by Gore and Associated Co. [21].
- In 1985, scientific studies on MD were revealed. Hydrophobic membrane (PP, PTFE) with 80% porosity and 50 μm thickness was created. Better understanding and explanation for temperature and concentration polarization and their effect on MD were published [22–24].

3.2 MD configuration and modules

In this section, different MD configurations that have been utilized to separate an aqueous feed solution using a microporous hydrophobic membrane will be addressed. In terms of permeate collection and driving force generation, MD technology can be classified into four categories: (1) direct contact membrane distillation (DCMD); (2) air gap membrane distillation (AGMD); (3) sweeping gas membrane distillation (SGMD); and (4) vacuum membrane distillation (VMD). In all configurations, the feed solution is in indirect contact with the membrane side surface. On the other hand, the other membrane side (permeate side) consists of condensing pure water (DCMD). Alternatively, the water vapor can be condensed over a metallic surface (AGMD), transferred by a sweep gas (SGMD), or extracted by vacuum (VMD).

3.2.1 *Direct contact membrane distillation*

DCMD (Fig. 3.1) is the simplest MD configuration. It is able to produce reasonably high flux. In this configuration, the hot solution (feed) is in direct contact with the hot membrane side surface. Therefore, evaporation takes place at the feed-membrane surface. The vapor is moved by the pressure difference across the membrane to the permeate side and condenses inside the membrane module. Because of the hydrophobic characteristic, the feed cannot penetrate the membrane (only the gas phase exists inside the membrane pores). This configuration is widely employed in desalination processes and concentration of aqueous solutions in food industries [25–29], boron

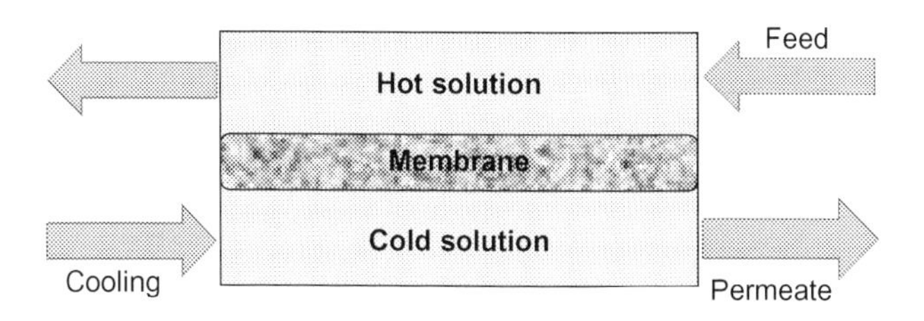

Fig. 3.1 Direct contact membrane distillation (DCMD).

removal [30], or acids manufacturing [31]. The main drawback of this design is high heat loss by conduction, therefore, low-energy efficiency.

3.2.2 Air gap membrane distillation

The schematic representation of the AGMD is shown in Fig. 3.2. The feed solution is in direct contact with the hot side of the membrane surface only. Stagnant air is introduced between the membrane and the condensation surface. The vapor crosses the air gap to condense over the cold surface inside the membrane cell. The flux achieved by AGMD is generally low due to mass transfer resistance; however, it has the highest energy efficiency. This configuration can be widely employed for most MD applications. For example, it is suitable for desalination [11,32,33]. In addition, this configuration has been used to treat highly concentrated solutions such as NaCl, $MgCl_2$, Na_2CO_3, and Na_2SO_4 [34] and also to remove volatile compounds from aqueous solutions [12,35,36]. Moreover, Alkhudhiri et al. [13] used this configuration to treat produced water.

3.2.3 Sweeping gas membrane distillation

In SGMD, as the schematic diagram in Fig. 3.3 shows, inert gas is used to sweep the vapor at the permeate membrane side to condense outside the membrane module. There is a gas barrier, like in AGMD, to reduce the heat loss; but this is not stationary, which enhances the mass transfer coefficient. This configuration is useful for removing volatile compounds from aqueous solution [32]. The main disadvantage of this configuration is that a small volume of permeate diffuses in a large sweep gas volume, requiring a large condenser.

It is worthwhile stating that AGMD and SGMD can be combined in a process called thermostatic sweeping gas membrane distillation (TSGMD) [37,38]. The inert gas in this case is passed through the gap between the membrane and the condensation surface. Part of the vapor is condensed over the condensation surface (AGMD) and the remainder is condensed outside the membrane cell by an external condenser (SGMD).

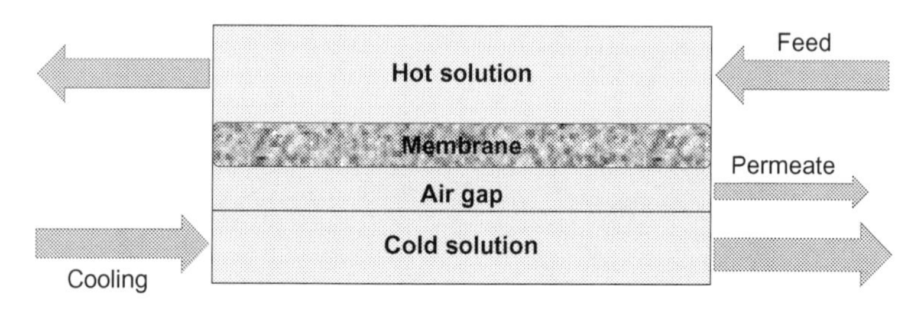

Fig. 3.2 Air gap membrane distillation (AGMD).

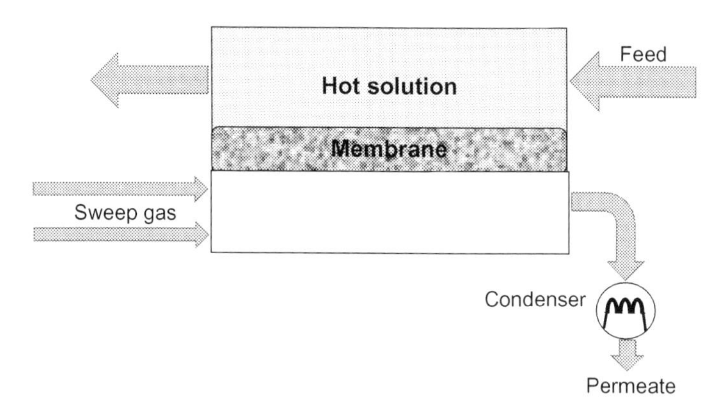

Fig. 3.3 Sweeping gas membrane distillation (SGMD).

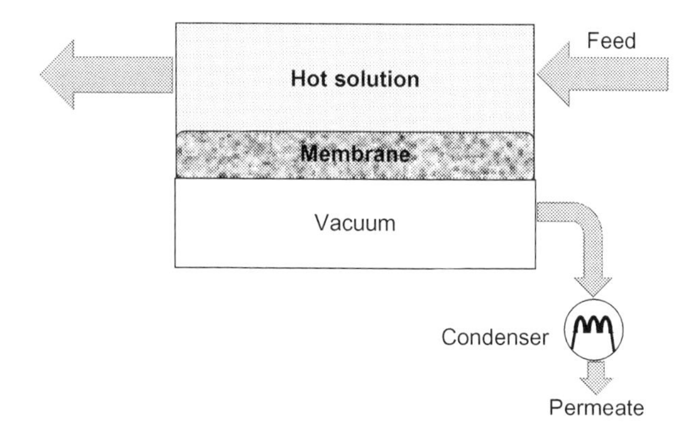

Fig. 3.4 Vacuum membrane distillation (VMD).

3.2.4 Vacuum membrane distillation

The schematic diagram of this module is shown in Fig. 3.4. In the VMD configuration, a pump is used to create a vacuum in the permeate membrane side. The vapor is transferred continuously from the vacuum chamber to the external condenser. Because of the vapor pressure on the permeate membrane side can be reduced to almost zero, the VMD configuration provides the highest driving force. The heat lost by conduction is negligible, which is considered a great advantage [39]. This type of MD is used to separate aqueous volatile solutions [40–42].

3.3 MD modules

Many types of membrane modules can be used in MD. These modules should provide high thermal stability, heat recovery, and low-pressure drop. Cath et al. [43] specified that the membrane module design and MD confirmation could simultaneously increase the permeate flux and reduce temperature polarization. It can be classified as the following.

3.3.1 Plate and frame

The membrane and the spacers are layered together between two plates (e.g., flat sheet). The feed flow is tangential to the flat sheet membrane surface. The flat sheet membrane configuration is widely used on the laboratory scale, because it is easy to clean and replace. However, the packing density, which is the ratio of the membrane area to the packing volume, is low. It varies between 100 and $400\,m^2/m^3$ [21]. Membrane support is required in this type of module. It can be seen that the flat sheet membrane is used widely in MD applications, such as desalination and water treatment.

3.3.2 Spiral wound membrane

In this type, flat sheet membrane and spacers are enveloped and rolled around a perforated central collection tube. The feed moves across the membrane surface in an axial direction, while the permeate flows radially to the center and exits through the collection tube. The spiral wound membrane has good packing density, which is greater than in the plate frame, average tendency to fouling, and acceptable energy consumption. A PTFE spiral wound membrane was employed by Koschikowski et al. [44] for desalination. The energy consumption was about $140–200\,kWh/m^3$. Moreover, Gore [21] used a spiral wound membrane for desalination purposes too.

3.3.3 Hollow fiber, capillary, and tubular

Hollow fiber, capillary, and tubular modules are similar in concept. However, they differ in diameters; therefore, they differ in packing density. The diameter of the hollow fiber is lower than 1 mm. On the other hand, the diameter of capillary membranes ranges from 1 to 3 mm and for the tubular membrane between 5 and 25 mm. Hollow fiber and capillary modules are self-supporting; however, the tubular module is not.

The hollow fiber module, which has been used in MD, has thousands of hollow fibers bundled and sealed inside a shell tube. The feed solution flows through the hollow fiber and the permeate is collected on the outside of the membrane fiber (inside-outside), or the feed solution flows from outside the hollow fibers and the permeate is collected inside the hollow fiber (outside-inside) [45]. For instance, Laganà et al. [46] and Fujii et al. [47] implemented a hollow fiber module (DCMD configuration) to concentrate apple juice and alcohol, respectively. Also, saline wastewater was treated successfully in a capillary polypropylene membrane [48]. The main advantages of the hollow fiber module are very high packing density and low-energy consumption. On the other hand, it has high tendency to fouling and is difficult to clean and maintain.

A capillary module consists of a large number of membrane capillaries, which are placed in parallel in shell and tube configuration. The packing density is in between the tubular and hollow fiber module. The packing density of this type is about $1200\,m^2/m^3$ [49].

In tubular membrane modules, the membrane is tube-shaped and inserted between two cylindrical chambers (hot and cold fluid chambers). In the commercial field, the tubular module is more attractive, because it has low tendency to fouling, is easy to clean, and has a high effective area. However, the packing density of this module is low, which is about $300\,m^2/m^3$ and it has a high operating cost [21,49]. Tubular membranes are also utilized in MD. Tubular ceramic membranes were employed in three MD configurations: DCMD, AGMD, and VMD to treat NaCl aqueous solution where salt rejection was >99% [50].

3.4 MD design and modifications

A heat recovery system is essential to increase MD efficiency; however, the total cost for the MD system will be raised. Effective module configuration design and better membrane characteristics are needed to minimize the heat loss and maximize the permeate flux. There are limited studies on new, efficient MD processes such as Fraunhofer ISE, Memstill, and multieffect membrane distillation (Memsys).

Fraunhofer ISE improved a solar-driven AGMD desalination plant to produce $10\,m^3$/day. A spiral wound module with a counter-current heat exchanger was used in this process. The technical design for spiral wound MD-modules was 5–14 m^2 total membrane area, channel length between 3.5 and 10 m, and 0.7 m channel height. The thermal energy requirement was reduced to $130\,kW/m^3$ by recycling of the internal energy [51] (Fig. 3.5).

The Memstill process was developed by TNO (Netherlands Organization for Applied Scientific Research). A feed solution (e.g., seawater) is pumped and heated by heat of condensation through a nonpermeable channel. An external heat source (waste heat) is used to heat the feed by 2–5°C; then it reenters the membrane evaporator, which consists of a microporous hydrophobic membrane in counter-current flow. Water vapor is transferred through the membrane and condensed on the other side. Low-energy consumption and low total cost are the best advantages of this process. Three pilot plants used Memstill technology. The first pilot plant was installed in

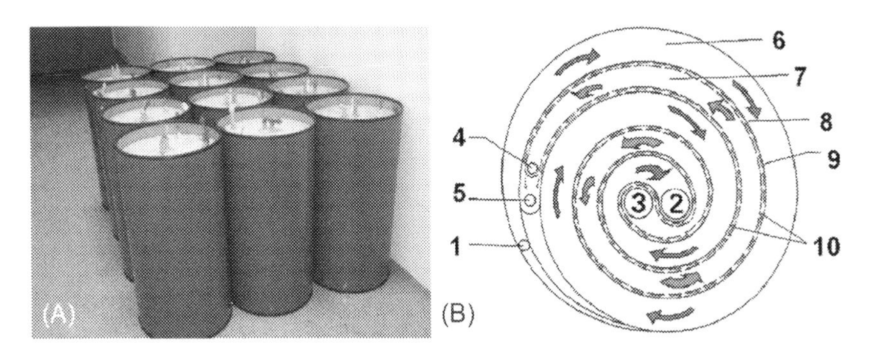

Fig. 3.5 Spiral wound module with schematic diagram for hot and cold channel. Concept: (1) condenser inlet, (2) condenser outlet, (3) evaporator inlet, (4) evaporator outlet, (5) distillate outlet, (6) condenser channel, (7) evaporator channel, (8) condenser foil, (9) distillate channel, and (10) hydrophobic membrane.

Singapore for seawater desalination. The second and third pilot plants were installed in the Netherlands for brackish water treatment. The thermal energy needed by Memstill to produce 1 m^3 was 56–100 kWh [52] (Fig. 3.6).

The Memsys (multieffect membrane distillation) process combines the benefits of VMD and multieffects. Each stage recovers the heat from the previous stage. Consequently, efficient heat recovery can be achieved. The capacity production varies because it depends on the number of stages and the size of each stage. Waste heat and applied vacuum pressure are used to produce water vapor in the first stage. The generated vapor passes through the hydrophobic membrane and condenses on a foil. This foil transfers the energy to the next effect. The distillate is produced in each stage and in the condenser [53]. The Memsys has successfully commercialized. The distillate flow for two stages can reach 33.5 L/h [54] (Fig. 3.7).

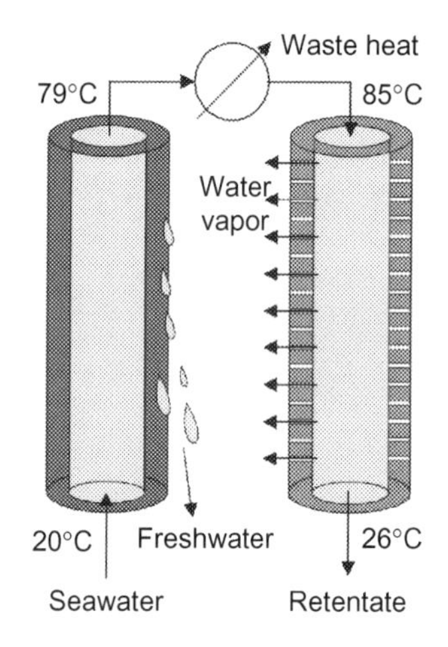

Fig. 3.6 Principle of the Memstill process [52].

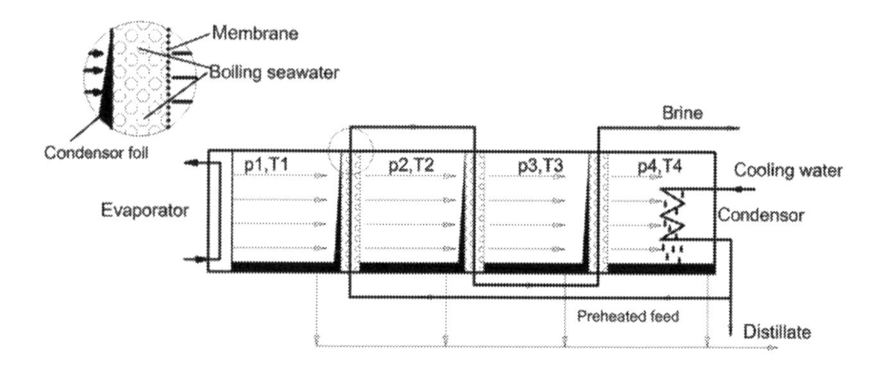

Fig. 3.7 Principle of Memsys V-ME-membrane distillation [54].

3.5 MD membrane

Hydrophobic (nonwetting) microporous membranes are used in the MD process. These membranes are located between the feed and permeate sides as a physical barrier when the mass and heat transfer tack place simultaneously. The permeate flux will be transported through the membrane by vapor pressure difference. The MD membrane properties will not be involved in mass transfer; however, it will be involved in heat transfer.

3.5.1 Membrane materials

The common materials used for MD membranes are made from polytetrafluoroethylene (PTFE), polypropylene (PP), or polyvinylidene fluoride (PVDF). In general, the membrane used in the MD system should have low resistance to mass transfer and low thermal conductivity to prevent heat loss across the membrane. In addition, the membrane should have good thermal stability in extreme temperatures, and high resistance to chemicals, such as acids and bases [5].

Some membranes are laminated to materials for additional structural support. They can be laminated to felts, woven, and nonwoven substrates from a variety of materials, including polyolefins, polyesters, and fluoropolymers. Support materials are selected according to the chemical, thermal, and mechanical requirements of the application [55].

PTFE has the highest hydrophobicity, excellent thermal, and chemical stability; however, it has the highest conductivity value. PVDF shows a good hydrophobicity, thermal resistance, and mechanical strength. PP also exhibits good thermal and chemical resistance.

Recently, several studies have made improvements to membrane properties. Feng et al. [56] prepared two microporous membranes made from PVDF and a modified PVDF (poly (vinyliden fluoride-*co*-tetrafluroethylene)). The mechanical performance and hydrophobicity of the modified PVDF membrane were better than the normal PVDF membrane. The modified PVDF membrane was used successfully in DCMD, where the rejection was almost 100%. Furthermore, the hydrophilic microporous membrane can be used in MD when the membrane surface is modified to become hydrophobic. For example, the surfaces of cellulose acetate and cellulose nitrate (hydrophilic membranes) were modified by radiation grafting polymerization and plasma polymerization to become hydrophobic.

Instead of polymeric membranes, Hengl et al. [57] made two flat metallic (stainless steel) hydrophobic membranes, where silicone was deposited on the top surface. The pore size of those membranes was 2.6 and 5 μm, respectively, and for 200 min, the flux was stable. A carbon nanotube membrane has been also utilized in DCMD. This membrane has a 113 degrees contact angle, 90% porosity, and low thermal conductivity (2.7 $kW/m^2 h$). Lawson et al. [58] also studied the influence of membrane compaction on membrane permeability. They found that the flux increased by 11% compared to a noncompacted membrane.

A dual-layer hydrophilic/hydrophobic membrane, which shows a different manner to water penetration and wetting, was utilized in MD. The hydrophobic side is contacted to the feed solution while the hydrophilic side is contacted to the permeate. Consequently, the hydrophilic side will be wet which improves the water vapor condensation and reduces the temperature polarization phenomenon. The hydrophilic/ hydrophobic membrane is proposed to control the heat loss [59].

There are several ways to create dual hydrophilic/hydrophobic membranes. For example, the hydrophilic membrane surface can be changed to a hydrophobic surface with a water contact angle above 120 degrees by plasma modification [60]. Moreover, the hydrophobicity of PVDF was improved to 130 degrees (superhydrophobic membrane) by TiO_2 coating at low temperature followed by fluoro-silanization of the surface. As a consequence, the membrane surface energy was reduced and there was 50% increase in the liquid entry pressure (LEP) of water [61].

Two ceramic membranes, zirconia (Zr50) and titania (Ti5), have been grafted, Cerneaux et al. [50]. The pore size of the zirconia membrane and the titania membrane were 50 and 5 nm, respectively, and the LEP was 9 bars for both membranes. In comparison, the permeation flux of zirconia membrane was slightly higher than that of the titania membrane when aqueous salts solution was used. In addition, a phase inversion/sintering method has been used to fabricate a hydrophilic alumina hollow fiber membrane by adding ceramic powders (Al_2O_3) into the polymer solution [62]. The wall thickness and porosity of the grafted membranes were 0.2 mm and 42.8%, respectively.

3.5.2 Membrane properties and characteristics

3.5.2.1 Liquid entry pressure (wetting pressure)

Liquid entry pressure (LEP) is a significant membrane characteristic. LEP is defined as the minimum transmembrane pressure that is required for the feed solution to penetrate the large pore size. Therefore, the hydrostatic pressure should be lower than LEP to avoid membrane wetting. This is a characteristic of each membrane and should be as high as possible. The feed liquid must not penetrate the membrane pores; so the pressure applied should not exceed the limit, or LEP, where the liquid (i.e., aqueous solution) penetrates the hydrophobic membrane (Table 3.1). LEP depends on the maximum pore size and the membrane hydrophobicity. It is directly related to feed concentration and the presence of organic solutes, which usually reduce the LEP. For example, LEP linearly decreases when ethanol concentration increases in the solution [63]. In addition, García-Payo et al. [64] indicated that LEP is strongly dependent on the alcohol type, alcohol concentration in the aqueous solution, and solution temperature.

According to Franken et al. [65], LEP can be estimated from the equation:

$$\triangle P = P_F - P_P = \frac{-2B\gamma_l \cos\theta}{r_{\max}} \tag{3.1}$$

Table 3.1 **Commercial membranes used in MD**

Trade name	Manufacturer	Material	Support[a]	Mean pore size (μm)	LEP_W (kPa)
TF200	Gelman	PTFE	PP	0.20	282
TF450	Gelman	PTFE	PP	0.45	138
TF1000	Gelman	PTFE	PP	1.00	48
Emflon	Pall	PTFE	PET	0.02	1585
Emflon	Pall	PTFE	PET	0.2	551
Emflon	Pall	PTFE	PET	0.45	206
FGLP	Millipore	PTFE	PE	0.20	280
FHLP	Millipore	PTFE	PE	0.50	124
Gore	Gore	PTFE	PP	0.2	368
GVHP	Millipore	PVDF	None	0.22	204
HVHP	Millipore	PVDF	None	0.45	105

LEP_W, membrane liquid entry pressure of water.
[a]Polytetrafluoroethylene (PTFE) supported by polypropylene (PP), polyester (PET), or polyethylene (PE).
Modified from Camacho LM, et al. Advances in membrane distillation for water desalination and purification applications. Water 2013;5(1):94–196.

where P_F and P_P are the hydraulic pressure on the feed and permeate side, B is a geometric pore coefficient (equal to 1 for cylindrical pores), γ_l is liquid surface tension, θ is the contact angle, and r_{max} is the maximum pore size. The contact angle of a water droplet on a Teflon surface varies from 108 to 115 degrees; it is 107 degrees for PVDF [45,66] and 120 degrees for PP [45]. It is worthwhile indicating that a flat ceramic membrane made by [67] had a contact angle varying from 177 to 179 degrees. Moreover, ceramic zirconia and titania hydrophobic membranes were prepared with a 160 degrees contact angle [50]. All these ceramic membranes were used for desalination purposes.

With regard to surface tension, Zhang et al. [68] studied the impact of salt concentration (NaCl) on the water surface tension and found that:

$$\gamma_{l\_new} = \gamma_l + 1.467\, c_f \tag{3.2}$$

γ_l stands for pure water surface tension at 25°C, which is 72 mN/m.

As a result, membranes that have a high contact angle (high hydrophobicity), small pore size, low surface energy, and high surface tension for the feed solution possess a high LEP value [69]. Typical values for surface energy for some polymeric materials are reported in Table 3.2. He et al. [70] suggested that the PEFE membrane appears more suitable for the MD process than PP or PVDF because of the high value of PTFE hydrophobicity and high permeate flux. The maximum pore size to prevent wetting should be between 0.1 and 0.6 μm [71,72]. Moreover, the possibility of liquid penetration in VMD is higher than other MD configurations; so a small pore size is recommended [39,73].

Table 3.2 Surface energy of some polymers

Polymer	Surface energy ($\times 10^3$ N/m)
PTFE (polytetrafluoroethylene)	19.1
PVDF (polyvinylidenefluride)	30.3
PP (polypropylene)	30.0
PE (Polyethylene)	33.2

Modified from Mulder M. Basic principles of membrane technology. 2nd ed. Netherlands: Kluwer Academic Publishers; 2003.

3.5.2.2 Mean pore size and pore size distribution

Regarding the MD membrane, two types of characteristics can be analyzed, the structural characteristic and the actual separation parameters (permeation) [37,74–76].

The pore size of the membrane and pore size distribution are important characteristic for MD membranes. Membranes with pore size between 100 nm and 1 μm are usually used in MD systems [39,75]. The maximum pore size is desirable to be close to the mean pore size; therefore, pore size distribution will be as narrow as possible. The permeate flux increases with increasing membrane pore size [75]. The mechanism of mass transfer can be determined, and the permeate flux calculated, based on the membrane pore size and the mean free path through the membrane pores taken by transferred molecules (water vapor). Generally, the mean pore size is used to determine the vapor flux. A large pore size is required for high permeate flux, while the pore size should be small to avoid liquid penetration. As a result, the optimum pore size should be determined for each feed solution and operating condition.

In fact, the membrane does not have a uniform pore size so more than mass transfer mechanisms occur simultaneously (depending on the pore size). Several investigations examined the importance of pore size distribution in MD flux [77–81]. Khayet et al. [78] reported that care must be taken when mean pore size is utilized to calculate the vapor transfer coefficient instead of pore size distribution. However, Martínez et al. [80] obtained a similar vapor transfer coefficient when the mean pore size and pore size distribution were used. Better understanding of membrane morphology such as pore size, pore size distribution, porosity, and thickness directs to have an accurate mass and heat transfer modeling. Woods et al. [82] used a numerical model to estimate the effect of pore size distribution on MD flux. They reported that the influence of pore size distribution is small in DCMD. However, it is much larger in VMD.

3.5.2.3 Membrane porosity and tortuosity

Membrane porosity refers to the void volume fraction of the membrane (defined as the volume of the pores divided by the total volume of the membrane). It should be as high as possible. Higher porosity membranes have a larger evaporation surface area. Two types of liquids are used to estimate membrane porosity. The first penetrates the membrane pores (e.g., isopropyl alcohol, IPA), while the other, like water, does not.

In general, a membrane with high porosity has a higher permeate flux and lower conductive heat loss because the thermal conductivity of air which is trapped inside the membrane is lower than the membrane material. The porosity (ε) can be determined by the Smolder-Franken equation,

$$\varepsilon = 1 - \frac{\rho_m}{\rho_{\text{pol}}} \tag{3.3}$$

where ρ_m and ρ_{pol} are the densities of membrane and polymer material, respectively.

According to El-Bourawi et al. [75], membrane porosity in the MD system varies from 30% to 85%.

Tortuosity (τ) is the deviation of the pore structure from the cylindrical shape. As a result, the higher the tortuosity value, the lower the permeate flux. The most successful correlation was suggested by Macki-Meares [83], where

$$\tau = \frac{(2-\varepsilon)^2}{\varepsilon} \tag{3.4}$$

3.5.2.4 Membrane thickness

Membrane thickness is a significant characteristic in the MD system. There is an inversely proportional relationship between the membrane thickness and the permeate flux. The permeate flux is reduced as the membrane becomes thicker, because the mass transfer resistance increases, while heat loss is also reduced as the membrane thickness increases. Membrane morphology, such as thickness and pore size distribution, has been studied theoretically by Laganà et al. [46]. They concluded that the optimum membrane thickness lies between 30 and 60 μm. It is worth noting that the effect of membrane thickness in AGMD can be neglected, because the stagnant air gap represents the predominant resistance to mass transfer.

3.5.2.5 Thermal conductivity

The thermal conductivity of the membrane is calculated based on the thermal conductivity of both polymer k_s and gas k_g (usually air). The thermal conductivity of the polymer depends on temperature, the degree of crystallinity, and the shape of the crystal. The thermal conductivities of most hydrophobic polymers are close to each other. For example, the thermal conductivities of PVDF, PTFE, and PP at 23°C are 0.17–0.19, 0.25–0.27, and 0.11–0.16 $\text{W m}^{-1}\,\text{K}^{-1}$, respectively [84]. The thermal conductivity of PTFE can be estimated by Sperati and DuPont de Nemours [85]

$$k_s = 4.86 \times 10^{-4} T + 0.253 \tag{3.5}$$

The thermal conductivity of the MD membrane is usually taken a volume average of both conductivities k_s and k_g as follows:

$$k_m = (1-\varepsilon)k_s + \varepsilon k_g \tag{3.6a}$$

However, Phattaranawik et al. [84] suggested that the thermal conductivity of an MD membrane is better based on the volume average of both resistances ($1/k_g$ and $1/k_s$), that is,

$$k_m = \left[\frac{\varepsilon}{k_g} + \frac{(1-\varepsilon)}{k_s}\right]^{-1} \tag{3.6b}$$

For, the thermal conductivity values for air and water vapor at 25°C are of the same order of magnitude. For instance, the thermal conductivity of air at 25°C is $0.026\,Wm^{-1}\,K^{-1}$ and for water vapor, it is $0.020\,Wm^{-1}\,K^{-1}$. As a result, the assumption of one component gas present inside the pores is justified.

Khayet et al. [86] suggested some ways to reduce the heat loss by conduction through the membrane: using membrane materials with low thermal conductivities, using a high porosity membrane, using thicker membrane, and minimizing heat losses. It is also suggested that the permeability can be enhanced by using a composite porous hydrophobic/hydrophilic membrane. In this case, the top layer is a very thin hydrophobic layer to stop liquid penetration, followed by a thick hydrophilic layer. Both layers reduce the heat losses through the membrane.

3.6 Transport phenomena

3.6.1 Mass transfer

A dusty gas model has been formulated and employed to illustrate gas transport through porous media by Evans and Mason between 1961 and 1967. The kinetic theory of gases is implemented to this model. The dusty gas model considered mass transfer mechanisms through porous media such as membranes. Moreover, it can be used for a multicomponent mixture of gases through porous media. According to this model, gas transport through porous media (e.g., membrane) can be classified into four modes: (1) surface diffusion; (2) Knudsen diffusion which is in the free molecular region with minimal molecule-molecule collisions and dominated by molecule-wall collisions; Knudsen diffusion considers the influence of the porous medium; (3) ordinary diffusion which is due to concentration, temperature gradients, or external forces; it is dominated by molecular-molecular collisions; and (4) viscous flow is linearly dependent on the pressure gradient which is dominated by molecular-molecular collisions [87–89].

The dusty gas model is used to describe the mass transfer in the MD system. This model relates the mass transport with collisions between molecules, and/or molecules with the membrane wall. In the Knudsen diffusion model, collision between the molecules and the inside walls of the membrane suitably expresses the mass transport and the collision between molecules can be ignored. The molecular diffusion model comes into play when the molecules move corresponding to each other under the influence of concentration gradients. In the viscous flow model (Poiseuille flow), the gas molecules act as a continuous fluid driven by a pressure gradient [87].

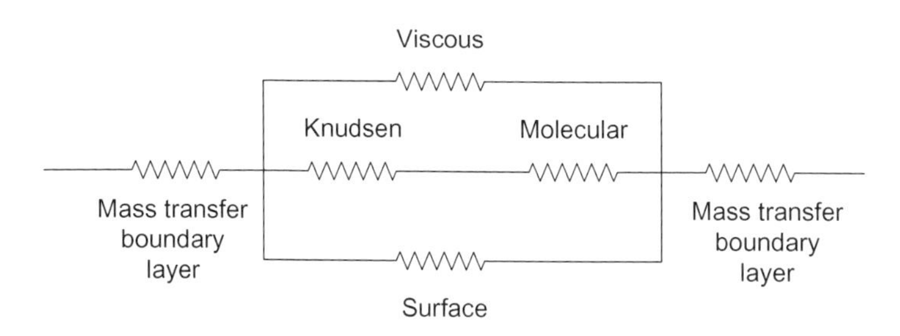

Fig. 3.8 Mass transfer resistances in MD.

It is worth noting that combined mass transfer mechanisms may be considered when Knudsen number is located in the transition region. The MD configuration, type of membrane, and the MD operating conditions should be considered [5]. Furthermore, the mass transfer boundary layer is generally negligible [45]. Similarly, the surface diffusion is insignificant, because of the membrane hydrophobicity and the surface area of the MD is small compared to the pore area [5] (Fig. 3.8).

The key parameter utilized to designate the mass transport mechanism inside the membrane pores is the Knudsen number. The Knudsen number (Kn), defined as the ratio of the mean free path (λ) of transported molecules to the membrane pore size (d_p), provides a guideline of which mechanism is active inside the membrane pore.

$$Kn = \frac{\lambda}{d_p} \tag{3.7}$$

According to the kinetic theory of gases, the molecules are assumed to be hard spheres with diameter d_e and are involved in binary collisions only. It is worth noting that the collision diameters for water vapor and air are about 2.64×10^{-10}, and 3.66×10^{-10}, respectively [85]. The average distance traveled by molecules to make collisions (λ) is defined as

$$\lambda = \frac{k_B T}{\sqrt{2}\pi P d_e^2} \tag{3.8}$$

k_B, T, and P are Boltzman's constant, absolute temperature, and average pressure within the membrane pores, respectively.

The mean free path value of water vapor at 60°C was estimated to be 0.11 μm [90].

3.6.1.1 Knudsen diffusion model

When the mean free path of water vapor molecules is large compared to the membrane pore size, the molecule-pore wall collisions are dominant over molecule-molecule collisions. In other words, $d_p < \lambda$ ($Kn > 1$). Ding et al. [91] proposed that the Knudsen diffusion model takes place when the pore size is too small.

The membrane permeability for a single pore can be computed such that [37,92]

$$C_{Kn} = \frac{1}{6}\frac{d_p^3}{\tau\delta}\left(\frac{2\pi}{M_w RT}\right)^{1/2} \tag{3.9}$$

where τ, d_p, δ, and M_w are pore tortuosity, membrane pore size, membrane thickness, and molecular weight of water vapor, respectively.

3.6.1.2 Ordinary diffusion (molecular) model

When $k_n < 0.01$ or $d_p > 100\lambda$ (continuum region), the ordinary molecular diffusion model represents the diffusion of the vapor flux through a stationary air film (the air which exist inside the membrane pores); ordinary molecular diffusion is used to describe the membrane permeability through a single pore [37,92]:

$$C_D = \frac{\pi}{4RT}\frac{PD}{P_{\text{air}}}\frac{d_p^2}{\tau\delta} \tag{3.10}$$

where P_{air} is the air pressure within the membrane pore, D is the diffusion coefficient, and P is the total pressure inside the pore which is equal to the partial pressure of air and water vapor.

The diffusivity of water vapor through the stagnant air inside the pores is given by Phattaranawik et al. [79]

$$PD = 1.895 \times 10^{-5} T^{2.072} \tag{3.11}$$

In addition, the Fuller equation, which is a common equation to predict binary gas diffusion, can be used [89,93]:

$$D = 1 \times 10^{-7} \frac{T^{1.75}\left(\frac{1}{Mw_a} + \frac{1}{Mw_b}\right)^{\frac{1}{2}}}{P\left[\left(\sum v_a\right)^{\frac{1}{3}} + \left(\sum v_b\right)^{\frac{1}{3}}\right]^2} \tag{3.12}$$

where $\sum v$ represents the diffusion volume, T is the temperature in Kelvin, and P is the pressure in atmospheres. The diffusion volumes of air and water are 20.1 and 12.7, respectively.

3.6.1.3 Viscous flow model

When the membrane pore size is larger than the mean free path ($d_p > 100\lambda$) and hydrostatic pressure (pressure gradient) is used like VMD configuration, the molecular-molecular collision dominates and the membrane permeability through a single pore can be expressed by Poisseuille flow (viscous), such that [37,92]

$$C_v = \frac{\pi}{2^7 \mu_i} \frac{P_{avg}}{RT} \frac{d_p{}^4}{\tau\delta} \tag{3.13}$$

where μ_i is the viscosity of species i, and P_{avg} is the average pressure in the pore.

3.6.1.4 Combined model

Combined mass transfer mechanisms are considered when the Knudsen number is located in the transition region $0.01 < k_n < 1$ ($\lambda < d_p < 100\lambda$). In the case of air presence in the membrane pores and absence of transmembrane hydrostatic pressure like the DCMD configuration, the water vapor molecules collide with each other and also diffuse through the air film. Consequently, the membrane permeability through a single pore can be expressed by both the Knudsen/ordinary diffusion mechanism, where [37,92]:

$$C_{Kn/D} = \frac{\pi}{RT} \frac{1}{\tau\delta} \left[\left(\frac{2}{3} \left(\frac{RT}{\pi M_w} \right)^{\frac{1}{2}} d_p{}^3 \right)^{-1} + \left(\frac{1}{4} \frac{PD}{P_a} d_p{}^2 \right)^{-1} \right]^{-1} \tag{3.14}$$

Nevertheless, when trapped air is removed in the membrane pores by the deaeration of the feed solution and continuous vacuum in the permeate side (VMD configuration), the ordinary molecular diffusion resistance is neglected and both molecular-molecular/molecular-wall collisions should be considered. As a result, the membrane permeability through a single pore can be represented by the Knudsen/viscous flow model such that [37,92]

$$C_{Kn/v} = \frac{\pi}{RT\delta\tau} \left[\frac{2}{3} \left(\frac{RT}{\pi M_{w_i}} \right)^{\frac{1}{2}} d_p{}^3 + \frac{P_{avg}}{2^7 \mu_i} d_p{}^4 \right] \tag{3.15}$$

where μ_i is the viscosity of species i and P_{avg} is the average pressure in the pore.

Table 3.3 shows DCMD membrane coefficients as reported by some researchers.

Lawson and Lloyd [39] stated that the molecule-pore wall collisions (Knudsen diffusion) and molecule-molecule collisions (molecular diffusion) take place simultaneously for pore sizes <0.5 μm. Moreover, Guijt et al. [96] point out that the flux can be expressed by molecular diffusion only for large pores. Furthermore, the vapor flux across the membrane can be expressed by Knudsen diffusion and the Poiseuille (viscous) flow model for deaerated DCMD [22,97]. On the other hand, Ding et al. [91] studied the effect of the Poiseuille flow mechanism in mass transfer through the membrane. They found that the Poiseuille flow should be considered as one of the mechanisms of mass transfer model in a large pore size membrane.

When the membrane permeability is computed, the permeate flux for MD can be calculated.

The mass flux (J) in MD is assumed to be proportional to the vapor pressure difference across the membrane, and is given by:

$$J = C \left[P_f - P_p \right] \tag{3.16}$$

Table 3.3 **DCMD membrane coefficients**

Ref.	Membrane type	Pore size (μm)	Membrane coefficient (kg/m^2 S Pa)	Feed solution
[64]	PTFE	0.2	14.5×10^{-7}	Distilled water
		0.45	21.5×10^{-7}	
[81]	PVDF	0.22	3.8×10^{-7}	Distilled water
[14]	Enka (pp)	0.1	4.5×10^{-7}	Distilled water
	Enka (pp)	0.2	4.3×10^{-7}	
	PVDF	0.45	4.8×10^{-7}	
[80]	GVHP	0.22	4.919×10^{-7}	Distilled water
	HVHP	0.45	6.613×10^{-7}	
[94]	PVDF	0.22	6.6×10^{-7}	NaCl solution
[95]	GVHP22	0.16	5.6×10^{-7}	Sucrose solution

where C is the membrane coefficient, and P_f and P_p are the vapor pressure at the membrane feed and permeate surfaces, respectively, which can found from the Antoine equation [56,71,98–100]. Therefore, Eq. (3.16) can be rewritten in terms of temperature difference across the membrane surfaces when the separation process is for pure water or very diluted solution, and the temperature difference across the membrane surfaces is less than or equal to 10°C [12,28,39,78]. Hence:

$$J = C\frac{dP}{dT}\left(T_{f,m} - T_{p,m}\right) \tag{3.17}$$

The vapor pressure and temperature relationship can be expressed by the Clausius-Clapeyron equation, as follows:

$$\frac{dP}{dT} = \left[\frac{\triangle H_v}{RT^2}\right] P^*(T) \tag{3.18}$$

However, Schofield et al. [23] adapted Eq. (3.17) for more concentrated solutions, such that

$$J = C\frac{dP}{dT}\left[\left(T_{f,m} - T_{p,m}\right) - \triangle T_{th}\right]\left(1 - x_m\right) \tag{3.19}$$

where $\triangle T_{th}$ is the threshold temperature, given by

$$\triangle T_{th} = \frac{RT^2}{M_w \triangle H_v} \frac{x_{f,m} - x_{p,m}}{1 - x_m} \tag{3.20}$$

where $x_{f,\ m}$, $x_{p,\ m}$, and x_m, represent the mole fraction of dissolved species at the hot membrane surface side, from the permeate membrane surface side and inside the membrane, and R and $\triangle H_v$ represent the universal gas constant and the latent heat of vaporization, respectively.

For low concentration solutions, the Antoine equation can be utilized to determine the vapor pressure, because it can be assumed that the vapor pressure is a function of temperature only, that is, dropping vapor pressure dependence on solution concentration. Martínez and Rodríguez-Maroto [101] and Godino et al. [102] estimated the effect of both concentration and temperature on the vapor pressures by considering the water activity at the feed and permeate sides, such that

$$P(T,x) = P^*(T)a_w(T,x) \tag{3.21}$$

where $a_w(T,x)$ is the water activity as a function of temperature and concentration, and $P^*(T)$ is the vapor pressure of pure water at a given temperature. Raoult's law has also been used to estimate the vapor pressure [29,36], where

$$P(T,x) = P^*(T)(1-x) \tag{3.22}$$

The molecular diffusion theory is used also to describe the transfer of vapor molecules through the membrane and the air gap. A stagnant gas film (air) is assumed to lie inside the membrane at the air gap side.

Kurokawa et al. [100] computed the flux by considering the diffusion in one direction through both membrane and air gap (l), where the air gap is below 5 mm:

$$J = \frac{PM_w}{RTP}\left(\frac{D}{\frac{\delta}{\varepsilon^{3.6}}+l}\right)\triangle P \tag{3.23}$$

where $\triangle P$ is the water vapor pressure difference between the feed on the membrane and the condensation surface and P is the partial pressure of water.

Moreover, Stefan diffusion was employed to describe the diffusion through a stagnant gas film. It can be represented mathematically as [103]

$$N = -\frac{cD}{1-y}\frac{dy}{dz} \tag{3.24}$$

where D, y, c, and z are the diffusion coefficient, the mole fraction of the vapor phase, the molar concentration and the diffusion length, respectively.

The Stefan equation was solved by [35]

$$N = \frac{cD}{z}\ \ln\frac{1-y_f}{1-y_m} \tag{3.25}$$

where y_m and y_f represent the mole fraction of vapor at the membrane and the condensation film, respectively.

However, Jönsson et al. [104] solved the same equation by neglecting the effect of temperature and concentration polarization. They suggested that the value of cD for water vapor and air at around 40°C to be calculated using this equation:

$$cD = 6.3 \times 10^{-5}\sqrt{T} \tag{3.26}$$

In addition, the molar concentration can be calculated from the ideal gas law:

$$c = \frac{P}{RT} \tag{3.27}$$

3.6.2 Heat transfer

MD is a nonisothermal process. Two main heat transfer mechanisms occur in the MD system: latent heat and conduction heat transfer.

The heat transfer, which occurs in MD, can be divided into three regions (Fig. 3.9) [23,28,45,75,83,91,94,101,105–111]:

Heat transfer by convection in the feed boundary layer (Table 3.4):

$$Q_f = h_f\left(T_f - T_{f,m}\right) \tag{3.28}$$

Heat transfer through the membrane by conduction and by movement of vapor across the membrane (latent heat of vaporization). The influence of mass transfer on the heat transfer can be ignored [75,97,107–109,114]

$$Q_m = \frac{k_m}{\delta}\left(T_{f,m} - T_{p,m}\right) + J\triangle H_v \tag{3.29}$$

$$Q_m = h_m\left(T_{f,m} - T_{p,m}\right) + J\triangle H_v \tag{3.30}$$

where h_m represents the heat transfer coefficient of the membrane.

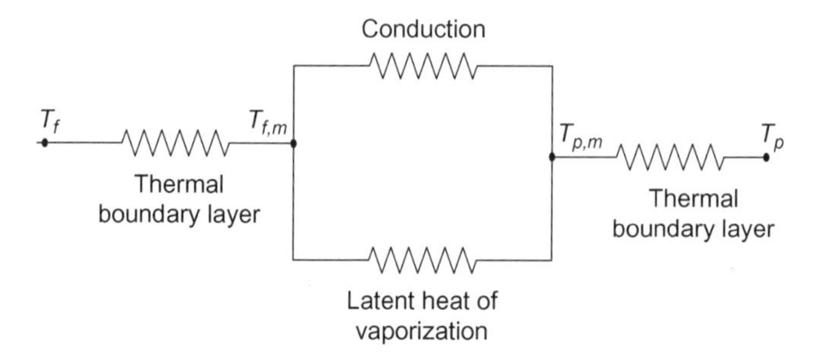

Fig. 3.9 Heat transfer resistances in the MD system.

Table 3.4 Heat transfer coefficients at different feed temperatures and concentrations

Ref.	Membrane type	Pore size (μm)	Feed temperature T_f (°C)	Feed type	Heat transfer coefficient h_f (W/m^2 K)
[112]	PVDF	0.2	26	Pure water	90
[112]	PVDF	0.2	42	Pure water	160
[94]	PVDF	0.2	60	NaCl solution	14,018
[113]	PTFE	–	50	NaCl solution	3457
[106]	PVDF	0.16	45	NaCl solution	13,500
[106]	PVDF	0.16	45	Sucrose solution	6300

It is worth mentioning that h_m can be rewritten for pure water or very diluted solution, and where temperature difference across the membrane surfaces is less than or equal to 10°C by substituting Eq. (3.17) into Eq. (3.29) [23,29,97]:

$$Q_m = \frac{k_m}{\delta}\left(T_{f,m} - T_{p,m}\right) + \left[C_m \frac{dP}{dT}\left(T_{f,m} - T_{p,m}\right)\right] \triangle H_v \tag{3.31}$$

$$Q_m = \left[\frac{k_m}{\delta} + \left(C_m \frac{dP}{dT}\right) \triangle H_v\right]\left(T_{f,m} - T_{p,m}\right) \tag{3.32}$$

$$Q_m = h_m\left(T_{f,m} - T_{p,m}\right) \tag{3.33}$$

For the permeate side, the convection heat transfer takes place in the permeate boundary layer

$$Q_p = h_p\left(T_{p,m} - T_p\right) \tag{3.34}$$

At the steady state, the overall heat transfer flux through the membrane is given by

$$Q = Q_f = Q_m = Q_p \tag{3.35}$$

$$h_f\left(T_f - T_{f,m}\right) = \frac{k_m}{\delta}\left(T_{f,m} - T_{p,m}\right) + J \triangle H_v = h_p\left(T_{p,m} - T_p\right) \tag{3.36}$$

$$Q = U\left(T_f - T_p\right) \tag{3.37}$$

where U represents the overall heat transfer coefficient.

It is worth pointing out that the heat conduction can be neglected for a non-sported thin membrane [108] and for high operating temperature as well [97,108]. Moreover, the heat transfer by convection is ignored in the MD process, except in AGMD [45].

The surface temperature of both sides of the membrane cannot be measured experimentally, or calculated directly. Therefore, a mathematical iterative model has been designed to estimate these temperatures [94]:

$$T_{f,m} = T_f - \frac{J \triangle H_v + \frac{k_m \left(T_{f,m} - T_{p,m}\right)}{\delta_m}}{h_f} \tag{3.38}$$

$$T_{p,m} = T_p - \frac{J \triangle H_v + \frac{k_m \left(T_{f,m} - T_{p,m}\right)}{\delta_m}}{h_p} \tag{3.39}$$

The value of H_v should be evaluated at the average membrane temperature. However, Phattaranawik and Jiraratananon [108] evaluated at the logarithmic average membrane temperature.

Lawson and Lloyd [39] pointed out that the $(T_{m,f} - T_{m,p})$ is about 0.1°C at low flux and does not exceed 0.5°C at high flux.

Gryta et al. [115] studied the presence of free and forced convection in laminar flow in DCMD and suggested the following equation to calculate the heat transfer coefficient:

$$Nu = 0.74\, Re^{0.2} (GrPr)^{0.1} Pr^{0.2} \tag{3.40}$$

In the AGMD configuration, the heat transfer through the AGMD was represented as in DCMD, except for the heat transfer across the air gap, which occurs by conduction and vapor (mass transfer) [12,116,117]:

$$\begin{aligned} h_f \left(T_f - T_{f,m}\right) &= J \triangle H_v + \frac{k_m}{\delta}\left(T_{f,m} - T_{p,m}\right) = J \triangle H_v + \frac{k_g}{l}\left(T_{p,m} - T_{\text{film}}\right) \\ &= h_d (T_{\text{film}} - T_5) \end{aligned} \tag{3.41}$$

Guijt et al. [96], Banat and Simandl [36], and Kimura et al. [35] suggested the following equation to calculate the heat transfer coefficient for the condensate film (pure vapor) on a vertical wall:

$$h_d = \frac{2}{3}\sqrt{2}\left(\frac{k_{\text{film}}{}^3 \rho^2 g \triangle H_v}{\mu L \left(T_{\text{film}} - T_5\right)}\right)^{\frac{1}{4}} \tag{3.42}$$

In addition, free convection heat transfer between two vertical plates is also used to describe the heat transfer phenomenon in the air gap region, when the air gap distance is over 5 mm [100]:

$$Nu = c(PrGr)^n \left(\frac{l}{L}\right)^{\frac{1}{9}} \quad (3.43)$$

where

$$10^5 < Gr < 10^7, \quad c = 0.07 \text{ and } n = \frac{1}{3}$$

$$10^4 < Gr < 10^5, \quad c = 0.2 \text{ and } n = \frac{1}{4}$$

Bouguecha et al. [118] designed a mathematical model for laminar flow, in which the heat and mass transfer are considered. The temperature profiles at different air gap thicknesses in two dimensions are plotted. The heat transfer by convection starts to change to natural convection at 5-mm air-gap thickness, and this dominates the heat transfer mechanism at a wide air gap.

Furthermore, the average membrane temperature was used by Kimura et al. [35] instead of the membrane surface temperature. They concluded that the sensible heat for the MD system can be neglected, because it has a very small magnitude compared to the heat of vaporization:

$$Q = J \triangle H_v \quad (3.44)$$

The heat transfer by conduction through the membrane is ignored in the VMD configuration [39,73]; so the heat transfer across the VMD can be written as

$$h_f\left(T_f - T_{f,m}\right) = J \triangle H_v \quad (3.45)$$

The heat transfer coefficients of the boundary layers can be estimated by the Nusselt correlation (Table 3.5). Its empirical form is

$$Nu = \text{Constant}\, Re^a Pr^b \quad (3.46)$$

Consequently, the heat transfer coefficient h can be calculated using Reynolds and Prandtl numbers (Re and Pr), that is,

$$\text{Reynolds number} = R_e = \frac{vdp}{\mu} \quad (3.47)$$

$$\text{Prandtl number} = Pr = \frac{c_p\mu}{k} \quad (3.48)$$

Table 3.5 Correlations used to estimate heat transfer coefficient

Ref.	Equations	Notes
[39]	$Nu = 0.023 Re^{0.8} Pr^{\frac{1}{3}} \left(\frac{\mu}{\mu_s}\right)^{0.14}$	Turbulent flow ($2500 < Re < 1.25 \times 10^5$)
[83]	$Nu = 0.023\ Re^{0.8} Pr^n$ $n = 0.4$ for heating, $n = 0.3$ for cooling	Turbulent flow ($2500 < Re < 1.25 \times 10^5$) $0.6 < Pr < 100$
[78]	$Nu = 0.027 Re^{\frac{4}{5}} Pr^n \left(\frac{\mu}{\mu s}\right)^{0.14}$ $n = 0.4$ for heating, $n = 0.3$ for cooling	Turbulent region
[79,84,94]	$Nu = 0.023\left(1 + \frac{6d}{L}\right) Re^{0.8} Pr^{\frac{1}{3}}$	The most suitable heat transfer correlation for turbulent flow
[79,84,94]	$Nu = 3.36 + \dfrac{0.036 RePr\frac{d}{L}}{1 + 0.0011\left(RePr\frac{d}{L}\right)^{0.8}}$	The best correlation to compute the heat transfer coefficient for the laminar flow
[83,107]	$Nu = 1.86\left(RePr\frac{d}{L}\right)^{0.33}$	For laminar flow ($Re < 2100$) Recommended for flat sheet module
[119,120]	$Nu = 1.86 Re^{\frac{1}{3}} Pr^{\frac{1}{3}} \left(\frac{d}{L}\right)^{\frac{1}{3}} \left(\frac{\mu}{\mu_s}\right)^{\frac{1}{7}}$	Laminar flow
[121]	$Nu = 1.86 Re^{0.96} Pr^{\frac{1}{3}} \left(\frac{d}{L}\right)^{\frac{1}{3}}$	Laminar flow
[35]	$Nu = 1.62\left[RePr\frac{d}{L}\right]^{\frac{1}{3}}$	Laminar flow
[115]	$Nu = 0.298\ Re^{0.646} Pr^{0.316}$ $Nu = 0.74\ Re^{0.2} (Gr\ Pr)^{0.1} Pr^{0.2}$	Laminar flow $150 < Re < 3500$ The best correlation for plate and frame module in laminar flow
[45]	$Nu = 0.036 Re^{0.8} Pr^{0.33} \left(\frac{d}{L}\right)^{0.055}$	Turbulent flow
[121]	$Nu = 0.036 Re^{0.96} Pr^{0.33} \left(\frac{d}{L}\right)^{0.055}$	Turbulent flow in tube and $10 \le \frac{L}{d} \le 400$
[122]	$Nu = 1 + 1.44\left(1 - \frac{1708}{Re}\right) + \left[\left(\frac{Re}{5830}\right)^{\frac{1}{3}} - 1\right]$	Not mentioned

$$\text{Grashoff number} = G_r = \frac{g\beta\Delta TL^3\rho^2}{\mu^3} \tag{3.49}$$

where v, ρ, μ, c_p, g, β, L, and k are fluid velocity, density, viscosity, heat capacity, gravity acceleration, thermal expansion coefficient, height, and thermal conductivity, respectively.

It is worth mentioning that the mass transfer and the heat transfer can be related, as proposed by García-Payo et al. [12], by

$$ShSc^{\frac{-1}{3}} = NuPr^{\frac{-1}{3}} \tag{3.50}$$

3.7 MD evaluation and assessment

In most lab-scale or pilot plant studies, three important parameters are used to evaluate and assessment the MD energy consumption; they are the thermal efficiency (Π), gain output ratio (GOR), and water production cost (WPC).

3.7.1 Thermal efficiency and energy consumption

The thermal efficiency Π in MD can be specified as the ratio of latent heat of vaporization to the total (latent and conduction) heat. Therefore, the heat transfer by conduction should be as low as possible to accomplish a high thermal efficiency process.

The thermal efficiency Π can be expressed as [123]

$$\Pi = \frac{J \triangle H_v}{J \triangle H_v + \frac{k_m}{\delta}\left(T_{f,m} - T_{p,m}\right)} \tag{3.51}$$

Around 50%–80% of the total heat flux across the membrane is considered to be latent heat; whereas 20%–40% of heat is lost by conduction through the membrane [39,97].

Thermal efficiency can be increased by optimizing the membrane properties such as porosity, thickness, and polymer thermal conduction. There is an inversely proportional relationship between the membrane thickness and the permeate flux; however, heat loss is reduced as the membrane thickness increases [124]. A composite hydrophobic/hydrophilic membrane was fabricated to solve this issue. The thermal conductivity of the hydrophobic layer should be as low as possible; in contrast, the hydrophilic layer should be as high as possible.

Al-Obaidani et al. [90] pointed out that the thermal efficiency can be improved by increasing the feed temperature due to the exponential increases of permeate flux and decreases of heat lost by conduction through the membrane. Feed flow rate and membrane thickness will affect positively the thermal efficiency. In contrast, it decreases when the concentration for salt solution increases.

For pure water, Bandini et al. [125] commented that the characteristics of the membrane, such as porosity and tortuosity, determine the thermal efficiency, with no dependence on membrane thickness.

Fane et al. [97] pointed out that there are three forms for heat transfer to be lost in the DCMD system. The first form is due to the presence of air within the membrane. Secondly, heat loss occurs through the membrane by conduction, and finally by temperature polarization. They suggested some solutions to minimize heat loss in the DCMD, such as deaeration of the feed solution, increasing the membrane thickness, creating an air gap between the membrane and the condensation surface, and operating within a turbulent flow regime.

Alklaibi and Lior [126] observed that by increasing the feed temperature from 40°C to 80°C, the thermal efficiency increased by 12%, whereas the salt concentration has a marginal effect on the thermal efficiency.

With regard to energy consumption, hybrid RO/MD becomes the best choice when an external energy source is available [7]. In addition, heat transfer to the cooling side by heat conduction and by heat of condensation can be used (recovered) to preheat the feed solution, which minimizes the heat requirement and improves the operation cost. The percentage of heat recovery depends on the heat exchanger area. Schneider et al. [72] indicated that the MD performance rises by 8% when heat recovery is used. Kurokawa and Sawa [8] reported that the heat input declines with increasing heat exchanger and membrane areas. They optimized the value of both heat exchanger and membrane areas for a plate and frame cell and PTFE membrane (0.2 μm pore size); this was 0.2 m^2. Likewise, Ding et al. [127] emphasized that the heat exchanger capacity should be optimized with membrane area, in order to get high production flux for a solar powered MD system.

It worth mentioning that several researchers utilized the energy efficiency which is the total energy input including both thermal and electrical energy instead of the thermal efficiency only [128] (Table 3.6).

Table 3.6 **Specific energy consumption for different MD systems**

Ref.	Process	Specific energy (kWh/m^3)
[129]	AGMD with waste energy	1.25
[7]	Hybrid NF-RO-DCMD	13
[7]	Hybrid NF-RO-DCMD (thermal energy is available)	2.58
[7]	Hybrid RO-DCMD	15
[7]	Hybrid RO-DCMD(thermal energy is available)	2.25
[44]	Solar driven AGMD	117
[130]	Geothermal AGMD	30.8
[131]	Hybrid NF-NF-MCr-RO-MD without energy recovery	2.05
[131]	Hybrid NF-NF-MCr-RO-MD with energy recovery	1.6

3.7.2 Gained output ratio

GOR is used to compute energy efficiency for MD. GOR is defined as the ratio of heat by movement of vapor across the membrane (latent heat of vaporization) associated with mass flux to the energy input. The GOR is a dimensionless parameter which reveals how much energy is consumed for water production:

$$\mathrm{GOR} = \frac{JA \triangle H_v}{E_{in}} \tag{3.52}$$

The higher the GOR value is the better is the performance of the system.

Several factors can help to increase the GOR such as: using heat recovery devices, improved module designs, effective insulation to the environment, optimized piping system, and multistaged operation [97,132].

GOR can be improved by optimizing the permeate flux and heat recovery. Heat recovery can be increased by designing a system with large membrane area, low flow velocity, and more recovery stages; however; the permeation flux will decline due to temperature polarization and concentration polarization impact. The GOR value varies in the literature between 0.3 and 8.1 because of different system design and operating conditions [133].

3.7.3 Water production cost

WPC is defined as the total water production to the total investment.

WPC can be estimated as [133]

$$\mathrm{WPC} = \frac{C_{\mathrm{total}}}{fM365} \tag{3.53}$$

where f is plant availability, assumed to be 90%; M is daily capacity kg/day; and C_{total} is the total cost. The total cost represents the capital cost which includes the direct and indirect capital and the annual operating cost.

The major factors of desalination plants cost are the capital cost and annual operating costs. Capital cost is represented by construction, membrane module, auxiliaries (such as heat exchanger, pumps), solar collectors and photovoltaic systems, and land and installation cost. Annual operating and maintenance cost is represented by the total annual costs of operation and maintenance of a desalination plant such as membrane replacement and chemical pretreatment.

Yang et al. [133] reported that the process equipment purchase (mainly on membrane modules) will represent the major cost. The installation and building is assumed to be around 25% of the equipment cost. In addition, indirect capital cost and annual operating cost are approximately calculated to be 10% of the direct capital cost. Table 3.7 illustrates cost and performance evaluation for MD with and without heat recovery system.

Table 3.7 Cost and performance evaluation for MD with and without heat recovery (HR) system [90]

	MD without HR	MD with HR
Total capital cost ($)	27,149,780	28,321,033
Membrane replacement ($/year)	2,246,256	2,246,256
Electricity ($/year)	10,515	10,515
Chemicals ($/year)	141,912	141,912
Spares ($/year)	260,172	260,172
Labor ($/year)	236,520	236,520
Total annual O&M costs ($)	7,476,680	6,932,981
Total water cost ($/m^3)	1.23	1.17
Total water cost ($/m^3) when using low-grade heat energy	0.64	0.66
Performance ratio	12.0	13.7
Specific heat consumption (kJ/kg)	162	143

3.8 Temperature and concentration polarization

Since the vaporization phenomenon occurs at the membrane hot surface and condensation at the other side of membrane, thermal boundary layers are established on both sides of the membrane. The temperature difference between the liquid-vapor interface and the bulk is called temperature polarization, ψ [23,94,97,105,134], which is defined as

$$\psi = \frac{T_{m,f} - T_{m,p}}{T_f - T_p} \tag{3.54}$$

Lawson and Lloyd [39] represented ψ with slight difference for VMD as

$$\psi = \frac{T_f - T_{m,f}}{T_f - T_p} \tag{3.55}$$

The effect of heat transfer boundary layer on the total heat transfer resistance of the system is measured by temperature polarization. Camacho et al. [53] pointed out that when the temperature difference between the feed and the coolant bulk temperatures is 10°C, the actual temperature difference across the membrane is only 3.2°C.

When the thermal boundary layer resistance is reduced, the temperature difference between the liquid-vapor interface and the bulk temperature diminishes and, consequently, ψ approaches 1, which means a typical system. On the other side, zero ψ means a high degree of concentration polarization is taking place, and the system is controlled by large boundary layer resistance. Usually, the value of ψ lies between 0.4 and 0.7 for DCMD [39,45,105,108,111]. Termpiyakul et al. [94] pointed out that temperature polarization becomes important at high concentration, high temperature, and low feed velocity. Pangarkar et al. [135] reported that the impact of concentration polarization is much less than the effect of temperature polarization.

Spacers can be employed to reduce temperature and concentration polarization. Spacers enhance the permeate flux by increasing the heat transfer coefficient. Martínez-Díez et al. [136] reported that flux has been improved by 31%–41% when DCMD with spacer-filled channels was used. In addition, he observed that the temperature polarization coefficients are noticeably improved and approach unity when spacers are utilized in the channels.

Concentration polarization, Φ, is defined as the increase of solute concentration on the membrane surface (c_m) to the bulk solute concentration (c_f):

$$\Phi = \frac{c_m}{c_f} \tag{3.56}$$

In order to estimate the concentration of the solute (mole fraction) on the membrane surface, Martínez [137], Martínez-Díez and Gonzalez [107,111] suggested the following relation:

$$c_m = c_f \exp\left(\frac{j}{\rho K}\right) \tag{3.57}$$

where ρ is the liquid density and K is the mass transfer coefficient.

Hwang et al. [138] reported that the effect of salt concentration on the flux showed a decrease greater than the vapor pressure difference and was attributed to the polarization layers on the membrane. Moreover, Yun et al. [110] studied the influence of high concentration on the mass transfer coefficient and distilled flux. Pure water and high concentration of NaCl (17.76%, 24.68%) are used as feed. They found that the viscosity, density of the feed, solute diffusion coefficient, and the convective heat transfer coefficient are directly related to the concentration and temperature. They noted that the solute accumulated on the membrane surface during the desalination process; as a consequence, a diffusive flow back to the feed was generated. Therefore, the concentration polarization and fouling must be considered in modeling, and the permeate flux cannot be predicted by Knudsen, molecular and Poiseuille flow, because the properties of the boundary layer at the membrane surface vary from the bulk solution.

3.9 Fouling

The fouling problem is significantly lower than that encountered in conventional pressure-driven membrane separation. Shirazi et al. [139] and Heru [140] pointed out that membrane fouling by inorganic salt depends on the membrane properties, module geometry, feed solution characteristic, and operating conditions. There are several types of fouling, which may block the membrane pores such as inorganic scaling and biofouling, colloidal fouling, and organic fouling. The main difference between these types is the kind of the particles that cause the fouling and the type of membrane performance impact [141].

Colloids such as clay minerals and suspended particles cover a wide size range, from a few nanometers to a few micrometers. During membrane filtration, colloids accumulate on the membrane surface or inside membrane pores (pore plugging). The colloidal fouling on the membrane surface creates an additional layer (cake layer). This cake layer makes an additional resistance to mass transfer across the membrane. The permeate flux and product quality will be influenced negatively.

Inorganic scaling including salt crystallization is a common problem in thermal/membrane desalination. The major scaling ions are calcium, magnesium, bicarbonate, and sulfate. Salt solubility and crystallization widely differ over temperature. For instance, the salt solubility of NaCl increases with temperature; however, the solubility of $CaCO_3$ and $CaSO_4$ decrease with temperature [141].

Biological fouling is growth on the surface of the membrane (by bacteria). Kullab and Martin [142] pointed out that fouling and scaling lead to blocking the membrane pores, which reduces the effective membrane, and therefore the permeate flux obviously decreases. These may also cause a pressure drop, and a higher temperature polarization effect. Gryta [143] indicated that the deposits formed on the membrane surface lead to the adjacent pores being filled with feed solution (partial membrane wetting). Moreover, additional thermal resistance will be created by the fouling layer, which is deposited on the membrane surface. As a result, the overall heat transfer coefficient is changed. For DCMD at a steady state, Gryta [144] specified:

$$h_f\left(T_f - T_{f,\text{fouling}}\right) = \frac{k_{\text{fouling}}}{\delta_{\text{fouling}}}\left(T_{f,\text{fouling}} - T_{f,m}\right) = \frac{k_m}{\delta}\left(T_{f,m} - T_{p,m}\right) + J\triangle H_v$$
$$= h_p\left(T_{p,m} - T_p\right) \tag{3.58}$$

where k_{fouling}, δ_{fouling}, and $T_{f,\text{fouling}}$ are the fouling layer thermal conductivity, thickness, and fouling layer temperature, respectively.

Tun et al. [120] examined the effect of high concentration of NaCl and Na_2SO_4 on the permeate flux. The flux gradually decreases during the MD process, until the feed concentration reaches the supersaturation point, and then the flux decreases sharply to zero. Afterwards, the membrane was completely covered by crystal deposits. Furthermore, Yun et al. [110] arrived at the same result, and concluded that when the membrane surface concentration reaches saturation, the properties of the boundary layer will differ from the bulk solution properties.

Warsinger et al. [141] indicated that scaling tendency could be decreased by acidifying the feed solution (below pH 7) and dissolved of CO_2. Moreover, biofouling could be decreased by increasing the operating temperature. Generally, most organisms will not grow at 60°C.

In addition, pretreatment and membrane cleaning are the main techniques to control fouling. Alklaibi and Lior [69] investigated the influence of fouling by preparing three different solutions: water pretreated by microfiltration, seawater, and 3% NaCl.

They concluded that the pretreatment process increased the flux by 25%, which means that the pretreatment process is important, in order to enhance the permeate flux. Hsu et al. [28] used the ultrasonic irradiation technique to clean fouling from the membrane. Moreover, pure water for 2 h, followed by 0.1 M NaOH was used to clean a membrane, which was utilized to filter a mixture of $CaCl_2$ and humic acid. The permeate flux was about 87% of the initial flux [145]. Gryta [144] proposed that the fouling intensity can be limited by operating at low temperature (feed temperature), and increasing the feed flow rate.

3.10 Application of membrane distillation

3.10.1 MD stand-alone process

Several studies have been applied to produce fresh water or treat wastewater by various MD configurations. New MD applications were considered in environmental protection and wastewater treatment. DCMD is the simplest MD configuration, and is widely employed in desalination processes and concentration of aqueous solutions in food industries [25–29] and boron removal [30]. Juice solutions are sensitive to high temperature. As a result, MD represents a competitive process to increase the quality of juice concentration. DCMD was successfully tested in the concentration of many juices such as orange, apple, and sugarcane juice. Furthermore, MD has a great potential application in the dairy industry process. DCMD has been applied to concentrate different dairy streams like whole milk, skim milk, and whey. Membrane performance and fouling mechanisms have been also examined in this study [146]. MD is a promising technology towards heavy metal ions and small molecule contaminates removal. DCMD was implemented to remove nickel sulfate from wastewater. The rejection factor was close to unity [14]. DCMD has been also used as a treatment method for liquid radioactive wastes produced from the nuclear industry [15,76].

Produced water was treated by MD [13]. They state that AGMD is a promising technology for the treatment of produced water. In addition, it is expected to be a cost-effective process for produced water treatment. Moreover, oily saline wastewater from a gas refinery has been treated by solar-powered AGMD [147]. The average permeate production rate was 1.3 L/m^2 day with 91 ppm total dissolved solid. In addition, this configuration has been used to treat highly concentrated solutions such as NaCl, $MgCl_2$, Na_2CO_3, and Na_2SO_4 [148].

Ammonia is the main pollutant in agriculture wastewater. SGMD has been implemented to remove ammonia [147]. It was concluded that the flux increased when the feed temperature raised while the selectivity decreased. Also, a negligible impact of sweeping gas temperature was noticed.

VMD is used for desalination where the feed concentration was varied between brackish, seawater, and brine (high concentration) [149–151]. In addition, VMD is widely used to remove volatile organic compounds from dilute aqueous solutions. It was employed to remove arsenic [152], acetone, isopropanol, and MTBE from water [153].

Table 3.8 summarize some of the MD application. Recently, most of the MD usage is at the laboratory stage. In spite of this, there are some pilot plants that have been developed currently to produce fresh water [32,142]. For instance, a DCMD pilot plant was operated successfully for 3 months by Song et al. [160]. Moreover, Memstill and AGMD Scarab AB systems with heat recovery were tested and used worldwide.

Table 3.8 Reported MD applications used by some researchers

Ref.	MD process	Membrane type	Thickness (μm)	Pore size (μm)	Feed solution
[154]	DCMD	TF200	178	0.2	Pure water and
		PVDF	125	0.22	humic acid
[145]	DCMD	PVDF	125	–	Humic acid/NaCl
[94]	DCMD	PVDF	126	0.22	Pure water, NaCl, brackish and seawater
[26]	DCMD	PVDF	–	0.45	Apple juice
[28]	DCMD	PTFE	175	0.2	Seawater and NaCl
	AGMD			0.5	
[91]	DCMD	PTFE	60	0.1	Pure water
		PTFE	60	0.3	
		PVDE	100	0.2	
[155]	DCMD	PVDF	–	0.4	Pure water, NaCl, and sugar
[156]	DCMD	PTFE	55	0.198	Olive mill wastewaters
[29]	DCMD	PVDF	140	0.11	Orange juice
[110]	DCMD	PVDF	120	0.22	Pure water, NaCl
			125	0.2	
[83]	DCMD	PVDF	125	0.22	Pure water and humic acid
[14]	DCMD	Not mentioned	120	0.25	Heavy metals waste
[157]	DCMD	PTFE	55	0.8	Pure water, NaCl,
			90		bovine plasma, and bovine blood
[13]	AGMD	PTFE	175	0.2	Produced water
				0.45	
				1.0	
[100]	AGMD	PTFE	–	0.2	LiBr and H_2SO_4
[35]	AGMD	PTFE	80	0.2	NaCl,H_2SO_4, NaOH, HCl, and HNO_3

Table 3.8 **Continued**

Ref.	MD process	Membrane type	Thickness (μm)	Pore size (μm)	Feed solution
[40]	VMD	PTFE	–	0.2	Acetone, ethanol, isopropanol, and MTBE
[41]	VMD	PTFE	60	0.2	Pure water, ethanol, and degassing water
[42]	VMD	3MC 3MB 3MA	76 81 91	0.51 0.4 0.29	Pure water and ethanol
[158,159]	SGMD	PTFE PTFE	178 178	0.2 0.45	NaCl

3.10.2 MD hybrid separation processes

The cost of the pretreatment process in pressure-driven technology represents the main desalination cost which might reach about 60% of overall costs. Cost reduction might be possible, if integrated membrane is introduced to the pretreatment process [161].

The MD system has the feasibility to be combined with other separation processes to create an integrated separation system, such as UF, nanofiltration (NF), and RO. The MD process is less affected by the high salt concentration compared to other separation processes. Consequently, a significant enhancement in the total water recovery and product quality can be achieved if MD is integrated to the existing desalination process (thermal or pressure-driven process). In addition, the potential impact on the environment can be minimized too when the concentrated brine is treated by MD.

De Andrés et al. integrated MD with an existing multieffect distiller unit [162]. The hot brine, which is rejected from the multieffect distiller, is used as a feed for MD. The pure water production and GOR of the whole system were improved to 7.5% and 10%, respectively.

An integrated membrane system was created, which included UF for pretreatment and VMD for the concentrated RO brine treatment [163]. The performance of VMD for treated and untreated RO brine was studied. The initial flux for RO brine, which was pretreated by hardness removal and UF, was 25.6 kg/m^2 h and decreased to 17.8 kg/m^2 h.

A hybrid desalination process composed of freeze desalination and direct contact membrane distillation (FD-DCMD) processes was developed [164]. Regasification of liquefied natural gas (LNG) was used as waste cold energy. The concentrated brine from the FD was further treated in the DCMD. A high water recovery of 71.5% was obtained from both processes.

An integrated membrane distillation/crystallizer (MD-C) process was proposed and used to achieve a higher water recovery from seawater desalination and wastewater treatment. MD is used to treat the feed solution to a desired concentration; then, the salt crystals can be easily precipitated by the external crystallizer [165]. It worth mentioning that the membrane crystallization process is still under evaluation at the lab scale.

Gryta et al. [6] proposed an integrated UF/MD system to treat oily wastewater. They stated that the permeate flux achieved via the UF process generally contains about 5 ppm of oil. A further purification for the UF permeate by MD results in a complete removal of oil from wastewater and a very high reduction of the total organic carbon (99.5%) and total dissolved solids (99.9%).

Scaling formation on the membrane surface is the main concern when MD is combined with current desalination processes. Inorganic salts in the brine such as calcium, magnesium, and sulfate ions will cause severe scaling in the MD process. The salt crystals can rip the membrane and allow the brine to penetrate the membrane. Qu et al. [166] combined an accelerated precipitation softening process to avoid calcium scaling with DCMD.

3.11 Operating parameters

In this section, the influence of feed temperature, concentration, and air gap will be reviewed and major findings will be cited and discussed.

3.11.1 Feed temperature

As can be seen in Table 3.9, the feed temperature has a strong influence on the distilled flux. According to the Antoine equation, the vapor pressure increases exponentially with temperature. Therefore, the operating temperature has an exponential effect on the permeate flux [33,69]. At constant temperature difference between the hot and the cold fluid, the permeate flux increases when the temperature of the hot fluid rises, which means the permeate flux is more independent of the hot fluid temperature. Alkhudhiri et al. [10], Qtaishat et al. [109], Gunko et al. [26], and Chen et al. [168] pointed out that increasing the temperature gradient between the membrane surfaces will affect the diffusion coefficient positively, which leads to increased vapor flux. Similarly, Srisurichan et al. [83] believed that there is a direct relation between diffusivity and temperature, so that working at high temperature will increase the mass transfer coefficient across the membrane. Moreover, temperature polarization decreases with increasing feed temperature [84,108].

In terms of coolant temperature, a noticeable change takes place in the permeate flux when the cold side temperature decreases [26,29,33]. In addition, more than double permeate flux can be achieved compared to a solution, at the same temperature difference [69]. Banat and Simandl [11] and Matheswaran et al. [169], however, found that the effect of the cold side temperature on the permeate flux is neglected at fixed hot side temperature, because of low variation of vapor pressure at low temperatures.

Table 3.9 **Effect of temperature on permeate flux**

Ref.	MD type	Membrane type	Pore size	Solution	Feed velocity (m/s)	T_{in} (°C)	Permeate (kg/m^2 h)
[11]	AGMD	PVDF	0.45	Artificial seawater	5.5 L/min	40–70	≈1–7
[84]	DCMD	PVDF	0.22	Pure water	0.1	40–70	≈3.6–16.2
[167]	DCMD	PTFE	0.2	NaCl (2 mol/L)	16 cm^3/s	17.5–31	≈2.88–25.2
[79]	DCMD	PTFE	0.2	Pure water	–	40–70	≈5.8–18.7
[155]	DCMD	PVDF	0.4	Sugar	0.45	61–81	≈18–38
[110]	DCMD	PVDF	0.4	Pure water	0.145	36–66	≈5.4–36
				NaCl (24.6 wt%)	0.145	43–68	≈6.1–28.8
[42]	VND	3MC	0.51	Pure water	–	30–75	≈0.8–8.8 mol/m^2 s
[83]	DCMD	PVDF	0.22	Pure water	0.23	40–70	≈7–33 L/m^2 h
[158]	SGMD	PTFE	0.45	Pure water	0.15	40–70	≈4.3–16.2
[29]	DCMD	PVDF	0.11	Orange juice	2.5 kg/min	25–45	30×10^3–108×10^3
[28]	DCMD	PTFE	0.2	NaCl (5%)	3.3 L/min	5–45	1–42
[28]	AGMD	PTFE	0.2	NaCl (3%)	3.3 L/min	5–45	0.5–6

3.11.2 The concentration and solution feature

There is a noticeable fall in the flux product when feed concentration increases due to decreasing vapor pressure [34,137] and increasing temperature polarization [107]. Likewise, Izquierdo-Gil et al. [116] concluded that the reduction in product flux is linear with time. Furthermore, Tomaszewska et al. [31,119] studied the influence of acid concentration on the permeate flux. They found that there is a reduction in the permeate when the acid concentration increase. Moreover, Sakai et al. [157] found a noticeable reduction in the water vapor permeability when the protein concentration of bovine plasma increases. On the other hand, Banat and Simandl [71], Qtaishat et al. [109], and Alklaibi and Lior [126] concluded that the permeate flux decreases slightly with increasing feed concentration. About 12% reduction in permeate flux happened when the feed (NaCl) increased from 0 to 2 M concentration [109]. This decrease in the permeate flux amount is due to the reduction in the water vapor pressure. Lawson and Lloyd [39] studied the reasons for decreasing product flux when the concentration of NaCl increases. They found three reasons for this reduction; (1) water activity, which is a function of temperature, decreases when the concentration increases, (2) the mass transfer coefficient of the boundary layer at the feed side decreases due to increased influence of concentration polarization, and (3) the heat transfer coefficient decreases as well at the boundary layer, because of the reduction in the surface membrane temperature. Therefore, the vapor pressure of the feed declines, which leads to reduced performance of MD. Schofield et al. [155] studied the impact of molecular weight fraction and viscosity on the flux by preparing sugar (30 wt%) and NaCl (25 wt%) solutions. Under the same conditions, they found that the sugar solution has less flux reduction than salt solution. They concluded that the viscosity is an important factor in flux reduction. The heat transfer coefficient decreases due to the reduced Reynolds number. The effect of thermal conductivity and heat capacity on flux reduction is negligible. Furthermore, the impact of density on flux production is important for salt solutions. Three aqueous solutions of methanol, ethanol, and isopropanol at different concentrations were studied by García-Payo [12]. They found that the type of alcohol is strongly related to the amount of flux. Isopropanol has the highest vapor pressure; consequently, isopropanol solution has the highest flux, while methanol solution has the lowest.

The influence of high concentration, such as in NaCl solutions, was reported by Yun et al. [110], who found that there is variation in the permeate flux with time (Table 3.10), and that it is difficult to calculate the permeate flux using existing models.

3.11.3 Recirculation rate

Table 3.11 summarizes the effect of recirculation rate. Working at a high recirculation rate minimizes the boundary layer resistance and maximizes the heat transfer coefficient. As a result, higher flux can be achieved [83]. Chen et al. [168] and Alkhudhiri et al. [33] indicated that the increasing volumetric flow rate will enhance the permeate flux. The fluid velocity rises when the volumetric flow rates increases, so that the

Table 3.10 **Effect of concentration and solution feature on permeate flux**

Ref.	MD type	Membrane type	Pore size (μm)	Solution	Concentration (g/L)	T_{in} (°C)	Permeate (kg/m^2 h)
[12]	AGMD	PVDF	0.22	Methanol/water	≈30–200	50	≈3.9–4.6
				Ethanol/water	≈30–150		≈3.95–4.9
				Isopropanol/ water	≈10–95		≈4.0–5.0
[167]	DCMD	PTFE	0.2	NaCl	0–116.8	31	≈32.4–25.2
[155]	DCMD	PVDF	0.4	NaCl	0–30 wt%	81	≈44–63
[110]	DCMD	PVDF	0.22	NaCl	0–24.6 wt%	68	≈36–28.8
[169]	AGMD	PTFE	0.22	HNO_3	2–6 M	80	≈0.9–2.1 L/m^2 h
[170]	VMD	PP	0.2	NaCl	100–300	55	10.7–7.0

Table 3.11 **Effect of recirculation rate on permeate flux**

Ref.	MD type	Membrane type	Pore size (μm)	Solution	T_f (°C)	Flow rate (m/s)	J (kg/m^2 h)
[158,159]	SGMD	TF200	0.2	NaCl (1 M)	50	0.07–0.21	≈3.24–3.96
		TF450	0.45				≈5.4–5.76
[94]	DCMD	PVDF	0.22	NaCl (35 g/L)	60	1.85–2.78	31–38
[11]	AGMD	PVDF	0.45	Artificial seawater	60	1–5.7 L/min	≈2.7
[137]	DCMD	PTFE	0.2	Sucrose (40 wt%)	39	5–16 cm^3/s	5.7–9.0
[155]	DCMD	PVDF	0.4	Sugar (30 wt%)	81	0.45–0.9	≈38–55
[110]	DCMD	PVDF	0.22	NaCl (17.7 wt%)	68	0.056–0.33	≈25.9–29.5
[42]	VMD	3MC	0.51	Pure water	74	37–63 cm^3/s	≈6.4–8.7 mol/m^2 s
[83]	DCMD	PVDF	0.22	Pure water	50	1.8–2.3	≈18–20 L/m^2 h
[40]	VMD	PTFE	0.2	Acetone (5 wt%)	30	0.1–2.6 L/min	12.6–21.6
[170]	VMD	PP	0.2	NaCl (300 g/L)	55	0.015–0.03 L/s	7–9.1

convective heat transfer coefficient develops and the thermal boundary layer thickness decreases. As a result, the temperature polarization effect reduces. Moreover, Martínez-Díez and Vázquez-González [105,107] found significant change in temperature polarization, when the rate of recirculation changes. This is because the recirculation rate enhances the heat transfer, which leads to rise in the product flux and decline in temperature polarization. Izquierdo-Gil et al. [116] and Xu et al. [171] investigated the influence of increasing the flow rate on the flux product. The concentration and temperature polarization decreased due to the higher flow rate.

With regard to coolant flow rate, Banat and Simandl [11] and Kubota et al. [172] studied the effect of flow rate on the cold side. Kubota et al. [172] found that the cold side flow rate has a noticeable effect on the permeate flux; however, Banat and Simandl [11] found negligible influence on the permeate flux, when the cold side flow rate was enhanced. It is worth pointing out that at a certain flow rate level, increasing the flow rate has a limited impact on the permeate flux production and it is not sensitive as before. This result can be attributed to the fact that the heat transfer in the boundary layer is no longer the controlling step [173].

3.11.4 The air gap

Lawson and Lloyd [39] pointed out that the flux declines linearly with $\frac{1}{l}$. Likewise, Izquierdo-Gil et al. [116] found a linear relation between the distillate flux and the number of gaskets removed. They studied the effect of opening the upper side of the gap to the atmosphere as well. A slight reduction in distillate flux happened compared to closing the gap. Alklaibi and Lior [69] reported that reducing the air gap will double the flux product, with a more significant effect when the gap is <1 mm. Banat and Simandl [11] concluded that reducing the air gap width will increase the temperature gradient within the gap, which leads to increased permeate flux. Table 3.12 summarizes the air gap effect on the permeate flux.

3.11.5 Membrane type

The membrane permeation flux is proportional to the porosity, and inversely proportional to the membrane thickness and tortuosity [175]. Izquierdo-Gil et al. [116] observed that for a larger pore size membrane, higher permeate flux is obtained. In addition, higher flux is achieved using a membrane without support, compared to the same membrane pore size with support [12]. Likewise, Izquierdo-Gil et al. [116] and Alklaibi and Lior [69] concluded that for a more efficient MD process, low thermal conductivity material should be used (unsupported membrane).

3.11.6 Long operation

Izquierdo-Gil et al. [116] did not observe any change in distillate flux during a month of operation. On the other hand, Schneider et al. [72] reported a 20% decline in permeate flux after 18 weeks, when tap water was used in a DCMD. Lawson and

Table 3.12 Air gap effect on the permeate flux

Ref.	Membrane type	Pore size (μm)	Solution	T_{in} (°C)	Air gap (mm)	Permeate (kg/m^2 h)
[11]	PVDF	0.45	Artificial seawater	60	1.9–9.9	≈5–2.1
[12]	PTFE	0.2	Isopropanol	50	1.62–0.55	≈5.1–6.3
[116]	PVDF	0.22	Sucrose	25.8	1–4	≈0.8–1.7 L/m^2 h
[35]	PTFE	0.2	NaCl	60	0.3–9	≈19–1.5
[169]	PTFE	0.22	HNO_3	80	0.5–2	≈5.3–4.25 L/m^2 h
[174]	PTFE	0.45	HCl/water	60	4–7	≈3.7–2.4
[174]	PTFE	0.2	Propionic acid/water	60	4–7	≈7.4–4.6

Lloyd [39] suggested that the decline in permeate flux was due to membrane fouling, or the pores being wetted. Banat and Simandl [71] analyzed the effect of long operation (2 months) on the permeate flux for tap water (297 μS/cm). They found that the flux increased during the first 50 h, and then fell for 160 h before reaching a steady state. The permeate conductivity was steady at about 3 μS/cm. For seawater, the experiment was conducted for 10 days. The flux declined until it reached a steady state [11]. Drioli and Wu [176] reached the same result, when 1 M NaCl had been used for 6 days.

3.12 Cost evaluation

There are abundant studies focusing on cost estimates and energy consumption of thermal and membrane desalination processes. Nevertheless, there are a few studies reported on energy consumption and cost estimation of MD process. This can be attributed to the fact that most MD applications were carried out at the lab scale and not fully applied to the commercial scale. Consequently, the total capital cost, MD performance, and the optimum flow conditions information varies for four different MD configurations. For example, the MD production cost could range from 0.26 to 130 $\$/m^3$, while total energy consumption could vary from 1 to 9000 kWh/m^3, depending on the types and size of the MD systems, operating conditions, energy sources and recovery approaches, and cost estimation procedures [133].

MD can be attractive for industry when free/cheap energy is available or MD can be implemented in extreme areas where other technologies are either not possible or too expensive [140]. For instance, Kesieme et al. [177] reported that the production cost of an MD desalination plant of 30,000 m^3/d (capacity plant) is reduced from 2.2 to 0.66 $\$/m^3$ when waste heat is utilized which is compared to the cost of an RO plant which is 0.8 $\$/m^3$. Furthermore, a solar powered MD plant was employed by Hogan et al. [181] to provide fresh water with a capital cost around 3500 \$. They estimated the optimum solar collector area to be around 3 m^2, a membrane area of 1.8 m^2, and heat exchanger area of 0.7 m^2 for plant size of 50 L/day.

Khayet [178] and Walton [32] reported that the WPC by MD could be competitive to RO if waste heat is employed. Khayet [178] pointed out that renewable energy source, energy recovery, and hydride system should be used to achieve the optimum MD design.

The impact of operating parameters and MD module design on GOR was studied. The GOR value is less than unity for a simple stage system that indicates low performance and poor heat recovery. On the other hand, the GOR value is higher than unity for multistages. For example, a numerical simulation for 10-stages DCMD module was created. The predicted GOR value for that system was 12 [133].

The impact of heat recovery for the MD system was examined. An economics study including unit production cost for a DCMD system with and without heat recovery was performed by Al-Obaidani et al. [90].They found that the unit production cost without heat recovery was 1.23 and 1.17 $\$/m^3$ with heat recovery. Moreover, the GOR for DCMD with a recovery system was about 13.7%. The WPC for DCMD with heat recovery is comparable to the cost of a conventional thermal process. The total capital cost was increased by 4% due to the additional cost of a heat exchanger for heat recovery. In contrast, the thermal energy needed was reduced by 11%. Furthermore, a modeling study for a DCMD plant with a capacity of 24,000 m^3/day revealed that a minimum WPC of 1.23 \$/m^3 can be achieved without heat recovery, while, the WPC for the system with heat recovery was 1.17 \$/m^3. The WPC continued to drop to 0.64 \$/m^3 if low grade waste heat was used [133].

A small MD solar-powered plant with a capacity of 100 L/day was implemented to treat brackish water. Also, a stand-alone MD powered completely by solar energy was employed for a capacity ranging from 0.2 to 20 m^3/day [44]. The GOR for a feed flow rate of 350 L/h at an inlet temperature of 75°C was computed to be around 5.5 with an energy consumption of 117 kWh/m^3. A large solar-powered MD plant was utilized to treat seawater. Spiral AGMD modules were employed in both plants. The WPCs for both plants were 15 and 18 \$/m^3 for the small and large system, respectively. Furthermore, they conclude that increasing the plant lifetime may reduce the WPC significantly [179].

An economic study was conducted by Saffarini et al. [180] for three MD configurations: DCMD, AGMD, and VMD connected to a solar heater and a heat recovery system. The WPC was 12.7, 18.6, and 16 \$/m^3 for DCMD, AGMD, and VMD, respectively. DCMD was the most cost-effective compared to AGMD and VMD. Moreover, they found that the WPC for AGMD could be lower when air gap and feed channel depth decreases and/or feed temperature increases. Saffarini et al. [180] and Banat and Jwaied [179] reported that the cost of a solar heater represents about 70% of the total cost. Therefore, using waste heat as a source of thermal energy is better or multistages of solar powered for more efficient heat used. Moreover, Hogan et al. [181] observed that the capital cost is very sensitive to heat recovery, because the heat exchanger is the most expensive item in a solar-powered MD plant. They optimized the solar collector area, the membrane area and heat recovery to achieve low capital cost and high flux. It was noted from the literature that the long-term performance for a solar-powered MD plant was not confirmed. Moreover, the thermal parabolic and spherical collectors are not tested.

Schwantes et al. [182] indicated that using waste heat as a source of thermal energy (e.g., combustion engines) is a promising and a cheaper alternative (saving the investment costs) compared to solar energy. For example, an integrated MD desalination plant (capacity of 36.6 m^3/h) with a power plant was employed. The WPC was 1.13 \$/m^3. A commercial study for AGMD was carried out by Memstill. They claimed that a WPC of 0.26 \$/m^3 can be obtained, which is competitive to RO when cheap industrial waste heat is utilized [183].

A geothermal resource was utilized for an MD plant size of 17 L/day to reduce the unit production cost. The estimated cost was 13 \$/m^3 [184]. In addition, geothermal water was filtered by VMD [185]. The economic study showed that the geothermal energy could reduce the WPC by 59% compared to a plant operated without geothermal energy. In spite of that geothermal energy is less competitive in terms of cost compared to other renewable energies [133]. Consequently, more intensive research is required to improve the process efficiency and minimize the WPC.

A hybrid NF-RO-MD system was utilized to treat seawater [131]. The total recovery was 76% and 310 ppm permeate concentration. The production cost was 0.92 \$/m^3.

Moreover, an energetic and exergetic study was performed by Criscuoli and Drioli [7] on a hybrid NF/RO/MD system which showed that 13 kWh/m^3 are required to produce 1 m^3 of fresh water for the hybrid system; however, this value falls down to 2.58 kWh/m^3 if low-grade thermal energy is available. In this case, the total operating costs are 0.56 \$/m^3.

A one-stage MED/DCMD hybrid system was used [162]. The hot brine of the MED is reused as feed in shell-and-tube DCMD modules. The water production of the MED/MD hybrid system was increased by 7.5% and the GOR improved by practically 10% compared to stand-alone MED.

As stated earlier, renewable energies such as solar, salt gradient solar ponds, and geothermal energy, which are considered as alternative low-grade energy resources, can be utilized to reduce the energy cost and then the total production cost. These alternatives can be integrated into the MD system to produce clean water, especially for arid and rural areas.

3.13 Conclusion and future directions

Primary energy consumption is higher in MSF plants than in MED plants. For a water production capacity of 30,000 m^3/day, the cost of MD is about 1.72 \$/m^3; however, the MSF cost is about 2.1 \$/m^3, MED cost is 1.48 \$/m^3, and RO cost is 0.69 \$/m^3 [177].

With the present state-of-the-art desalination technology (4 kWh/m^3), RO is the lowest pollutant, and its environmental load is one order of magnitude less than thermal desalination (MSF and MED) [186].

A continuous improvement in terms of membrane fabrication and module engineering on MD technology was shown. The MD process has several advantages and possibility for industrial application. For instance, MD showed a great potential in seawater desalination and water treatment when it was integrated with conventional seawater desalination systems such as MSF, MED, and RO. The brine of a conventional desalination plant can be treated to saturation and then sent to the membrane crystallizer for salt recovery. Furthermore, small-scale stand-alone MD, which is powered by renewable energy, can be utilized in isolated or remote areas such as offshore platforms. Produced water, which is the largest waste stream associated with oil or gas production and represents a potential risk to the environment, can be treated by MD.

However, few studies were performed for a large scale in long-term MD performance, membrane fouling and scaling, and WPC, which are considered the main MD challenges. More research is required using different membranes types and modules as well as different types of feed aqueous solutions and wastewaters. Moreover, integrated MD systems to other separation processes and renewable energy sources are needed.

To achieve these aims, more intensive research is necessary to improve membrane characterization and membrane module to minimize the energy consumption and enhance the permeate flux quality and quantity. In addition, new membrane materials or composite membranes are required to increase the permeate flux and selectivity.

References

[1] Danoun R. Desalination plants: potential impacts of brine discharge on marine life. Sydney: The University of Sydney; 2007.

[2] Ghaffour N. The challenge of capacity-building strategies and perspectives for desalination for sustainable water use in MENA. Desalin Water Treat 2009;5(1–3):48–53.

[3] IDA. I.D.A. desalination by the numbers, Available from: http://idadesal.org/desalination-101/desalination-by-the-numbers/. Accessed 12 September 2015.

[4] Ghaffour N, Missimer TM, Amy GL. Technical review and evaluation of the economics of water desalination: current and future challenges for better water supply sustainability. Desalination 2013;309:197–207.

[5] Alkhudhiri A, Darwish N, Hilal N. Membrane distillation: a comprehensive review. Desalination 2012;287:2–18.

[6] Gryta M, Karakulski K, Morawski AW. Purification of oily wastewater by hybrid UF/MD. Water Res 2001;35(15):3665–9.

[7] Criscuoli A, Drioli E. Energetic and exergetic analysis of an integrated membrane desalination system. Desalination 1999;124(1–3):243–9.

[8] Kurokawa H, Sawa T. Heat recovery characteristics of membrane distillation. Heat Transfer—Jpn Res 1996;25:135–50.

[9] Blanco Gálvez J, García-Rodríguez L, Martín-Mateos I. Seawater desalination by an innovative solar-powered membrane distillation system: the MEDESOL project. Desalination 2009;246(1–3):567–76.

[10] Alkhudhiri A, Darwish N, Hilal N. Treatment of saline solutions using air gap membrane distillation: experimental study. Desalination 2013;323:2–7.

[11] Banat FA, Simandl J. Desalination by membrane distillation: a parametric study. Sep Sci Technol 1998;33(2):201–26.

[12] García-Payo MC, Izquierdo-Gil MA, Fernández-Pineda C. Air gap membrane distillation of aqueous alcohol solutions. J Membr Sci 2000;169(1):61–80.

[13] Alkhudhiri A, Darwish N, Hilal N. Produced water treatment: application of air gap membrane distillation. Desalination 2013;309:46–51.

[14] Zolotarev PP, et al. Treatment of waste water for removing heavy metals by membrane distillation. J Hazard Mater 1994;37(1):77–82.

[15] Zakrzewska-Trznadel G, Harasimowicz M, Chmielewski AG. Concentration of radioactive components in liquid low-level radioactive waste by membrane distillation. J Membr Sci 1999;163(2):257–64.

[16] Bodell, B., Silicone rubber vapor diffusion in saline water distillation. United States Patent Serial, (285,032); 1963.

[17] Weyl PK. Recovery of demineralized water from saline waters; 1967. Google Patents.

[18] Findley M. Vaporization through porous membranes. Ind Eng Chem Process Des Dev 1967;6(2):226–30.

[19] Findley M, et al. Mass and heat transfer relations in evaporation through porous membranes. AICHE J 1969;15(4):483–9.

[20] Van Haute A, Henderyckx Y. The permeability of membranes to water vapor. Desalination 1967;3(2):169–73.

[21] Sanjay N, Ganapathi P, Raghavarao K. Membrane distillation in food processing. In: Handbook of membrane separations. New York: CRC Press; 2008. p. 513–51.

[22] Schofield RW, Fane AG, Fell CJD. Gas and vapour transport through microporous membranes. I. Knudsen-Poiseuille transition. J Membr Sci 1990;53(1–2):159–71.

[23] Schofield RW, Fane AG, Fell CJD. Heat and mass transfer in membrane distillation. J Membr Sci 1987;33(3):299–313.
[24] Schofield RW, Fane AG, Fell CJD. Gas and vapour transport through microporous membranes. II. Membrane distillation. J Membr Sci 1990;53(1–2):173–85.
[25] Alves VD, Coelhoso IM. Orange juice concentration by osmotic evaporation and membrane distillation: a comparative study. J Food Eng 2006;74(1):125–33.
[26] Gunko S, et al. Concentration of apple juice using direct contact membrane distillation. Desalination 2006;190(1–3):117–24.
[27] Godino MP, et al. Water production from brines by membrane distillation. Desalination 1997;108(1–3):91–7.
[28] Hsu ST, Cheng KT, Chiou JS. Seawater desalination by direct contact membrane distillation. Desalination 2002;143(3):279–87.
[29] Calabro V, Jiao BL, Drioli E. Theoretical and experimental study on membrane distillation in the concentration of orange juice. Ind Eng Chem Res 1994;33(7):1803–8.
[30] Hou D, et al. Boron removal from aqueous solution by direct contact membrane distillation. J Hazard Mater 2010;177(1–3):613–9.
[31] Tomaszewska M, Gryta M, Morawski AW. Study on the concentration of acids by membrane distillation. J Membr Sci 1995;102:113–22.
[32] Walton J, et al. Solar and waste heat desalination by membrane distillation. Texas: College of Engineering University of Texas at El Paso; 2004.
[33] Alkhudhiri A, Darwish N, Hilal N. Treatment of saline solutions using air gap membrane distillation: experimental study. Desalination 2013;323:2–7.
[34] Alkhudhiri A, Darwish N, Hilal N. Treatment of high salinity solutions: application of air gap membrane distillation. Desalination 2012;287:55–60.
[35] Kimura S, Nakao S-I, Shimatani S-I. Transport phenomena in membrane distillation. J Membr Sci 1987;33(3):285–98.
[36] Banat FA, Simandl J. Membrane distillation for dilute ethanol: separation from aqueous streams. J Membr Sci 1999;163(2):333–48.
[37] Khayet M. Membranes and theoretical modeling of membrane distillation: a review. Adv Colloid Interf Sci 2011;164(1–2):56–88.
[38] García-Payo MC, et al. Separation of binary mixtures by thermostatic sweeping gas membrane distillation: II. Experimental results with aqueous formic acid solutions. J Membr Sci 2002;198(2):197–210.
[39] Lawson KW, Lloyd DR. Membrane distillation. J Membr Sci 1997;124(1):1–25.
[40] Bandini S, Sarti GC. Heat and mass transport resistances in vacuum membrane distillation per drop. AIChE J 1999;45:1422–33.
[41] Bandini S, Gostoli C, Sarti GC. Separation efficiency in vacuum membrane distillation. J Membr Sci 1992;73(2–3):217–29.
[42] Lawson KW, Lloyd DR. Membrane distillation. I. Module design and performance evaluation using vacuum membrane distillation. J Membr Sci 1996;120(1):111–21.
[43] Cath TY, Adams VD, Childress AE. Experimental study of desalination using direct contact membrane distillation: a new approach to flux enhancement. J Membr Sci 2004;228(1):5–16.
[44] Koschikowski J, Wieghaus M, Rommel M. Solar thermal-driven desalination plants based on membrane distillation. Desalination 2003;156(1–3):295–304.
[45] Curcio E, Drioli E. Membrane distillation and related operations—a review. Sep Purif Rev 2005;34(1):35–86.
[46] Laganà F, Barbieri G, Drioli E. Direct contact membrane distillation: modelling and concentration experiments. J Membr Sci 2000;166(1):1–11.

[47] Fujii Y, et al. Selectivity and characteristics of direct contact membrane distillation type experiment. I. Permeability and selectivity through dried hydrophobic fine porous membranes. J Membr Sci 1992;72(1):53–72.
[48] Gryta M. Concentration of saline wastewater from the production of heparin. Desalination 2000;129(1):35–44.
[49] García-Fernández L, Khayet M, García-Payo MC. Membranes used in membrane distillation: preparation and characterization. In: Pervaporation, vapour permeation and membrane distillation. Oxford: Woodhead Publishing; 2015. p. 317–59.
[50] Cerneaux S, et al. Comparison of various membrane distillation methods for desalination using hydrophobic ceramic membranes. J Membr Sci 2009;337(1–2):55–60.
[51] Winter D, Koschikowski J, Wieghaus M. Desalination using membrane distillation: experimental studies on full scale spiral wound modules. J Membr Sci 2011;375 (1–2):104–12.
[52] Meindersma GW, Guijt CM, de Haan AB. Desalination and water recycling by air gap membrane distillation. Desalination 2006;187(1–3):291–301.
[53] Camacho LM, et al. Advances in membrane distillation for water desalination and purification applications. Water 2013;5(1):94–196.
[54] Heinzl W, Büttner S, Lange G. Industrialized modules for MED desalination with polymer surfaces. Desalin Water Treat 2012;42(1–3):177–80.
[55] Wikol M, et al. Expanded polytetrafluoroethylene membranes and their applications. Drug pharm Sci 2008;174:619.
[56] Feng C, et al. Preparation and properties of microporous membrane from poly(vinylidene fluoride-co-tetrafluoroethylene) (F2.4) for membrane distillation. J Membr Sci 2004;237 (1–2):15–24.
[57] Hengl N, et al. Study of a new membrane evaporator with a hydrophobic metallic membrane. J Membr Sci 2007;289(1–2):169–77.
[58] Lawson KW, Hall MS, Lloyd DR. Compaction of microporous membranes used in membrane distillation. I. Effect on gas permeability. J Membr Sci 1995;101(1–2):99–108.
[59] Peng P, Fane AG, Li X. Desalination by membrane distillation adopting a hydrophilic membrane. Desalination 2005;173(1):45–54.
[60] Chiam C-K, Sarbatly R. Vacuum membrane distillation processes for aqueous solution treatment—a review. Chem Eng Process Process Intensif 2013;74:27–54.
[61] Razmjou A, et al. Superhydrophobic modification of TiO_2 nanocomposite PVDF membranes for applications in membrane distillation. J Membr Sci 2012;415–416:850–63.
[62] Fang H, et al. Hydrophobic porous alumina hollow fiber for water desalination via membrane distillation process. J Membr Sci 2012;403–404:41–6.
[63] Gostoli C, Sarti GC. Separation of liquid mixtures by membrane distillation. J Membr Sci 1989;41:211–24.
[64] García-Payo MC, Izquierdo-Gil MA, Fernández-Pineda C. Wetting study of hydrophobic membranes via liquid entry pressure measurements with aqueous alcohol solutions. J Colloid Interface Sci 2000;230(2):420–31.
[65] Franken ACM, et al. Wetting criteria for the applicability of membrane distillation. J Membr Sci 1987;33(3):315–28.
[66] Tomaszewska M. Preparation and properties of flat-sheet membranes from poly(vinylidene fluoride) for membrane distillation. Desalination 1996;104(1–2):1–11.
[67] Khemakhem S, Amar RB. Grafting of fluoroalkylsilanes on microfiltration Tunisian clay membrane. Ceram Int 2011;37(8):3323–8.
[68] Zhang J, et al. Identification of material and physical features of membrane distillation membranes for high performance desalination. J Membr Sci 2010;349(1–2):295–303.

[69] Alklaibi AM, Lior N. Membrane-distillation desalination: status and potential. Desalination 2005;171(2):111–31.
[70] He K, et al. Production of drinking water from saline water by direct contact membrane distillation (DCMD). J Ind Eng Chem 2011;17(1):41–8.
[71] Banat FA, Simandl J. Theoretical and experimental study in membrane distillation. Desalination 1994;95(1):39–52.
[72] Schneider K, et al. Membranes and modules for transmembrane distillation. J Membr Sci 1988;39(1):25–42.
[73] Khayet M, Khulbe KC, Matsuura T. Characterization of membranes for membrane distillation by atomic force microscopy and estimation of their water vapor transfer coefficients in vacuum membrane distillation process. J Membr Sci 2004;238(1–2):199–211.
[74] Mulder M. Basic principles of membrane technology. 2nd ed. Netherlands: Kluwer Academic Publishers; 2003.
[75] El-Bourawi MS, et al. A framework for better understanding membrane distillation separation process. J Membr Sci 2006;285(1–2):4–29.
[76] Khayet M, Mengual JI, Zakrzewska-Trznadel G. Direct contact membrane distillation for nuclear desalination. Part I: review of membranes used in membrane distillation and methods for their characterisation. Int Nuclear Desalination 2005;1(4):435–49.
[77] Imdakm AO, Matsuura T. Simulation of heat and mass transfer in direct contact membrane distillation (MD): the effect of membrane physical properties. J Membr Sci 2005;262(1–2):117–28.
[78] Khayet M, Velázquez A, Mengual JI. Modelling mass transport through a porous partition: effect of pore size distribution. J Non-Equilib Thermodyn 2004;29(3):279–99.
[79] Phattaranawik J, Jiraratananon R, Fane AG. Effect of pore size distribution and air flux on mass transport in direct contact membrane distillation. J Membr Sci 2003;215(1–2):75–85.
[80] Martínez L, et al. Estimation of vapor transfer coefficient of hydrophobic porous membranes for applications in membrane distillation. Sep Purif Technol 2003;33(1):45–55.
[81] Imdakm AO, Matsuura T. A Monte Carlo simulation model for membrane distillation processes: direct contact (MD). J Membr Sci 2004;237(1–2):51–9.
[82] Woods J, Pellegrino J, Burch J. Generalized guidance for considering pore-size distribution in membrane distillation. J Membr Sci 2011;368(1–2):124–33.
[83] Srisurichan S, Jiraratananon R, Fane AG. Mass transfer mechanisms and transport resistances in direct contact membrane distillation process. J Membr Sci 2006;277(1–2):186–94.
[84] Phattaranawik J, Jiraratananon R, Fane AG. Heat transport and membrane distillation coefficients in direct contact membrane distillation. J Membr Sci 2003;212(1–2):177–93.
[85] Sperati CA, DuPont de Nemours EI. In: Brandrup J, Immergut EH, editors. Polymer handbook. 2nd ed. New York: John Wiley; 1975.
[86] Khayet M, et al. Design of novel direct contact membrane distillation membranes. Desalination 2006;192(1–3):105–11.
[87] Webb SW. Gas-phase diffusion in porous media: evaluation of an advective-dispersive formulation and the dusty-gas model including comparison to data for binary mixtures. California: Sandia National Laboratories; 1996.
[88] He W, et al. Gas transport in porous electrodes of solid oxide fuel cells: a review on diffusion and diffusivity measurement. J Power Sources 2013;237:64–73.
[89] Cussler E. Diffusion mass transfer in fluid systems. 2nd ed. Cambridge: Cambridge University Press; 1997.
[90] Al-Obaidani S, et al. Potential of membrane distillation in seawater desalination: thermal efficiency, sensitivity study and cost estimation. J Membr Sci 2008;323(1):85–98.

[91] Ding Z, Ma R, Fane AG. A new model for mass transfer in direct contact membrane distillation. Desalination 2003;151(3):217–27.

[92] Essalhi M, Khayet M. Fundamentals of membrane distillation. In: Pervaporation, vapour permeation and membrane distillation. Oxford: Woodhead Publishing; 2015. p. 277–316.

[93] Geankoplis C. Transport process and separation principles. 4th ed. Upper Saddle River, NJ: Prentice Hall; 2003.

[94] Termpiyakul P, Jiraratananon R, Srisurichan S. Heat and mass transfer characteristics of a direct contact membrane distillation process for desalination. Desalination 2005;177 (1–3):133–41.

[95] Pangarkar BL, et al. Status of membrane distillation for water and wastewater treatment—a review. Desalin Water Treat 2014;52(28–30):5199–218.

[96] Guijt CM, et al. Modelling of a transmembrane evaporation module for desalination of seawater. Desalination 1999;126(1–3):119–25.

[97] Fane AG, Schofield RW, Fell CJD. The efficient use of energy in membrane distillation. Desalination 1987;64:231–43.

[98] Drioli E, Calabro V, Wu Y. Microporous membranes in membrane distillation. Pure Appl Chem 1986;58(12):1657–62.

[99] Drioli E, Wu Y, Calabro V. Membrane distillataion in the treatment of aqueous solutions. J Membr Sci 1987;33(3):277–84.

[100] Kurokawa H, et al. Vapor permeate characteristics of membrane distillation. Sep Sci Technol 1990;25(13):1349–59.

[101] Martínez L, Rodríguez-Maroto JM. On transport resistances in direct contact membrane distillation. J Membr Sci 2007;295(1–2):28–39.

[102] Godino P, Peña L, Mengual JI. Membrane distillation: theory and experiments. J Membr Sci 1996;121(1):83–93.

[103] Brid R, Stewart W, Lightfoot E. Transport phenomena. 2nd ed. New York: John Wiley; 2001.

[104] Jönsson AS, Wimmerstedt R, Harrysson AC. Membrane distillation—a theoretical study of evaporation through microporous membranes. Desalination 1985;56:237–49.

[105] Martínez-Díez L, Vázquez-Gonzàlez MI. Temperature polarization in mass transport through hydrophobic porous membranes. AlChE J 1996;42:1844–52.

[106] Martínez L, Rodríguez-Maroto JM. Effects of membrane and module design improvements on flux in direct contact membrane distillation. Desalination 2007;205 (1–3):97–103.

[107] Martínez-Díez L, Vázquez-González MI. Temperature and concentration polarization in membrane distillation of aqueous salt solutions. J Membr Sci 1999;156(2):265–73.

[108] Phattaranawik J, Jiraratananon R. Direct contact membrane distillation: effect of mass transfer on heat transfer. J Membr Sci 2001;188(1):137–43.

[109] Qtaishat M, et al. Heat and mass transfer analysis in direct contact membrane distillation. Desalination 2008;219(1–3):272–92.

[110] Yun Y, Ma R, Fane AG. Direct contact membrane distillation mechanism for high concentration NaCl solutions. Desalination 2006;188(1–3):251–62.

[111] Martínez-Díez L, Gonzalez MIV. Effects of polarization on mass transport through hydrophobic porous membranes. Ind Eng Chem Res 1998;37(10):4128–35.

[112] Wirth D, Cabassud C. Water desalination using membrane distillation: comparison between inside/out and outside/in permeation. Desalination 2002;147(1–3):139–45.

[113] Khayet M, Imdakm AO, Matsuura T. Monte Carlo simulation and experimental heat and mass transfer in direct contact membrane distillation. Int J Heat Mass Transf 2010;53 (7–8):1249–59.

[114] Gryta M, Tomaszewska M. Heat transport in the membrane distillation process. J Membr Sci 1998;144(1–2):211–22.
[115] Gryta M, Tomaszewska M, Morawski AW. Membrane distillation with laminar flow. Sep Purif Technol 1997;11(2):93–101.
[116] Izquierdo-Gil MA, García-Payo MC, Fernández-Pineda C. Air gap membrane distillation of sucrose aqueous solutions. J Membr Sci 1999;155(2):291–307.
[117] Liu GL, et al. Theoretical and experimental studies on air gap membrane distillation. Heat Mass Transf 1998;34(4):329–35.
[118] Bouguecha S, Chouikh R, Dhahbi M. Numerical study of the coupled heat and mass transfer in membrane distillation. Desalination 2003;152(1–3):245–52.
[119] Tomaszewska M, Gryta M, Morawski AW. A study of separation by the direct-contact membrane distillation process. Sep Technol 1994;4(4):244–8.
[120] Tun CM, et al. Membrane distillation crystallization of concentrated salts–flux and crystal formation. J Membr Sci 2005;257(1–2):144–55.
[121] Izquierdo-Gil MA, Fernández-Pineda C, Lorenz MG. Flow rate influence on direct contact membrane distillation experiments: different empirical correlations for Nusselt number. J Membr Sci 2008;321(2):356–63.
[122] Sarti GC, Gostoli C, Matulli S. Low energy cost desalination processes using hydrophobic membranes. Desalination 1985;56:277–86.
[123] Fan H, Peng Y. Application of PVDF membranes in desalination and comparison of the VMD and DCMD processes. Chem Eng Sci 2012;79:94–102.
[124] Zhang Y, et al. Review of thermal efficiency and heat recycling in membrane distillation processes. Desalination 2015;367:223–39.
[125] Bandini S, Gostoli C, Sarti GC. Role of heat and mass transfer in membrane distillation process. Desalination 1991;81(1–3):91–106.
[126] Alklaibi AM, Lior N. Transport analysis of air-gap membrane distillation. J Membr Sci 2005;255(1–2):239–53.
[127] Ding Z, et al. Analysis of a solar-powered membrane distillation system. Desalination 2005;172(1):27–40.
[128] Bui VA, Vu LTT, Nguyen MH. Simulation and optimisation of direct contact membrane distillation for energy efficiency. Desalination 2010;259(1–3):29–37.
[129] Carlsson L. The new generation in sea water desalination SU membrane distillation system. Desalination 1983;45(2):221–2.
[130] Bouguecha S, Hamrouni B, Dhahbi M. Small scale desalination pilots powered by renewable energy sources: case studies. Desalination 2005;183(1–3):151–65.
[131] Macedonio F, Curcio E, Drioli E. Integrated membrane systems for seawater desalination: energetic and exergetic analysis, economic evaluation, experimental study. Desalination 2007;203(1–3):260–76.
[132] Yu H, et al. Numerical simulation of heat and mass transfer in direct membrane distillation in a hollow fiber module with laminar flow. J Membr Sci 2011;384 (1):107–16.
[133] Yang X, Fane AG, Wang R. Membrane distillation: now and future. In: Desalination. New Jersey: John Wiley & Sons Inc.; 2014. p. 373–424.
[134] Calabrò V, Drioli E, Matera F. Membrane distillation in the textile wastewater treatment. Desalination 1991;83(1–3):209–24.
[135] Pangarkar B, et al. Review of membrane distillation process for water purification. Desalin Water Treat 2014;1–23.
[136] Martínez-Díez L, Vázquez-González MI, Florido-Díaz FJ. Study of membrane distillation using channel spacers. J Membr Sci 1998;144(1–2):45–56.

[137] Martínez L. Comparison of membrane distillation performance using different feeds. Desalination 2004;168:359–65.
[138] Hwang HJ, et al. Direct contact membrane distillation (DCMD): experimental study on the commercial PTFE membrane and modeling. J Membr Sci 2011;371(1–2):90–8.
[139] Shirazi S, Lin C-J, Chen D. Inorganic fouling of pressure-driven membrane processes—a critical review. Desalination 2010;250(1):236–48.
[140] Heru S. Towards practical implementations of membrane distillation. Chem Eng Process Process Intensif 2011;50(2):139–50.
[141] Warsinger DM, et al. Scaling and fouling in membrane distillation for desalination applications: a review. Desalination 2015;356:294–313.
[142] Kullab A, Martin A. Membrane distillation and applications for water purification in thermal cogeneration plants. Sep Purif Technol 2011;76(3):231–7.
[143] Gryta M. Long-term performance of membrane distillation process. J Membr Sci 2005;265(1–2):153–9.
[144] Gryta M. Fouling in direct contact membrane distillation process. J Membr Sci 2008;325 (1):383–94.
[145] Srisurichan S, Jiraratananon R, Fane AG. Humic acid fouling in the membrane distillation process. Desalination 2005;174(1):63–72.
[146] Hausmann A, et al. Fouling mechanisms of dairy streams during membrane distillation. J Membr Sci 2013;441:102–11.
[147] Shirazi MMA, Kargari A. A review on application of membrane distillation (MD) process for wastewater treatment. J Membr Sci Res 2015;1:101–12.
[148] Alkhudhiri A, Hilal N. Air gap membrane distillation: a detailed study of high saline solution. Desalination 2017;403:179–86.
[149] Li J-M, et al. Microporous polypropylene and polyethylene hollow fiber membranes. Part 3. Experimental studies on membrane distillation for desalination. Desalination 2003;155(2):153–6.
[150] Cabassud C, Wirth D. Membrane distillation for water desalination: how to chose an appropriate membrane? Desalination 2003;157(1–3):307–14.
[151] Li B, Sirkar KK. Novel membrane and device for vacuum membrane distillation-based desalination process. J Membr Sci 2005;257(1–2):60–75.
[152] Criscuoli A, Bafaro P, Drioli E. Vacuum membrane distillation for purifying waters containing arsenic. Desalination 2013;323:17–21.
[153] Bandini S, Saavedra A, Sarti GC. Vacuum membrane distillation: experiments and modeling. AIChE J 1997;43:398–408.
[154] Khayet M, Velázquez A, Mengual JI. Direct contact membrane distillation of humic acid solutions. J Membr Sci 2004;240(1–2):123–8.
[155] Schofield RW, et al. Factors affecting flux in membrane distillation. Desalination 1990;77:279–94.
[156] El-Abbassi A, et al. Concentration of olive mill wastewater by membrane distillation for polyphenols recovery. Desalination 2009;245(1–3):670–4.
[157] Sakai K, et al. Effects of temperature and concentration polarization on water vapour permeability for blood in membrane distillation. Chem Eng J 1988;38(3):B33–9.
[158] Khayet M, Godino P, Mengual JI. Nature of flow on sweeping gas membrane distillation. J Membr Sci 2000;170(2):243–55.
[159] Khayet M, Godino P, Mengual JI. Theory and experiments on sweeping gas membrane distillation. J Membr Sci 2000;165(2):261–72.
[160] Song L, et al. Pilot plant studies of novel membranes and devices for direct contact membrane distillation-based desalination. J Membr Sci 2008;323(2):257–70.

[161] Drioli E, et al. Integrated membrane operations in desalination processes. Desalination 1999;122(2):141–5.
[162] de Andrés MC, et al. Coupling of a membrane distillation module to a multieffect distiller for pure water production. Desalination 1998;115(1):71–81.
[163] Chunrui W, et al. Membrane distillation and novel integrated membrane process for reverse osmosis drained wastewater treatment. Desalin Water Treat 2010;18 (1–3):286–91.
[164] Wang P, Chung T-S. A conceptual demonstration of freeze desalination–membrane distillation (FD–MD) hybrid desalination process utilizing liquefied natural gas (LNG) cold energy. Water Res 2012;46(13):4037–52.
[165] Edwie F, Chung T-S. Development of simultaneous membrane distillation–crystallization (SMDC) technology for treatment of saturated brine. Chem Eng Sci 2013;98:160–72.
[166] Qu D, et al. Integration of accelerated precipitation softening with membrane distillation for high-recovery desalination of primary reverse osmosis concentrate. Sep Purif Technol 2009;67(1):21–5.
[167] Martínez-Díez L, Florido-Díaz FJ, Vázquez-González MI. Study of evaporation efficiency in membrane distillation. Desalination 1999;126(1–3):193–8.
[168] Chen T-C, Ho C-D, Yeh H-M. Theoretical modeling and experimental analysis of direct contact membrane distillation. J Membr Sci 2009;330(1–2):279–87.
[169] Matheswaran M, et al. Factors affecting flux and water separation performance in air gap membrane distillation. J Ind Eng Chem 2007;13(6):965–70.
[170] Safavi M, Mohammadi T. High-salinity water desalination using VMD. Chem Eng J 2009;149(1–3):191–5.
[171] Xu Z, Pan Y, Yu Y. CFD simulation on membrane distillation of NaCl solution. Front Chem Eng China 2009;3(3):293–7.
[172] Kubota S, et al. Experiments on seawater desalination by membrane distillation. Desalination 1988;69(1):19–26.
[173] Zuo G, et al. Energy efficiency evaluation and economic analyses of direct contact membrane distillation system using Aspen Plus. Desalination 2011;283:237–44.
[174] Udriot H, Araque A, von Stockar U. Azeotropic mixtures may be broken by membrane distillation. Chem Eng J Biochem Eng J 1994;54(2):87–93.
[175] Xiuli Z, et al. Mathematical model of gas permeation through PTFE porous membrane and the effect of membrane. Chin J Chem Eng 2003;11(4):383–7.
[176] Drioli E, Wu Y. Membrane distillation: an experimental study. Desalination 1985;53 (1–3):339–46.
[177] Kesieme UK, et al. Economic analysis of desalination technologies in the context of carbon pricing, and opportunities for membrane distillation. Desalination 2013;323:66–74.
[178] Khayet M. Solar desalination by membrane distillation: dispersion in energy consumption analysis and water production costs (a review). Desalination 2013;308:89–101.
[179] Banat F, Jwaied N. Economic evaluation of desalination by small-scale autonomous solar-powered membrane distillation units. Desalination 2008;220(1–3):566–73.
[180] Saffarini RB, et al. Economic evaluation of stand-alone solar powered membrane distillation systems. Desalination 2012;299:55–62.
[181] Hogan PA, et al. Desalination by solar heated membrane distillation. Desalination 1991;81(1–3):81–90.
[182] Schwantes R, et al. Membrane distillation: solar and waste heat driven demonstration plants for desalination. Desalination 2013;323:93–106.

[183] Hanemaaijer JH, et al. Memstill membrane distillation—a future desalination technology. Desalination 2006;199(1):175–6.
[184] Bouguecha S, Dhahbi M. Fluidised bed crystalliser and air gap membrane distillation as a solution to geothermal water desalination. Desalination 2003;152(1–3):237–44.
[185] Sarbatly R, Chiam C-K. Evaluation of geothermal energy in desalination by vacuum membrane distillation. Appl Energy 2013;112:737–46.
[186] Raluy RG, et al. Life-cycle assessment of desalination technologies integrated with energy production systems. Desalination 2004;167:445–58.

Membrane properties in membrane distillation

Sahar Zare, Ali Kargari
Amirkabir University of Technology (Tehran Polytechnic), Tehran, Iran

4.1 Introduction

Membrane distillation (MD) is a nonisothermal membrane operation that has received much attention in recent years, especially for salty water desalination [1–4]. In the MD process, only vapor molecules are allowed to pass through the hydrophobic membrane pores. The pressure difference resulting from the temperature difference is applied between the surfaces of the hydrophobic membrane, so the separation takes place [2,5].

The MD process has unique advantages compared to other processes. Operating pressure is lower compared to other processes, which increases process safety and reduces the equipment cost. It also has a low operating temperature because it is not essential that the liquid temperature becomes as high as its boiling temperature. Feed temperature in MD is from 45°C to 85°C [6–9]; therefore, renewable energy sources such as solar or geothermal energies, or even waste heats can be used for the MD process which would significantly reduce the energy costs [10]. Complete rejection (mineral ions, colloids, cells, macromolecules, and other nonvolatile organic compounds) is another advantage of the MD process. Since the membranes for MD are highly hydrophobic, fouling and scaling of the MD membranes are not as serious as those for hydrophilic membranes. Therefore, pretreatment of the feeds with high concentration is not as strict as RO and also the pretreatment costs are low [2,8].

The MD process is used in various fields, especially when industrial products are temperature-sensitive. Fields of MD application include the food industry, pharmaceutical industry, textile industry, water treatment plants and radioactive wastes, removal of volatile materials from aqueous solutions, and desalination [2,11,12]. A summary of MD applications is shown in Table 4.1. Most MD applications are still carried out at laboratory scale or small-scale pilot plants [38]. Among the mentioned applications, the most focused areas are desalination, water, and wastewater treatments, respectively.

Despite the advantages and a wide range of applications that were introduced to MD, there are numbers of factors and limitations that have prevented large-scale MD process designs, so these factors have limited its commercialization [39].

The MD process has lower flux than other desalination processes. Uncertainty of energy and economic costs and concentration polarization are the limiting factors of the MD process. Wetting of the membrane pores reduces the stability and durability

Emerging Technologies for Sustainable Desalination Handbook. https://doi.org/10.1016/B978-0-12-815818-0.00004-7

Table 4.1 **Some applications of MD process [9,13–37]**

MD configuration	Membranes type	Feed solution	Target	Ref.
DCMD	PTFE	Simulated water	Chromium removal	[13]
DCMD	TF200	Pure water and humic acid	Treatment of humic acid solutions	[14]
DCMD	Hydrophobic ally modified FS PS or PES	Synthetic radioactive wastewater	Removal of radioactive elements	[15]
VMD	PP	Aqueous solution of *N*-methyl-2-pyrrolidone	Concentration of *N*-methyl-2-pyrrolidone solution	[16]
DCMD	PTFE	Seawater of Persian Gulf	Desalination	[17]
DCMD	PP	Zabłocka Thermal Brine	Brine concentration	[18]
DCMD	PVDF	Seawater	Boron removal	[19]
Multi effect MD	PP	Aqueous H_2SO_4 solution	Concentration of H_2SO_4 solution	[20]
DCMD	PTFE	Olive mill wastewaters	Concentration of olive mill wastewater	[21]
DCMD	FS PP	Cooling tower blow down water	Desalination	[22]
DCMD	Accurel PP hollow fiber	Glycerol fermentation broth	Separation of acetic acid from the broth	[23]
DCMD	PP	Model lactose solution	Ethanol production	[24]
DCMD	FS PTFE	Synthetic solution of trace OC	Removal of complex trace organic compounds	[25]
VMD	PP	Dying solution	Treatment of dye solutions	[26]
SGMD	FS PTFE	Dilute glycerol wastewater	Concentration of glycerol	[9]
SGMD	PP	Pure water, aqueous of sodium chloride	Desalination	[27]
DCMD	PTFE	Produced water	Desalination	[28]
DCMD	PVDF	Hot dying solution	Treating hot dyeing solution	[29]

DCMD	FS PTFE and PP	Water from great salty lake	Recovery of minerals	[30]
DCMS	FS PTFE	Olive oil waste mill water	Concentration of phenolic compounds	[31]
AGMD	FS PTFE	Produced water	Desalination	[32]
DCMD	PVDF	Retentate of NF and RO	Improvement of water RF and salt crystallization	[33]
SGMD	FS PTFE	Ethanol-water mixture	Ethanol separation	[34]
DCMD and MDCMD	PVDF	Aqueous ammonia solution	Removal of ammonia	[35]
VMD	PP and FS PVDF	Arsenic containing wastewater	Removal of arsenic	[36]
VMD	PTFE	Human urine	Water regeneration	[37]

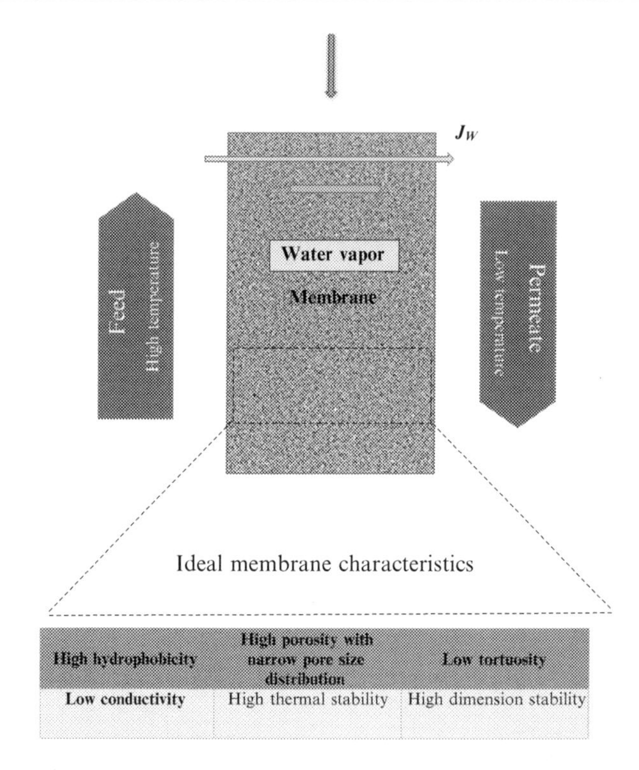

Fig. 4.1 Schematic presentation of a MD process and the ideal membrane characteristics [40].

of process performance. The membrane performance declines over the time, so the MD process industrialization has become limited.

The fabrication of suitable membranes for MD is the Achilles' heel for a large-scale operation and commercialization of this new type of membrane operation (Fig. 4.1). Pore wetting, scaling, and low permeation flux are the major shortcomings of the presently available MD membranes. Most of the membranes which are used for MD processes have been fabricated for other pressure-driven membrane processes such as microfiltration which have their own requirements that are different from MD. Nearly five decades of research on the MD process and membranes give us important information about the requirements and characteristics for the MD membranes [11,17,41]. Since MD is not a pressure-driven membrane operation, the mechanical strength of the MD membranes has less importance in comparison to other pressure-driven membranes, but the transport process in MD membranes is completely complicated. Simultaneous heat and mass transfer occurs in the porous MD membranes ration through nonwetted pores; therefore, besides the membrane and pore geometries, characters such as hydrophobicity, surface topology, thermal conductivity, membrane thickness, and bulk properties affect the permeation flux and solute rejection.

In this regard, this chapter gives a comprehensive overview of the desired properties for MD membranes. In addition, achievements regarding the MD membranes such as new materials and fabrication techniques, multilayer membranes with different hydrophobicities, as well as the surface modified macromolecules (SMM) membranes, hydrophobic/hydrophilic membranes, mixed matrix membranes (MMM), electrospun fibrous membranes, and chemically surface modified membranes have been comprehensively reviewed.

4.2 MD membranes

4.2.1 Membrane materials

Membranes used in MD process should be porous and hydrophobic, should not be wetted, and also exhibit suitable thermal stability at high temperatures. Thus, polymeric materials that are used for the production of MD membranes should be inherently hydrophobic and must have low thermal conductivity to prevent heat loss. The materials used for the manufacture of membrane MD are divided into two general categories. These materials are listed below.

4.2.1.1 Polymeric materials

The most commonly reported polymeric materials for the fabrication of the MD membranes are polytetrafluoroethylene (PTFE), polypropylene (PP), polyvinylidene fluoride (PVDF), and polyethylene (PE), respectively. Other applied materials such as polysulfone (PSf), polyethersulfone (PES), poly (ethylene chloro trifluoroethylene) (ECTFE), and polyetherimide (PEI) are also used in the manufacture of MD membranes as a based polymer but with surface modification [42–47]. Among the materials listed, most research studies have focused on PVDF because it has high mechanical strength, good chemical resistance, suitable thermal stability, and excellent aging resistance. Its thermal conductivity is 0.17–0.21 $W\ m^{-1}\ K^{-1}$. In addition, it offers acceptable processability for the production of flat sheet, hollow fiber, and tubular membranes. PVDF is soluble in common solvents such as *N*-methyl-2-pyrrolidone (NMP), dimethylformamide (DMF), and dimethylacetamide (DMAC). The widely applied fabrication method for PVDF membranes is nonsolvent induced phase separation (NIPS) while in limited researches the electrospinning method has been used [42,43,48]. PTFE has been the most efficient membrane material for MD membranes. It is highly hydrophobic and has very good chemical resistance and thermal stability. The thermal conductivity of PTFE is 0.25–0.29 $W\ m^{-1}\ K^{-1}$. It is difficult to prepare a PTFE membrane with a desired pore size because it is nearly insoluble in most solvents. The methods for the fabrication of PTFE membranes are limited to stretching of thin films and sintering of the fine PTFE powders [44,49]. Polypropylene (PP) is a crystalline polymer with high hydrophobicity and a thermal conductivity of 0.1–0.22 $W\ m^{-1}\ K^{-1}$. PP is nearly insoluble in most solvents at ambient temperature. Even at high temperatures, only limited solvents could dissolve PP such as boiling xylenes or paraffin. PP membranes are not as hydrophobic as PTFE and are not as easy to fabricate as PVDF membranes. These have caused its application to be less than PVDF membranes for MD application. PP fabrication techniques have been limited to thermally induced phase separation (TIPS) and stretching [42,44,48].

PSf does not have many applications. Its thermal conductivity is about 0.26 W m^{-1} K^{-1} and is mainly used when modified via fluorinated SMMs to increase the hydrophobicity since its hydrophobicity is less than that of PVDF. Nevertheless, it is used in the MD process to improve its hydrophobicity and this is the main reason for using it in MD. PSf is soluble in NMP, DMF, tetrahydrofuran (THF), DMAC solvent, etc. A suitable solvent is selected depending on the structure of the membrane. Phase inversion is a method for fabrication of this membrane [47,50,51]. PES and PES-based membranes with thermal conductivities of 0.13–0.18 W m^{-1} K^{-1} have numerous applications in separation processes since they have excellent chemical stability, good mechanical strength, and suitable thermal resistance. There are etheric bonds in chains of PES, so these membranes also have hydrophilic properties and this is one of their drawbacks. This is because it causes an increase of fouling and a tendency to swell in water and also a reduction of mechanical strength and rejection. Thus, PES and PES-based membranes cannot be used for MD and their surface should be modified for use in MD and to improve their performance. Polymer concentration, solvent type, additives, temperature of the solution, as well as nonsolvent and coagulation bath are effective on the structure of PES membranes. These solvents include DMF, NMP, and DMAC. The solubility of these membranes is higher than that of PSf. Phase inversion is a manufacturing method of PES and PES-based membranes [51–55]. ECTFE is a novel hydrophobic polymer material which is applied in the manufacture of membranes such as MD membranes, and it is taken into consideration because of its excellent properties. The thermal conductivity of ECTFE is 0.151–0.157 W m^{-1} K^{-1}. Excellent mechanical properties in a wide temperature range, hydrophobicity, durability in the environment and under ambient conditions, when organic solvents and chemicals, chlorine, strong acids, and oxidizers exist, are considered its distinguishing features. So it indicates high stability in extreme conditions. ECTFE has very low solubility in common solvents at room temperature, which limits fabrication of membranes from this polymer. Its solvents include NMP, glycerol triacetate (GTA), triphenylphosphite (TPP), diethyl phthalate (DEP), dioctyladipate (DOA), dibutylsebacate (DBS), bis (2-ethylhexyl) adipate (DEHA), etc. A construction method of ECTFE membranes is TIPS [56–58]. A summary of the membrane materials that are used in the MD process is shown in Table 4.2.

4.2.1.2 Inorganic materials

Now, a briefing will be given about polymeric materials for the production of MD membranes. In addition to polymeric materials, mineral materials are used in the manufacture of MD membranes. Polymeric materials cannot be applied to membranes at very high temperatures and harsh chemical conditions due to their limited chemical resistance and thermal stability. In general, minerals have higher chemical and thermal stability than polymeric materials. Four types of minerals used for membranes include metals, zeolites, carbons, and ceramics. Membranes that are made from these materials are different in terms of pore diameter and are used in various applications. Ceramics are the main and most important group of minerals to make these membranes, especially in the MD. Ceramics are made from ceramic membranes by

Table 4.2 **The membrane materials that are used in MD [17,46,58–75]**

Type	Configuration	Method
PVDF/Ultem	VMD	Co-extrusion
PVDF/Ultem with Al_2O_3	VMD	Co-extrusion
PDMS/PMMA	DCMD	Electrospinning
PTFE/PVA	VMD	Sintering electrospun
F2.4 with LiCl	DCMD	Phase inversion
F2.4 with $LiClO_4 \cdot 3H_2O$	DCMD	Phase inversion
PTFE/TiO_2	DCMD	Vacuum filtration
PVDF/SiO_2	VMD	Phase inversion
E-PDMS	DCMD	Electrospinning
PES modified	DCMD	Plasma polymerization
PVDF/calcium phosphate	DCMD	Phase inversion
PVDF	AGMD	Electrospinning
PP	DCMD	
PVDF/PVP	VMD	Dry/wet spinning
PVDF/PTFE	VMD	Electrospinning
e-PTFE modified	DCMD	Plasma treatment
PS	DCMD	VIPS
PES	DCMD	Phase inversion
PVDF	DCMD	Phase inversion
PP	VMD	Melt-spinning and stretching
ECTFE	VMD	TIPS
PEI/SMM	DCMD	Phase inversion

combining a metal such as aluminum, titanium, and zirconium with a nonmetal in the form of oxide, nitride, or carbide, the most notable of them being alumina (aluminum oxide), zirconia (zirconium oxide), and titania (titanium oxide). Their melting and boiling are shown in Table 4.3. Ceramic membranes have recently been highly regarded and are usually made with the sol-gel process. The mentioned materials are inherently hydrophilic due to the hydroxyl groups present on the surface and can readily be linked with water molecules. The advantages of ceramic membranes over polymeric membranes are: higher mechanical strength, chemical inertness, having nonswelling behavior, higher thermal and chemical stability, high resistance to

Table 4.3 **Melting and boiling points of ceramics materials [76]**

Material	Melting point (°C)	Boiling point (°C)
Zirconia	2715	4300
Alumina	2072	2977
Titania	1843	2972

corrosive environments, long lifetime, and ease of cleaning. Hydrophilic properties of ceramic materials can be changed to hydrophobic. Nowadays, changes in the properties of hydrophilic to hydrophobic are highly regarded and take place by the grafting of hydrophobic molecules on the surface. Hydrophobic ceramic membranes are used in the MD process [52–58,76–80].

4.2.2 Membrane fabrication techniques in MD

In this section, various techniques for making MD membranes are studied. The type of polymer, the desired structure, and application of the membrane are determinant and effective parameters in the selection of the appropriate method for making MD membranes. In general, methods suitable for making polymeric membranes used in the MD process include: phase inversion, sintering, stretching, and electrospinning [81].

4.2.2.1 Phase inversion

Phase inversion is a demixing process. In this method, a homogeneous polymer solution is immersed in a coagulation bath and is converted into two phases. This conversion is done in several ways that are listed below [82,83]:

- Immersion precipitation

There are three components in this method, polymer, solvent, and nonsolvent [84]. The polymer is dissolved in a suitable solvent. The resulting polymer solution is casted on a suitable support and is immersed in the coagulation bath containing a nonsolvent (usually water). The solvent penetrates into the nonsolvent and the nonsolvent penetrates in the polymer solution. This exchange continues until demixing occurs. As a result, a homogeneous polymer solution turns into two phases. One of them is the polymer-rich phase, which is a membrane with an asymmetric structure, and the other is the liquid-rich phase [76,85]. Factors that are effective on the structure and pore size of the MD membrane, are as follows [86]:

- Polymer concentration. If the polymer concentration increases, the pore size and porosity of the membrane are reduced and, as a result, the permeate flux decreases and the membrane is willing to form a sponge-like structure.
- Solvent and nonsolvent. The choice of an appropriate solvent and nonsolvent has a great effect on the morphology and properties of the membrane. If the polymer is dissolved in a solvent in small quantities, a membrane is formed that will be nonporous. However, the high solubility of the polymer in solvents causes a membrane with high porosity to be obtained.
- The type of additive. Sometimes minerals or organic materials with high molecular weight are added as additives to the casting solution. These additives can improve morphology and properties of the membrane and the fabricated membranes have better performance. Also, they accelerate the immersion precipitation process and increase the viscosity of the solution.
- Precipitation time.
- Bath temperature. By increasing the bath temperature, the membrane pore size increases [81,87].

- TIPS

TIPS (thermally induced phase separation) is a method in which polymers that cannot be solved at room temperature, such as PP, are dissolved in solvents at high temperatures and the resulting polymer solution is casted on the support. Then the temperature is reduced for demixing to occur. Notably, to remove the solvent in TIPS, evaporation, extraction, and freeze-drying are used. Unique advantages of TIPS are as follows: process simplicity, high reproducibility, low trend to create defeats, high porosity, and the ability to create microstructures that have a narrow PSD. In addition, by using solvents and factors of the process, it can handle polymer polymorphism [88,89].

- EIPS

In EIPS (evaporation induced phase separation), a polymer solution, which contains a polymer and a solvent or mixture of solvents and a volatile nonsolvent, is casted on a porous support with a doctor blade method. The solvent can be evaporated, so demixing occurs and a thin porous membrane is formed on the support. To be able to control the morphology of the membrane, solvents with different boiling points are used. The EIPS method is known as the solution casting method.

- VIPS

In the VIPS (vapor induced phase separation) technique, a polymer solution is placed in an environment that contains a nonsolvent (usually exposed to air that contains water vapor as a nonsolvent). The nonsolvent is absorbed by the polymer solution and, as a result, demixing occurs and the membrane is formed.

The above techniques for the construction of MD membranes, immersion precipitation, and TIPS are more prevalent [81,90].

4.2.2.2 Stretching

Since 1970, the stretching method was used for construction of the polymer membrane for MF, UF, and MD processes [91]. In this method, the polymer is heated to a temperature above the melting point and is extruded into thin film forms with a fast drawdown until the polymer film is formed. After that, by stretching, the films are made porous. Thus, it is the appropriate method for making microporous membranes [84,88,92]. Stretching does not need solvents, so it is considered an economic approach [93]. This method is suitable for polymers that have high crystallinity. Crystals in the polymer are aligned in the direction of drawing and create strength. The amorphous part of the polymer causes the porous structure to be formed [92]. Stretching is done in two stages, cold stretching and hot stretching. Cold stretching is initially used to nucleate the microporousity in the formed film and then used for hot stretching to increase/control the ultimate porous structure of the membrane. After the above steps, a mechanical stress is applied in the direction perpendicular to the direction of drawing. Membranes that are made by this method have a relatively uniform porous structure and their porosity is about 90% [88,93]. The porous structure and properties of the membrane are under control with the physical properties of the membrane material, such as crystallinity, melting point, tensile strength, and processing parameters applied [81].

4.2.2.3 Sintering

In sintering, polymer particles are powdered and the resulting powder is pressed into a plate or a film. Then it is sintered at the temperature below the melting point of the polymer. Membranes that are produced by this method have an irregular PSD and a microporous structure. Their porosity range is 10%–40%. The pore size of these membranes depends on the size of polymer particles used and is variable from 0.2 to 20 mm [44,88,94].

4.2.2.4 Electrospinning

Electrospinning is a relatively new and varied method which is used for the production of the nanofibrous polymer membrane. These membranes are used in filtration, desalination, and wastewater-treatment processes [95]. Electrospinning is done in four stages that are shown in Fig. 4.2. As can be seen, it starts from a high voltage and ends with the solidification of a jet into nanofibers [96].

Between the droplet of the polymer solution and the metal collector enters a high voltage of about 14–16 kV. With this technique, different morphologies can be made. For this purpose, electrospinning parameters and properties of the polymer should be changed [95,97]. The processing parameters affecting the electrospinning process are shown in Table 4.4.

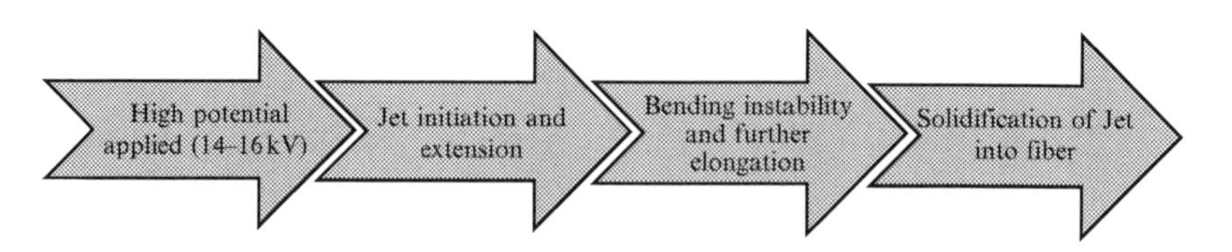

Fig. 4.2 Schematic flow chart of production of fibers in electrospinning technique [96].

Table 4.4 Process parameters for electrospinning [97]

Solution parameters	Process parameters	Environmental conditions
Concentration	Electrostatic potential	Temperature
Viscosity	Electric field strength	Humidity
Surface tension	Electrostatic field shape	Local atmosphere flow
Conductivity	Working distance	Atmospheric composition
Dielectric constant	Feed rate	Pressure
Solvent volatility	Orifice diameter	

4.2.3 Membrane characteristics

Previously, most researches and studies on the MD process were about designing and modeling systems. While there are no cheap and suitable membranes for MD, industrialization of the MD process would face obstacles. Membranes that can show a good performance in the MD process require a series of attributes [98]. In addition to having low mass transfer resistance, good thermal stability at high temperatures, and high resistance to chemicals such as acids and alkalis, these membranes must contain the following features [99,100].

4.2.3.1 Liquid entry pressure

The liquid (feed solution) should not penetrate into the pores of the membrane. For this reason, the applied pressure should be lower than liquid entry pressure (LEP) [101]. LEP is the minimum required pressure of the liquid for entering the membrane pores and must be as high as possible to prevent the occurrence of pore wetting. It depends on the maximum pore size and membrane hydrophobicity. This means that to obtain a high LEP, the materials should be used for membranes that have high hydrophobicity, low energy levels, and small maximum pore size [102]. It should be noted that smaller maximum pore size leads to reduction of the permeability of the membrane [103].

Therefore, to achieve high permeability and high LEP, the appropriate maximum pore size should be considered. For this purpose, the range of maximum pore sizes is reported between 0.1 and 0.6 μm [104]. LEP can be acquired from [100,105]

$$\Delta P = P_f - P_p = \frac{-2\beta\gamma\cos\theta}{r_{\max}} \tag{4.1}$$

in which P_f and P_p are the hydraulic pressure of the feed and permeate sides, respectively. β is the pore geometric coefficient (for cylindrical pores equal to one). γ is the surface tension of the liquid. θ is the contact angle between the liquid and permeate and r_{max} is the maximum pore size. LEP is reduced by the existence of organic materials and feed concentration. For example, when ethanol concentration increases in the solution, LEP is reduced [106]. Garcia et al. [101] showed that the solution temperature, the type of alcohol, and the alcohol concentration in the aqueous solution are effective in LEP.

4.2.3.2 Membrane thickness

Membrane thickness is an important parameter in membrane performance. It is inversely related to the permeate flux in the MD process like any other membrane processes. On the other hand, a thicker membrane has lower permeate flux. The increased thickness increases the mass transfer resistance and reduces heat loss as well [107,108]. In the single layer, the membrane should be considered an optimum thickness with increased flux and reduced mass transfer resistance [108]. Lagana et al. [109] have reported the optimal thickness range as 30–60 μm. However, in the multilayer membrane (a hydrophobic layer and a hydrophilic layer), a hydrophobic layer that is the top layer should be thin as much as possible to reduce the mass transfer resistance. Also, the total thickness of the membrane (which contains two layers) should be as thick as possible to prevent heat loss [42].

4.2.3.3 Membrane porosity and tortuosity

The volume fraction of the pores of the membrane is called membrane porosity (the ratio of pores volume to the total volume of the membrane) [107]. Membranes that have higher porosity provide greater surface area for evaporation. So higher permeate flux and less heat loss are achieved. In other words, the porosity of the membrane is directly related to permeate flux and has an inverse relationship with the conductive heat loss. Heat transfer coefficient and porosity of the membrane are closely related to Eq. (4.2) [110]

$$h_m = \varepsilon h_{mg} + (1-\varepsilon)h_{ms} \tag{4.2}$$

in which ε is membrane porosity, h_{mg} is the heat transfer coefficient of the gas, and h_{ms} is the heat transfer coefficient of the membrane solid. El-Bourawi et al. [111] reported ranges of membrane porosity from 30% to 85%. To calculate the porosity of the membrane, two types of liquid are used; one of them penetrates into the pores, such as isopropyl alcohol (IPA), and the other, such as water, cannot pass through the pores. Membrane porosity is calculated with the Smolder-Franken equation [102,107]:

$$\varepsilon = 1 - \frac{\rho_m}{\rho_{Pol}} \tag{4.3}$$

in which ρ_m is the density of the membrane and ρ_{Pol} is the density of the polymer material. In all configurations of MD, high porosity increases the flux.

Deviation of pore structure from the cylindrical state is named tortuosity. Tortuosity is inversely related to the permeate flux. It is noteworthy that tortuosity is dependent on the membrane geometric [42,112]. For this reason, according to the membrane structure, it is divided into two types [107]:

(a) In loose-packed domains:

$$\tau = \frac{1}{\varepsilon} \tag{4.4}$$

(b) In distances between closed domains:

$$\tau = \frac{(2-\varepsilon)^2}{\varepsilon} \tag{4.5}$$

Since the measurement of membrane tortuosity is difficult, flux estimation and modeling assumes that tortuosity is 2 [110]. It is necessary to mention that tortuosity is an effective factor in the mass transfer mechanism [112].

4.2.3.4 The mean pore size and pore size distribution

Pore size of the MD membrane is between 0.1 and 1 μm and has a direct relationship with the permeate flux [111,113]. In the MD process, pore size is proportional to the mass transfer mechanism and, depending on the size of the pores, it takes place in a type of mass transfer mechanism. For example, when the pore size is small, a mass transfer mechanism is the Knudsen diffusion and when the pore size is large, the Knudsen-viscose mechanism is established. The mean pore size is used to estimate the vapor flux [110]. From one side, the pore size should be small so that the liquid cannot penetrate into the pores, in which case the flux is reduced. On the other hand, the pore size should be large, so a high permeate flux is reached. In this case, even if the membrane is very hydrophobic, wetting occurs in pores, and thus the membrane selectivity decreases [98,99]. So for each MD performance, an optimum pore size should be considered. Choosing an optimum pore size depends on feed solution and operating conditions [99,107]. In general, MD membranes do not have a uniform pore size; instead, they have a distribution of various pore sizes. For this reason, more than one mechanism of heat and mass transfer occurs at the same time. It should be noted that PSD will have little effect on MD performance. Khayet et al. [114] noted that to calculate the heat transfer coefficient, mean pore size should be used instead of PSD, while Martinez et al. [115] found that the heat transfer coefficient is obtained by assuming a uniform pore size very close to the heat transfer coefficient and is calculated by assuming PSD.

4.2.3.5 Thermal conductivity

Thermal conductivity of the membrane should be low; otherwise, heat transfer across the membrane increases. On the other hand, since the interface temperature difference is low, heat flux is reduced [110]. The thermal conductivity of commercial membranes has been reported from 0.04 to 0.06 W m^{-1} K^{-1} [42].

The thermal conductivity of the membrane is obtained according to the thermal conductivity of the polymer (K_p) and gas thermal conductivity (K_g) that is usually air. The thermal conductivity of the polymer depends on the degree of crystallinity, crystal shape, and temperature. Most hydrophobic polymers have similar thermal conductivity [38,116]. According to Phattaranawik et al. [116] a parallel model or Iso-strain is the best model to estimate the thermal conductivity of the MD membrane and it is a suitable model because the tortuosity value is close to one. The model assumes that the heat flows through the air and membrane material are parallel. For example, for PTFE with a tortuosity of about 1.1, this model is used to calculate thermal conductivity. Thermal conductivity is obtained from [108,117]

$$K_m = (1 - \varepsilon)K_p + K_g \tag{4.6}$$

Thermal conductivities of water vapor, air, and some polymers are shown in Table 4.5.

Table 4.5 **Thermal conductivity of different materials [110]**

Thermal conductivity	Range ($W m^{-1} K^{-1}$)
Water vapor	0.026 at 298 K 0.03 at 348 K
Air	0.020 at 298 K 0.022 at 348 K
PVDF	0.17–0.19 at 296 K 0.21 at 348 K
PTFE	0.25–0.27 at 296 K 0.29 at 348 K
PP	0.11–0.16 at 296 K 0.2 at 348 K

4.2.4 *Membrane characterization*

The shape and size of pores, PSD, and membrane porosity are effective parameters in MD performance. The MD membrane can be characterized by specifying the mean pore size, PSD, pore density, and effective porosity. A suitable membrane can be elected by understanding the basic properties of the membrane for any special performance in the MD process. Also, a membrane can be designed and produced with the desired properties. The following methods are used for membrane characterization [118,119].

4.2.4.1 *Scanning electron microscopy*

The morphology of the porous membrane can be observed with scanning electron microscopy (SEM) (top and bottom surfaces of flat sheets, internal and external surfaces of capillary and cross section of hollow fiber membranes). Also, SEM is used to determine PSD, pore size, and surface porosity. In SEM, a narrow high-energy electron beam hits the atoms that are on the sample surface. As a result, electrons with low energy are released from the sample surface. Thus, the micrograph picture is determined. SEM images are shown in Fig. 4.3.

Khemakhem and Amar [168] used SEM in order to investigate the morphology, surface quality, and thickness of the top layer of hydrophobic ceramic membranes. Garcia et al. [121] used SEM to study the cross section of PVDF-HFP hollow fibers. It should be noted that the membrane sample should be put in liquid nitrogen before the test to be frozen. Then it is broken and the broken pieces are used for testing. Before using the sample, the surface should be covered with gold and this is SEM's major drawback. This is because it causes damage to the membrane surface and the pore size is determined less accurately. Determining the pore size is difficult in SEM [85,99,122–124].

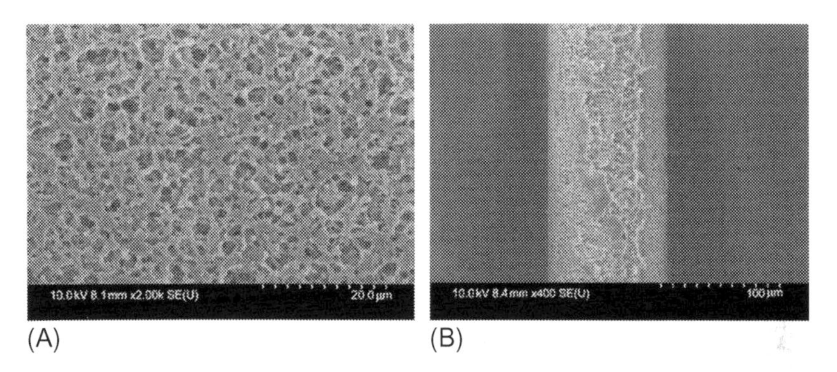

Fig. 4.3 SEM of PVDF membrane (A) top surface and (B) cross section [120].

4.2.4.2 Atomic force microscopy

Studying the surface morphology of organic and inorganic membrane surfaces in less than a nanometer scale is possible by atomic force microscopy (AFM) which has high resolution (Fig. 4.4). AFM was developed by Binnig et al. The search for polymer membrane surfaces was applied for the first time by Albrecht and Quate in 1988. In AFM, the topography of the surface is derived by scanning a sharp tip on a surface. When the tip is close to the surface, the van der Waals force alters/diverts vibration frequency of the tip. When the vibrating frequency or deviation of the tip appears, a three-dimensional image of the surface topography is achieved. Unlike SEM, AFM is used in aqueous solution. Sample preparation is not required before the test. The sample size is 0.5 to 0.5 cm. AFM is used in several modes: contact mode, noncontact mode,

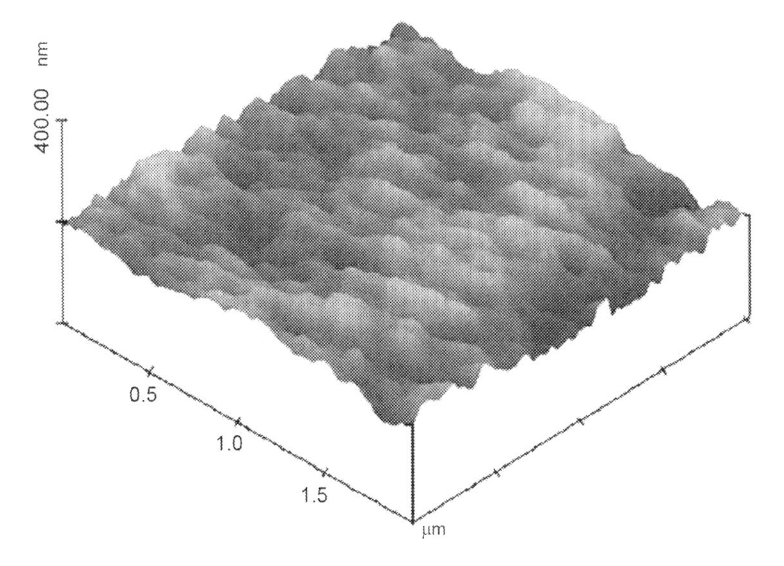

Fig. 4.4 AFM image of the internal surface PVDF/HFP membrane [42].

and trapping mode. In MF, UF, NF, RO, gas separation, and MD processes, in order to check the morphology of the membrane surface, pore size, nodule size, pore density, porosity, and roughness, AFM is used. The electrical properties of the membrane surface, its fouling potential, and performance of the filtration as a function of roughness will also be discussed. Thus, the fouling tendency is predictable. In the field of surface properties of membranes, AFM has attracted the attention of many researchers [122,125–129].

4.2.4.3 Contact angle method

The hydrophobic membrane is an important factor in the MD membrane, and it is related to the contact angle between the membrane surface and the liquid. The contact angle is measured by a goniometer. For this purpose, the sessile drop is often used. In this method, about 10–20 μL of deionized water is placed on the surface of the dry membrane by a micro-syringe, and within a few seconds, the contact angle between the membrane surface and water will be measured. Contact angle measurement is performed in different parts of the sample. To prevent the membrane wetting, the contact angle of both sides of the membrane should be reduced. Contact angle measurement is difficult in porous membranes. Its reasons are: capillary force into the pores, contraction in the dry state, heterogeneity, roughness, and surface reconstruction. It is noteworthy that the membrane surface free energy can be determined by measuring the contact angle [119,130–133].

4.2.4.4 Wet/dry flow method

This method is a combination of the bubble point and gas permeation methods, and it is used to determine the max pore size, the mean pore size, and MD membrane PSD [88].

Initially, a brief description of the bubble point and gas permeation methods will be discussed. In the bubble point technique, necessary pressure should be measured to blow air via membrane with pores that have been full of the liquid and this is inversely related to the radius. As the pressure increases gradually, air bubbles pass through the membrane pores. The first hole that is empty of the liquid is the biggest hole and it is the indicated bubble point [119,126,134]. The gas permeation method was presented for the first time by Yasud and Tsai. The mean pore size and effective porosity (the ratio of porosity to the effective length of the pores) of the membrane were calculated by this method. At room temperature, permeability will be measured at various pressures and its chart is achieved as a straight line. The pore size and effective porosity are achieved by the slope and intercept of the chart. In this way, various gases such as air and nitrogen are used, but nitrogen is often considered as a standard gas. Permeate pressure can be less than or equal to atmospheric pressure [46,88,99,135]. The total gas permeation flow rate (J) in the MD membrane is given by the following equation:

$$J = BA_m \Delta P \tag{4.7}$$

B is the gas permeance, A_m is the total membrane area, and ΔP is the pressure difference. The gas permeance for a porous medium is included in viscous term and diffusive term. Owing to the applied pressure, the contribution of each term is determined. The equation related to it is presented by Carmen.

$$B = \frac{4}{3}\left(\frac{2}{\pi MRT}\right)^{0.5}\frac{r\varepsilon}{L_p} + \frac{P_m}{8\mu RT}\frac{r^2\varepsilon}{L_p} \quad (4.8)$$

R is the gas constant, T is the absolute temperature, M is M_w of the gas, μ is the gas viscosity, P_m is the mean pressure into the pores of the membrane, r is the pore radius, ε is the porosity, and L_p is the effective pore length [46].

In the wet/dry flow method, first in different pressures, the gas permeation is measured for a dry membrane and its chart is drawn that is a straight line. Then the membrane is soaked with a liquid such as IPA, so that all of the membrane pores are full of the liquid. The gas permeation of the wet membrane will be measured at different pressures. When the pressure is very low, all of the pores are full of the liquid. As the pressure is increased, flux is further improved because the pores will be empty of the liquid by applying pressure. Thus, it can be said that the pressure range should be applied to be empty of the liquid and all the membrane pores depend on the pore size of the membrane. This process persists until all pores are empty of the liquid and the flux of the wet membrane is equal to the flux of the dry membrane. The chart of the wet membrane is nonlinear and its touch point with the dry membrane is indicated by the equality of both membranes' flux. The above steps should be carried out at room temperature. Various liquids can be used to fill the pores and, as a result, different results are obtained. The rate of pressure increase and the length of the pores affect the results. If water is used to fill the pores, the pores can be measured down to a few nanometers [88,99,119].

4.2.5 Commercial membranes used in MD

In the MD process, to achieve high efficiency, selection of suitable materials for making MD membranes has an important role. Commercial membranes on the market that are used in the MD process are made from hydrophobic polymers such as PP, PTFE, and PVDF (have a low energy level) [136–138]. When the membrane is placed in contact with an aqueous solution, it should not be wet and only vapor and noncondensable gases must pass from the pores [42].

The energy level of PP is $0.03\,N\,m^{-1}$ [76]. It has high crystallinity and good resistance to solvents. There are few solvents to dissolve PP and this is one of the disadvantages of PP. Stretching and thermal phase inversion are commercial membrane fabrication techniques of PP [44,139].

The energy level of PTFE is $0.009\,N\,m^{-1}$[76,100]. It has high chemical resistance and thermal stability. The first membrane that was used for the MD process in the early 1980s was the PTFE membrane. The disadvantages of PTFE are low solubility in common solvents and difficult processability. PTFE membranes are made by sintering and stretching methods [44,100,139].

Table 4.6 Flat sheet commercial membranes commonly used in MD [42]

Trade name	Manufacturer	Material	δ (μm)	d_p (μm)	ε (%)	LEP (kPa)
TF200	Gelman	PTFE/PP	178	0.20	80	282
TF1000	Gelman	PTFE/PP	178	1.00	80	48
TS22	Osmonics	PTFE/PP	175	0.22	70	–
TS45	Osmonics	PTFE/PP	175	0.45	70	–
GVSP	Millipore	PVDF	108	0.22	80	–
FGLP	Millipore	PTFE/PE	130	0.20	70	280
Gore	Gore	PTFE	64	0.20	90	368

Table 4.7 Capillary and hollow fiber commercial membranes commonly used in MD [42]

Trade name	Manufacturer	Material	δ (μm)	d_i (μm)	d_p (μm)	ε (%)
MD080CO2N	Enka Microdyn	PP	650	1.5	0.2	70
MD020PT2N	Enka Microdyn	PP	1550	5.5	0.2	70
Capillary	Membrana GmbH,	PP	510	1.79	0.2	75
UPE test fiber	Millipore	PE	250	0.2	0.2	–
PTFE	Sumitomo Electric	Poreflon	550	0.9	0.8	62
PTFE	Gore-tex	TA001	400	1	2	50

PVDF with an energy level of $0.303\,N\,m^{-1}$ has a high chemical and thermal resistance. PVDF is soluble at room temperature and in many solvents such as DMF, DMAC, and TEP. Phase inversion is a method of PVDF membrane fabrication [8,76,100,137].

Tables 4.6 and 4.7 show the list of commercial flat sheet and capillary membranes that have been frequently used in the MD processes, including the name of manufacturers and their important characteristics.

4.2.6 Fabricated membranes for MD

The number of laboratory studies on the structure of a membrane that is just for the MD process is extremely low compared to the membranes made for other processes such as RO, NF, MF, UF, pervaporation, and gas separation. The structure and properties of the MD membrane should be such that the needs of the MD can be resolved.

Therefore, it is necessary to perform further studies on the properties of the MD membranes.

In the past few years, numerous studies have been conducted in this regard that has led to the development and industrialization of the MD process. In the following sections, some important points on the design and construction of MD membranes that proposed by the researchers are discussed [42].

4.2.6.1 Single layer hydrophobic flat sheet membranes

Flat-sheet membrane configuration is very useful on a laboratory scale because it is easy to clean and replace. Flat sheet membranes require support, because the packing density (the ratio of membrane area to packing volume) is very low. Flat sheet membrane has been used very much in the MD process such as desalination and water treatment. Several flat sheet single hydrophobic layer membranes are designed and built in the MD (Table 4.8) [99]. Khayet et al. [145] applied the fractional factorial design of experiments to construct a flat sheet asymmetric membrane by phase inversion. Membranes were constructed from poly (vinylidene fluoride-*co*-hexafluoropropylene) (PVDF-HFP) that were different in terms of properties. These membranes were applied in the DCMD system for desalination. In this report, the concentration of copolymer PVDF-HFP, additives concentration (polyethylene glycol, PEG) in casting solution, coagulation bath temperature, and evaporation time of the solvent were considered as input parameters. A linear model was considered to predict the effects of variables on the DCMD permeate flux and salt rejection coefficient. This model was generated with respect to the fractional factorial. The results were obtained such that the salt rejection coefficient increases by increasing the amount of polymer, coagulation bath temperature, and evaporation time of the solvent, while the permeate flux of DCMD is reduced. However, if the concentration of PEG increases in casting solution, a reverse effect is applied on the responses. The conditions of membrane manufacturing were optimized by Lagrange multipliers using factorial models. These conditions were as follows, the concentration of PVDF-HFP in casting solution is 19.1 wt%, the

Table 4.8 Permeate flux of different types of fabricated and modified flat sheet membranes for MD applications [72,102,140–144]

Type	Configuration	J (kg m^{-2}h^{-1})	Fabrication method
PVDF-unsupported	VMD	16.56	Phase inversion
PVDF-supported	VMD	14.076	Phase inversion
PVDF/CTFE	DCMD	20.65	NIPS
PSf	DCMD	30	NIPS
PVDF/CTFE	DCMD	62.09	NIPS
SMM/PEI (20M)	DCMD	17.28	Phase inversion
PVDF nanofiber	AGMD	11.592	Electrospinning
PVDF/CTFE	DCMD	62.09	Dry/wet phase inversion

concentration of PEG in casting solution is 4.99 wt%, the coagulation bath temperature is 35°C and solvent evaporation time is 102 s. The results that were obtained under the above conditions include salt rejection coefficient of 99.95%, which was higher compared to the commercial membranes. The permeate flux of DCMD was gained at 4.41 L m^{-2} h^{-1}. The results indicated that the membrane of the above conditions has the most performance in the DCMD system.

Feng et al. [146] made a flat sheet microporous membrane of poly(vinylidene fluoride-*co*-tetrafluoroethylene) (F2.4) via phase inversion for the MD process. DMAC was regarded as a solvent and TMP as an additive. The effects of factors such as composition of casting solution, exposure time before coagulation, and precipitation bath temperature were examined on the structure of the F2.4 membrane. Properties of the membrane structure and permeate applications are controlled by the parameters such as the action of the pore-forming additive, the composition of casting solution, the precipitation bath temperature, and exposure time. The F2.4 membrane has special properties compared to the PVDF membrane that are mechanical performances and hydrophobicity via a series of examinations. These membranes have higher stress-at-break and strength stress compared to the PVDF membrane. Also, their elongation percentage at break is an eightfold PVDF membrane. In addition, hydrophobicity of the F2.4 membrane was higher than that of the PVDF membrane. Therefore, these membranes could be used successfully in the MD. The permeate flux of PVDF membranes prepared with LiCl was lower than the flux of the F2.4 membrane in the same condition. On the basis of the above results, it can be seen that F2.4 can be considered as a membrane material to manufacture the MD membrane.

Nejati et al. [147] fabricated a flat sheet hydrophobic microporous membrane with an asymmetric structure for the MD process by NIPS. PVDF with a concentration range of 7%–15% was dissolved in TEP and was used as a casting solution with a water coagulation bath and comprising 2-propanol. The composition of the coagulation bath was highly effective on top surface properties of the membrane made. Increasing the concentration of a soft coagulation agent in the coagulation bath increased surface porosity and top surface roughness. Water vapor flux performance of the membrane is different according to the membrane orientation. The results for the partial wetting of the more open bottom surface of the membrane by the lower tension surface hot feed stream were obtained. It was also observed that water vapor flux was strongly influenced by the thickness and porosity of the membrane. Water vapor flux was increased with increasing porosity and thickness reduction. It was concluded that it would build a specific membrane for MD with the desired performance using NIPS.

4.2.6.2 Hollow fiber single hydrophobic layer membranes

Unlike the flat sheet configuration, in this configuration the packing density is high and its energy consumption is low. It is difficult to clean and maintain and there is the possibility of fouling due to its high affinity (Table 4.9) [99]. The porous PVDF asymmetric hollow fiber membrane was made and was characterized by Wang et al. [139] by applying a small molecular nonsolvent (water, ethanol, and 1-propanol) or a mixture of water/LiCl, ethanol/LiCl, and LiCl/1-propanol as additives. High water

Table 4.9 **Permeate flux of fabricated and modified hollow fiber membranes for MD applications [8,144,148–150]**

Type	Configuration	J ($kg m^{-2} s^{-1}$)	Fabrication method
PP	VMD	0.23	Melt-extruded/cold-stretched
PE	DCMD	0.864	Melt-extruded/cold-stretched
PVDF/CTFE	DCMD	62.09	Dry/jet wet spinning
PVDF	DCMD	26.4	Phase inversion
PVDF/PTFE	DCMD	40.39	Dry/jet wet phase inversion
PVDF (M4)	DCMD	40.5	Phase inversion

permeability observed in wet fibers and high gas permeability, good mechanical strength, and excellent hydrophobicity were observed in dry fibers. When a mixture of LiCl/1-propanol was used as additives, more circular nodules emerged in the membrane made. However, for a hollow fiber made with LiCl/water as an additive, there is an interconnecting sponge-like structure in the outer layer of the skin. When the PVDF hollow fiber membrane is made just of a small molecular nonsolvent, it has low permeability and weak fiber integrity because the rate of coagulation is very quiet. To prepare a hollow fiber membrane that has a circular lumen and high permeability, water should be used as an internal coagulant. If ethanol or a mixture of ethanol–water is used as an internal coagulant, it would terminate in noncircular lumen hollow fibers. The membrane permeability is increased somewhat by increasing the temperature of the coagulation bath. The dry water-wetted PVDF hollow fiber makes a great shrinkage. If methanol, ethanol, and 1-propanol are used instead of water in a wet membrane, gas permeability of the dry membrane increases and shrinkage decreases.

For the first time, Maab et al. [151] made a porous hollow fiber membrane from fluorinated polyoxadiazol and polytriazol via the dry/wet spinning method. This membrane was made for desalination of red sea water in different temperatures that was tested in DCMD. Information about this membrane was evaluated in comparison to the PVDF commercial membrane. The hollow fiber membrane was characterized in terms of porosity by SEM, LEP, and pore diameter measurements. Salt rejection was obtained higher than 99.95% and water fluxes were reported for the polyoxadiazol hollow fiber membrane $35 L m^{-2} h^{-1}$ and for polytriazol hollow fiber membrane $41 L m^{-2} h^{-1}$. Feed and permeate temperatures were 80°C and 20°C, respectively. It was observed that, by comparing the performance of the PVDF commercial hollow fiber membrane and polyazol hollow fiber membrane in the same condition, the polyazol hollow fiber membrane has a higher and better performance.

Drioli et al. [152] prepared the microporous PVDF hydrophobic hollow fiber membrane via phase inversion in different processing conditions and by changing the composition of the polymeric dope. They used various additives such as water, PVP, and AMAL to achieve fibers with different morphologies. Also, they applied different concentrations and molecular weights (15%–18%, medium, and high) of PVDF in the solution of the dope. Characteristics of fibers were evaluated in terms of thickness, porosity, improved morphology of the membrane, and the transmembrane water vapor flux in MD. There is higher porosity, reduced mechanical strength, and enhanced flux

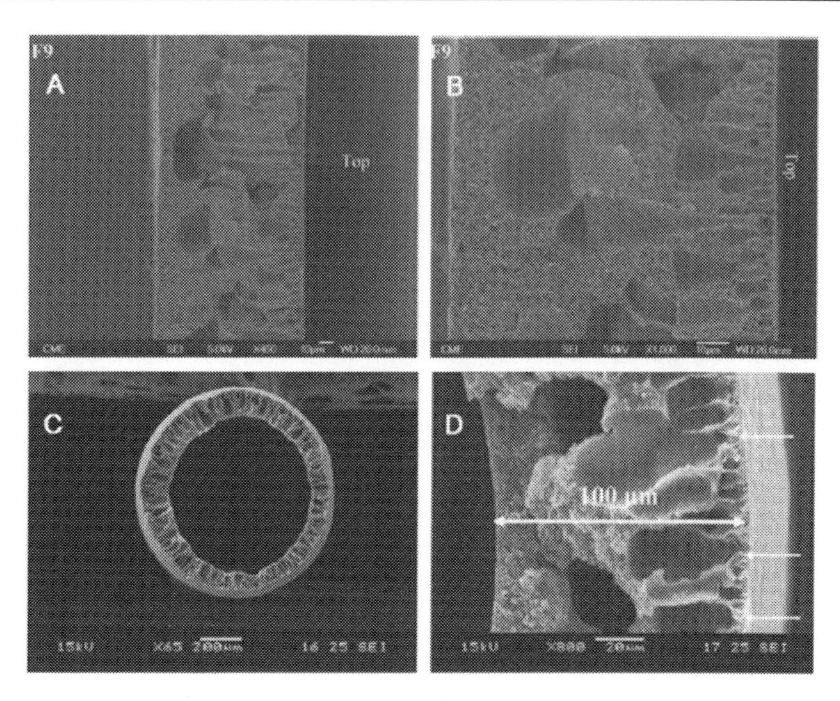

Fig. 4.5 SEM images of flat sheet and hollow fiber membranes. A and B are the images of cross-section of a flat sheet membrane with two different magnifications. C and D are the images of the a hollow fiber membrane cross-section with two different magnifications [40].

in the fiber structure that indicated macro-voids. When a weak nonsolvent such as NMP or ethanol solution was used as the bore liquid and external coagulant, the best mechanical properties were observed. It is visible that when the introduction of more pore forming additives (AMAL) with a lower composition of polymer in the dope solution is combined (fiber type M3), the membrane structure has the most transmembrane flux in VMD and DCMD. The flux of VMD in 50°C is 41.78 $kg\,m^{-2}\,s^{-1}$ and the flux of DCMD in 70°C is 21.78 $kg\,m^{-2}\,s^{-1}$. When the M3 membrane is used, the VMD flux will be 90% greater than the flux of the commercial membrane. This is due to the asymmetric structure of the M3 membrane in which the sponge layer has reduced thickness and it is supported by an open finger structure. In Fig. 4.5 the SEM images of flat sheet and hollow fiber membranes are shown. In Fig. 4.5A and 4.5B represent the SEM images of cross-section of flat sheet membrane with magnification of x450 and x1000, respectively. Fig. 4.5C and 4.5D show the SEM images of the hollow fiber membrane cross-section with magnification of x65 and x800, resectively.

4.2.6.3 Composite bi- and multilayered membranes

The hydrophobic porous membrane that is used for the MD process can be a membrane with a hydrophobic layer or a membrane as a hydrophilic/hydrophobic porous composite layer. For the first time, Cheng and Wiersma presented a report to use the composite membrane in the MD process. In recent years, a group of Spaniards/Canadians expanded the hydrophilic/hydrophobic composite flat sheet membranes and examined the DCMD system for desalination [42]. The flat sheet PVDF/fabric composite membrane was made via casting and the wet phase inversion process by Huo et al. [137] to use in MD. The membrane was composed from a PVDF porous membrane and a layer of fabric. For the manufacture of the PVDF porous membrane, the casting solution was 7%–12% PVDF and 3% LiCl and DMF as a solvent. The thin polyester filament woven fabric with water and an oil repellent finish was applied as a support in the construction of the composite membrane. Construction and properties of the composite membrane are greatly influenced by factors such as PVDF concentration in casting solution, functional finishing of fabric, and fabric texture. These parameters have the greatest effect on laboratory results. Experimental results showed that the composite membrane made has the best performance in the tensile strength, peeling strength, and permeability to water vapor. When the concentration of PVDF is 10%–12% in casting solution, the support fabric has 435 warps 10 cm^{-1} and 273 wefts 10 cm^{-1} and the area weights of 7 g m^{-2} was ended with 2 g L^{-1} water and oil repellent agent FK-501. Under the above terms, the obtained composite membrane has a mean pore size of 0.63 μm and an overall porosity of 57.6%. Thus, this method is suitable for the MD and can help to realize the industrialization of the MD process.

Lu et al. [153] made an amphiphobic PVDF composite membrane via dynamically functionalizing perfluorooctyltrichlorosilane (PFTS) coated with modified SiO_2 nanoparticles on the surface of the membrane. The water contact angle of modified membranes was 167.3 degrees and contact angle of diiodomethane was 140.9 degrees that stemmed from microfluorinated SiO_2 particles coated on the membrane surface. Good mechanical and thermal resistances were observed in DCMD by the dynamically formed SiO_2-PFTS/PVDF-2 membrane. As a result of the surface modification, the LEP soared from 160 to 250 MPa without any adverse effect on porosity and the pore size. The pristine PVDF and modified SiO_2-PFTS/PVDF-2 membranes were tested in DCMD. In this system, there were various foulants. This test was carried out in order to observe the antifouling and antiwetting behavior of the two membranes. Fouling caused by three types of foulants (hydrophilic, hydrophobic, and amphiprotic) led to a sharp decrease in permeation flux and the salt/foulant penetration in the pristine PVDF membrane had devastating effects on the performance of this membrane. However, in modified membranes, the permeate flux and high salt rejection were kept stable because they had high resistance to wetting and fouling in comparison to the three types of foulants. A fluoro-saline layer is established by the fluorinated SiO_2 that greatly increases the excretion of different organic material from the feed solution. Therefore, Lu et al. were able to build a new membrane that indicated a good resistance against fouling and can be used in various applications for water treatment.

Teoh et al. [154] prepared the PVDF/PTFE composite hollow fiber membrane for desalination of seawater and tested it in DCMD. The hollow fiber membrane was manufactured by the dry-jet wet-spinning method. To construct the bilayer membrane, the polymer solution was fixed at 15/70/15 wt% PVDF/NMP/EG for the inner layer but the amount of PTFE particles was variable in the polymeric solution to the outer layer.

The formation of macro-voids and increased hydrophobicity of the outer surface could be affected by the incorporation of PTFE particles in the formulated dope solution. A double-layer hollow fiber with a suitable macro-void-free morphology is obtained by blending 30 wt% PTFE particles in the outer large dope. These hollow fibers have a relatively thin outer layer with a thickness of 13 μm. Porosity and contact angle of the hollow fiber were gained at 81.5% and 114.5 degrees, respectively. If this membrane is compared to the single-layer hollow fiber membrane with 30 wt% PTFE particles, it can be seen that the flux is increased at about 24% by the dual-layer hollow fiber. Thus, the inner layer mass transfer resistance is reduced. If the dual-layer hollow fiber has lower wall thickness and larger inner and outer diameters, its water vapor transmission rate will be more than is favorable for desalination. The permeate flux of the hollow fiber membrane at 80°C is obtained at 50.9 $kg\,m^{-2}\,h^{-1}$, which was relatively high and has more content by reducing the thickness of the fibers. They found that in both single-layer and double-layer membranes after 100 hours of continuous testing, long-term stability was good and also salt rejection was 100%. The thermal conductivity of both membranes at 80°C was reported as 80%. In addition, Teoh et al. stated that the hollow fiber membrane can be designed and produced using dual-layer spinning technology which has better performance and morphology.

The highly hydrophobic Hyflon AD60/PVDF composite hollow fiber membrane is fabricated by Tong et al. [155] via coating a copolymer from TFE (tetrafluoroethylene) and TTD (2,2,4-trofluoro-5-trifluoromethoxy-1,3-dioxole) on PVDF membranes. Tong and his colleagues examined the effects of Hyflon concentration, coating time, heat treatments temperature, and time of heat treatments on the structure and performance of the Hyflon AD60/PVDF membrane composite hollow fiber in VMD. They found that coating Hyflon AD60 causes the hydrophobicity of the PVDF hollow fiber membrane to increase significantly. The optimum conditions were obtained as follows: Hyflon concentration was 0.1 wt%, coating time was 10–20 min, heat treatment temperature was 40–50°C, and heat treatment time was 9 hours. The composite hollow fiber membranes that were used in VMD had salt rejection higher than 99.9% and a water flux of 10 $kg\,m^{-2}\,h^{-1}$. Of course, these results were achieved despite conditions such as salt solution with a concentration of 35 $g\,L^{-1}$ under a vacuum of 0.09 MPa and feed temperature of 70°C. This membrane creates an increased contact angle of 138 degrees and LEP of 0.696 MPa. As well as the mechanical strength of PVDF, the hollow fiber membrane was increased after coating Hyflon AD60. Therefore, Tong reported the Hyflon AD60/PVDF composite hollow fiber membrane as a suitable membrane in the VMD.

The hydrophobic microporous cellulose nitrate membrane surface was modified by Kong et al. [156] via plasma polymerization. Two monomers, OFCB (octafluorocyclobutane) and VTMS/CF4 (vinyltrimethylsilicon/carbon tetrafluoride), were used for the polymerization. They fabricated a composite membrane with a hydrophilic layer that was sandwiched between two hydrophobic layers. Then this membrane is tested in the DCMD system and an acceptable performance was observed. SEM, X-ray microscopic analysis, and XPS were applied to evaluate the effects of polymerization conditions such as glow-discharge powder and deposition time on the structure and performance of the composite membrane. To obtain a hydrophobic composite membrane that has good performance in the MD process, the polymerization condition should be moderate. Hydrophobic composite membranes prepared and membranes that were made from PVDF, PP, and PTFE have similar MD behaviors. A modified composite CN membrane via plasma polymerization of VTMS/CF4 has better performance in comparison to the modified composite CN membrane by plasma polymerization of OFCB. When the discharge time was longer, salt rejection increased and permeate flux decreased. These changes were made in both membranes. The reason for this is the gradual reduction of the pore size. Since the discharge time increases, the pore size decreases and this process continues until a dense layer is created on the membrane surface.

Qtaishat et al. [47] created a new composite membrane for the MD via blending hydrophilic polysulfone with SMMs hydrophobic. In this experiment, several SMMs were used. SMMs were synthesized and characterized for the amount of fluorine molecular weights and glass transition temperature. The composite membrane was manufactured by phase inversion in one casting stage. They characterized the membrane by measuring the contact angle, gas permeation test, LEP, and SEM. Then the membrane was tested in the DCMD system for desalination. They examined the various conditions to construct a membrane which was effective on the morphology and structure of the membrane as well as its performance. The type of SMMs, the concentration of polysulfone, type of solvent, and nonsolvent concentration in casting solution were also explored. They also tried to find the relationship between the morphology of the membrane and its performance in the DCMD system. Qtaishat et al. concluded that the permeate flux of the hydrophilic/hydrophobic composite membrane is reduced if it increases the concentration of the polymer or nonsolvent concentration. Since LEP increased and the ratio of the membrane pore size time and porosity to effective length of the pore is reduced, composite membranes that have higher LEP have less permeate flux. The PTFE commercial membrane was compared to the polysulfone membrane blended SMM in terms of performance. It was observed that the flux of the composite membrane is higher. Their flux was 43% higher than the commercial membrane flux and the salt separation factor was 99.9%.

4.2.6.4 Mixed matrix membranes (MMMs)

Mixed matrix membrane is the incorporation of a solid phase in a continuous polymer matrix. The application of these membranes is a good way to reach contributory effects between the polymeric matrix and solid particles. In MMMs, solid particles are added to the polymer dope and hollow fiber or the flat sheet membranes are formed by the phase inversion method. The composites mixed matrix PVDF hollow fiber membrane was constructed by Wang et al. [157] via NIPS for the MD. These membranes involved PVDF/NMP/Cloisite 20A/EG and were tested in DCMD. In this experiment, water was applied as internal and external coagulants. The void fraction of the membrane was higher than 90%, so the thermal insulation of fibers was improved and resistance to vapor transmission was reduced. Pores in the outer surface of the fibers were extremely small, approximately less than 50 nm in diameter. However, the inner surface pores of fibers were larger, less than 1 μm in diameter. In addition, the inner surface of fibers has a large number of streaky pores. Therefore, it can be concluded that pores of the membranes that have been made on the nanoscale had high water vapor permeation flux and salt rejection was 100%. For instance, in the PVDF/clay composite hollow fiber membrane, the permeation flux was gained at $79.2\,kg\,m^{-2}\,h^{-1}$ when the temperature at outside/inside of the fibers were on the fibers outer diameter at internal temperatures of 81.5°C/17.5°C. This membrane was used for desalination with 3.5 wt% salt solution. According to the membrane performance, it can be found that decay does not occur within 220 h continuous operation. Adding clay particles to fibers could increase the tensile modulus and lead to improving long-term stability in comparison to fibers that do not have clay particles.

4.2.6.5 Electrospun nanofibrous membranes

A membrane that is prepared by electrospinning is called an electrospun nanofibrous membrane (ENM) and has attractive features. For this reason, it has attracted a lot of attention recently. These features include: high porosity, pore size ranging from tens of nanometers to a few micrometers, high permeability for gases, interconnected open pore structures, and a large surface area per unit volume [158]. The electrospinning technique was used to increase the hydrophobicity of the MD membrane by Lee et al. [159]. To this end, TiO_2 functionalized with 1H,1H,2H,2H-perfluorooctyltriethoxysilane was added to the dope solution. Varying amounts of TiO_2 (1%, 5%, and 10%) were applied for fiber production and different concentrations (10%, 15%, and 20%) from PH (PVDF-HFP) were used in order for electrospinning to take place. The presence of TiO_2 not only increases the hydrophobicity of the membrane, but also the membrane pore size was reduced because of the reduction of the fiber's size and it affects the membrane performance. The highest hydrophobicity was in ENM that was made from 20% PH with 10% TiO_2 because it had a good scattering of TiO_2. Its contact angle was reported at about 149 degrees. However, ENM which was made from 10% PH with 10% TiO_2 had the highest LEP because its pore size was decreased. On the basis of the results, they found that ENMs with 10% TiO_2 had better flux and salt rejection stability compared to the commercial membrane and the membranes that do not have TiO_2. When TiO_2 concentration is the same in the two membranes, then the morphology of TiO_2 on fiber surfaces influences the concentration of the polymer, the mass of particles, and volatility of solvents. It was observed that there was no wetting in EMN constructed from 20% PH with 10% TiO_2 after one week of operation, though the feedwater contained high concentrations of salt (7 wt% NaCl). The flux of this membrane was reported at $40\,L\,m^{-2}\,h^{-1}$.

A nanofiber PVDF membrane was constructed and characterized for use in the MD process by Liao et al. [160] via electrospinning method. They were able to optimize the structure and properties of the membrane made with the control of polymeric dope composition and parameters related to spinning. Therefore, the performance of the membrane was improved in DCMD. In addition, they evaluated the effect of the heat-press posttreatment on the performance of the MD membrane. Based on the laboratory results, they reported that the membrane structure is highly influenced by factors such as electrospinning process parameters and properties of the dope solution. They were controlled polymer concentration and adding appropriate additives in the dope solution to manufacture the nanofiber with small diameter and formation of a membrane with small pore sizes by nanofibers. In addition, the sprayer moving speed can be slowed for membranes with a small pore size and reduced humidity in the spinning chamber. Rough surface with a high hydrophobicity was verified for electrospun PVDF membranes by measuring the contact angle of the surface. The heat-press post-treatment is essential to be able to increase the permeate flux of water, avoid wetting in membrane pores, and improve fresh nanofiber membrane integrity. The permeate flux of the post-treated PVDF nanofiber membrane was reported at $21\,kg\,m^{-2}\,h^{-1}$ after a 15h test that was higher than the flux of the untreated fresh membrane. Ultimately, PVDF EMN was introduced as a suitable membrane with a high potential for the MD process by Liao. SEM and AFM images of the PVDF nanofiber-based membrane are shown in Fig. 4.6.

Self-sustained PVDF ENMs were manufactured by Essalhi and Khayet [161] for desalination. These membranes were tested in DCMD. PVDF ENMs were constructed with various thicknesses of 144.4–1529.3 μm and different electrospinning times. For the first time, Essalhi examined the effect of membrane thickness on DCMD performance. SEM was used to study the surface and cross section of ENMs. In addition, the mean size of the fibers and its distribution were calculated. To determine the water

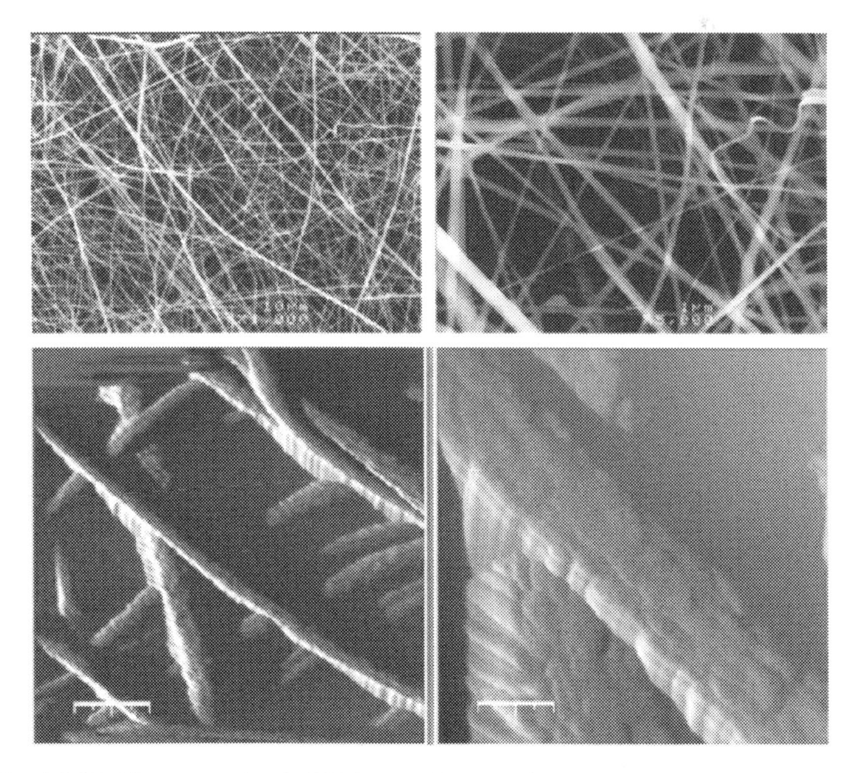

Fig. 4.6 SEM images (top) and AFM images (bottom) of PVDF nanofiber-based membrane for AGMD desalination application [7].

contact angle, the interfiber space, the void volume fraction, and LEP of water in the space between the fibers, different methods were used for characterization. Increasing the electrospinning time caused an increase in the thickness and LEP and reduced the mean size of the interfiber space. However, it had a slight effect on the diameter of electrospun fibers (1–1.3 μm), volume fraction of the void (0.85–0.93), and water contact angle (137.4–141.1 degrees). It is noteworthy that the size of the interfiber space is not uniform in all parts of ENM thickness. Essalhi considered different temperatures for feed and salt feed aqueous solutions with concentrations higher than $60\,g\,L^{-1}$ to observe the effects of ENM thickness on DCMD performance. Decreasing changes of the permeate flux is not uniform with ENM thickness and this is due to reduced energy lost via heat conduction. The heat conduction is related to ENMs that have an increase in thickness. The permeate flux was reported at $54.72\,kg\,m^{-2}\,h^{-1}$ for PVDF ENM in the DCMD system. Temperatures of feed and permeate were 81°C and 21°C, respectively. The salt rejection factor was higher than 99.39%. When distilled water was considered as feed, changes in its permeate flux were less than 5% before and after desalination tests. There was no wetting after 25-h operation in DCMD. When the time of electrospinning is longer, the permeate flux of ENMs is lower. However, the permeate flux obtained from this experiment that was done by Essalhi and Khayet was the highest reported permeate flux for PVDF ENMs so far.

4.3 Surface modification methods in MD

Membrane surface modification is one of the methods to improve and develop membranes for the MD process. It is also a good technique to achieve functional polymers by controlling their surface properties. Surface modification is used for reducing maximum pore size, narrowing down PSD, increasing the flux and/or selectivity, removing existing defects in the membrane, and improving chemical resistance, such as resistance to solvents, swelling, and also fouling and wetting. In surface modification techniques, deeper layers are not modified and the modification is done only for the first layer. There are various methods of MD membrane surface modification that are discussed in this section [136,162–165].

4.3.1 Surface coating

Surface coating is a simple and inexpensive method for surface modification of the membrane and it can be done easily in industrial operations and also on a large scale. In this way, the membrane surface is covered by a coated layer. There is a possibility of penetration of the coated layer into the pores of the membrane, so it should be used in very high molecular weight polymers for the coated layer. The coated layer can be controlled by adjusting the operating parameters. Membranes with a coated layer have a good stability in the long term without being wet. The use of these membranes reduces cost and power consumption because there is less heat loss surface between them. The modified membranes by surface coating can be used in all of the MD configurations (DCMD, AGMD, VMD, and SGMD) [88,162,166]. The coated layer on the membrane surface is unstable, so it is the major problem in the surface coating method. This is because there is a relatively weak interaction between the membrane surface and the coated layer and the coated layer is separated from the membrane surface during the process or during washing operations. In order to anchor the coated layer, chemical treatments such as sulfonation or cross-linking on the membrane surface can be used in some cases. The surface coating is applied for the surface modification of both hydrophilic and hydrophobic surfaces. If the MD membrane (which is hydrophobic) is in contact with a solution containing hydrophobic particles, it is better to have hydrophilic modification. This is because the hydrophilic modification reduces the pore size of the membrane and also prevents the occurrence of fouling. Peng et al. [136] coated a PVA/PEG layer on the surface of the PVDF membrane and they fabricated a composite membrane for the MD process. Owing to the increased hydrophilic surface, permeability of the modified membrane is greater than that of the unmodified membrane [90,136]. As mentioned, one of the major problems of the MD membrane is membrane wetting. In addition to wetting the membrane, membrane wetting has disadvantages on mass transfer. Thus, the coated layer can be used to increase the hydrophobicity and to prevent wetting. Yang et al. [43] modified the PVDF membrane surface by the surface coating method and they strongly increased hydrophobicity. Fig. 4.7 shows the surface morphology of the coated and noncoated membranes. The noncoated membrane has an open and rough surface structure (Fig. 4.8A), but the coated membrane is thicker and shows a different structure (Fig. 4.8B) [162].

4.3.2 Surface grafting

One of the surface modification methods is surface grafting that is considered a chemical treatment. In this way, the surface is modified by covalent bonding interaction. In other words, chains of membrane material are activated by a chemical reaction or high-energy radiation, and then they are grafted. Grafted macromolecular chains

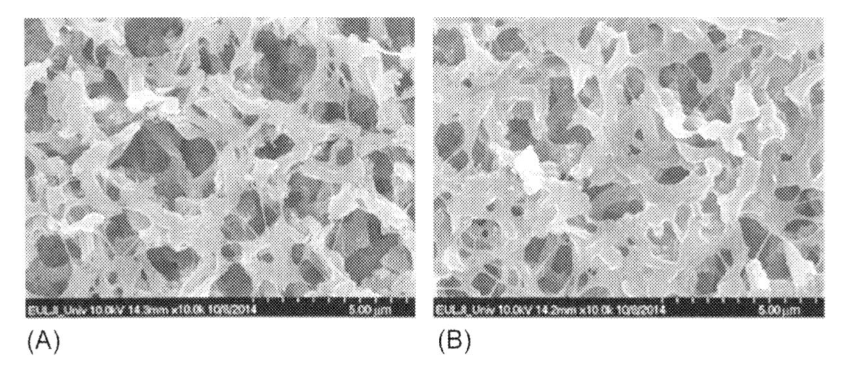

Fig. 4.7 Surface SEM images of modified PA membrane. (A) Base membrane and (B) membrane coated by WRC [162].

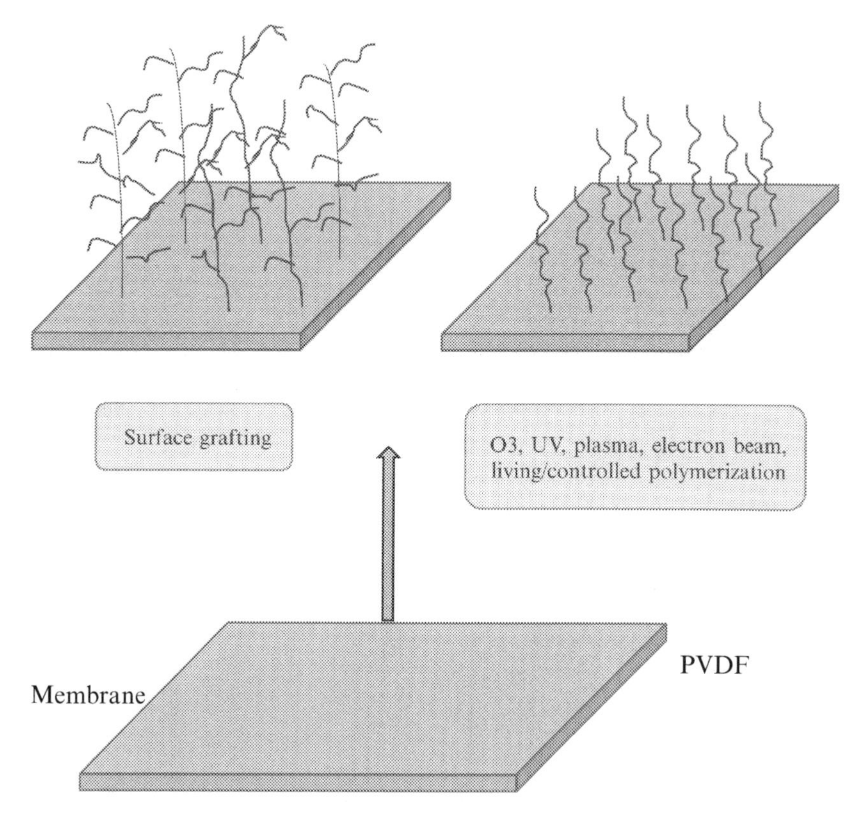

Fig. 4.8 Schematic representation of surface modification of PVDF membrane by surface grafting [90].

are established as covalent bonding with the membrane surface (Fig. 4.8). Created covalent attachments on the surface created a long-term chemical stability and they prevented surface delamination. Surface grafting improves the membrane surface properties with no effect on the membrane bulk. It can be done with UV radiation, plasma treatments, high-energy radiation, O_3/O_2 preactivation, living/controlled polymerization, defluorination-sulfonation, etc. According to the chemical structure of the membrane and desired properties after membrane modification, one of these methods is selected. Since the chemical modification is unlike physical modification, there is a covalent bonding between modifiers and the membrane surface, and improvement of the hydrophobic membrane is not temporary. However, the chains of the membrane material on the surface may be demolished because of chemical modification [43,90,167–170]. Surface grafting based on the type of monomers can be divided into two categories: grafting with a monomer that can be done in one step and grafting with a mixture of two or more monomers which may be simultaneously (one step) or may be consecutive (multistage). It is done in both aqueous and organic solutions. Monomers on the membrane surface can be polymerized via various starting processes that is called "grafting." Polymers can be placed directly on the membrane surface without moving by the coupling reaction, which in this case is called "grafting to." "Grafting from" has more options to control the degree of grafting, length of chains and their structure, temperature, solvent, concentration, monomers, additives, and other reaction conditions [90,136].

4.3.3 Plasma polymerization

Plasma polymerization leads to the least damage compared to other methods of surface modification and it is an electrical ionization of monomers that generates different active particles. These particles are combined on the surface and form a cross-linked structure (to form a polymer). The deposited polymer layer on the surface provides an additional mass transfer resistance and reduces the water flux of the membrane. The cross-linking degree of the plasma polymers is higher than that of the conventional polymers. They also have good thermal stability and stronger adhesion to the substrate and are not solved in organic solvents. The process of forming the polymer in the plasma polymerization is different from that in conventional polymerization. Plasma polymerization is a single-step method and the monomer or combination of monomers is used in chemical modification of the surface layers. Of course, this depends on the monomer that is used. In plasma polymerization, the volatile molecules can form the polymer, and even those do not have functional groups such as vinyl and cyclic groups. In this way, the composition of the polymer depends on the reaction chamber geometry and operating conditions such as flow rate of the monomer, operating pressure, and power discharge [170–173]. At first, plasma polymerization was used for electrical insulation and protective coatings but now it is focused on its applications for preparation of high-performance membranes [88]. Kong et al. [156] modified the surface of the hydrophilic microporous cellulose nitrate membrane. They used plasma polymerization of OFCB (octafluorocyclobutane) and VTMS/CF4 (vinyltrimethylsilicon/carbon tetrafluoride) for modification and the resulting membranes were tested in the DCMD system. More details are shown in reference [156]. Membranes that are made or modified using plasma polymerization have a good performance in the MD process. In other words, plasma polymerization is an efficient and effective method to create a hydrophobic porous membrane from a hydrophilic membrane [88].

4.3.4 Surface modifying macromolecules (SMMs)

Surface modifying macromolecules is a simple method for membrane modification and is defined as active additives that can migrate to the surface and change the surface chemistry. However, the bulk properties of the membrane will be unchanged. There is a driving force in the membrane surface for the migration of the additive chemical groups that makes interfacial energy to reach the minimum value. In this procedure, small amounts of additives are required for surface modification (less than 5 wt% of the base polymer) and this is considered an advantage of SMMs [103,142,163,174]. On this basis, the membrane can be made through the phase inversion method in a casting step using a polymer solution. Polymer solution includes a host hydrophilic polymer and SMM with/without other additives [88]. The porous and dense composite membrane can be made by fluorinated SMM. Most commercial fluoropolymers have difficult processability to surface fluorination. In copolymers, the fluorinated segments increase and accumulate on the surface. Thus, surface modifying macromolecules improve and develop that are oligomeric fluoropolymers which have been synthesized via polyurethane chemistry and appropriated with fluorinated end groups. It has an amphiphatic structure and theoretically includes a main polyurethane chain terminated with two low polarity polymer chains (such as fluorine segments) [163]. The SMM blended membranes have low surface energy, high chemical resistance, high mechanical strength, less fouling, ability to delete chloroform from water by pervaporation, less water and gas permeation flux because of high hydrophobicity, smaller pore size, and ability to enhance LEP of the porous hydrophilic membranes. A schematic representation of the SMMs' blended membrane is shown in Fig. 4.9.

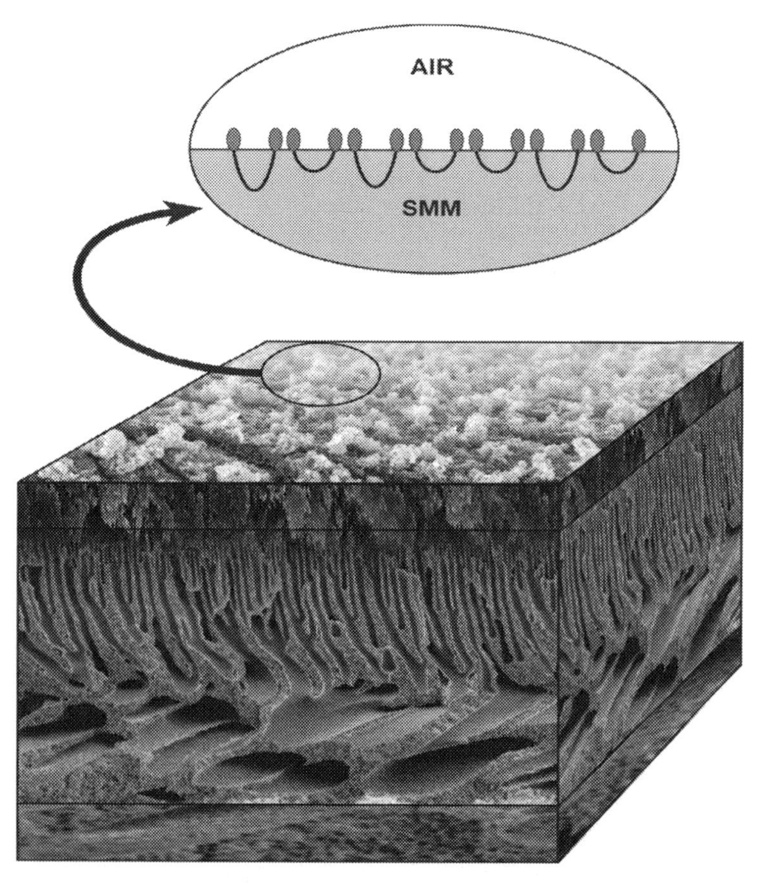

Fig. 4.9 An art diagram illustrating SMM-blended membranes: dumbell-shaped tail contains functional end groups, and the bent line contains polymer chain [175].

These membranes are also widely useful for medical applications, pharmaceutical, food industry, and industrial chemical processes. There have been many studies in order to carry out the synthesis of several fluorinated surface modifying macromolecules blended with various polymers such as polyurethane, PVDF, and PES [103,142,175,176]. Khayet and Matsuura [124] used blended fluorinated surface modifying macromolecules for surface modification of the PEI flat sheet membrane. As a result, membrane hydrophobicity increased and it was suitable for the MD process.

4.4 MD fouling and control methods

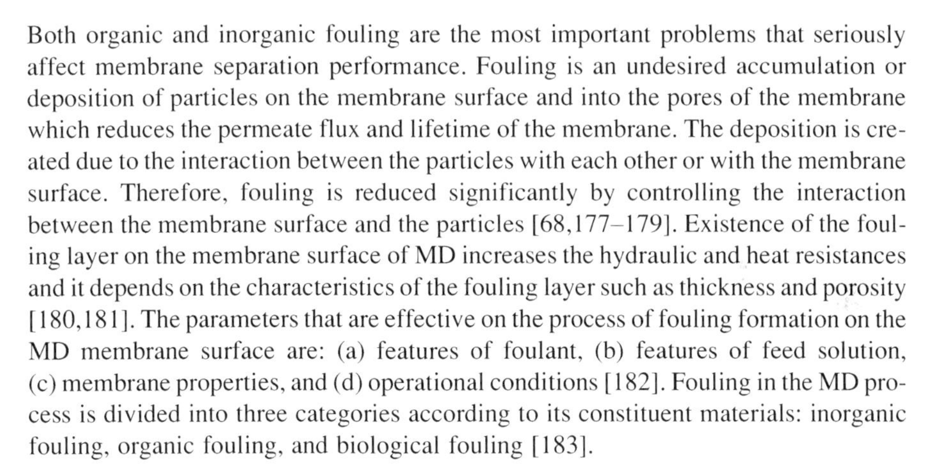

Both organic and inorganic fouling are the most important problems that seriously affect membrane separation performance. Fouling is an undesired accumulation or deposition of particles on the membrane surface and into the pores of the membrane which reduces the permeate flux and lifetime of the membrane. The deposition is created due to the interaction between the particles with each other or with the membrane surface. Therefore, fouling is reduced significantly by controlling the interaction between the membrane surface and the particles [68,177–179]. Existence of the fouling layer on the membrane surface of MD increases the hydraulic and heat resistances and it depends on the characteristics of the fouling layer such as thickness and porosity [180,181]. The parameters that are effective on the process of fouling formation on the MD membrane surface are: (a) features of foulant, (b) features of feed solution, (c) membrane properties, and (d) operational conditions [182]. Fouling in the MD process is divided into three categories according to its constituent materials: inorganic fouling, organic fouling, and biological fouling [183].

- *Inorganic fouling:* Inorganic fouling or scaling refers to deposits of hard minerals from feed solution. Inorganic colloidal particles like silt, clays, corrosion products, silica, etc. cause inorganic fouling. This type of fouling is one of the key problems of the MD process to desalination. In the MD process, since the larger pores have a greater tendency to wettability than the pores with a smaller size, initially fouling happens in the largest pores. Also, in the MD process, the most popular scales are silicates, $CaCO_3$, calcium phosphate, $CaSO_4$, iron oxides, $BaSO_4$, $MgSO_4$, aluminum oxide, $MgCl_2$, and $SrSO_4$ [177,184,185].
- *Organic fouling:* Adsorption or deposition of unresolved and organic colloidal particles on the membrane surface like humic acid (HA), extracellular polymeric substances (EPS), carboxylic acids, proteins and polysaccharides, etc. form organic fouling. The membrane surface cleaning from organic sediment is difficult, if that not use the chemicals. NOMs are the most common reason to organic fouling [186,187].
- *Biological fouling:* Accumulation and growth of biological material on the membrane surface is known as biological fouling or biofouling. In the MD process, biofouling occurs less frequently than in other types of fouling. In other words, it can be stated that presence of microorganisms is one of the most important reasons for biofouling. In the MD process, feed solution has high salinity and the operating temperature of MD is higher than the growth temperature of microorganisms, so microorganisms cannot grow enough [188].

Techniques that are applied to control and reduce fouling in the MD process are as follows:

- *Surface modification:* Membrane surface modification, which is the simplest way to control and mitigate membrane fouling, has been carried out by many researchers to improve hydrophobicity and increase antifouling attributes of the membrane in the MD. Applying diverse superhydrophobic coating on variant layers demonstrates that the formation of biofouling on coated layers is less than that on uncoated substrates. According to the work done, it can be seen that hydrophilization of the membrane surface is considered to be one of the ways to control organic fouling [189,190].
- *Pretreatment:* Feed pretreatment is a convenient and efficient method for reducing fouling in membrane processes and it improves the quality of permeation. The MD process compared to other membrane processes has less need for pretreatment, because there is no high flux in the MD. However, pretreatment has great importance in this process, since it is considered one of the main ways to protect the membrane, particularly in industrial desalination. In MD application, pretreatment is applied to reduce the three types of fouling that were listed above. So the feed solutions that include organic and inorganic colloidal particles and microorganisms should pass through a pretreatment step before entering the MD system to remove macromolecules and microorganisms at first. Generally, when seawater is considered as feed for the MD process, pretreatment is applied [191–193].
- *Gas bubbling:* This is one way to reduce fouling in which a two-phase (gas-liquid) flow is formed and makes the secondary flow increase whereby shear stress on the surface reaches the maximum value. Therefore, the rising shear rate in the membrane surface reduces the formation of deposits on the surface. Mechanisms of the gas bubbling method in control of fouling are as follows: (a) bubble caused by the secondary flow, (b) physical movement of concentration polarization layer, (c) pressure pulsing induced passing bubbles, and (d) increase in superficial cross-flow velocity [194–198].
- *Use of antiscalants:* Antiscalants are one of the ways to reduce fouling in the MD. In this method, chemical additives that intervene with the formation reaction of scales and sediments and diminish the scales cohesion to the surface of the membrane. Use of antiscalants in the MD process is an effective way to reduce inorganic salts scaling. However, it has the following limitations: (a) achieving optimal dosage is essential because the high dosage of antiscalants can be considered as foulants, (b) according to some reports, it has been found that some antiscalants will increase the biological growth of a number of microorganisms, and thus biofouling is formed, (c) reaction of some antiscalants with used chemical additives causes fouling to form and/or decreases returns of antiscalants, (d) The reaction of some metal ions and antiscalants leads to the formation of fouling, and (e) applying this method is complicated and difficult [199–202].
- *Chemical cleaning:* alkalis, surfactants, metal chelating agents, acids, enzymes, and oxidizing agents are chemical cleaning agents which are used to reduce fouling. Generally, potent or impotent acids are applied to clean, particularly those that are deposited with $CaCO_3$. To reduce inorganic fouling, washing with an acid like HCl is suitable, but washing with a basic/alkali solution is usually used to remove organic fouling. Biocides persistent dosage is the basic way in debarment of biofouling. If the scales accumulate and deposit on the membrane surface, washing can significantly enhance primary flux according to the place of scales [68,203–205].
- *Application of magnetic field or microwave irradiation:* Magnetic field is exerted to reduce the scaling in the heat exchanger and cooling water-fouling application. A nonchemical water treatment way using microwave irradiation is applied for mitigation of scaling formation in the MD process. Microwave irradiation has four mechanisms: (a) water molecules may be destroyed by microwave energy, so the exit of molecules from the bulk solution occurs faster, (b) influence of molecules gets faster due to increasing the activity of the polar

structure of the membrane material by microwave, (c) thermal effects are induced by microwave radiation and this causes the temperature polarization in the MD process to be reduced, and (d) the thermal effect can be reduced by microwave irradiation, so temperature polarization also decreases [206–208].

- *Membrane flushing:* A method that reduces membrane scaling in DCMD is studied by Nghiem and Cath via regular membrane flushing. According to the results, they found that membrane scaling arising from $CaCO_3$ or silicate was less than that from $CaSO_4$. Feed and distillate temperatures were 20°C and 40°C, respectively, and owing to the experimental conditions it was observed that scaling of $CaSO_4$ took place just after a lengthy enough induction time (that is, up to 25 h). Increase in feed temperature caused a reduction in the induction time and the size of $CaSO_4$ crystals was increased. Membrane flushing was very effective to control $CaSO_4$ scaling [209].
- *Back-pulsing:* In order to recover the overall rate of filtration, decrease fouling in membranes, and prolong a time-out of cleaning (it is time between two cleaning steps of successive membranes), transmembrane pressure pulsing or back-pulsing (BP) has been introduced as an effective and impressive method for reducing fouling. It is also an in situ way to clean the membranes via the transmembrane pressure of periodically returning. Permeate liquid traced back to the feed by membrane if the transmembrane pressure was inverted. Given that BP reduces the filtration impressive time, it should be noted that the application time of BP is important. If the BP way is applied at the proper time, the positive effects of this method will be noticeable on the permeate flux. Back-pulse duration (BD) and back-pulse interval (BI) are the two factors that are related to BP. The operation time of the filtration system under negative pressure is BD and the time between two persistent pulses is considered as BI. This is a high-performance method to remove deposits from the surface of the membrane [210–212].
- *Spacers:* One way to reduce fouling in the MD process is applying a mechanical device, called a "spacer," to increase the turbulence in the system which is placed in the membrane channels and operates as a mixer. Since the amount of flux is dependent on the feed flow rate, spacers reduce fouling and thus increase the permeate flux by increasing turbulence in the feed side. Geometry of the spacers affects the fouling of membranes and study of spacer geometry is currently an attractive subject for many researchers. In the MD process, according to the needs of a process like the highest flux, the lowest pressure drop, and the highest flux at a given pressure drop in the process, several spacers are used. Access these results conducted by both laboratory methods and CFD. As mentioned, spacers increase permeate flux with increasing turbulence, but it should be noted that the spacer causes stagnant areas to be created in the membrane channels. These areas increase fouling and thus reduce permeability of the membranes [213–216].
- *Back-flushing:* Back-flushing is a way to reduce fouling in the MD system. In this method, channels of feed and permeate sides were returned. In other words, the feed is introduced to the permeate side and the permeate is taken from the feed side. This method is a new and effective way for fouling mitigation that was reported recently. Back-flushing improved flux of the membrane due to reduced fouling [217].

4.5 Economics in MD

A few studies reported the cost estimates of MD process [88]. The economics of MD process for desalination and water treatment are evaluated according to parameters such as source of energy, plant capacity, salinity, and design features. From the thermodynamic point of view, the minimum energy for desalination of seawater is about 0.8 $kWh\,m^{-3}$. This is true for all types of desalination methods. The difference between various desalination processes is the source and the form of energy. In the MD method, nearly all of the energy required is thermal energy that could be supplied from low grade or waste sources but in the RO method, only electrical energy is required, which is more expensive. Owing to the nonideality of the separation processes and energy loss, the required energy for desalination is higher than that based on the thermodynamic minimum. It is reported that the required thermal energy for MD, RO, and MED is 100, 0, and 60 $kWh\,m^{-3}$, respectively. The RO process requires the most electrical energy (3.5 $kWh\,m^{-3}$), followed by MD and MED (2 $kWh\,m^{-3}$). The impact of energy source and plant capacity is very important on the economics of the overall process [218]. The cost of seawater desalination is shown in Fig. 4.10 as a function of plant size when the required thermal energy is supplied only by fresh steam ($0.0078 kg^{-1}). When the production capacity is 30,000 $m^3\,day^{-1}$, the cost of MD is $1.72 m^{-3}, the cost of MED is $1.48 m^{-3}, while the cost of RO is just $0.69 m^{-3}. Therefore, in this situation, RO is the best choice and MD is not economical. Unlike the RO process, which only requires electrical energy, MD and MED processes need heat energy in addition to electrical energy. MD cost is higher because it needs high temperature to heat water and it is supplied by steam. Heat energy requirement accounts for 90% of the total energy. This energy could be supplied from the waste heat or solar energy, which has a low price. If solar energy is used as the energy source and energy is recovered constantly, it can be stated that in terms of energy, the MD process is an economic technology. If the waste heat is used as an energy source, the

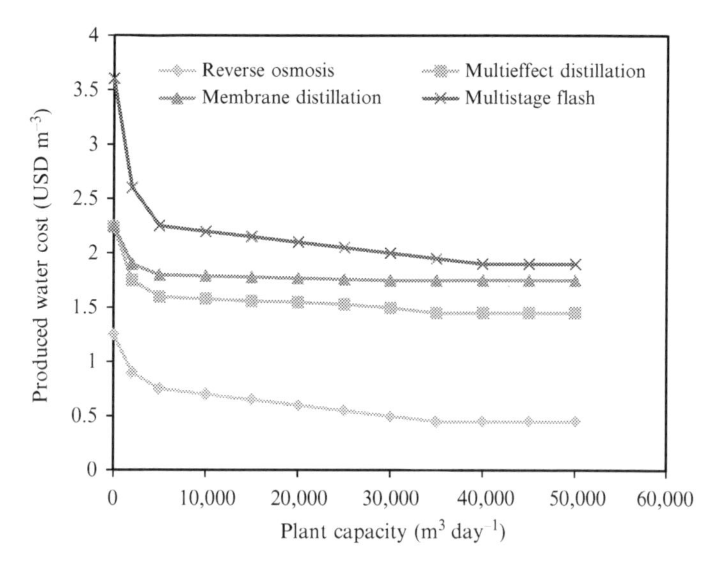

Fig. 4.10 Water production cost of MD, RO, MED, and MSF driven by steam and/or electricity for seawater desalination [219].

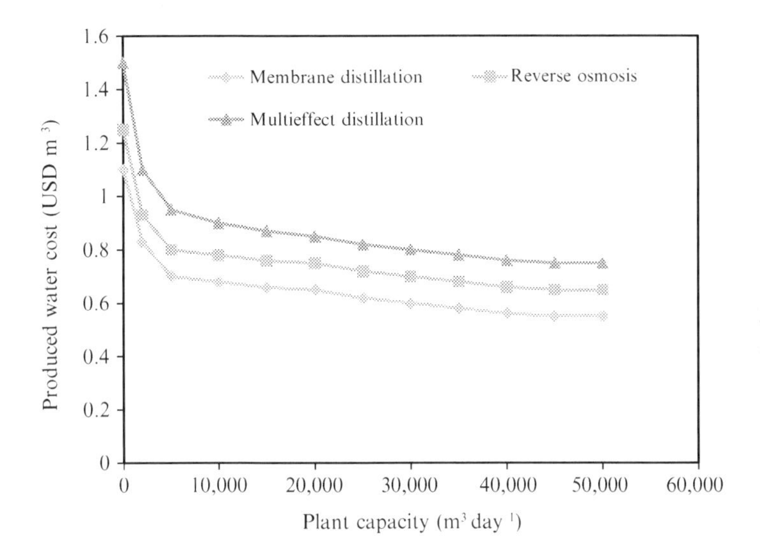

Fig. 4.11 Cost of MD, RO, and MED for seawater desalination when high temperature waste heat is available for both MD and MED [219].

cost of the MD process will significantly reduce for desalination and it is considered to be an economic technology compared to the RO process. Fig. 4.11 shows a comparison between MD, RO, and MED seawater desalination processes costs when waste heat is used as a thermal energy source (which has been assumed to have a cost of 10% of fresh steam cost). The cost of water produced for a plant capacity of $30{,}000\,m^3\,day^{-1}$ is reported to be $\$0.61\,m^{-3}$ for MD, $\$0.81\,m^{-3}$ for MED, and $\$0.69\,m^{-3}$ for RO. It could be seen that in this situation MD is cheaper than RO and is more competitive. When there is a low-cost waste thermal heat source, the MD process will be very suitable for desalination [219–222]. The economics of different MD configurations were investigated by Saffarini et al. [223]. The costs were compared at constant operating conditions and identical membrane area. They observed that while the heat loss through conduction is high from the feed to permeate side in the DCMD system, in terms of the cost, DCMD with a heat recovery system is more economical than AGMD and VMD. In addition, they found that 70% of the total cost of the system is spent for the solar heater. Thus, it is better to use an alternative energy source such as waste heat [223]. Finally, it must be stated that for the commercialization of the MD process in the industry, the MD process should be modernized economically [221].

4.6 Conclusion

In this chapter, membrane materials, membrane characteristics, commercial membranes for MD, membrane characterization, fabricated membranes for MD, methods for membrane fabrication and surface modification in MD, and the economics of MD were reviewed. This study indicates that membrane performance is heavily dependent on the structure and properties of the membrane, especially in the long run. As a result, the membrane should have a structure and properties so that the obtained membrane could be devoted to the MD process and can resolve requirements of the MD process, such as increasing the mass transfer, reducing conductive heat transfer, preventing fouling/scaling and wetting, achieving high flux and energy efficiency, and the ability for long-time applications without loss of efficiency. Thus, it can be stated that the MD membrane should have a small pore size, and the surface energy of membrane materials should be low which makes increased hydrophobicity, and must be porous with an optimum thickness, membrane tortuosity should be low, it should have narrow PSD, water permeability and LEP should be high and it should also be inexpensive.

MD is a process that will surely become one of the most promising desalination methods with the following perspectives intensified in the future:

- Unfortunately, this technology is not yet applied on a large scale and has not reached the commercialization stage despite its potential in many applications. Thus, the membrane must be built with new materials and properties for MD to improve membrane performance. Thus, focus on this issue should be more in future research.
- The MD process is a combination of simultaneous momentum, heat, and mass transfer. Most of the studies have considered the membrane as only an inert porous medium for transferring the vapor phase from feed to permeate side like in most of the microfiltration or even ultrafiltration operations, while the chemistry of membrane material is critical for a successful MD operation. For example, the effects of functional groups and surface energy on the membrane flux and rejection have not been considered yet. It seems that we should take a fresh look on the effect of membrane material on the MD performance such as we made for gas separation membranes. Therefore, the MD process, besides its simplicity in operation, is a very mysterious separation operation and wider studies and investigations should be done for understanding this process.
- The use of CFD and molecular dynamics before fabrication of the membrane can help to design the membrane with appropriate properties and structure for the MD. In addition, CFD has the potential to design a surface patterned membrane with different patterns and can predict the membrane performance as well as investigation of the polarization effect on the membrane. If it is acceptable, an appropriate membrane can be fabricated MD, in order to obtain the best results and to improve the performance of the MD membrane and its industrialization. Meanwhile, predicting the properties of the membrane via CFD also saves time and cost.

References

[1] Lin S, Nejati S, Boo C, Hu Y, Osuji CO, Elimelech M. Omniphobic membrane for robust membrane distillation. Environ Sci Technol Lett 2014;1(11):443–7.

[2] Al-Obaidani S, Curcio E, Macedonio F, Di Profio G, Al-Hinai H, Drioli E. Potential of membrane distillation in seawater desalination: thermal efficiency, sensitivity study and cost estimation. J Membr Sci 2008;323(1):85–98.

[3] Bilad MR, Guillen-Burrieza E, Mavukkandy MO, Al Marzooqi FA, Arafat HA. Shrinkage, defect and membrane distillation performance of composite PVDF membranes. Desalination 2015;376:62–72.

[4] Shirazi MM, Kargari A, Ismail AF, Matsuura T. Computational fluid dynamic (CFD) opportunities applied to the membrane distillation process: state-of-the-art and perspectives. Desalination 2016;377:73–90.

[5] Bouguecha S, Chouikh R, Dhahbi M. Numerical study of the coupled heat and mass transfer in membrane distillation. Desalination 2003;152(1):245–52.

[6] Koo J, Lee S, Choi JS, Hwang TM. Theoretical analysis of different membrane distillation modules. Desalin Water Treat 2015;54(4-5):862–70.

[7] Kargari A, Shirazi MM. Water desalination: solar-assisted membrane distillation. In: Encyclopedia of energy engineering and technology. Boca Raton: CRC Press; 2014.

[8] Hou D, Wang J, Qu D, Luan Z, Ren X. Fabrication and characterization of hydrophobic PVDF hollow fiber membranes for desalination through direct contact membrane distillation. Sep Purif Technol 2009;69(1):78–86.

[9] Shirazi MM, Kargari A, Tabatabaei M, Ismail AF, Matsuura T. Concentration of glycerol from dilute glycerol wastewater using sweeping gas membrane distillation. Chem Eng Process Process Intensif 2014;78:58–66.

[10] Khayet M, Payo CG. Progress on membrane distillation and related technologies. J Membr Sci Res 2013;2(4):161–2.

[11] Shirazi MM, Bastani D, Kargari A, Tabatabaei M. Characterization of polymeric membranes for membrane distillation using atomic force microscopy. Desalin Water Treat 2013;51(31–33):6003–8.

[12] Tomaszewska M. Membrane distillation-examples of applications in technology and environmental protection. Pol J Environ Stud 2000;9(1):27–36.

[13] Bhattacharya M, Dutta SK, Sikder J, Mandal MK. Computational and experimental study of chromium (VI) removal in direct contact membrane distillation. J Membr Sci 2014;450:447–56.

[14] Khayet M, Velázquez A, Mengual JI. Direct contact membrane distillation of humic acid solutions. J Membr Sci 2004;240(1):123–8.

[15] Khayet M. Treatment of radioactive wastewater solutions by direct contact membrane distillation using surface modified membranes. Desalination 2013;321:60–6.

[16] Shao F, Hao C, Ni L, Zhang Y, Du R, Meng J, et al. Experimental and theoretical research on N-methyl-2-pyrrolidone concentration by vacuum membrane distillation using polypropylene hollow fiber membrane. J Membr Sci 2014;452:157–64.

[17] Shirazi MM, Kargari A, Bastani D, Fatehi L. Production of drinking water from seawater using membrane distillation (MD) alternative: direct contact MD and sweeping gas MD approaches. Desalin Water Treat 2014;52(13–15):2372–81.

[18] Gryta M. The concentration of geothermal brines with iodine content by membrane distillation. Desalination 2013;325:16–24.

[19] Hou D, Dai G, Wang J, Fan H, Luan Z, Fu C. Boron removal and desalination from seawater by PVDF flat-sheet membrane through direct contact membrane distillation. Desalination 2013;326:115–24.

[20] Li X, Qin Y, Liu R, Zhang Y, Yao K. Study on concentration of aqueous sulfuric acid solution by multiple-effect membrane distillation. Desalination 2012;307:34–41.

[21] El-Abbassi A, Hafidi A, García-Payo MD, Khayet M. Concentration of olive mill wastewater by membrane distillation for polyphenols recovery. Desalination 2009;245(1):670–4.

[22] Yu X, Yang H, Lei H, Shapiro A. Experimental evaluation on concentrating cooling tower blowdown water by direct contact membrane distillation. Desalination 2013;323:134–41.

[23] Gryta M, Markowska-Szczupak A, Bastrzyk J, Tomczak W. The study of membrane distillation used for separation of fermenting glycerol solutions. J Membr Sci 2013;431:1–8.

[24] Tomaszewska M, Białończyk L. Production of ethanol from lactose in a bioreactor integrated with membrane distillation. Desalination 2013;323:114–9.
[25] Wijekoon KC, Hai FI, Kang J, Price WE, Cath TY, Nghiem LD. Rejection and fate of trace organic compounds (TrOCs) during membrane distillation. J Membr Sci 2014;453:636–42.
[26] Criscuoli A, Zhong J, Figoli A, Carnevale MC, Huang R, Drioli E. Treatment of dye solutions by vacuum membrane distillation. Water Res 2008;42(20):5031–7.
[27] Khayet M, Godino MP, Mengual JI. Theoretical and experimental studies on desalination using the sweeping gas membrane distillation method. Desalination 2003;157(1–3):297–305.
[28] Singh D, Sirkar KK. Desalination of brine and produced water by direct contact membrane distillation at high temperatures and pressures. J Membr Sci 2012;389:380–8.
[29] Mokhtar NM, Lau WJ, Ismail AF. The potential of membrane distillation in recovering water from hot dyeing solution. J Water Process Eng 2014;2:71–8.
[30] Hickenbottom KL, Cath TY. Sustainable operation of membrane distillation for enhancement of mineral recovery from hypersaline solutions. J Membr Sci 2014;454:426–35.
[31] Xie M, Nghiem LD, Price WE, Elimelech M. A forward osmosis–membrane distillation hybrid process for direct sewer mining: system performance and limitations. Environ Sci Technol 2013;47(23):13486–93.
[32] Alkhudhiri A, Darwish N, Hilal N. Produced water treatment: application of air gap membrane distillation. Desalination 2013;309:46–51.
[33] Tun CM, Groth AM. Sustainable integrated membrane contactor process for water reclamation, sodium sulfate salt and energy recovery from industrial effluent. Desalination 2011;283:187–92.
[34] Shirazi MM, Kargari A, Tabatabaei M. Sweeping gas membrane distillation (SGMD) as an alternative for integration of bioethanol processing: study on a commercial membrane and operating parameters. Chem Eng Commun 2015;202(4):457–66.
[35] Qu D, Sun D, Wang H, Yun Y. Experimental study of ammonia removal from water by modified direct contact membrane distillation. Desalination 2013;326:135–40.
[36] Criscuoli A, Bafaro P, Drioli E. Vacuum membrane distillation for purifying waters containing arsenic. Desalination 2013;323:17–21.
[37] Zhao ZP, Xu L, Shang X, Chen K. Water regeneration from human urine by vacuum membrane distillation and analysis of membrane fouling characteristics. Sep Purif Technol 2013;118:369–76.
[38] Kullab A, Martin A. Membrane distillation and applications for water purification in thermal cogeneration plants. Sep Purif Technol 2011;76(3):231–7.
[39] Ramakrishna S, Shirazi MM. Electrospun membranes: next generation membranes for desalination and water/wastewater treatment. Desalination 2013;308:198–208.
[40] Tijing LD, Choi JS, Lee S, Kim SH, Shon HK. Recent progress of membrane distillation using electrospun nanofibrous membrane. J Membr Sci 2014;453:435–62.
[41] Shirazi MM, Kargari A, Tabatabaei M, Ismail AF, Matsuura T. Assessment of atomic force microscopy for characterization of PTFE membranes for membrane distillation (MD) process. Desalin Water Treat 2015;54(2):295–304.
[42] Khayet M. Membranes and theoretical modeling of membrane distillation: a review. Adv Colloid Interface Sci 2011;164(1):56–88.
[43] Kang GD, Cao YM. Application and modification of poly (vinylidene fluoride)(PVDF) membranes—a review. J Membr Sci 2014;463:145–65.
[44] Curcio E, Drioli E. Membrane distillation and related operations—a review. Sep Purif Rev 2005;34(1):35–86.

[45] Wang KY, Chung TS, Gryta M. Hydrophobic PVDF hollow fiber membranes with narrow pore size distribution and ultra-thin skin for the fresh water production through membrane distillation. Chem Eng Sci 2008;63(9):2587–94.
[46] Qtaishat M, Rana D, Khayet M, Matsuura T. Preparation and characterization of novel hydrophobic/hydrophilic polyetherimide composite membranes for desalination by direct contact membrane distillation. J Membr Sci 2009;327(1):264–73.
[47] Qtaishat M, Khayet M, Matsuura T. Novel porous composite hydrophobic/hydrophilic polysulfone membranes for desalination by direct contact membrane distillation. J Membr Sci 2009;341(1):139–48.
[48] Zhang J, Dow N, Duke M, Ostarcevic E, Gray S. Identification of material and physical features of membrane distillation membranes for high performance desalination. J Membr Sci 2010;349(1):295–303.
[49] Huang LT, Hsu PS, Kuo CY, Chen SC, Lai JY. Pore size control of PTFE membranes by stretch operation with asymmetric heating system. Desalination 2008;233(1):64–72.
[50] Kim JY, Lee HK, Baik KJ, Kim SC. Liquid-liquid phase separation in polysulfone/solvent/water systems. J Appl Polym Sci 1997;65(13):2643–53.
[51] Madaeni SS, Rahimpour A. Effect of type of solvent and non-solvents on morphology and performance of polysulfone and polyethersulfone ultrafiltration membranes for milk concentration. Polym Adv Technol 2005;16(10):717–24.
[52] Barzin J, Feng C, Khulbe KC, Matsuura T, Madaeni SS, Mirzadeh HA. Characterization of polyethersulfone hemodialysis membrane by ultrafiltration and atomic force microscopy. J Membr Sci 2004;237(1):77–85.
[53] Zhao C, Xue J, Ran F, Sun S. Modification of polyethersulfone membranes–a review of methods. Prog Mater Sci 2013;58(1):76–150.
[54] Rastegarpanah A, Mortaheb HR. Surface treatment of polyethersulfone membranes for applying in desalination by direct contact membrane distillation. Desalination 2016;377:99–107.
[55] Boussu K, Van der Bruggen B, Vandecasteele C. Evaluation of self-made nanoporous polyethersulfone membranes, relative to commercial nanofiltration membranes. Desalination 2006;200(1):416–8.
[56] Roh IJ, Ramaswamy S, Krantz WB, Greenberg AR. Poly (ethylene chlorotrifluoroethylene) membrane formation via thermally induced phase separation (TIPS). J Membr Sci 2010;362(1):211–20.
[57] Santoro S, Drioli E, Figoli A. Development of novel ECTFE coated PP composite hollow-fiber membranes. Coatings 2016;6(3):40.
[58] Pan J, Xiao C, Huang Q, Liu H, Hu J. ECTFE porous membranes with conveniently controlled microstructures for vacuum membrane distillation. J Mater Chem A 2015;3(46):23549–59.
[59] Zuo J, Chung TS, O'Brien GS, Kosar W. Hydrophobic/hydrophilic PVDF/Ultem® dual-layer hollow fiber membranes with enhanced mechanical properties for vacuum membrane distillation. J Membr Sci 2017;523:103–10.
[60] Ren LF, Xia F, Shao J, Zhang X, Li J. Experimental investigation of the effect of electrospinning parameters on properties of superhydrophobic PDMS/PMMA membrane and its application in membrane distillation. Desalination 2017;404:155–66.
[61] Huang Y, Huang QL, Liu H, Zhang CX, You YW, Li NN, et al. Preparation, characterization, and applications of electrospun ultrafine fibrous PTFE porous membranes. J Membr Sci 2017;523:317–26.
[62] Feng C, Shi B, Li G, Wu Y. Preparation and properties of microporous membrane from poly (vinylidene fluoride-co-tetrafluoroethylene)(F2. 4) for membrane distillation. J Membr Sci 2004;237(1):15–24.

[63] Fan Y, Chen S, Zhao H, Liu Y. Distillation membrane constructed by TiO 2 nanofiber followed by fluorination for excellent water desalination performance. Desalination 2017;405:51–8.

[64] Baghbanzadeh M, Rana D, Lan CQ, Matsuura T. Effects of hydrophilic silica nanoparticles and backing material in improving the structure and performance of VMD PVDF membranes. Sep Purif Technol 2016;157:60–71.

[65] An AK, Guo J, Lee EJ, Jeong S, Zhao Y, Wang Z, et al. PDMS/PVDF hybrid electrospun membrane with superhydrophobic property and drop impact dynamics for dyeing wastewater treatment using membrane distillation. J Membr Sci 2017;525:57–67.

[66] Wei X, Zhao B, Li XM, Wang Z, He BQ, He T, et al. CF 4 plasma surface modification of asymmetric hydrophilic polyethersulfone membranes for direct contact membrane distillation. J Membr Sci 2012;407:164–75.

[67] Agyemang FO, Sheikh FA, Appiah-Ntiamoah R, Chandradass J, Kim H. Synthesis and characterization of poly (vinylidene fluoride)–calcium phosphate composite for potential tissue engineering applications. Ceram Int 2015;41(5):7066–72.

[68] Gryta M. Influence of polypropylene membrane surface porosity on the performance of membrane distillation process. J Membr Sci 2007;287(1):67–78.

[69] Simone S, Figoli A, Criscuoli A, Carnevale MC, Rosselli A, Drioli E. Preparation of hollow fibre membranes from PVDF/PVP blends and their application in VMD. J Membr Sci 2010;364(1):219–32.

[70] Dong ZQ, Ma XH, ZL X, You WT, Li FB. Superhydrophobic PVDF–PTFE electrospun nanofibrous membranes for desalination by vacuum membrane distillation. Desalination 2014;347:175–83.

[71] Lai CL, Liou RM, Chen SH, Huang GW, Lee KR. Preparation and characterization of plasma-modified PTFE membrane and its application in direct contact membrane distillation. Desalination 2011;267(2):184–92.

[72] Peng Y, Dong Y, Fan H, Chen P, Li Z, Jiang Q. Preparation of polysulfone membranes via vapor-induced phase separation and simulation of direct-contact membrane distillation by measuring hydrophobic layer thickness. Desalination 2013;316:53–66.

[73] Khayet M. Treatment of radioactive wastewater solutions by direct contact membrane distillation using surface modified membranes. Desalination 2013;321:60–6.

[74] Hou D, Dai G, Wang J, Fan H, Luan Z, Fu C. Boron removal and desalination from seawater by PVDF flat-sheet membrane through direct contact membrane distillation. Desalination 2013;326:115–24.

[75] Shao F, Hao C, Ni L, Zhang Y, Du R, Meng J, et al. Experimental and theoretical research on N-methyl-2-pyrrolidone concentration by vacuum membrane distillation using polypropylene hollow fiber membrane. J Membr Sci 2014;452:157–64.

[76] Mulder M. Basic principles of membrane technology. Dordrecht, The Netherlands: Kluwer Academic Publishers; 1996.

[77] Kujawski W, Kujawa J, Wierzbowska E, Cerneaux S, Bryjak M, Kujawski J. Influence of hydrophobization conditions and ceramic membranes pore size on their properties in vacuum membrane distillation of water–organic solvent mixtures. J Membr Sci 2016;499:442–51.

[78] Krajewski SR, Kujawski W, Bukowska M, Picard C, Larbot A. Application of fluoroalkylsilanes (FAS) grafted ceramic membranes in membrane distillation process of NaCl solutions. J Membr Sci 2006;281(1):253–9.

[79] Larbot A, Gazagnes L, Krajewski S, Bukowska M, Kujawski W. Water desalination using ceramic membrane distillation. Desalination 2004;168:367–72.

[80] Kujawa J, Kujawski W, Koter S, Jarzynka K, Rozicka A, Bajda K, et al. Membrane distillation properties of TiO_2 ceramic membranes modified by perfluoroalkylsilanes. Desalin Water Treat 2013;51(7-9):1352–61.
[81] Lalia BS, Kochkodan V, Hashaikeh R, Hilal N. A review on membrane fabrication: structure, properties and performance relationship. Desalination 2013;326:77–95.
[82] Drioli E, Giorno L, editors. Membrane operations: innovative separations and transformations. Weinheim: Wiley-VCH Verlag GmbH & Co. KGaA; 2009.
[83] Ulbricht M. Advanced functional polymer membranes. Polymer 2006;47(7):2217–62.
[84] Trommer K, Morgenstern B. Nonrigid microporous PVC sheets: preparation and properties. J Appl Polym Sci 2010;115(4):2119–26.
[85] Tomaszewska M. Preparation and properties of flat-sheet membranes from poly (vinylidene fluoride) for membrane distillation. Desalination 1996;104(1):1–11.
[86] Smolders CA, Reuvers AJ, Boom RM, Wienk IM. Microstructures in phase-inversion membranes. Part 1. Formation of macrovoids. J Membr Sci 1992;73(2-3):259–75.
[87] Kim IC, Yoon HG, Lee KH. Formation of integrally skinned asymmetric polyetherimide nanofiltration membranes by phase inversion process. J Appl Polym Sci 2002;84 (6):1300–7.
[88] Khayet Souhaimi M, Matsuura T. Membrane distillation: principles and applications. Amsterdam: Elsevier; 2011.
[89] Kim JF, Kim JH, Lee YM, Drioli E. Thermally induced phase separation and electrospinning methods for emerging membrane applications: a review. AICHE J 2016;62(2):461–90.
[90] Liu F, Hashim NA, Liu Y, Abed MM, Li K. Progress in the production and modification of PVDF membranes. J Membr Sci 2011;375(1):1–27.
[91] Sarada T, Sawyer LC, Ostler MI. Three dimensional structure of Celgard® microporous membranes. J Membr Sci 1983;15(1):97–113.
[92] Sadeghi F. Developing of microporous polypropylene by stretching. Montreal: Ecole Polytechnique; 2007.
[93] Zhu W, Zhang X, Zhao C, Wu W, Hou J, Xu M. A novel polypropylene microporous film. Polym Adv Technol 1996;7(9):743–8.
[94] Dickey C, Mcdaniel J. Method of producing spherical thermoplastic particles. US Pat, 1975. 3 (896,196).
[95] Prince JA, Singh G, Rana D, Matsuura T, Anbharasi V, Shanmugasundaram TS. Preparation and characterization of highly hydrophobic poly (vinylidene fluoride)–clay nanocomposite nanofiber membranes (PVDF–clay NNMs) for desalination using direct contact membrane distillation. J Membr Sci 2012;397:80–6.
[96] Ray SS, Chen SS, Li CW, Nguyen NC, Nguyen HT. A comprehensive review: electrospinning technique for fabrication and surface modification of membranes for water treatment application. RSC Adv 2016;6(88):85495–514.
[97] Ahmed FE, Lalia BS, Hashaikeh R. A review on electrospinning for membrane fabrication: challenges and applications. Desalination 2015;356:15–30.
[98] Yang X, Fane AG, Wang R. Membrane distillation: now and future. In: Desalination: water from water. 2013. p. 373–424.
[99] Alkhudhiri A, Darwish N, Hilal N. Membrane distillation: a comprehensive review. Desalination 2012;287:2–18.
[100] Zhang J, Dow N, Duke M, Ostarcevic E, Gray S. Identification of material and physical features of membrane distillation membranes for high performance desalination. J Membr Sci 2010;349(1):295–303.

[101] García-Payo MD, Izquierdo-Gil MA, Fernández-Pineda C. Wetting study of hydrophobic membranes via liquid entry pressure measurements with aqueous alcohol solutions. J Colloid Interface Sci 2000;230(2):420–31.
[102] Khayet M, Matsuura T. Preparation and characterization of polyvinylidene fluoride membranes for membrane distillation. Ind Eng Chem Res 2001;40(24):5710–8.
[103] Khayet M, Matsuura T, Mengual JI, Qtaishat M. Design of novel direct contact membrane distillation membranes. Desalination 2006;192(1):105–11.
[104] Banat FA, Simandl J. Theoretical and experimental study in membrane distillation. Desalination 1994;95(1):39–52.
[105] Franken AC, Nolten JA, Mulder MH, Bargeman D, Smolders CA. Wetting criteria for the applicability of membrane distillation. J Membr Sci 1987;33(3):315–28.
[106] Gostoli C, Sarti G. Separation of liquid mixtures by membrane distillation. J Membr Sci 1989;41:211–24.
[107] Le My D. Membrane distillation application in purification and process intensification. M.Sc. Thesis in Environmental Engineering and Managemen,Thailand: Asian Institute of Technology School of Environment, Resources and Development; 2015.
[108] Lawson KW, Lloyd DR. Membrane distillation. J Membr Sci 1997;124(1):1–25.
[109] Laganà F, Barbieri G, Drioli E. Direct contact membrane distillation: modelling and concentration experiments. J Membr Sci 2000;166(1):1–11.
[110] Keyvani Fard AM. Membrane distillation desalination: water quality and energy efficiency analysis. M.Sc. Thesis in Environmental Engineering,Doha, Qatar: Qatar University; 2013.
[111] El-Bourawi MS, Ding Z, Khayet M. A framework for better understanding membrane distillation separation process. J Membr Sci 2006;285(1):4–29.
[112] Iversen SB, Bhatia VK, Dam-Johansen K, Jonsson G. Characterization of microporous membranes for use in membrane contactors. J Membr Sci 1997;130(1):205–17.
[113] Manawi YM. An investigation into the potential of industrial low grade heat in membrane distilaltion for freshwater production. M.Sc. Thesis in Environmental Engineering, Doha, Qatar: Qatar University; 2013.
[114] Khayet M, Velázquez A, Mengual JI. Modelling mass transport through a porous partition: effect of pore size distribution. J Non-Equilib Thermodyn 2004;29(3):279–99.
[115] Martínez L, Florido-Díaz FJ, Hernandez A, Pradanos P. Estimation of vapor transfer coefficient of hydrophobic porous membranes for applications in membrane distillation. Sep Purif Technol 2003;33(1):45–55.
[116] Phattaranawik J, Jiraratananon R, Fane AG. Heat transport and membrane distillation coefficients in direct contact membrane distillation. J Membr Sci 2003;212(1):177–93.
[117] Ibrahim SS, Alsalhy QF. Modeling and simulation for direct contact membrane distillation in hollow fiber modules. AICHE J 2013;59(2):589–603.
[118] Chakrabarty B, Ghoshal AK, Purkait MK. SEM analysis and gas permeability test to characterize polysulfone membrane prepared with polyethylene glycol as additive. J Colloid Interface Sci 2008;320(1):245–53.
[119] Guillen GR, Pan Y, Li M, Hoek EM. Preparation and characterization of membranes formed by nonsolvent induced phase separation: a review. Ind Eng Chem Res 2011;50(7):3798–817.
[120] Karanikola V, Corral AF, Jiang H, Sáez AE, Ela WP, Arnold RG. Effects of membrane structure and operational variables on membrane distillation performance. J Membr Sci 2017;524:87–96.
[121] García-Payo MD, Essalhi M, Khayet M. Effects of PVDF-HFP concentration on membrane distillation performance and structural morphology of hollow fiber membranes. J Membr Sci 2010;347(1):209–19.

[122] Wyart Y, Georges G, Deumie C, Amra C, Moulin P. Membrane characterization by microscopic methods: multiscale structure. J Membr Sci 2008;315(1):82–92.
[123] Nakao SI. Determination of pore size and pore size distribution: 3 Filtration membranes. J Membr Sci 1994;96(1-2):131–65.
[124] Mirtalebi E, Shirazi MM, Kargari A, Tabatabaei M, Ramakrishna S. Assessment of atomic force and scanning electron microscopes for characterization of commercial and electrospun nylon membranes for coke removal from wastewater. Desalin Water Treat 2014;52(34-36):6611–9.
[125] Bowen WR, Doneva TA. Artefacts in AFM studies of membranes: correcting pore images using fast fourier transform filtering. J Membr Sci 2000;171(1):141–7.
[126] Arvay A, Yli-Rantala E, Liu CH, Peng XH, Koski P, Cindrella L, et al. Characterization techniques for gas diffusion layers for proton exchange membrane fuel cells—a review. J Power Sources 2012;213:317–37.
[127] Hilal N, Bowen WR. Atomic force microscope study of the rejection of colloids by membrane pores. Desalination 2002;150(3):289–95.
[128] Bowen WR, Doneva TA, Stoton JA. The use of atomic force microscopy to quantify membrane surface electrical properties. Colloids Surf A Physicochem Eng Asp 2002;201(1):73–83.
[129] Shirazi MM, Kargari A, Bazgir S, Tabatabaei M, Shirazi MJ, Abdullah MS, et al. Characterization of electrospun polystyrene membrane for treatment of biodiesel's water-washing effluent using atomic force microscopy. Desalination 2013;329:1–8.
[130] Vatanpour V, Madaeni SS, Moradian R, Zinadini S, Astinchap B. Fabrication and characterization of novel antifouling nanofiltration membrane prepared from oxidized multiwalled carbon nanotube/polyethersulfone nanocomposite. J Membr Sci 2011;375 (1):284–94.
[131] Vrijenhoek EM, Hong S, Elimelech M. Influence of membrane surface properties on initial rate of colloidal fouling of reverse osmosis and nanofiltration membranes. J Membr Sci 2001;188(1):115–28.
[132] Arkhangelsky E, Kuzmenko D, Gitis V. Impact of chemical cleaning on properties and functioning of polyethersulfone membranes. J Membr Sci 2007;305(1):176–84.
[133] Qiu C, Nguyen QT, Ping Z. Surface modification of cardo polyetherketone ultrafiltration membrane by photo-grafted copolymers to obtain nanofiltration membranes. J Membr Sci 2007;295(1):88–94.
[134] Smolders K, Franken A. Terminology for membrane distillation. Desalination 1989;72 (3):249–62.
[135] Wang D, Li K, Teo WK. Preparation and characterization of polyvinylidene fluoride (PVDF) hollow fiber membranes. J Membr Sci 1999;163(2):211–20.
[136] Zuo G, Wang R. Novel membrane surface modification to enhance anti-oil fouling property for membrane distillation application. J Membr Sci 2013;447:26–35.
[137] Huo R, Gu Z, Zuo K, Zhao G. Preparation and properties of PVDF-fabric composite membrane for membrane distillation. Desalination 2009;249(3):910–3.
[138] Shirazi MM, Kargari A, Tabatabaei M. Evaluation of commercial PTFE membranes in desalination by direct contact membrane distillation. Chem Eng Process Process Intensif 2014;76:16–25.
[139] Wang D, Li K, Teo WK. Porous PVDF asymmetric hollow fiber membranes prepared with the use of small molecular additives. J Membr Sci 2000;178(1):13–23.
[140] Zheng L, Wang J, Li J, Zhang Y, Li K, Wei Y. Preparation, evaluation and modification of PVDF-CTFE hydrophobic membrane for MD desalination application. Desalination 2017;402:162–72.

[141] Zheng L, Wu Z, Wei Y, Zhang Y, Yuan Y, Wang J. Preparation of PVDF-CTFE hydrophobic membranes for MD application: Effect of LiCl-based mixed additives. J Membr Sci 2016;506:71–85.
[142] Khayet M, Matsuura T. Application of surface modifying macromolecules for the preparation of membranes for membrane distillation. Desalination 2003;158(1):51–6.
[143] Feng C, Khulbe KC, Matsuura T, Gopal R, Kaur S, Ramakrishna S, et al. Production of drinking water from saline water by air-gap membrane distillation using polyvinylidene fluoride nanofiber membrane. J Membr Sci 2008;311(1):1–6.
[144] Wang J, Zheng L, Wu Z, Zhang Y, Zhang X. Fabrication of hydrophobic flat sheet and hollow fiber membranes from PVDF and PVDF-CTFE for membrane distillation. J Membr Sci 2016;497:183–93.
[145] Khayet M, Cojocaru C, García-Payo MD. Experimental design and optimization of asymmetric flat-sheet membranes prepared for direct contact membrane distillation. J Membr Sci 2010;351(1):234–45.
[146] Feng C, Shi B, Li G, Wu Y. Preparation and properties of microporous membrane from poly (vinylidene fluoride-co-tetrafluoroethylene)(F2. 4) for membrane distillation. J Membr Sci 2004;237(1):15–24.
[147] Nejati S, Boo C, Osuji CO, Elimelech M. Engineering flat sheet microporous PVDF films for membrane distillation. J Membr Sci 2015;492:355–63.
[148] Li JM, Xu ZK, Liu ZM, Yuan WF, Xiang H, Wang SY, et al. Microporous polypropylene and polyethylene hollow fiber membranes. Part 3. Experimental studies on membrane distillation for desalination. Desalination 2003;155(2):153–6.
[149] Hou D, Wang J, Qu D, Luan Z, Zhao C, Ren X. Preparation of hydrophobic PVDF hollow fiber membranes for desalination through membrane distillation. Water Sci Technol 2009;59(6):1219–26.
[150] Teoh MM, Chung TS. Membrane distillation with hydrophobic macrovoid-free PVDF–PTFE hollow fiber membranes. Sep Purif Technol 2009;66(2):229–36.
[151] Maab H, Al Saadi A, Francis L, Livazovic S, Ghafour N, Amy GL, et al. Polyazole hollow fiber membranes for direct contact membrane distillation. Ind Eng Chem Res 2013;52(31):10425–9.
[152] Drioli E, Ali A, Simone S, Macedonio F, Al-Jlil SA, Al Shabonah FS, et al. Novel PVDF hollow fiber membranes for vacuum and direct contact membrane distillation applications. Sep Purif Technol 2013;115:27–38.
[153] Lu X, Peng Y, Ge L, Lin R, Zhu Z, Liu S. Amphiphobic PVDF composite membranes for anti-fouling direct contact membrane distillation. J Membr Sci 2016;505:61–9.
[154] Teoh MM, Chung TS, Yeo YS. Dual-layer PVDF/PTFE composite hollow fibers with a thin macrovoid-free selective layer for water production via membrane distillation. Chem Eng J 2011;171(2):684–91.
[155] Tong D, Wang X, Ali M, Lan CQ, Wang Y, Drioli E, et al. Preparation of Hyflon AD60/PVDF composite hollow fiber membranes for vacuum membrane distillation. Sep Purif Technol 2016;157:1–8.
[156] Kong Y, Lin X, Wu Y, Chen J, Xu J. Plasma polymerization of octafluorocyclobutane and hydrophobic microporous composite membranes for membrane distillation. J Appl Polym Sci 1992;46(2):191–9.
[157] Wang KY, Foo SW, Chung TS. Mixed matrix PVDF hollow fiber membranes with nanoscale pores for desalination through direct contact membrane distillation. Ind Eng Chem Res 2009;48(9):4474–83.
[158] Gopal R, Kaur S, Ma Z, Chan C, Ramakrishna S, Matsuura T. Electrospun nanofibrous filtration membrane. J Membr Sci 2006;281(1):581–6.

[159] Lee EJ, An AK, He T, Woo YC, Shon HK. Electrospun nanofiber membranes incorporating fluorosilane-coated TiO_2 nanocomposite for direct contact membrane distillation. J Membr Sci 2016;520:145–54.
[160] Liao Y, Wang R, Tian M, Qiu C, Fane AG. Fabrication of polyvinylidene fluoride (PVDF) nanofiber membranes by electro-spinning for direct contact membrane distillation. J Membr Sci 2013;425:30–9.
[161] Essalhi M, Khayet M. Self-sustained webs of polyvinylidene fluoride electrospun nanofibers at different electrospinning times: 1. Desalination by direct contact membrane distillation. J Membr Sci 2013;433:167–79.
[162] Jung J, Shin Y, Choi YJ, Sohn J, Lee S, An K. Hydrophobic surface modification of membrane distillation (MD) membranes using water-repelling polymer based on urethane rubber. Desalin Water Treat 2016;57(22):10031–41.
[163] Khayet M, Suk DE, Narbaitz RM, Santerre JP, Matsuura T. Study on surface modification by surface-modifying macromolecules and its applications in membrane-separation processes. J Appl Polym Sci 2003;89(11):2902–16.
[164] Wavhal DS, Fisher ER. Membrane surface modification by plasma-induced polymerization of acrylamide for improved surface properties and reduced protein fouling. Langmuir 2003;19(1):79–85.
[165] Inagaki N. Plasma surface modification and plasma polymerization. Boca Raton: CRC Press; 1996.
[166] Xi ZY, Xu YY, Zhu LP, Wang Y, Zhu BK. A facile method of surface modification for hydrophobic polymer membranes based on the adhesive behavior of poly (DOPA) and poly (dopamine). J Membr Sci 2009;327(1):244–53.
[167] Kochkodan V, Hilal N. A comprehensive review on surface modified polymer membranes for biofouling mitigation. Desalination 2015;356:187–207.
[168] Khemakhem S, Amar RB. Grafting of fluoroalkylsilanes on microfiltration Tunisian clay membrane. Ceram Int 2011;37(8):3323–8.
[169] Liu L, Shen F, Chen X, Luo J, Su Y, Wu H, et al. A novel plasma-induced surface hydrophobization strategy for membrane distillation: Etching, dipping and grafting. J Membr Sci 2016;499:544–54.
[170] Yang C, Li XM, Gilron J, Kong DF, Yin Y, Oren Y, et al. CF_4 plasma-modified superhydrophobic PVDF membranes for direct contact membrane distillation. J Membr Sci 2014;456:155–61.
[171] Zou L, Vidalis I, Steele D, Michelmore A, Low SP, Verberk JQ. Surface hydrophilic modification of RO membranes by plasma polymerization for low organic fouling. J Membr Sci 2011;369(1):420–8.
[172] Yasuda H, Gazicki M. Biomedical applications of plasma polymerization and plasma treatment of polymer surfaces. Biomaterials 1982;3(2):68–77.
[173] Inagaki N, Ohkubo J. Plasma polymerization of hexafluoropropene/methane mixtures and composite membranes for gas separations. J Membr Sci 1986;27(1):63–75.
[174] Prince JA, Rana D, Singh G, Matsuura T, Kai TJ, Shanmugasundaram TS. Effect of hydrophobic surface modifying macromolecules on differently produced PVDF membranes for direct contact membrane distillation. Chem Eng J 2014;242:387–96.
[175] Rana D, Matsuura T. Surface modifications for antifouling membranes. Chem Rev 2010;110(4):2448–71.
[176] Rana D, Matsuura T, Narbaitz RM. Novel hydrophilic surface modifying macromolecules for polymeric membranes: polyurethane ends capped by hydroxy group. J Membr Sci 2006;282(1):205–16.

[177] He F, Gilron J, Lee H, Song L, Sirkar KK. Potential for scaling by sparingly soluble salts in crossflow DCMD. J Membr Sci 2008;311(1):68–80.

[178] Derjaguin BV, Landau L. Theory of the stability of strongly charged lyophobic sols and of the adhesion of strongly charged particles in solutions of electrolytes. Prog Surf Sci 1993;43(1-4):30–59.

[179] Verweij EJ, Overbeek JT. Theory of stability of lyophobic colloids.

[180] Srisurichan S, Jiraratananon R, Fane AG. Humic acid fouling in the membrane distillation process. Desalination 2005;174(1):63–72.

[181] Curcio E, Ji X, Di Profio G, Fontananova E, Drioli E. Membrane distillation operated at high seawater concentration factors: role of the membrane on $CaCO_3$ scaling in presence of humic acid. J Membr Sci 2010;346(2):263–9.

[182] Tang CY, Chong TH, Fane AG. Colloidal interactions and fouling of NF and RO membranes: a review. Adv Colloid Interface Sci 2011;164(1):126–43.

[183] Meng F, Chae SR, Drews A, Kraume M, Shin HS, Yang F. Recent advances in membrane bioreactors (MBRs): membrane fouling and membrane material. Water Res 2009;43 (6):1489–512.

[184] Gryta M, Tomaszewska M, Karakulski K. Wastewater treatment by membrane distillation. Desalination 2006;198(1–3):67–73.

[185] Schneider K, Hölz W, Wollbeck R, Ripperger S. Membranes and modules for transmembrane distillation. J Membr Sci 1988;39(1):25–42.

[186] Gryta M. Fouling in direct contact membrane distillation process. J Membr Sci 2008;325 (1):383–94.

[187] Tijing LD, Pak BC, Baek BJ, Lee DH. A study on heat transfer enhancement using straight and twisted internal fin inserts. Int Commun Heat Mass Transfer 2006;33(6):719–26.

[188] Gryta M. The assessment of microorganism growth in the membrane distillation system. Desalination 2002;142(1):79–88.

[189] Zhang H, Lamb R, Lewis J. Engineering nanoscale roughness on hydrophobic surface—preliminary assessment of fouling behaviour. Sci Technol Adv Mater 2005;6(3):236–9.

[190] Privett BJ, Youn J, Hong SA, Lee J, Han J, Shin JH, et al. Antibacterial fluorinated silica colloid superhydrophobic surfaces. Langmuir 2011;27(15):9597–601.

[191] Huang H, Schwab K, Jacangelo JG. Pretreatment for low pressure membranes in water treatment: a review. Environ Sci Technol 2009;43(9):3011–9.

[192] Shon HK, Kim SH, Vigneswaran S, Aim RB, Lee S, Cho J. Physicochemical pretreatment of seawater: fouling reduction and membrane characterization. Desalination 2009;238(1-3):10–21.

[193] Macedonio F, Curcio E, Drioli E. Integrated membrane systems for seawater desalination: energetic and exergetic analysis, economic evaluation, experimental study. Desalination 2007;203(1–3):260–76.

[194] Teoh MM, Bonyadi S, Chung TS. Investigation of different hollow fiber module designs for flux enhancement in the membrane distillation process. J Membr Sci 2008;311 (1):371–9.

[195] Yang X, Wang R, Fane AG. Novel designs for improving the performance of hollow fiber membrane distillation modules. J Membr Sci 2011;384(1):52–62.

[196] Yeo AP, Law AW, Fane AT. The relationship between performance of submerged hollow fibers and bubble-induced phenomena examined by particle image velocimetry. J Membr Sci 2007;304(1):125–37.

[197] Ratkovich N, Chan CC, Berube PR, Nopens I. Experimental study and CFD modelling of a two-phase slug flow for an airlift tubular membrane. Chem Eng Sci 2009;64 (16):3576–84.

[198] Cui ZF, Chang S, Fane AG. The use of gas bubbling to enhance membrane processes. J Membr Sci 2003;221(1):1–35.
[199] Ghani S, Al-Deffeeri NS. Impacts of different antiscalant dosing rates and their thermal performance in Multi Stage Flash (MSF) distiller in Kuwait. Desalination 2010;250 (1):463–72.
[200] Amjad Z. Calcium sulfate dihydrate (gypsum) scale formation on heat exchanger surfaces: The influence of scale inhibitors. J Colloid Interface Sci 1988;123(2):523–36.
[201] Kavitha AL, Vasudevan T, Prabu HG. Evaluation of synthesized antiscalants for cooling water system application. Desalination 2011;268(1):38–45.
[202] Lyster E, Kim MM, Au J, Cohen Y. A method for evaluating antiscalant retardation of crystal nucleation and growth on RO membranes. J Membr Sci 2010;364(1):122–31.
[203] Lee H, Amy G, Cho J, Yoon Y, Moon SH, Kim IS. Cleaning strategies for flux recovery of an ultrafiltration membrane fouled by natural organic matter. Water Res 2001;35 (14):3301–8.
[204] Al-Amoudi A, Lovitt RW. Fouling strategies and the cleaning system of NF membranes and factors affecting cleaning efficiency. J Membr Sci 2007;303(1):4–28.
[205] Karakulski K, Gryta M. Water demineralisation by NF/MD integrated processes. Desalination 2005;177(1-3):109–19.
[206] Barrett RA, Parsons SA. The influence of magnetic fields on calcium carbonate precipitation. Water Res 1998;32(3):609–12.
[207] Ji Z, Wang J, Yin Z, Hou D, Luan Z. Effect of microwave irradiation on typical inorganic salts crystallization in membrane distillation process. J Membr Sci 2014;455:24–30.
[208] Ji Z, Wang J, Hou D, Yin Z, Luan Z. Effect of microwave irradiation on vacuum membrane distillation. J Membr Sci 2013;429:473–9.
[209] Nghiem LD, Cath T. A scaling mitigation approach during direct contact membrane distillation. Sep Purif Technol 2011;80(2):315–22.
[210] Larsson E. Microfiltration of milk with backpulsing. Lund: Tetra Pak Processing Systems AB; 2011.
[211] Sondhi R, Bhave R. Role of backpulsing in fouling minimization in crossflow filtration with ceramic membranes. J Membr Sci 2001;186(1):41–52.
[212] Kambarani M, Bahmanyar H, Mousavian MA, Mousavi SM. Crossflow filtration of sodium chloride solution by a polymeric nanofilter: minimization of concentration polarization by a novel backpulsing method. Iran J Chem Chem Eng 2016;35 (4):135–41.
[213] Shakaib M, Hasani SM, Ahmed I, Yunus RM. A CFD study on the effect of spacer orientation on temperature polarization in membrane distillation modules. Desalination 2012;284:332–40.
[214] Cath TY, Adams VD, Childress AE. Experimental study of desalination using direct contact membrane distillation: a new approach to flux enhancement. J Membr Sci 2004;228 (1):5–16.
[215] Martınez-Dıez L, Vazquez-Gonzalez M, Florido-Dıaz F. Study of membrane distillation using channel spacers. J Membr Sci 1998;144(1):45–56.
[216] Chernyshov M, Meindersma G, De Haan A. Comparison of spacers for temperature polarization reductionin air gap membrane distillation. Desalination 2005;183 (1–3):363–74.
[217] Prisciandaro M, Mazziotti di Celso G. Back-flush effects on superficial water ultrafiltration. Desalination 2010;256:22–6.
[218] Fath H, Abbas Z, Khaled A. Echno-economic assessment and environmental impacts of desalination technologies. Desalination 2011;266(1–3):263–73.

[219] Kesieme UK, Milne N, Aral H, Cheng CY, Duke M. Economic analysis of desalination technologies in the context of carbon pricing, and opportunities for membrane distillation. Desalination 2013;323:66–74.
[220] Kesieme UK, Milne N, Cheng CY, Aral H, Duke M. Recovery of water and acid from leach solutions using direct contact membrane distillation. Water Sci Technol 2014;69 (4):868–75.
[221] Pangarkar BL, Deshmukh SK, Sapkal VS, Sapkal RS. Review of membrane distillation process for water purification. Desalin Water Treat 2016;57(7):2959–81.
[222] Kesieme UK. Mine waste water treatment and acid recovery using membrane distillation and solvent extraction. Ph.D. Thesis,Melbourne, Australia: Institute for Sustainability and Innovation, School of Engineering & Science, Victoria University; 2015.
[223] Saffarini RB, Summers EK, Arafat HA. Economic evaluation of stand-alone solar powered membrane distillation systems. Desalination 2012;299:55–62.

Sustainable desalination by permeate gap membrane distillation technology

5

Farzaneh Mahmoudi, Aliakbar Akbarzadeh
RMIT University, Melbourne, VIC, Australia

Nomenclature

A_m	membrane surface area (m^2)
C_m	membrane mass transfer coefficient (kg/Pa m^2 s)
C_{pC}	condenser channel specific heat capacity (J/kg K)
C_{pE}	evaporator channel specific heat capacity (J/kg K)
D_h	hydraulic diameter (m)
E_i	total power input (W)
f	friction factor
GOR	gained output ratio
h_{fg}	specific heat of vaporization (J/kg)
h_C	heat transfer coefficient at the condenser channel (W/m^2 K)
h_E	heat transfer coefficient at the evaporator channel (W/m^2 K)
h_F	heat transfer coefficient at the impermeable polymeric film (W/m^2K)
h_{PG}	heat transfer coefficient at the permeate gap channel (W/m^2 K)
h_M	heat transfer coefficient at the membrane surface (W/m^2 K)
J_p	permeate flux (kg/m^2 s)
K_E	evaporator channel thermal conductivity (W/m K)
K_m	membrane thermal conductivity (W/m K)
K_{pg}	permeate gap thermal conductivity (W/m K)
L	module length (m)
l_s	orthogonal distance between net spacer filament (m)
$\dot{m}_f$	feed flow rate (kg/s)
$\dot{m}_{Ci}$	condenser channel inlet mass flow rate (kg/s)
$\dot{m}_{Ei}$	evaporator channel inlet mass flow rate (kg/s)
$\dot{m}_{Eo}$	evaporator channel outlet mass flow rate (kg/s)
$\dot{m}_{PG}$	permeate flow rate (kg/s)
$P_{v,w}$	pure water vapor pressure (Pa)
$P_{v,sw}$	saltwater vapor pressure (Pa)
Pr	Prandtl number
q_{STEC}	specific thermal energy consumption (kWh/m^3)
$\dot{q}_C$	convective heat flux from the condenser channel (W/m^2)
$\dot{q}_{cold}$	condenser channel heat flux (W/m^2)
$\dot{q}_E$	convective heat flux from the evaporator channel (W/m^2)

Emerging Technologies for Sustainable Desalination Handbook. https://doi.org/10.1016/B978-0-12-815818-0.00005-9

$\dot{q}_{hot}$ evaporator channel heat flux (W/m^2)
$\dot{q}_M$ heat transfer rate from the membrane surface (W/m^2)
Re Reynolds number
S salinity (g/kg)
S_{Ci} condenser channel inlet feed water salinity (g/kg)
S_{Ei} feed water salinity in the evaporator channel inlet (g/kg)
S_{Eo} feed water salinity in the evaporator channel outlet (g/kg)
T_{Ci} temperature of the fluid in the condenser inlet (°C)
T_{Co} temperature of the fluid in the condenser outlet (°C)
T_{Ei} temperature of the fluid in the evaporator inlet (°C)
T_{Eo} temperature of the fluid in the evaporator outlet (°C)
T_{Me} temperature of the fluid at the membrane surface in the evaporator side (°C)
T_{Mp} temperature of the fluid at the membrane surface in the permeate gap side (°C)
T_{PG} temperature of the fluid in the permeate gap channel (°C)
u_m feed velocity (m/s)
$\dot{V}_f$ feed flow rate (m^3/s)
$\dot{V}_p$ permeate flow rate (m^3/s)
α Antoine equation coefficient
β Antoine equation coefficient
γ Antoine equation coefficient
δ_F impermeable film thickness (m)
δ_{m} membrane thickness (m)
δ_{PG} permeate gap channel thickness (m)
ΔP pressure drop (Pa)
ρ feed density (kg/m^3)

5.1 Introduction

The need for freshwater is considered a critical global problem and, according to the World Water Council, 17% of the world population will have inadequate freshwater supply by 2020 [1]. Consequently, the demand for alternative sustainable water sources including ground water, desalinated water, and recycled water has increased over recent years and, as a result, the implementation of desalination plants is growing on a large scale. Freshwater can be derived from seawater by evaporation processes, for example, multistage flash (MSF), multieffect distillation (MED), or by membrane-based processes, including reverse osmosis (RO), electrodialysis (ED), and membrane distillation (MD).

MD is a separation process that involves phase change (liquid–vapor equilibrium) across a hydrophobic, highly porous membrane, which acts as an interface between the vapor and liquid phases. In contrast to most membrane separation processes, which are isothermal and have driving forces as transmembrane hydrostatic pressures, concentrations, and electrical or chemical potentials, MD is a nonisothermal process. If a temperature difference occurs across a non-wetting membrane, the created partial vapor pressure difference as a driving force leads to water molecules evaporating at the hot side, crossing the membrane in the vapor phase, and condensing at the cold side. Fig. 5.1 depicts the direct contact membrane distillation (DCMD) configuration.

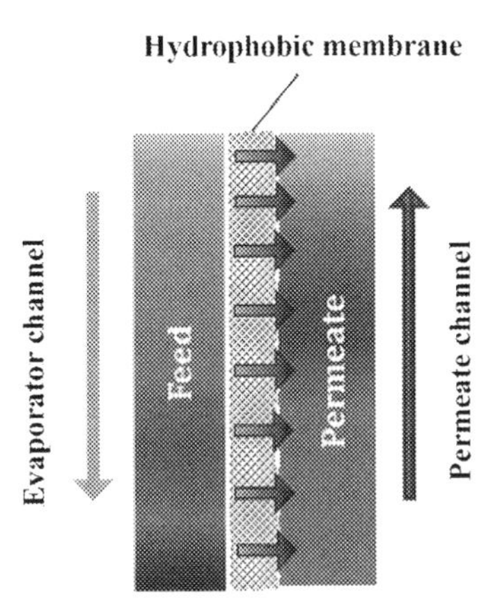

Fig. 5.1 Direct contact membrane distillation configuration.

Commercially developed RO technology is associated with high electrical energy consumption in the range of (6–12) kWh/m^3 with the electricity currently being generated from nonrenewable and polluting fossil fuels [2]. In contrast, MD is a thermal process which uses lower top temperature (80°C or less), compared with the traditional thermal desalination processes such as MSF and MED, making it suitable for using waste heat or solar heat. Table 5.1 provides a comparison between the most developed current desalination technologies in terms of specific thermal energy consumption (STEC), specific electrical energy consumption (SEEC), and operating temperature. In addition, an advantage of the MD process [4] is that aqueous solutions of salts with higher concentrations than seawater can be treated by MD, reducing discharge volumes and increasing the water recovery factor up to 95%, which considerably diminishes the environmental impact of brine disposal.

However, the MD process is still under research phase and the lack of experimental data has indicated that there is a need for more comprehensive research in this field, both experimentally and mathematically. The central issues are the external energy source for MD units, lack of MD membranes, and fabrication of modules for each MD configuration. There are also uncertain energetic and economic costs as well as difficulties with long-term operation and the possibility of membrane pore wetting

Table 5.1 **Comparison of the most developed desalination technologies [3]**

Technology	Plant capacity (m^3/day)	STEC (kWh/m^3)	SEEC (kWh/m^3)	Operation temperature (typical) (°C)
MSF	4000–450,000	55–220	4–6	90–120 (112)
MED	100–56,000	40–220	1.5–2.5	50–70 (70)
RO	0.01–360,000	–	2.8–12	≤40

and membrane fouling. Overall, optimization of MD plants is required in order to reach higher MD performance and to decrease energy consumption [5]. The reported values for permeate flux are relatively low and to overcome this issue, an appropriate redesign of the MD module is demanded in order to achieve mass transfer improvement and to increase the membrane surface area per module volume. In addition, with the exception of DCMD, which has been more widely studied, other MD configurations have not been properly investigated, so more focus on other MD configurations is required [6].

Moreover, the energy source of the MD process is an important issue for commercialization of this technology as a sustainable process. MD associated with renewable energy is considered a highly promising process, especially for situations where low-temperature solar, waste, or other heat is available. The STEC of MD systems varies based on the module configuration, setup scale, and operating condition. A wide dispersion of reported values is observed in the literature for STEC based on different MD configurations, with the STEC varying in a range of (1–9000) kWh/m^3. Moreover, the energy consumption of a small-scale installation is much higher than for pilot plants with higher effective membrane surface areas [7].

Additionally, too little information is available about the long-term performance of MD. Deposit formation on the membrane surface causes the closing of the evaporation surfaces, which leads to a decline of module efficiency during long-term operation. Furthermore, membrane pore wetting occurs in the areas where deposits form leading to liquid penetration of the membrane pores and decrease of the permeate flow rate and possibly permeate quality [8].

In this chapter, firstly a fundamental study is described concerning the PGMD system configuration, as a novel developed MD technology with its potential for use as a sustainable desalination technology. Then the theoretical study of PGMD modules is described and design parameters for a laboratory-scale PGMD setup, as a sample case study, are detailed. This sample case study provides a good estimation of the most important characteristics of an MD system, including permeate flux, STEC, GOR, and their dependence on the module design and the particular operating conditions. Furthermore, validation of the PGMD technology numerical model is provided by comparing the experimental data from this case study with the modeling results.

By reviewing studies carried out for the PGMD configuration, a more reliable and comprehensive understanding of the current status of this technology is provided and the possibility of developing it in commercial scale real plant applications is evaluated. In addition, a comprehensive literature survey is presented concerning published projects, which study the feasibility of developing a sustainable MD desalination plant by the use of solar energy to eliminate the requirement for external electrical energy.

In the last section, the general techno-economic feasibility of the PGMD system in comparison with other MD configurations is described. Overall, introducing an essentially new PGMD system and developing knowledge and technology in this area through describing fundamental concepts and through theoretical study aims to help optimize clean water production by desalination technology as a sustainable way to meet the future high demand of drinking water.

5.2 Permeate gap membrane distillation theory

5.2.1 PGMD separation technology fundamentals

Generally, there are four basic MD configurations, including: (a) DCMD; (b) air gap membrane distillation (AGMD); (c) vacuum membrane distillation (VMD); and (d) sweeping gas membrane distillation (SGMD);

DCMD, as depicted in Fig. 5.1, is the most studied MD configuration because of the simplicity of the MD configuration in terms of the channel arrangement, although the heat transferred by conduction through the membrane is higher than in the other MD configurations [9].

In DCMD, the membrane is in direct contact with the feed solution on one side and the permeate on the other side, with the temperature difference across the two sides of the membrane as the process driving force. A recently introduced configuration of DCMD is called PGMD or liquid gap membrane distillation (LGMD). In this configuration, the permeate is extracted from the highest module position, so that the gap between the membrane and the impermeable film fills with permeate during the operation (Fig. 5.2).

Also, in the DCMD configuration, both sides of the membrane are in direct contact with the feed and permeate fluid and the driving force is provided by temperature difference. However, in PGMD, by considering a third channel for produced freshwater with an impermeable film on the permeate side, the cold fluid in the condenser side separates from the permeate, and therefore it could be any other liquid like saline feedwater. Through introducing the permeate gap and impermeable film, there is an additional resistance to heat transfer across the membrane which leads to a reduction of the effective temperature difference, so that the permeate gap and film thickness should be minimized to decrease the thermal resistance [9]. Therefore, the DCMD configuration will have higher flux when compared with PGMD, but the main advantage of the PGMD configuration is providing internal heat recovery for cold salt feedwater by preheating through absorbing latent heat of condensation from the

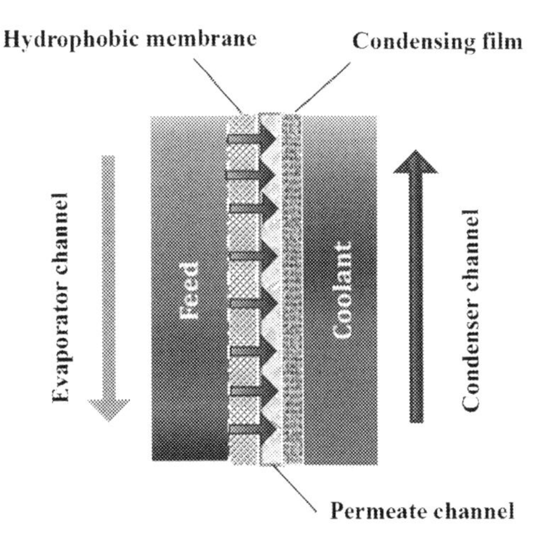

Fig. 5.2 PGMD module channel arrangement with internal heat recovery.

permeate channel. The DCMD channel configuration has low overall heat and mass transfer resistance compared to more complex channel configurations, which leads to the higher fluxes. However, the thin membrane layer causes high heat loss by conductive heat transfer.

In AGMD, the vapor molecules cross both the membrane pores and the air gap, driven by a temperature difference, and condense over a cold surface inside the membrane module. Similar to the PGMD channel configuration, the permeate is separated from the cooling fluid by an impermeable film on the permeate side of the membrane. In order to prevent the air gap from filling with liquid, the permeate is extracted at the bottom of the membrane module. The sensible heat recovery concept is applicable for DCMD, PGMD, and AGMD by absorbing the latent heat of condensation. For PGMD and AGMD, with an impermeable wall applied for separation of the permeate from the coolant and the preheating of the feed taking place within the condenser channel, no external heat exchanger is required for heat recovery (as opposed to DCMD) [9].

In VMD, vacuum or a low pressure is applied to the permeate side of the membrane module by means of a vacuum pump and the gap separates the condensing surface from the membrane with condensation happening outside the MD module [10]. In SGMD, a cold, inert gas sweeps the vapor molecules on the permeate side of the membrane and condensation, as in VMD, takes place outside the membrane module. Generally, DCMD and AGMD are best suited to applications where water is the distillate flux; however, SGMD and VMD are normally used to remove volatile organic or dissolved gases from an aqueous solution [11]. Table 5.2 provides an overall comparison between different MD configurations.

5.2.2 Mathematical model development in PGMD

In a PGMD module, water and volatile components evaporate at the membrane interfacial surface of the evaporator channel, diffuse through the microporous membrane structure, and are condensed and extracted from the membrane module at the permeate channel outlet. To predict the PGMD module performance, a single-node theoretical model is developed using some simplifying assumptions. The following assumptions are considered:

- Steady-state condition
- Stagnant air inside the membrane pore
- Fully filled permeate channels with pure water
- No total pressure difference across the membrane, so no mass transfer by viscous flow
- No heat loss by conduction to the environment
- Stagnant permeate in the permeate channel and so no heat transfer by convection in the permeate gap

To model the PGMD system, the above assumptions, heat and mass conservation laws in all channels, and the five main thermal resistances between evaporator and condenser channels, as shown in Fig. 5.3, are considered. In addition, by defining the known variables including: T_{Ci}, T_{Ei}, $\dot{m}_{Ci}$, $\dot{m}_{Ei}$, S_{Ci}, and S_{Ei} which are respectively temperature (°C), mass flow rate (kg/s) and salinity (g/kg) at the condenser and evaporator channels inlets, governing equations are developed.

Table 5.2 **Different configurations of MD technology [10]**

MD configurations	Main features
DCMD	Both feed and permeate fluids are in direct contact with the two sides of the membrane Simplest and the most studied MD variant High heat loss by conduction through the membrane Low overall heat and mass transfer resistance compared to more complex channel configurations
PGMD	A third channel is considered for the produced freshwater which provides internal heat recovery within the system The permeate extracts from the highest module position, so the gap is filled with permeate during operation Lower distillate flux and lower STEC with compare to DCMD
AGMD	The permeate extracts from the lowest module position, to prevent the air gap from filling up with liquid Low heat loss by conduction through the membrane
VMD	Vacuum or a low pressure is applied in the permeate side of the membrane module Condensation takes place outside the membrane module High risk of membrane pore wetting
SGMD	A cold, inert gas sweeps the vapor molecules in the permeate side of the membrane Condensation takes place outside the membrane module High-temperature polarization effect

It is also required to specify the membrane properties including pore size, porosity, thickness and thermal conductivity, impermeable film thickness and thermal conductivity, as well as bulk conditions and module geometry, in order to carry out numerical modeling.

The convective heat flux from the evaporator channel to the membrane surface is expressed as

$$\dot{q}_E = h_E\left(\frac{T_{Ei}+T_{Eo}}{2} - T_{Me}\right) \tag{5.1}$$

where $\dot{q}_E$ is the evaporator channel heat flux (W/m^2), h_E is the heat transfer coefficient at the evaporator channel (W/m^2 K), T_{Ei} and T_{Eo} are, respectively, the temperatures at the evaporator channel inlet and outlet, and T_{Me} is the temperature at the evaporator side of the membrane surface.

The heat transfer rate from the membrane surface $\dot{q}_M$ arises from the latent heat of the produced vapor flux and the heat transferred by conduction across both the membrane matrix and the gas-filled membrane pores [12].

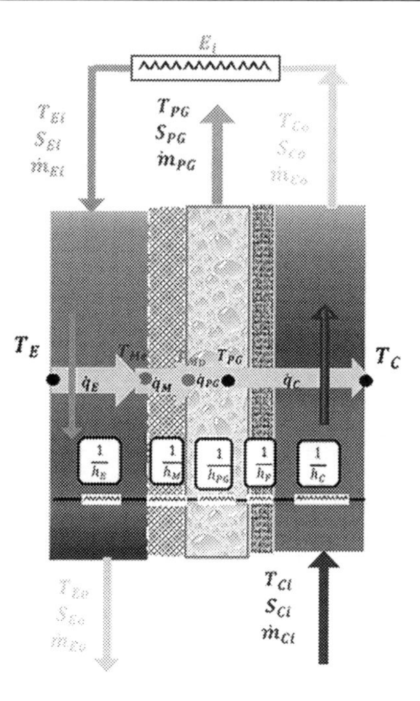

Fig. 5.3 Single-node approach for mathematical modeling of the PGMD module.

$$\dot{q}_M = h_M(T_{Me} - T_{Mp}) \tag{5.2}$$

In this equation, h_M is the heat transfer coefficient at the membrane (W/m^2 K) (its inverse is depicted in Fig. 5.3 as membrane resistance)

$$\dot{q}_M = \frac{K_m}{\delta_m}(T_{Me} - T_{Mp}) + J_p h_{fg} \tag{5.3}$$

where h_{fg} is the water vaporization enthalpy (J/kg), and K_m and δ_m are the membrane thermal conductivity (W/m K) and thickness (m), respectively. J_p (kg/m^2 s) (based on Eq. 5.4) is defined as a function of partial vapor pressure on the two sides of the membrane and of the membrane mass transfer coefficient (C_m). The membrane mass transfer coefficient can be derived by defining the main mass transfer mechanism via the Knudsen number, as comprehensively described by Essalhi et al. [10]. This coefficient is a function of membrane physical properties including pore size, porosity, thickness, and tortuosity besides the temperature and pressure inside the membrane pores.

$$J_p = C_m(P_{v,swe} - P_{v,swp}) \tag{5.4}$$

The partial vapor pressure at the membrane surface can be defined by the Antoine equation (Eq. 5.5) as a function of membrane surface temperatures in the evaporator and permeate gap sides, respectively (T_{Me}) and (T_{Mp}). In the Antoine equation, $P_{v,w}$ is vapor pressure of pure water (Pa), T is absolute temperature (K), and for water α, β, and γ are taken as 23.1964, 3816.44, and −46.13, respectively [12].

$$P_{v,w}(T) = \exp\left(\alpha - \frac{\beta}{\gamma + T}\right) \tag{5.5}$$

In desalination applications, for salt feedwater, the reduction of hot feedwater vapor pressure by the presence of salt ions must be taken into account and some empirical correlations for water vapor pressure of seawater are given in the literature. Sharqawy et al. [13] summarized the existing correlations and data with their ranges of validity and accuracy for estimating the thermophysical properties of seawater. As a primary estimation by assuming seawater as an ideal solution, the reduced vapor pressure may be calculated by Eq. (5.6), in which S is feedwater salinity in g/kg, as derived from Raoult's law.

The partial vapor pressure on the evaporator side $P_{v,swe}$ and the permeate gap side $P_{v,swp}$ in Eq. (5.4) may be derived from Eq. (5.6), by assuming ($S_{PG} = 0$), so that $P_{v,swp}$ would be equal to $P_{v,w}$.

$$P_{v,sw} = P_{v,w}\left(1 + 0.5735\left(\frac{S}{1000 - S}\right)\right)^{-1} \tag{5.6}$$

Emerson and Jamieson also introduced Eq. (5.7), based on experimental measurements in synthetic seawater for a (30–170) ppt salinity range and (100–180)°C temperatures range. The reduced vapor pressure $P_{v,sw}$ based on pure water vapor pressure $P_{v,w}$ may then be calculated using Eq. (5.7) [14].

$$P_{v,sw} = P_{v,w} 10^{\left(-2.1609 \times 10^{-4} S - 3.5012 \times 10^{-7} S^2\right)} \tag{5.7}$$

Fig. 5.4 compares these two correlations (Eqs. 5.6, 5.7) for different feed salinities in a range of (30–160) ppt and for similar operating conditions. In this graph, J_S/J_{sw} is the

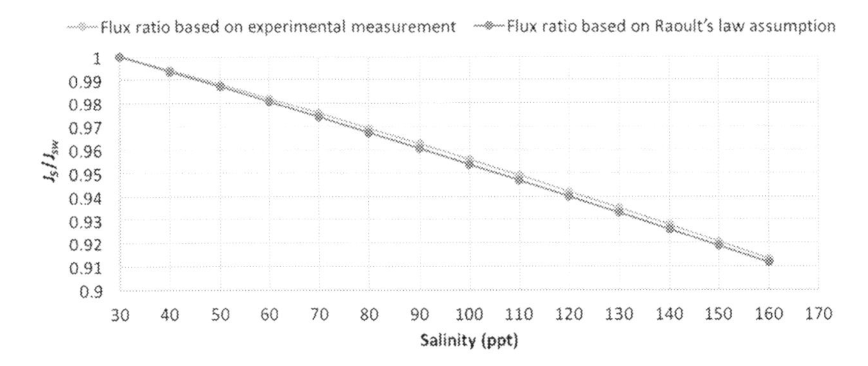

Fig. 5.4 Flux ratio (J_S/J_{sw}) at different feedwater salinity.

ratio of freshwater flux for different feedwater salinities (J_S) to freshwater flux for seawater salinity (J_{sw} at $S=30\,\text{ppt}$) at a constant hot and cold channel temperature difference and for identical membrane characteristics. It is evident from this graph that reduced vapor pressure calculation based on the assumption of Raoult's law (Eq. 5.6) has low accuracy compared to experimental measurement values (Eq. 5.7's data), especially for high concentrate solutions, since in Raoult's law theory, the interaction between different ions in seawater is ignored, which leads to high error for a non-dilute solution. Sharqawy et al. [13] concluded that Eq. (5.7), which is derived based on experimental measurement and suggested by Emerson and Jamieson, is the best correlation to describe seawater vapor pressure.

On the basis of the assumptions made, the produced permeate is considered stagnant in the permeate channel, so that the heat transfer from the gap takes place only in the form of conduction calculated by

$$\dot{q}_{PG}=h_{PG}\left(T_{Mp}-T_{PG}\right) \tag{5.8}$$

In this equation, h_{PG} is the heat transfer coefficient at the permeate gap (W/m K) (its inverse is depicted in Fig. 5.3 as permeate gap resistance), which is defined as Eq. (5.9). T_{PG} is the permeate temperature at the permeate gap.

$$\dot{q}_{PG}=\frac{1}{\dfrac{\delta_{PG}}{2K_{PG}}}\left(T_{Mp}-T_{PG}\right) \tag{5.9}$$

In this equation, K_{PG} (W/m K) and δ_{PG} (m) are the permeate gap thermal conductivity and thickness, respectively. Since it is assumed that the permeate gap is completely filled with distillate water, the permeate gap thermal conductivity is equivalent to the thermal conductivity of freshwater. Moreover, the heat is transferred from the permeate gap to the condenser channel by a series combination of thermal resistances including the permeate channel and impermeable polymeric film and condenser side thermal resistances. Therefore, the condenser channel heat flux ($\dot{q}_C$) is defined as shown below:

$$\dot{q}_C=\frac{1}{\dfrac{1}{h_{PG}}+\dfrac{1}{h_F}+\dfrac{1}{h_C}}\left(T_{PG}-\frac{T_{Ci}+T_{Co}}{2}\right) \tag{5.10}$$

Similarly, in this equation, h_F and h_C (W/m^2 K) are the heat transfer coefficients at the impermeable polymeric film and condenser channel, respectively. The inverse is depicted in Fig. 5.3 as permeate gap and condenser channel resistance, as described in the following equation:

$$\dot{q}_C=\frac{1}{\dfrac{\delta_{PG}}{2K_{PG}}+\dfrac{\delta_F}{K_F}+\dfrac{1}{h_C}}\left(T_{PG}-\frac{T_{Ci}+T_{Co}}{2}\right) \tag{5.11}$$

K_F (W/m K) and δ_F (m) are the impermeable polymeric film thermal conductivity and thickness, respectively, T_{Ci} and T_{Co} are the temperatures at the condenser channel inlet and outlet, respectively.

Considering the energy balance correlations for both evaporator and condenser channels and the total membrane surface area (A_m), Eqs. (5.12), (5.13) could be assumed, respectively, for the evaporator channel and condenser channel heat fluxes ($\dot{q}_{hot}$ and $\dot{q}_{cold}$).

$$\dot{q}_{cold} = \frac{\dot{m}_{Ci}\, C_{pC}(T_{Co} - T_{Ci})}{A_m} \tag{5.12}$$

$$\dot{q}_{hot} = \frac{\dot{m}_{Ei}\, C_{pE}(T_{Ei} - T_{Eo})}{A_m} \tag{5.13}$$

Given the assumption of a steady-state condition, it may be concluded that

$$\dot{q}_E = \dot{q}_M = \dot{q}_{PG} = \dot{q}_C = \dot{q}_{cold} = \dot{q}_{hot} \tag{5.14}$$

Solving the seven main Eqs. (5.1, 5.3, 5.4, 5.9, 5.11, 5.12, 5.13) using the Matlab equation solver, the seven unknown variables (T_{Me}, T_{MP}, T_{PG}, T_{Co}, T_{Eo}, J_P, and $\dot{q}$), as depicted in the mathematical modeling schematic, may be calculated.

Based on the J_P value and mass balance equation for saline feedwater on the evaporator side, $\dot{m}_{Eo}$ and S_{Eo} may be evaluated by Eqs. (5.15), (5.16). S_{Ei}, S_{Eo}, and $\dot{m}_{PG}$ describe, respectively, the feedwater salinity in the input and output of the evaporator channel and the produced distillate rate in the permeate channel.

$$\dot{m}_{Eo} = \dot{m}_{Ei} - \dot{m}_{PG} = \dot{m}_{Ei} - J_p A_m \tag{5.15}$$

$$S_{Eo} = \frac{\dot{m}_{Ei} S_{Ei}}{\dot{m}_{Eo}} \tag{5.16}$$

where h_C and h_E in Eqs. (5.1), (5.10) are similarly calculated by Eq. (5.17) for a spacer filled channel and for a situation in which spacers do not induce change in flow direction [15,16].

$$Nu = 0.644 Re^{0.5} Pr^{0.333} \left(\frac{2D_h}{l_s}\right)^{0.5} \tag{5.17}$$

Which in this equation, l_s (m) is orthogonal distance between the net spacer filaments and D_h (m) is the hydraulic diameter. Hence, the applied heat transfer correlations significantly effects on the whole modeling of MD system, considering correct and accurate heat transfer equations based on the system channel/spacer configuration are so essential to simulate the actual MD system condition. Phattaranawik et al. [17] and Winter [9] proposed to consider the complex hydraulic condition in a spacer filled channel to improve the accuracy of MD system modeling. Applying the basic forms of heat transfer correlation (as Eq. 5.18) and defining the exponents value and constants for Nu number via recorded experimental data and by considering each specific system geometry, is suggested as the more accurate approach in compares to predefined heat transfer correlations.

$$Nu = aRe^{b}Pr^{c}\left(\frac{d_h}{L}\right)^{d} \tag{5.18}$$

5.3 PGMD module experimental-based case studies

5.3.1 Laboratory-scale system design and experimental approach

In this section, the details, technical design, and performance characteristics of a research project on a laboratory-scale PGMD system at RMIT University are presented as a case study to provide comprehensive understanding and a reliable guideline to the design characteristics, system components and module performance, as well as efficiency and possible improvements for future research activities in this field.

A novel optimized experimental approach was based on a laboratory-scale plate-and-frame PGMD module with a $0.12\,m^2$ effective membrane area, which was constructed using two transparent polyacrylic sheets of 25 mm thickness for evaporator and condenser sides and the third sheet for the produced permeate water of 6 mm thickness (as shown in Fig. 5.5). As shown in Fig. 5.5A, two main cylindrical channels with 10 mm diameter holes were milled in each evaporator and condenser plate for the flow inlet and outlet manifolds. To improve flow distribution, a set of 11×3 mm distributor holes was drilled in each flow channel. The hydrophobic PTFE membrane with a 0.22 μm nominal pore size and (140–200) μm thickness on a polypropylene (PP) support, with an effective surface area $(760 \times 160)\,mm^2$, was applied.

The permeate channel was separated from the condenser channel by an impermeable 100 μm clear PP film. This plastic PP film had a thermal conductivity similar to that of the PTFE membrane. All channels are filled with diamond mesh plastic spacers which apply both as a mechanical support for membrane and condensing polymeric films and as turbulence promoters (as shown in Fig. 5.5B). The gap width in the condenser and evaporator channels was adjusted to 1.5 mm and the gap width in the permeate channel was 3 mm. Four rubber gasket frames were used for sealing and to provide the required connection between both the membrane and impermeable film to the polyacrylic sheets.

A schematic diagram of the laboratory-scale experimental setup is shown in Fig. 5.6. The feedwater was pumped from a 100-L storage tank using a small 12-V DC water pump, the feed flow rate to the module controlled either by adjusting the DC pump voltage or by an in-line control valve. A 100 μm pore size mechanical filter was installed before the pump to protect the module and pump from unwanted solids. To adjust the inlet feed temperature to the condenser channel, a laboratory cooling circuit was used.

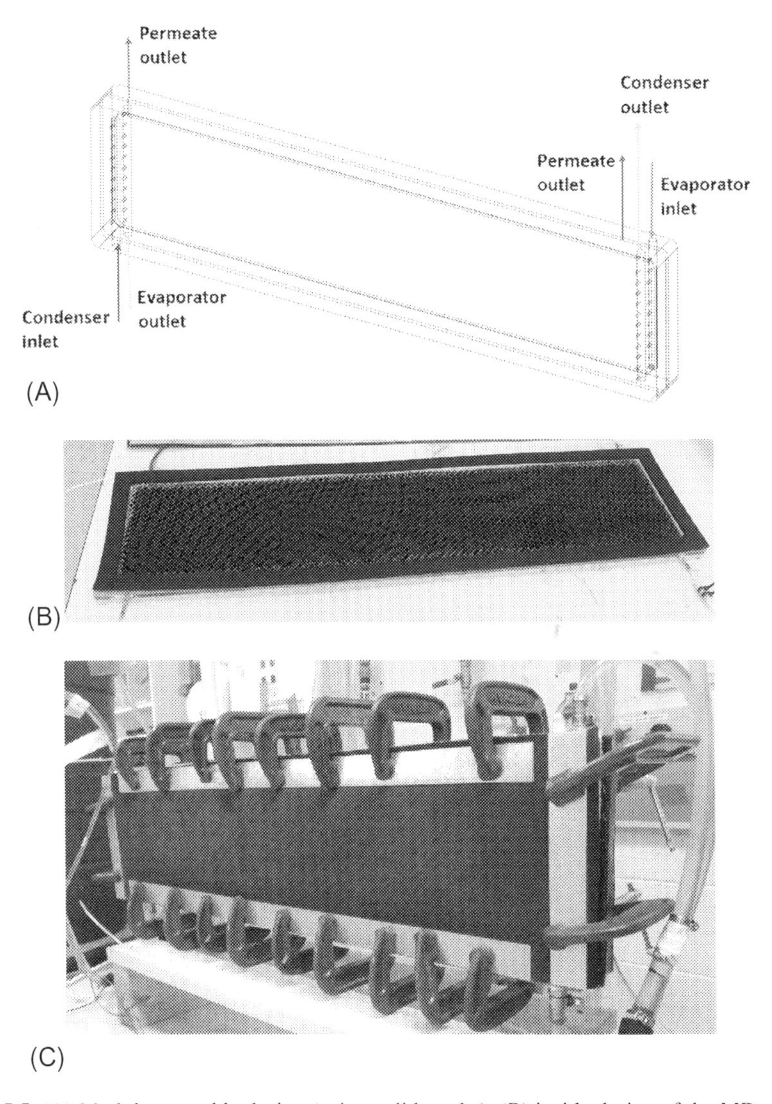

Fig. 5.5 (A) Module assembly design (using solid works), (B) inside design of the MD system with mesh spacer and rubber gasket, and (C) outside view of the manufactured system.

The feedwater gradually preheats while flowing through the condenser channel and using the latent heat of condensation and conduction via the PP condensing film. The condenser outlet temperature was increased by using an external electrical heat source immersed in an insulated electric water tank (2.4 kW), to provide the required evaporator channel inlet temperature for the PGMD module. In the evaporator channel, the hot water vaporizes at the membrane surface, diffuses through the hydrophobic PTFE membrane pores, and condenses on the permeate channel film. The evaporator

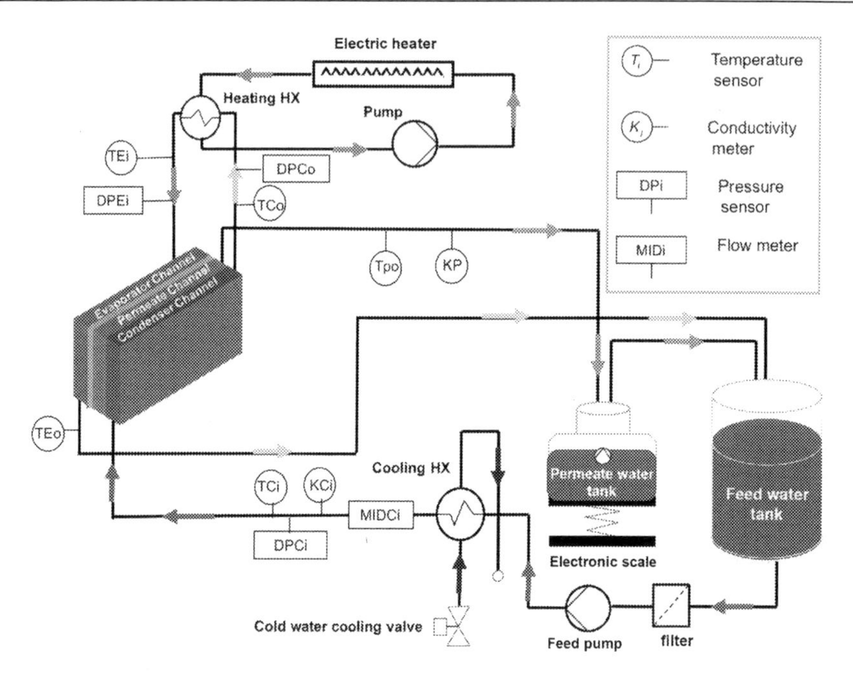

Fig. 5.6 Schematic diagram of the PGMD experimental setup.

channel outlet feed, with higher salinity than the inlet feed to the condenser channel, returns to the feed tank. In order to maintain the feed tank salinity at a constant level, freshwater is pumped from the permeate tank by applying a floating ball valve and electronic scale.

The produced freshwater exited from the top manifold of the permeate gap. The inlet and outlet water temperatures to the condenser and evaporator channels were continuously measured using four T-type thermocouples connected to a data acquisition system, Agilent 34970 A. Digital pressure sensors were used to measure the condenser channel inlet and outlet pressure and evaporator channel inlet pressure and two Salinity Meters (HI98192) were applied to monitor and measure the salinity of feed and distillate during all experimental runs. Each experiment ran for at least 90 min to reach a steady-state condition and the average values of the recorded data in the steady-state condition were used for analysis.

5.3.2 Laboratory-scale experimental and theoretical results

5.3.2.1 Hydraulic study and heat transfer results

In this section, the influence of feedwater flow rate on hydraulic system parameters including pressure drop, friction factor, and the applied pumping power in the condenser and evaporator flow channels is studied. Furthermore, the effects of different

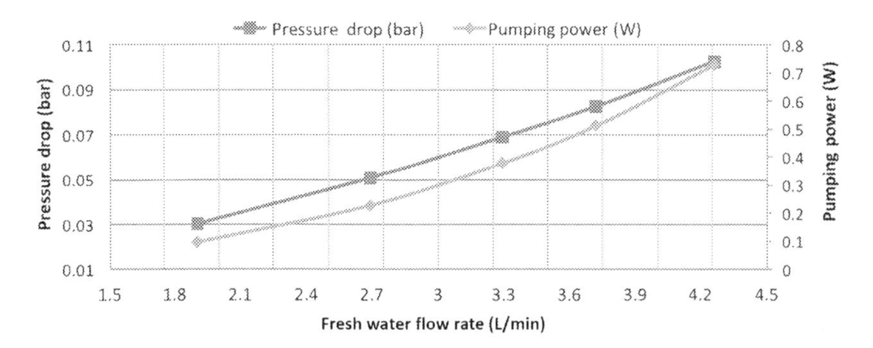

Fig. 5.7 Pressure drop and required pumping power in the condenser channel at different freshwater flow rate.

feedwater flow rates and salinities on internal heat recovery through the system are presented.

Fig. 5.7 illustrates the effect of feed flow rate on the pressure drop in the condenser channel. This experiment is done to study the hydrodynamic condition in the flow channels for cold freshwater feed. As is evident from these graphs, by increasing the feed flow rate, the pressure drops increase so that the required pumping power, which is a function of feed flow rate ($\dot{V}_f$) and pressure drop (ΔP), as described by Eq. (5.18), will also increase. As is shown by the hydraulic tests, the maximum pressure drop in the MD setup for the designed flow rate (<3 L/min) was less than 6 kPa in the condenser channel and less than 0.5-W pumping power was required for the designed system, within the experiment flow rate range. The selected feed flow rate range was chosen to have the optimum amount of feed residence time and therefore the optimum internal heat recovery, which will be explained in the next section. It is noted that the reported values are based on fresh feedwater (nearly 0% salinity feed); however, the pressure drop and pumping power would increase for saline feedwater.

$$\text{Pumping power} = \dot{V}_f \cdot \Delta P \tag{5.18}$$

Furthermore, the friction factor in the condenser and evaporator channels may be calculated on the basis of the Darcy equation:

$$f = \frac{-\Delta P D_h}{L \dfrac{\rho u_m^2}{2}} \tag{5.19}$$

where D_h is the hydraulic diameter (defined by Eq. 5.20), L is the module length, and u_m is the feed velocity. It is noted that by increasing the feed flow rate, the turbulence in the flow channel increases so the Reynolds number increases. Increasing the Reynolds number leads to improved heat transfer rate in the flow channels and reduction of the temperature polarization effect at the two sides of the membrane surface as well as promoting the temperature difference across the membrane surface causing the permeate flux to increase, which will be discussed subsequently. As described by Eq. (5.19), increasing the feed flow rate also leads to a lower friction factor in the module channels.

$$D_h = \frac{4A_c}{P} = \frac{4WH}{2(W+H)} \tag{5.20}$$

In Eq. (5.20), A_c is the flow cross-sectional area and P is the wetted perimeter. For the described laboratory-scale PGMD module, the effective surface area was (760×160) mm^2, the gap width was 1.5 mm, the channel cross-sectional width (W) was equivalent to 160 mm, and the channel cross-sectional gap height (H) was 1.5 mm.

The Reynolds number is defined as a function of fluid density (ρ) and viscosity (μ), hydraulic diameter (D_h), and feed velocity (u_m) (Fig. 5.8).

$$Re = \frac{\rho u_m D_h}{\mu} \tag{5.21}$$

The effect of feedwater salinity on internal heat recovery was also investigated and the results depicted in Fig. 5.9. As seen in this figure, by increasing the feedwater salinity from freshwater feed (nearly 0% salinity) to seawater salinity (around 30 ppt), the temperature rise in the condenser channel (internal heat recovery rate) decreased. An increase of feed flow rate leads to lower feed residence time in the flow channels, so there was less time for heat transfer between the hot and cold channels; hence, the temperature rise in the condenser channel was less significant [12,18,19]. In this figure also, the negative effect of salinity on internal heat recovery is shown, which led to a lower amount of released latent heat of condensation at higher salinity and thus less temperature rise in the condenser channel.

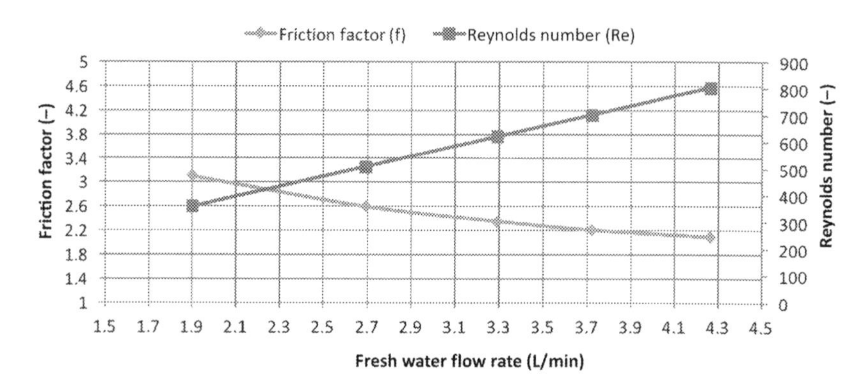

Fig. 5.8 Friction factor and Reynolds number of condenser channel at different freshwater flow rates.

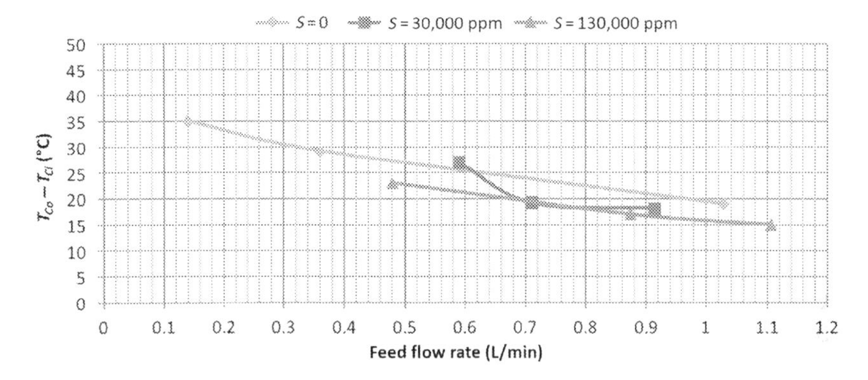

Fig. 5.9 Cold channel temperature rise ($T_{Co} - T_{Ci}$) by internal heat recovery through the membrane at different feed salinity, test condition: $T_{Ci} = 15°C$, $T_{Ei} = 82°C$.

5.3.2.2 Module performance in terms of distillate rate and energy consumption

In the MD operation, the most important characteristic values are permeate flux (J_P), STEC, and GOR which are defined as follows:

The permeate flux J_P of the MD module could be defined by dividing the permeate flow rate $\dot{m}_{PG}$ (kg/s) to the total membrane area (A_m):

$$J_p = \frac{\dot{m}_{PG}}{A_m} \tag{5.22}$$

The quantity q_{STEC} is the amount of total power input (E_i) to produce 1 m^3 of freshwater [20]. In this equation, $\dot{V}_p$ is the produced distillate rate (m^3/s).

$$q_{STEC} = \frac{E_i}{\dot{V}_p} = \frac{\dot{m}_{Ei} C_p (T_{Ei} - T_{Co})}{\dot{V}_p} \tag{5.23}$$

GOR is an indication of how well the total energy input to the system is utilized to produce freshwater:

$$GOR = \frac{\dot{m}_{PG} h_{fg}}{E_i} \tag{5.24}$$

The effect of the feed flow rate on permeate flux for a constant set of operating conditions ($T_{Ci} = 15°C$, $T_{Ei} = 82°C$) and for a range of feed salinities from 0 to 300 ppt is investigated and shown in Fig. 5.10. It is seen in this graph that by increasing the feed flow rate, the permeate flux increases.

For freshwater feed (assuming 0% salinity), increasing the feed flow rate from 0.14 to 1.03 L/min, led to an approximately 400% increase in the permeate flux. For a feed

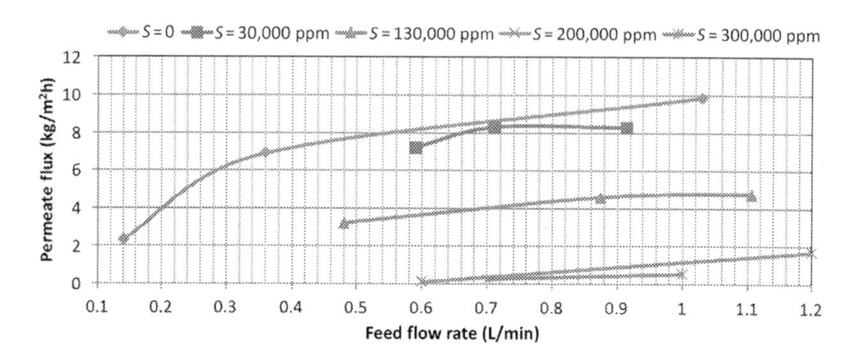

Fig. 5.10 Permeate flux at different feed flow rates for $S=0$, 30, 130, 200, and 300 g/kg, test condition: $T_{Ci}=15°C$ and $T_{Ei}=82°C$.

sample with 130 ppt (nearly four times seawater salinity), doubling the feed flow rate led to a similar upward trend on the produced distillate rate. That is, J_p increases from approximately 3–5 kg/m^2 h (a nearly 70% increase). This effect may be explained by the higher turbulence in the flow channel at the higher feed flow rate, associated with a higher value for the Reynolds number, which improved the heat transfer rate in flow channels and reduced the temperature polarization effect on the two sides of the membrane surface. As a result, the temperature difference across the membrane surface increased, leading to higher permeate flux.

Fig. 5.10 also presents the effect of salinity on permeate flux, which shows a decrease in permeate water flux for saline water compared to the freshwater feed case. As seen in Eq. (5.6) and according to Raoult's law, a higher concentration of solutes in the feed causes a decrease of vapor pressure above the solution and decrease of the MD process partial vapor pressure difference across the two sides of the membrane, eventually leading to lower permeate flux [12]. The produced distillate rate decreased significantly for high salt concentration (200–300) ppt, which is near the saturated state of saline feedwater. As seen in this graph, for a feed sample with 20 ppt salinity and at a nearly 1 L/min flow rate, the distillate flux decreased to <2 kg/m^2 h. Owing to the significant effect of salinity in the permeate flux rate, the permeate test only runs for a higher value of feed flow rate (>0.6 L/min) in (200–300) ppt of salinity range (as shown in Fig. 5.10) and the test did not perform for a very low feed flow rate because of the negligible permeate flux.

However, Dow et al. [21] reported an approximately 3 kg/m^2 h permeate flux over a 3-month operation of an MD plant pilot trial to treat the power station's wastewater, using a multilayered flat-sheet DCMD system with a 0.67 m^2 membrane area. Dow et al. reported that their desalination setup performed consistently without considerable flux loss even at high feed concentration factors, which was attributed to applying a series of prefiltration stages as an effective solution (at the end of their trial, the feed TDS value had reached approximately 70 ppt). Two mechanical pretreatment stages, including a 25 μm high capacity bag filter followed by a 5 μm cartridge filter in the feed cycle, at the highest temperature point in the MD module (before the evaporator channel inlet) is applied to remove the precipitating minerals and prevents scale formation on the membrane surface. These mechanical filter stages lead to high salt rejection and prevent pore wetting at the membrane surface area, which then leads to a longer life span of applied polymeric membrane, less frequent membrane replacement, and production of a higher quality distillate.

As a further stage in this project, in development of a pilot-scale PGMD module desalination rig with a 0.62 m^2 membrane area, a 22 μm screen pore size strainer with a plastic mesh basket followed by a 5 μm pore size plastic bag filter will be installed. The first filter will be located after the saline feed tank and the second one at the highest temperature position of the MD rig to filter salts with reverse solubility, which maintain the permeate flux quantity and quality within an acceptable range. It is necessarily expected that by applying these low mesh size filtration stages, there will be a higher-pressure drop along the module length, demanding higher pumping power and hence higher SEEC.

The effect of feed flow rates on STEC of the system for three different feed salinity values (0, 30, and 130) ppt, which are calculated on the basis of Eq. (5.23), is shown in Fig. 5.11. The results show that by increasing the feed flow rate, the required amount of thermal energy will increase. A higher feed flow rate is equivalent to a higher value for $\dot{m}_{Ei}$, resulting in higher demand for external energy input (E_i) to reach the constant T_{Ei} value in comparison with the lower feed flow rate. Also, as explained previously, a high feed flow rate leads to shorter feed residence time in the flow channel; therefore, a less efficient sensible heat recovery in the condenser channel is possible. As a result, for a similar operating condition, T_{Co} decreases and the amount of external heat demand (STEC) to reach the designed value for T_{Ei}, increases.

On the other hand, as seen in Fig. 5.10, by increasing the feed flow rate, the permeate flux rate increases (higher value of $\dot{V}_p$ in Eq. 5.23). However, the overall effect of higher energy demand (higher value of E_i in Eq. 5.23) is not completely compensated by higher permeate flux, so the STEC values increase at the higher feed flow rate. As is clear from this figure, at higher feed salinities, the STEC also increases because of the lower permeate output at higher salinities which is also confirmed by previous studies [22,23].

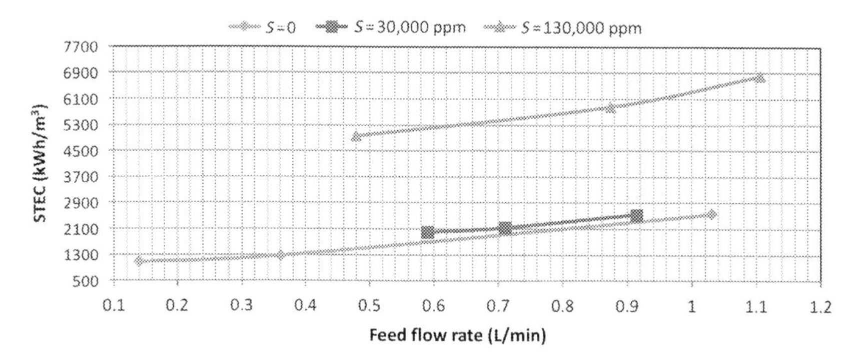

Fig. 5.11 Specific thermal energy consumption at different feed flow rates for $S=0$, 30 and 130 g/kg, test condition: $T_{Ci}=15°C$ and $T_{Ei}=82°C$.

The values determined for distillate flux rate and STEC are in good agreement with similar research by Cipollina et al. [23], and their experimental system is described in more detail in the next section. In this research, by applying a PGMD module configuration of a $0.04\,m^2$ area and with saline feedwater having 35 ppt salt concentration, a 0.6 L/min feed flow rate and for $T_{Ei}=80°C$, a value of $8\,kg/m^2h$ is reported for distillate rate. The reported result is comparable with the value for $S=30$ ppt under the test condition mentioned in this study, as depicted in Fig. 5.11 (a value of $8\,kg/m^2h$ was obtained at approximately 0.7 L/min). In addition, the effect of feed flow rate on produced distillate rate showed a similar trend in both researches.

Comparison of STEC values in the stated test condition shows that Cipollina et al. reported a value of $6000\,kWh/m^3$, which, as shown in Fig. 5.11, is much higher than the value from the present study (around $2100\,kWh/m^3$). This difference may be explained by the lower effective membrane surface area, which is nearly one-third of that in this study. Therefore, negligible sensible heat recovery and thermal integration along the system happened in this condition.

Gained output ratio is an alternative representation of the STEC and for analyzing the thermal efficiency of the desalination systems may be used to quantify the module's capability for internal heat recovery. As is explained by Winter et al. [24], in MD desalination systems, owing to heat loss by conduction through the MD module, all the thermal energy input cannot be applied to the evaporation process and GOR must be <1. However, in ideal MD systems with optimum internal heat recovery with a high surface area for heat transfer between hot and cold fluids, a higher value of GOR would be possible.

As investigated in this study and illustrated in Fig. 5.12, by increasing feed flow rate and feedwater salinity, GOR values show a downward trend. The maximum value achieved for GOR was approximately one, which confirmed the high heat loss rate through the laboratory-scale module and insufficient internal heat recovery by the developed PGMD module with the geometry described. A longer module flow channel with a higher membrane surface area will provide more efficient sensible heat recovery leading to a higher GOR value and so develop a more thermally efficient MD system [24].

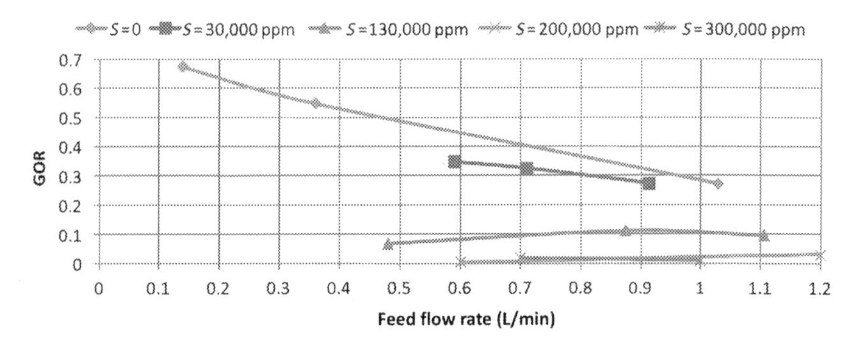

Fig. 5.12 Gained output ratio at different feed flow rates for $S=0$, 30, 130, 200, and 300 g/kg, Test condition: $T_{Ci}=15°C$ and $T_{Ei}=82°C$.

Winter et al. [24] reported GOR values of approximately 5 and 3.5, respectively, for freshwater feed and seawater feed (nominally 35 ppt) at 300 kg/h feed flow rate in a spirally wound MD module. They carried out a set of similar experiments, with a PGMD module arrangement but in a compact spiral-wound configuration and on a large scale, for which the system specification is explained briefly in the next section.

A comparison between reported STEC values from Winter et al. and the values from this study shows that the current laboratory-scale arrangement with a 0.1 m^2 membrane surface area has an order of magnitude higher thermal energy demand than the spiral-wound full-scale arrangement with a 10 m^2 membrane surface area. Therefore, in the PGMD configuration based on internal heat recovery characteristics, it would be achievable to reduce the amount of external thermal energy demand by providing more fluid residence time in the module channels. Therefore, providing longer module channels will provide a higher heat transfer rate between cold and hot channels, which will increase the temperature rise in the condenser channel and reduce the external STEC rate.

On the basis of the developed theoretical model described in Section 5.2.2, the influences of feed flow rate and salinity on the two important desalination systems characteristic of permeate flux (J_p) and internal heat recovery rate ($T_{Co} - T_{Ci}$) are plotted and the results are compared with the measured values. Figs. 5.13 and 5.14 show a good comparison between experimental results and theoretical values for produced freshwater rate and for internal heat parameters.

Comparison is made for feed flow rate in the range 0.1–1.1 L/min and with three inlet feed salinities of approximately (0, 30, and 130) ppt. As is evident from these graphs, values obtained from numerical modeling using heat and mass balance equations and using the Matlab Equation solver for a single-node model, as explained in the mathematical model development section, are in good agreement with experimental measured values. The developed theoretical model could be applied as a reliable tool, to design the geometrical configuration of an optimized PGMD system based on simulation of the system performance.

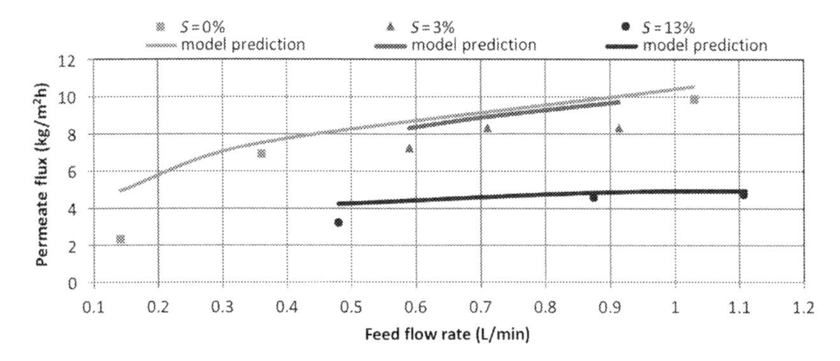

Fig. 5.13 Theoretical and experimental value comparison for influence of feed flow rate and salinity on permeate flux.

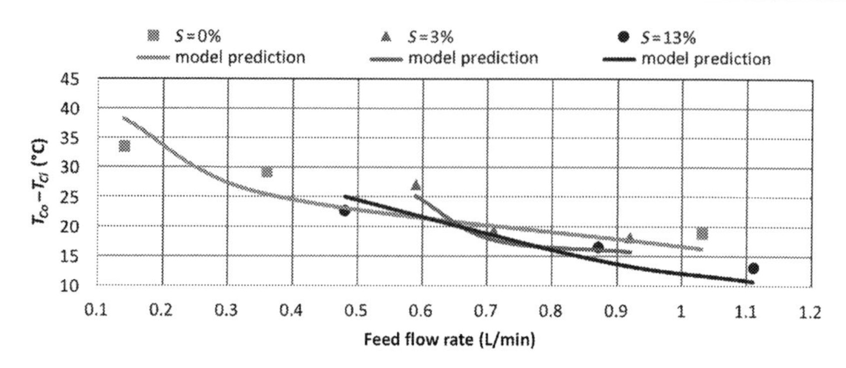

Fig. 5.14 Theoretical and experimental value comparison for influence of feed flow rate and salinity on internal heat recovery ($T_{Co} - T_{Ci}$).

The theoretical study provides a basis for developing a more efficient PGMD module by estimating the effects of system parameters including module length, feed flow rate, and temperature on the main output parameters including permeate flux rate and STEC. As a first step, the developed theoretical model is used for a numerical study of the PGMD pilot plant under construction, which is a new module design with a 0.6 m^2 membrane area to provide higher thermal efficiency and a lower STEC rate.

5.3.3 Commercial scale PGMD module rig case study

A commercial scale PGMD module configuration was developed by the Fraunhofer Institute for Solar Energy System and SolarSpring GmbH (Germany). Since 2001, Fraunhofer ISE has been developing independent desalination systems based on solar driven MD technology with different membrane areas and has installed some of the developed systems in different locations [22]. SolarSpring GmbH, started as a spin-off from Fraunhofer ISE, was founded in 2009 and is pursuing research activities in the field of sustainable autonomous desalination systems for remote areas [25]. The company developed the PGMD technology using the spiral-wound configuration and for different sizes. Fig. 5.15A shows the channel arrangement for their developed spiral-wound PGMD module configuration.

This module technology uses internal heat recovery over the length of the system, as cold salty feedwater firstly flows through the condenser channel at a temperature of nearly 20°C and by gaining latent and sensible heat of condensation reaches a temperature of 73°C. Then by applying a renewable source of energy such as thermal solar collectors as an external heat source, the temperature increases to 80°C at the entrance to the evaporator channel and the salty hot feed flows in a countercurrent flow direction with respect to the condenser channel and leaves the hot channel at a temperature of 27°C. In this configuration, an almost constant temperature difference (approximately 7°C) is established through the module channels. Fig. 5.15B shows this PGMD module arrangement in a compact spiral-wound configuration, to provide more fluid

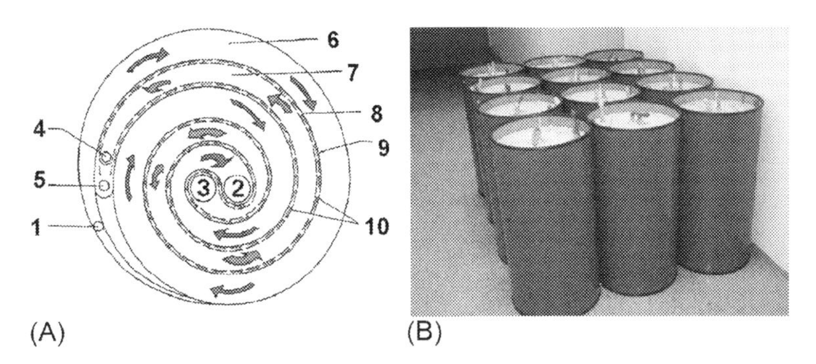

Fig. 5.15 (A) Schematic representation of the spiral-wound module concept:(1) condenser inlet, (2) condenser outlet, (3) evaporator inlet, (4) evaporator outlet, (5) distillate outlet, (6) condenser channel, (7) evaporator channel, (8) condenser foil, (9) distillate channel, and (10) hydrophobic membrane, (B) ready-to-use spiral-wound desalination modules with a 14 m^2 membrane area [22].

residence time in each channel and better internal heat recovery through the system. The designed spiral-wound PGMD modules have different membrane areas of 5, 10, and 14 m^2 with 3.5–7 and 10 m of channel lengths, respectively.

Winter et al. [22] performed a set of experiments with a system having a 10 m^2 total membrane area, 25°C and 80°C, respectively, at the condenser and evaporator channel inlets, and feed flow rates varied from 200 to 500 kg/h with different feed salinities.

For the freshwater source ($S = 0$ g/kg), increasing the feed flow rate from 200 to 500 kg/h led to the distillate rate significantly increasing from 10 to 25 kg/h, and the specific energy consumption rising from 130 to 207 kWh/m^3. However, for the higher values of feedwater salinity in a range from 0 to 105 g/kg (ppt), the distillate rate decreased with salt concentration by approximately 1 kg/h per 10 ppt of feedwater salinity. These results enable finding the optimum feed flow rates at each given salinity level to provide minimum specific energy demand. Increasing the condenser inlet temperature also causes a reduction in the temperature difference across the membrane surface, which leads to a lower distillate rate. An increment of 10°C in the condenser inlet temperature led to the distillate flux decreasing to approximately 2.7 kg/h for the entire salinity range. In this study, the effect of the condenser inlet temperature within a range of (20–40)°C at a constant evaporator inlet temperature of 80°C and a constant feed flow rate of 500 kg/h was also investigated. The condenser inlet temperature depends on the available cooling power and recirculation strategy.

It is also observed that the effect of increasing membrane surface area on reduction of specific energy consumption of the system is more dominant in comparison to increasing freshwater flux as it provides more contact time between feed and membrane and better heat transfer performance in the system. In addition, increasing feed salinity decreased the effect of membrane area and it is reported that in all the experiments the conductivity of the produced freshwater was low (3.5 μS/cm).

5.4 MD systems integrated with solar energy

As described earlier, for the MD process, it becomes possible to utilize directly low-temperature waste heat or solar energy as the process driver. As a result, in recent years some projects investigated the feasibility of coupling MD systems of different configurations to renewable sources. Next is a review of accomplished research activities, which evaluate the solar MD desalination process.

Cipollina et al. [23], developed a laboratory-scale plate-and-frame MD module for solar energy seawater desalination by applying a PTFE membrane with 0.2-μm pore size on a PP support and with the effective membrane surface area of $0.042\,m^2$. An impermeable PP film with a thickness of 770 μm was used to create an additional channel. Three different channel configurations were investigated during this research, including free air gap, permeate gap, and partial vacuum air gap.

Fig. 5.16 shows the geometrical layout of the designed module with three channels and the applied net spacer not only provides mechanical support but also is a turbulence promoter. Adding baffle layers may also help to increase the fluid velocity and increase heat and mass transfer coefficients in the flow channels. In this study, a conceptual design of a solar powered MD system using a thermal solar collector has been performed to characterize all the design parameters and the operating condition. In the proposed integrated MD system coupled with a solar system the use of electrically operated thermostatic valves and circulation pumps can make the system adapt to different configurations based on the environmental conditions and the solar radiation power (Fig. 5.17A and B).

As seen in Fig. 5.17A, which represents a cold hour of a sunny day, solar energy is directly transferred to the MD module to provide the thermal energy required by the system. However, in configuration B, which depicted a warm hour of a sunny day, the amount of solar radiation is higher. Therefore, in this configuration by applying a

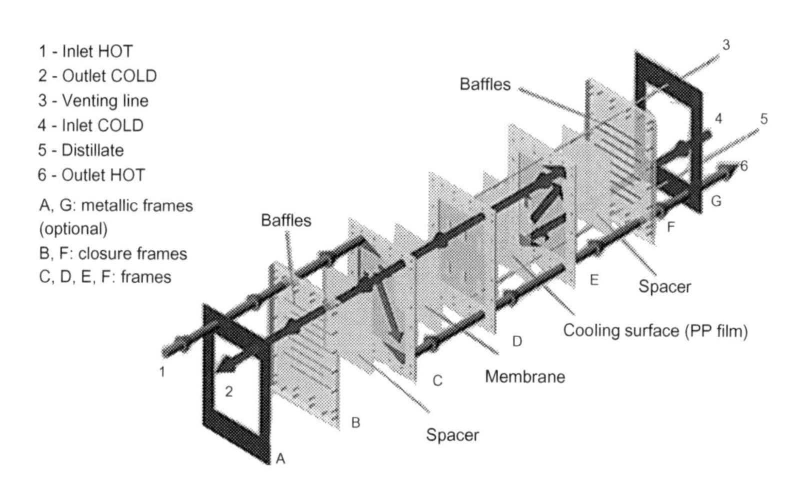

Fig. 5.16 Sketch of the laboratory-scale MD unit with planar geometry [23].

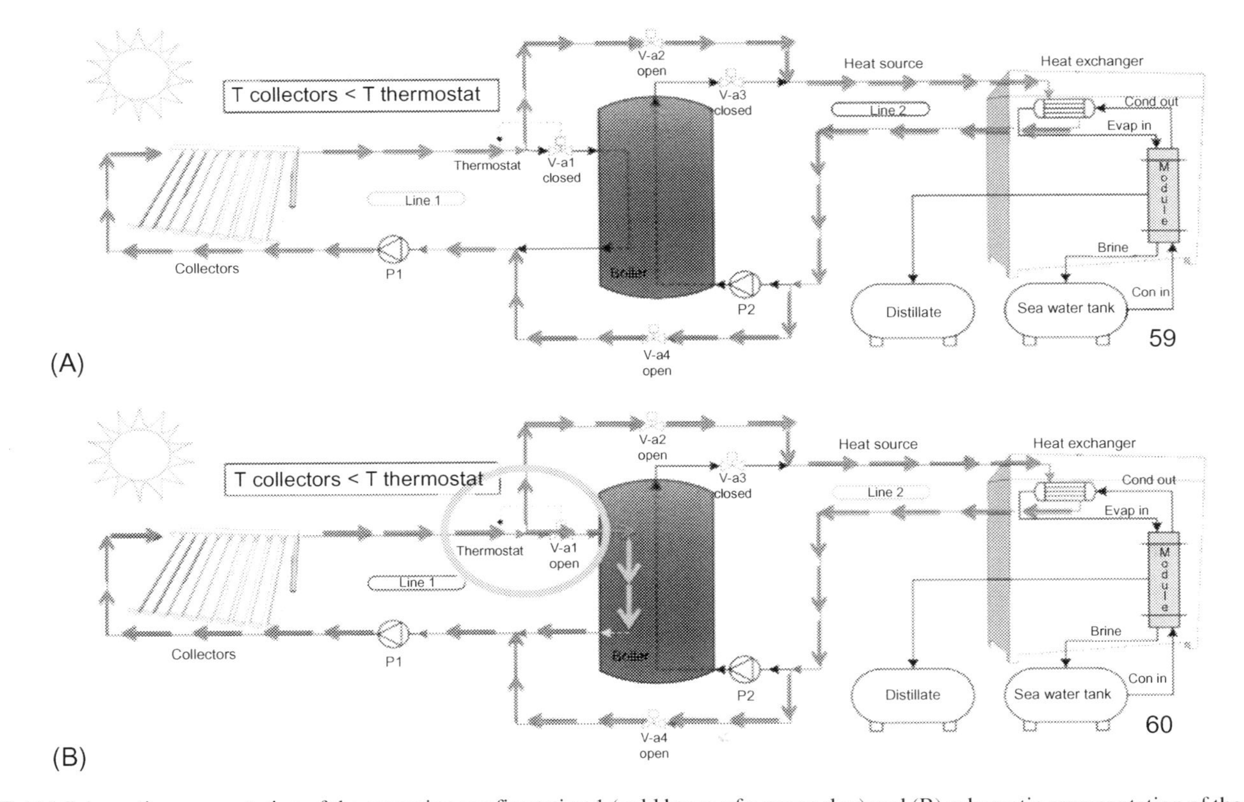

Fig. 5.17 (A) Schematic representation of the operating configuration 1 (cold hours of a sunny day) and (B) schematic representation of the operating configuration 2 (warm hours of a sunny day) [23].

thermostatic splitting valve, some hot fluid from the thermal solar collector may be stored in a heat storage tank and the remainder transferred to the MD module.

A simplified theoretical model of the MD system for AGMD and PGMD was also developed and validated by comparison with experimental data. The numerical model predictions were used for the design of a scaled-up multistage unit for integration in a solar powered MD system. It is generally determined that the PGMD configuration showed better performance in terms of distillate flux and STEC than with free AGMD. As mentioned previously, in this study, different values of feed flow rate and feed inlet temperature, respectively, in ranges of (0.4–1.2) L/min and (50–80)°C were investigated. The corresponding results show that the distillate flux varied in the range (2–12 L/m^2 h) with higher flux produced at a higher feed flow rate and temperature and the STEC varied in a range of approximately (5000–20,000) kWh/m^3.

An experimental investigation of coupling a solar pond with MD was accomplished by Nakoa et al. [26]. A plate-and-frame DCMD module using a PTFE membrane with a 0.22-μm pore size and 0.1074-m^2 surface area was used in saltwater desalination coupled with a salinity gradient solar pond (SGSP) to supply the thermal energy required for the process.

An SGSP is an ideal sustainable solar source to integrate with MD desalination. It comprises three layers, an upper convective zone (UCZ), a nonconvective zone (NCZ), and a lower convective zone (LCZ). As depicted in Fig. 5.18, the top layer (UCZ) consists of cold and low salinity water. The NCZ contains a salt gradient and acts as an insulation layer for the LCZ, which has the highest temperature and highest salt concentration. In this experiment, the integrated SGSP, which was used to provide heat to a DCMD module, had a 50 m^2 area and was 2.05 m deep and was located at RMIT University, Australia. The performance of the coupled DCMD/SGSP

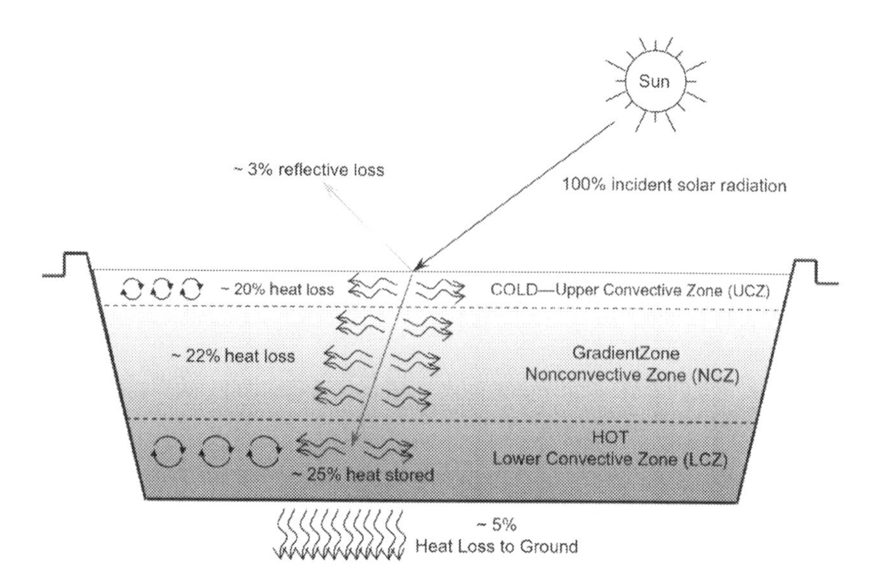

Fig. 5.18 Salinity gradient solar pond with different layers [26].

system is estimated, particularly in terms of produced freshwater flux, energy consumption of the system, and amount of heat extracted from the SGSP.

The feed stream of 1.3% salinity was firstly heated by the SGSP and circulated through the DCMD module and then injected into an evaporation pond. In addition, a thermal energy system was used to recover heat from the outlet brine stream of the DCMD for use in preheating the inlet feedwater stream. The feedwater with a salinity of approximately 13 g/L was provided from the evaporation pond which was located next to the solar pond. Referring to Fig. 5.19, the feed passed through the in-pond heat exchanger and extracted heat from the NCZ and the LCZ before entering the DCMD module, with a temperature close to the LCZ temperature. The test facility included two auxiliary heat exchangers, the first used to preheat the saline feedwater from the evaporation pond, a stainless steel heat exchanger used to exchange heat between the produced freshwater which was warm and the cold inlet feedwater. The second heat exchanger was also used to preheat the feedwater before entering the in-pond heat exchanger pipe which was fixed to the wall of the pond. This method of heat extraction had an efficiency of 30% for the solar pond described, in recovering the heat from the MD module's hot feedwater outlet.

The experimental results show that feed temperature to the MD module has a significant effect on freshwater mass flux, especially at a higher feed temperature. The mass flux changed in a range of 2–6 L/m^2 h and the thermal energy consumption of the system varied in the range from 13,000 to 6000 kJ/kg. The required thermal energy could be provided by the solar pond and the heat exchangers. The auxiliary heat exchangers can supply approximately 50% of the total thermal energy supplied from the solar pond, which is a function of the inlet feed temperature. It is estimated that, by assuming 30% efficiency and an average daily radiation of 310 W/m^2, the average rate of thermal energy which could be delivered by this 50 m^2 solar pond would be approximately 4.7 kW, where this estimate applies to the end of the summer season.

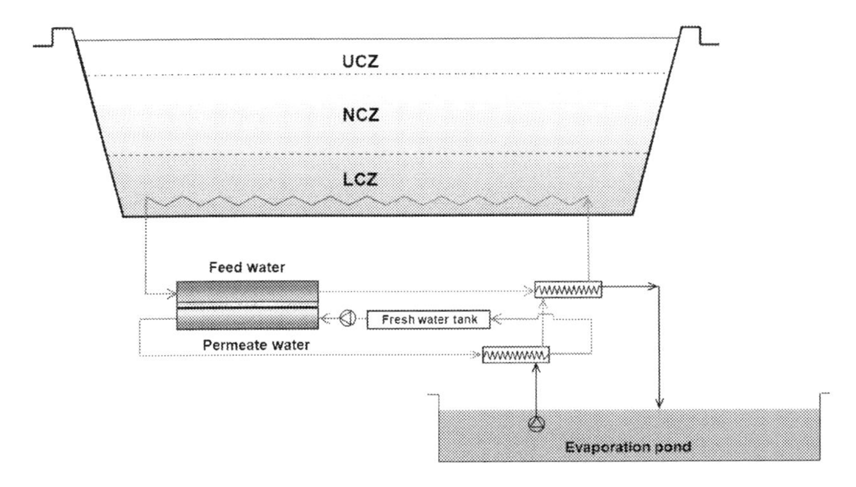

Fig. 5.19 The schematic diagram of a laboratory-scale coupled DCMD/SGSP setup [26].

Nakoa et al. followed up their experimental research on an integrated DCMD system with a solar pond by developing a sustainable zero liquid discharge desalination system. The hot saline water (nearly 50°C) with a high salt concentration (16% salinity) was extracted from the NCZ of the SGSP and the brine disposal of the MD module, which had higher salinity than the feed inlet, and was discharged to the LCZ. With this configuration, it is possible to produce salt from the remaining brine leading to zero liquid discharge desalination. The initial trials, which extracted saline hot water as a feed from the NCZ and discharged the brine module outlet to the corresponding NCZ level, caused significant stratification of the NCZ layer and disturbed the density profile of the NCZ, as shown in Fig. 5.20. By extracting brine from the NCZ and discharging to the LCZ, this issue was overcome. The top layer of the solar pond was also used as a cooling system for the process by locating plastic pipes near the pond surface to cool the process fluid as depicted in Fig. 5.20. The developed system could produce 52 L/day of freshwater for each square meter of membrane area coupled with the SGSP, consuming almost 11 kW/m^2 of renewable thermal energy.

The Fraunhofer Institute for Solar Energy Systems (ISE) has been working on the development of solar energy desalination systems since 2001. In that research, Schwantes et al. [28] studied three different desalination plants in different locations. They applied different sources of salty water and used various process driver sources including a thermal solar collector and waste heat supplying a parallel multi MD-module system with different surface areas and module lengths as a sustainable MD desalination demonstration plant. Each system's performance was investigated in terms of energy efficiency and the plant distillate output rate. No chemical pretreatment of the feedwater unit was required for the installed solar desalination plants; hence, resistance to scaling and fouling is considered to be one of the most beneficial characteristics of MD.

One of the demonstration plants was in Namibia, Africa, which used ground water with salinity in the range of nearly (1.5%–3%) as the plant feed source. PV panels and solar thermal collectors supplied the thermal and electrical energy for the system. This system, as shown in Fig. 5.21, was designed to produce freshwater with a capacity of up to 4 m^3/day for 24 h operation. The 100 flat plate collectors with a total aperture area of 232 m^2 were used to supply the thermal energy demand of the system. During the daytime operation of the system, excess heat energy was stored in a tank and,

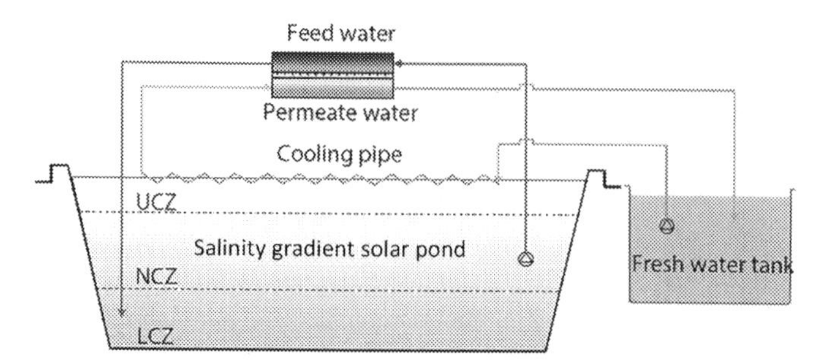

Fig. 5.20 The schematic diagram of a zero liquid discharge desalination setup by integrated MD to SGSP [27].

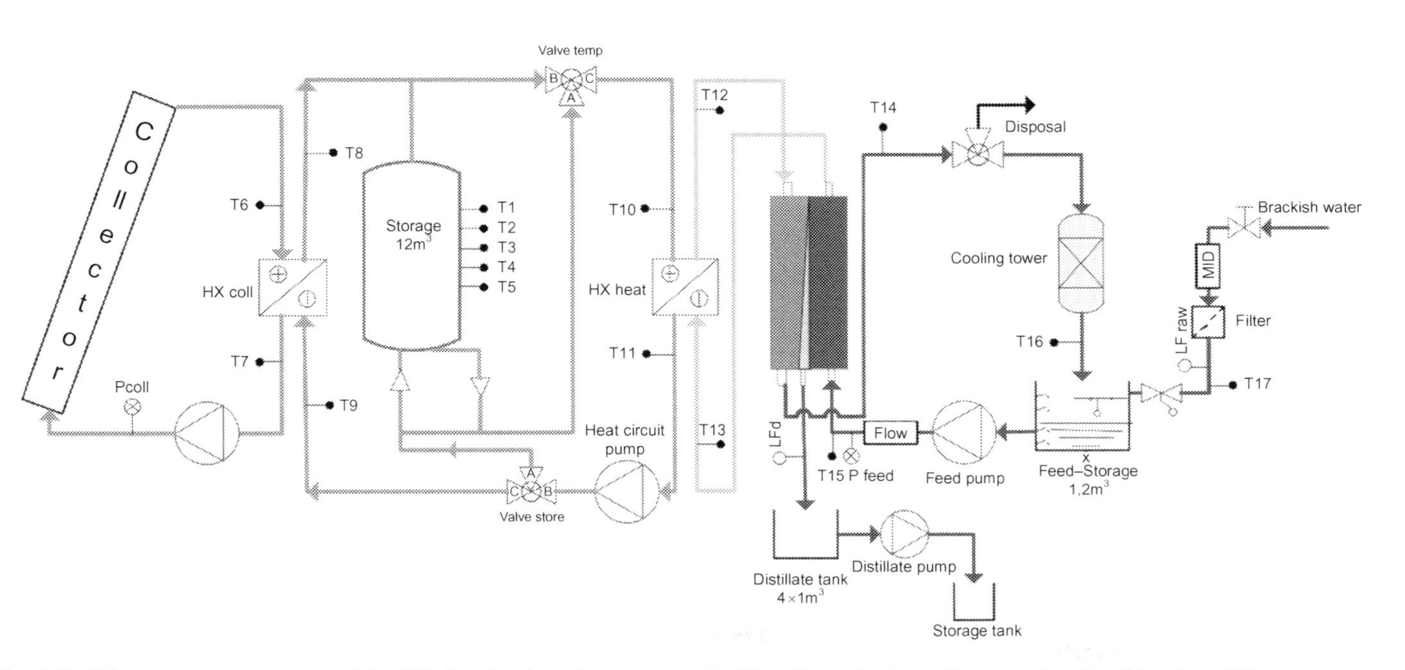

Fig. 5.21 Schematic representation of the MD desalination plant integrated with a thermal solar collector at the Namibia plant [28].

during the evening and nighttime, the excess energy in the storage tank was extracted to sustain the operation.

The average operating temperatures in the evaporator channel and condenser channel were respectively approximately 70°C and 65°C, corresponding to a small temperature difference (nearly 5°C) between the cold and hot channels. This small temperature difference resulted from applying a longer MD module, which led to more heat transfer surface and better internal heat recovery. Therefore, plants with a longer MD channel length have better performance in terms of specific energy consumption.

Comparison of the operation of three desalination plants showed that some operational faults, including defective thermal collectors and high heat loss from them, led to non-efficient operation of the desalination plant for which at full capacity, as shown in Fig. 5.22, only 29% of the heat supplied by radiation reached the heat circuit of the MD plant. Heat losses happened through each of the heat exchangers transferring heat from the solar collectors to the MD system. It is reported that 26 kW of losses occurred in this plant from piping and heat exchangers, in addition to the heat loss through the storage tank. The heat losses reduced the overall plant energy efficiency.

The other plant investigated in this study was located in the Pantelleria island site [28] in which waste heat from a diesel generator (combustion engine), as a constant source of energy, was used to supply the process. It was confirmed that the waste heat from the combustion engine could provide a suitable temperature level for MD, so by having a constant waste heat source, the cost of the solar collector array would be eliminated. It was also observed that, by providing better insulation and an optimum size of heat storage tank, it would be possible to improve the system's energy efficiency. In addition, the long-term performance of the solar powered MD units in all locations under study showed that the quality of freshwater produced was sufficient enough to be used directly for drinking, irrigation, and industrial purposes without any posttreatment.

In another similar project, a solar MD demonstration plant in the PozoIzquierdo-Gran Canaria Island was studied by Raluy et al. [29] with a PGMD module of (8.5–10) m^2 effective area from the Franhaufer Institute with (7–20) L/h distillate flux capacity coupled with a 6.96 m^2 solar thermal collector field. The feedwater first entered the condenser channel and gained heat by internal heat recovery; the preheated saltwater then entered the solar collector to reach the required temperature of the evaporator channel. An almost constant temperature difference was established across the membrane.

The system developed as shown in Fig. 5.23 was completely self-running since the electrical energy required for pumping and electrical systems was provided by PV panel solar collectors, so that the operation was only possible during daytime. The higher mean temperature difference between condenser and evaporator channels caused higher flux so that, if more heat becomes available from the solar collector, a higher feed flow rate could be transferred to the module. Higher feed salinities caused a lower vapor pressure of water and lower driving force for MD, as a result of the lower distillate flux. The STEC of the unit during a 5-year investigation was (140–350) kWh/m^3 with a distillate conductivity of (20–200) μs/cm. The PGMD system

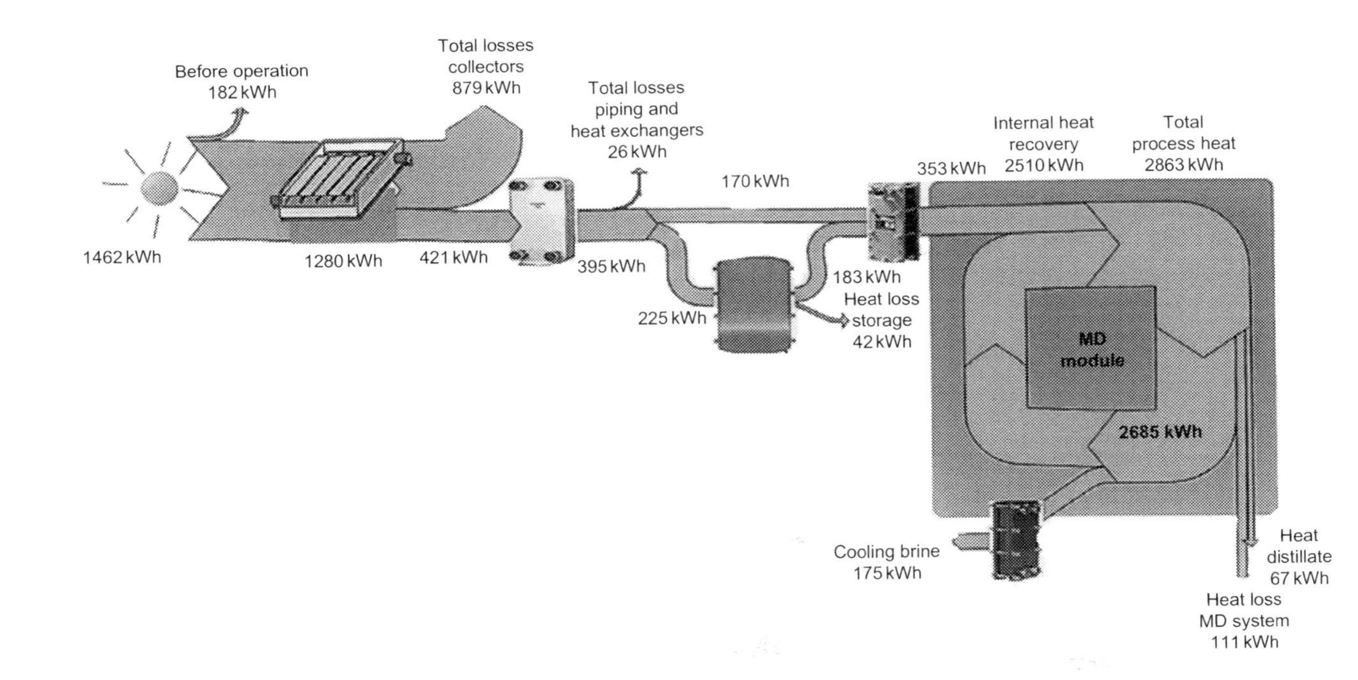

Fig. 5.22 Schematic representation of heat flows and losses on a sunny day in the Namibia plant [28].

Fig. 5.23 Fraunhofer solar MD compact system installed in Pozo Izquierdo [29].

described operated at a low temperature (80°C) which facilitated coupling with solar sources. It was also mentioned that future improvements could enable lower STEC, reduced specific investment, and lower final water cost.

In their research, Shim et al. [30] did an experimental study around a small-scale solar energy assisted DCMD module with a surface area of $0.06\,m^2$ for seawater desalination. In this configuration, a solar heating system with an effective heat collection area of $4.7\,m^2$, which ran continuously for >150 days, was applied. In this configuration, >77.3% of the required heating energy was supplied by solar energy during daytime. A schematic representation of the MD system coupled to thermal collectors for home applications is shown in Fig. 5.24.

The STEC of the system was estimated at between (890 and 1450) kWh/m^3 with a GOR of (0.44–0.7). The results showed that the proposed system was more energy intensive than the traditional desalination system such as RO; however, since it could directly use heat from solar collectors and could operate at a lower temperature compared to MSF and MED, it would be a more sustainable and economical choice. In the developed system, it would be possible to produce $40.9\,L/m^2\,h$ under the operating condition of feed temperature of 60°C, permeate temperature of 20°C, and feed flow rate of 4.5 L/min.

Shim et al. [30] also developed a two-dimensional mathematical model to simulate unsteady conditions during the long-time operation of a module. For instance, to investigate the flux change over a long-time operation, a time factor in the heat transfer equation and flux prediction was applied to simulate the operating condition change related to fouling phenomena. The model showed that the experimental values and modeling data were in good agreement at lower temperatures but under high feed temperature condition (70°C), the difference was higher. The difference was attributed to higher heat loss at a higher temperature, which was ignored in the modeling Also, the discrepancy could be related to a concentration polarization effect, which is more significant at higher temperature because of a higher flux rate.

The optimum condition in terms of lowest STEC and highest GOR was reported for feed temperature 60°C, permeate temperature 20°C, and feed flow rate of 4.5 L/min.

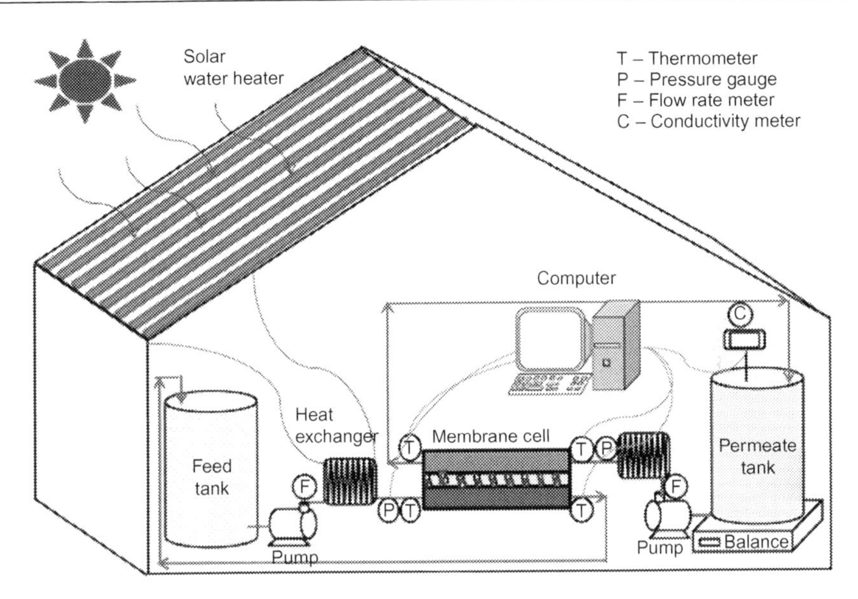

Fig. 5.24 A schematic diagram of the DCMD module integrated with a solar heating system for domestic application [30].

The experimental results show that after 150 days of operation, the permeate flux decreased from 28.48 to 26.50 L/m^2 h. The feed temperature and flow rate are significant factors affecting the flux reduction in that, at the lower feed temperature and feed flow rate, more particles remain at the membrane surface and the fouling effect increases.

Another research study which investigated the MD process coupled with waste heat recovery was performed by Dow et al. [21]. A pilot trial of an MD plant, using a multilayered flat-sheet DCMD system with a 0.67 m^2 membrane area, driven by low-grade waste heat (<40°C) from a power station, was developed in this project to treat the power station's wastewater. The main source of waste heat for MD from the power station was the main steam condenser. The permeate flux over a 3-month trial operation was estimated at approximately 3 kg/m^2 h and varied with waste heat temperature. The module reported water recovery up to 92.8% and the distillate produced had high quality in terms of total dissolved solid.

The STEC for the trial period was reported at approximately 1500 kWh/m^3, which increased by the end of the trial, because of the fouling layer at the membrane surface, which reduced the overall mass transfer coefficient (C_m) and the permeate flux. It was also confirmed that, by applying an MD plant with internal heat recovery characteristics, it would be possible to reduce the value of the STEC of the system, which is a more important consideration when thermal energy is supplied from an expensive heat source. However, in this study, abundant low-grade waste heat was available and the DCMD system showed better performance in terms of the produced distillate quantity and the required membrane surface area because the applied MD setup was without internal heat recovery. Therefore, in this trial situation, achieving a high distillate flux and smaller plant size had higher priority than high STEC (or low GOR) because of the availability of excess low-cost waste heat.

In recent years, some commercial MD modules in different configurations have been developed. Zaragoza et al. studied the efficiency of the use of solar thermal energy in small membrane desalination systems for decentralized water production [19,31]. A comparison, especially in terms of energy analysis during several months of operation, was made between the most advanced commercial MD module prototype technologies in different configurations (air gap, permeate gap, and vacuum) and different structures, including (plate-and-frame and spiral wound) integrated with solar energy. Each module was investigated in terms of STEC by considering different operating conditions including feed temperature and feed salinities. The solar thermal energy installation studied compromised parabolic solar collectors and flat-plate collectors. The solar collectors were connected to a thermal storage system to provide a steady state heat supply. The investigated commercial MD modules, as shown in Fig. 5.25 are listed:

- Flat-sheet AGMD applying a PTFE membrane with a 2.8 m^2 membrane area from a Swedish company (SC module)
- Two different MD modules from a Singaporean company (KeppleSeghers) licensed from Memstill including, firstly, a single flat-sheet PGMD with a total surface area of 9 m^2 (M33 module) and, secondly, three MD modules connected in series, each of them with a 3 m^2 area for each membrane to increase the latent heat recovery of condensation (PT5 module)
- Spiral-wound PGMD from a German company (Solar Spring) with a 10 m^2 PTFE membrane area. There was an additional channel for the permeate. The system was designed to maximize internal heat recovery (Oryx 150 module).
- A flat-sheet vacuum multieffect MD (V-MEMD) system with multiple stages and a 5.76 m^2 total surface area of the PTFE membrane was built by Aquaver (licensed by the Memsys Company). The stage low pressure allowed the feedwater to boil at a reduced temperature. However, the specific energy consumption to create and maintain vacuum in the module should be considered in the assessment (WTS-40A module).

A comparison of performance of these commercial modules in terms of produced distillate quality and quantity and energy efficiency was performed. Table 5.3 shows the obtained value for maximum distillate flux (F_d) and distillate quality (σ_d) at two different feed conductivities (σ_F) which are approximately equivalent to 0.1% and 3.5% salinities under the test operating condition. In this table, ΔT is the temperature difference between hot and cold inputs and f_f is the feed flow rate in each case. With respect to F_d, as is shown in Table 5.2, it was of the same order of magnitude for all MD technologies which were investigated in this research including AGMD, PGMD, and V-MEMD but varied with operational conditions like feed flow rate and feed temperature. The increasing feed flow rate showed a detrimental effect on SEEC, because of increased heat loss at a higher flow rate and higher turbulence; however, a higher distillate flux would also be obtained. It is also observed that the PGMD module in the spiral-wound structure had a higher latent heat of condensation recovery compared with all the single-effect configurations, leading to the minimum value of 210 kWh/m^3 for the STEC [31].

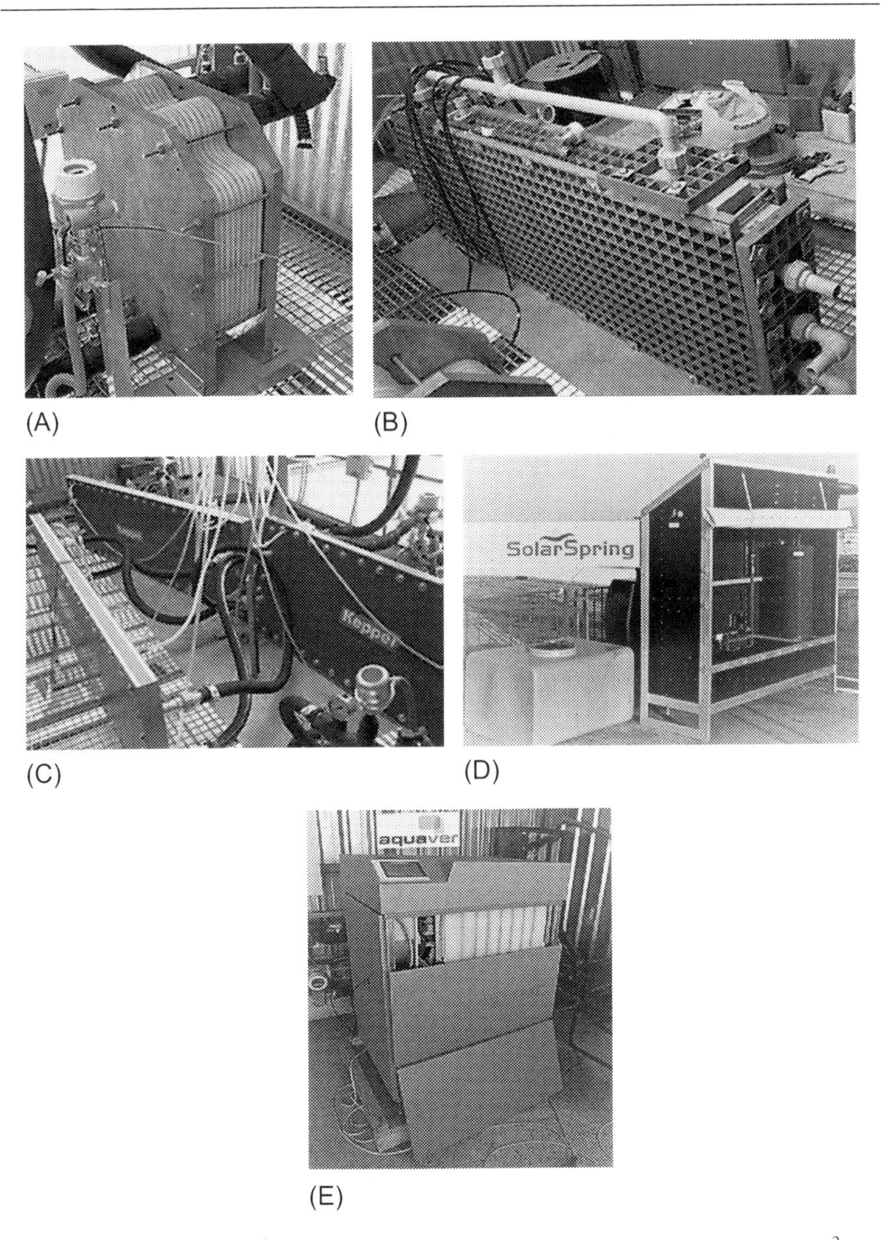

Fig. 5.25 (A) Scarab 2.8 m^2 flat-sheet AGMD module; (B) Keppel Seghers' single 9-m^2 flat-sheet PGMD module; (C) Keppel Seghers' PT5 model; (D) Aquaver's WTS-40A V-MEMD prototype; and (E) Solar Spring's Oryx 150 spiral-wound LGMD module [31].

Table 5.3 Distillate quality and productivity for each module at two different feed conductivities (σ_f) [31]

Module	σ_f (mS/cm)	σ_d (μS/cm)	Max. F_d (L/m^2 h)	Conditions
SC	2	<10	6.5	$\Delta T = 65°C$ $f_f = 1200$ L/h
	48	40–60	5.5	$\Delta T = 65°C$ $f_f = 1200$ L/h
M33	2	2–5	3.5	$\Delta T = 50°C$ $f_f = 1560$ L/h
	48	2–5	3.1	$\Delta T = 50°C$ $f_f = 1560$ L/h
PT5	48	>5	5.0	$\Delta T = 50°C$ $f_f = 1020$ L/h
Oryx 150	2	2–5	3.4	$\Delta T = 55°C$ $f_f = 600$ L/h
	46	2–5	3.2	$\Delta T = 55°C$ $f_f = 600$ L/h
WTS- 40A	2	4–6	7.1	$\Delta T = 85°C$ $f_f = 70$ L/h
	46	4–6	7	$\Delta T = 85°C$ $f_f = 70$ L/h

The results of this project confirmed that the maximum distillate flux is obtained at the maximum feed flow rate because of higher turbulence at the higher feed flow rate and the lower temperature polarization effect. In addition, it is observed that the obtained distillate quantity was less than half of the scaled lab values, with this difference attributed to the macroscopic effects on a large scale, which changed the optimum temperature and flow rate distribution profile along the larger membrane surface area compared to the lab situation.

Table 5.4 provides a summary of literature surveys accomplished in the available publications, and focuses on recent advances, and improvements in MD technologies and commercial MD modules operation. Referring to documented research projects provides a good understanding of current MD technology status and the latest developments and optimization of MD units' performance.

5.5 Techno-economic feasibility of the PGMD system

Membrane distillation as an emerging desalination technology needs analysis in terms of technological and economical features in order to be compared with the current conventional desalination systems. Few studies are reported in terms of energy consumption (EC) and water production cost (WPC) of MD setups and a wide variation in energy consumption is noted.

Khayet [7] presented a review of energy consumption analysis and WPCs of solar desalination by MD. The reported values show the EC changes between 1 and 9000 kWh/m^3 and the WPC varies from \$0.3/m^3 to \$130/m^3. This wide dispersion in the energy consumption rate is attributed to the different factors, including: the MD process did not completely study in the full-scale application, so the investment cost in terms of membrane module varies widely; besides, there is not a good estimation of other cost related items including pretreatment processes, optimum flow conditions, long-term MD performance, fouling, and membrane life time. At present, the

Table 5.4 **Novel research in desalination by MD technology**

Paper title	Main objective or observation
Desalination using membrane distillation: Flux enhancement by feedwater deaeration on spiral-wound modules [24]	Investigation of flux enhancement by removal of air (deaeration) from membrane pores of a spiral-wound MD module The air in the membrane pore is one of the main resistances to molecular diffusion and one effective method to remove air from the membrane pore is by applying the deaerated feedwater For a wide range of feed salinity, deaerations caused a higher distillate flux and higher GOR By increasing the deaeration degree, the distillate output increased and the module thermal efficiency increased due to reduction of heat loss by conduction
Evaluation of MD process performance: Effect of backing structures and membrane properties under different operating conditions [32]	The membrane process performance in a DCMD setup was investigated with the existence of backing with different designs in membrane structure and without it over a wide range of operating conditions The backing application has an impact on an effective cross-sectional area for mass transfer and the formation of a complex network of thermal resistances, leading to a significant reduction in process performance Presence of backing material reduces the distillate flux and thermal efficiency Optimized backing design with small surface coverage and high porosity could improve the process performance
Evaluation of commercial PTFE membranes in desalination by direct contact membrane distillation [33]	Investigation of nine commercially PTFE membranes performance, with different specification including pore size, support layer, thickness, and porosity, in a 0.0169 m^2 area of a plate-and-frame DCMD module under different operating conditions It is concluded that the membrane pore size and pore size distribution have a significant effect on the overall performance of the MD module and generally a pore size in a range of (0.2–0.5) μm of the PTFE membrane is the best choice for MD application It is also concluded that the PTFE membrane with fiber HDPE (high-density

Continued

Table 5.4 **Continued**

Paper title	Main objective or observation
	polyethylene) as a support and 0.22 μm pore size showed the best results in terms of distillate flux
Commercial PTFE membranes for membrane distillation application: effect of microstructure and support material [34]	Six types of different commercial PTFE membranes (with support and without it) from three different suppliers were used in a laboratory-scale DCMD setup to investigate the effect of membrane characteristics on the module performance It is mentioned that support materials not only block some active pores, but also reduce the porosity, which cause a reduction in flux Increasing membrane thickness will increase the thermal conduction resistance, thus leading to much difference between T_{mf} and T_{mp} and higher flux in DCMD In this study, the spacer effect on the heat transfer coefficient is also studied, so based on the applied spacer porosity, a spacer factor is considered in the Reynold number correlation
Characterization and performance evaluation of commercially available hydrophobic membranes for direct contact membrane distillation [35]	Over 20 semicommercial hydrophobic membranes including PVDF, PTFE, and PP in a laboratory-scale DCMD module were compared, and the proper characterization methods for measuring porosity, contact angle, and pore size distribution were introduced. An overview of the optimal membrane properties also recommended, including the contact angle more than 90°, a minimum of 2.5 bar for LEP and membrane porosity, pore diameter and thickness respectively in a range of 80–90%, 0.1–1 μm, and 70–280 μm
Energy efficiency of the permeate gap and novel conductive gap membrane distillation [36]	A novel configuration for MD, conductive gap membrane distillation (CGMD) which could be considered as a modification of PGMD by considering a high conductive material (metal mesh spacer) instead of plastic ones The numerical study compares the PGMD performance with CGMD by applying a conductive spacer, including aluminum

Table 5.4 **Continued**

Paper title	Main objective or observation
	mesh spacers, copper plate with fins, etc. in the gap between the membrane and the condensing surface There is a more efficient internal heat recovery system in CGMD and the system has shown a two times higher GOR than PGMD The freshwater flux is higher in CGMD than in PGMD, since there is a larger temperature difference across the membrane
Effect of module inclination angle on air gap membrane distillation [37]	The effects of the flat plate AGMD module's inclination angle on permeate production and thermal efficiency were investigated for different cold and hot fluid temperatures and in different module angles In an extremely inclined module position, the condensate may fall on the membrane surface (flooding), which reduced the mass transfer resistance and increased the permeate production rate which caused the AGMD system to behave as a PGMD system, with desirable internal heat recovery This tilt angle variation has a hydrostatic effect, which affects the droplet flow and film thickness on the condenser surface of AGMD
Effect of operating parameters and membrane characteristics on air gap membrane distillation performance for the treatment of highly saline water [18]	The performance of 10 commercially hydrophobic membranes in an AGMD configuration with a 0.005 m^2 effective membrane area in terms of the effect of each membrane characteristic including thickness, support configuration, and pore size on permeate flux was investigated A stainless steel perforated plate, having a thickness of 0.5 mm and circular holes diameter of 6.5 mm, was used to support the membrane from possible damage and also act as a spacer to create turbulence near the membrane surface Higher feed temperature and flow rate lead to higher permeate flux, however the effect of feed flow rate on permeate flux is lower at lower feed temperature

Continued

Table 5.4 Continued

Paper title	Main objective or observation
	It is reported that the existence of support on the membrane matrix, which has a slight effect on permeate flux, just leads to a slight reduction of about 2.35% on permeate flux
Experimental and theoretical investigation on water desalination using air gap membrane distillation [38]	An experimental research on a laboratory-scale designed AGMD module with 2.29×10^{-3} m^2 membrane areas by applying two commercial PTFE membranes with 0.22 and 0.45 μm pore size is investigated Results showed that the system performance is highly affected by changes in both feed temperature and air gap width The permeate flux increased with the feed temperature and the feed flow rate. However, it decreases with increasing air gap width and the coolant temperature
Novel method for the design and assessment of direct contact membrane distillation modules [39]	The equivalent effectiveness-number of transfer units (e-NTU) method was applied for a flat-sheet DCMD module by considering simultaneous heat and mass transfer phenomena, which were validated against a MATLAB model and confirmed by experimental data This method is suitable for any primary calculation, to have an estimation regarding the required module area based on the required distillate water and output parameters

most expensive parts of MD systems are the MD membranes and modules. To have a more accurate estimation of these two parameters (WPC and EC), more efficient MD plants should be designed and developed on a pilot plant scale. It is predicted that over the next 10–20 years, MD systems will be developed further, because of many advances in terms of higher permeability membranes, more efficient modules with energy recovery options and in multistage configuration, application of renewable energy sources, and waste heat and brine disposal management technologies.

Camacho et al. provided a section in terms of economic aspects of MD desalination technology compared to other traditional desalination technologies [4]. They summarized the main parameters that affect the investment cost of desalination facilities in three main categories including capital, operation and maintenance (O & M), and other costs. In this category, the capacity of the plant, project location, quality of the source water and required water quality, co-location with the power generation plants, and environmental limitation are introduced as the main capital cost factors. In addition, energy consumption of the plant and cost, water quality, equipment replacement including MD module, filtration devices, and labor cost are considered to be the main factors which affect O & M costs.

Sirkar and Li [40] investigated the investment cost of a DCMD at the bench and pilot scales compared with the RO plant. The RO plant cost estimation was accomplished by assuming 1000 psi operating pressure and feed flow rate of $0.15\,m^3/s$ and a DCMD system, operating at 10 psi with a feed flow rate of $0.36\,m^3/s$ for a feed brine solution with 10% salinity and feed temperature in the range of (40–95)°C.

In this cost estimation study, the costs of some capital items including site development, water source, utilities, and some O & M costs such as membrane replacement, cartridge filters, and labor are considered identical for both RO and DCMD plants. Sirkar and Li concluded that by considering the availability of industrial waste heat, the evaluated cost of a DCMD setup would be lower than that of an RO plant. That is, the total capital cost for DCMD was estimated at \$0.85/Lday compared to \$1.19/L day for RO and the total production cost for DCMD is reported as \$0.96/m^3 in comparison with \$1.18/m^3 for RO. This difference results from lower operating pressure and a higher distillate production rate because of the non-fouling characteristics of hydrophobic membranes used in MD compared to RO systems. Fig. 5.26 depicts a breakdown cost comparison of the described DCMD and RO systems [40].

Winter also made an economic evaluation of MD desalination projects by presenting a new cost model and analyzing different MD configurations in small-scale plants [9]. Besides, the water unit cost in terms of two main categories and by considering the availability of waste heat or integrated solar sources to the MD setup is presented with details in this research [9]. In the cost model developed by Winter, the overall project costs were divided into two main categories: investment costs and operational costs. Investment costs generally included land cost, civil works, desalination unit, pre-, and posttreatment process, and brine disposal. The capacity and factorial method was the most common procedure for developing the fundamental investment cost and was introduced with more details in the model. The capacity model is a function of plant capacity in which by increasing the plant capacity the specific cost would decrease, whereby having a reference cost for a reference capacity, the price of any considered capacity could be estimated by means of a coefficient. However, in the factorial method, the overall cost of the plant is estimated from the costs of the main plant components multiplied by a factor.

For the traditional membrane-based desalination system such as RO developed on a commercial scale, the capacity model could provide an accurate cost estimation. However, regarding the investment cost of an MD plant, the membrane module cost depends on plant capacity and the degree of heat recovery in the system while for other system components including heat exchangers, deaeration systems, pre-treatment facilities, etc., costs are estimated mainly by only considering the plant capacity.

The operational costs, which define all the costs associated with plant operation, include the fixed operational costs which are independent of the plant operational

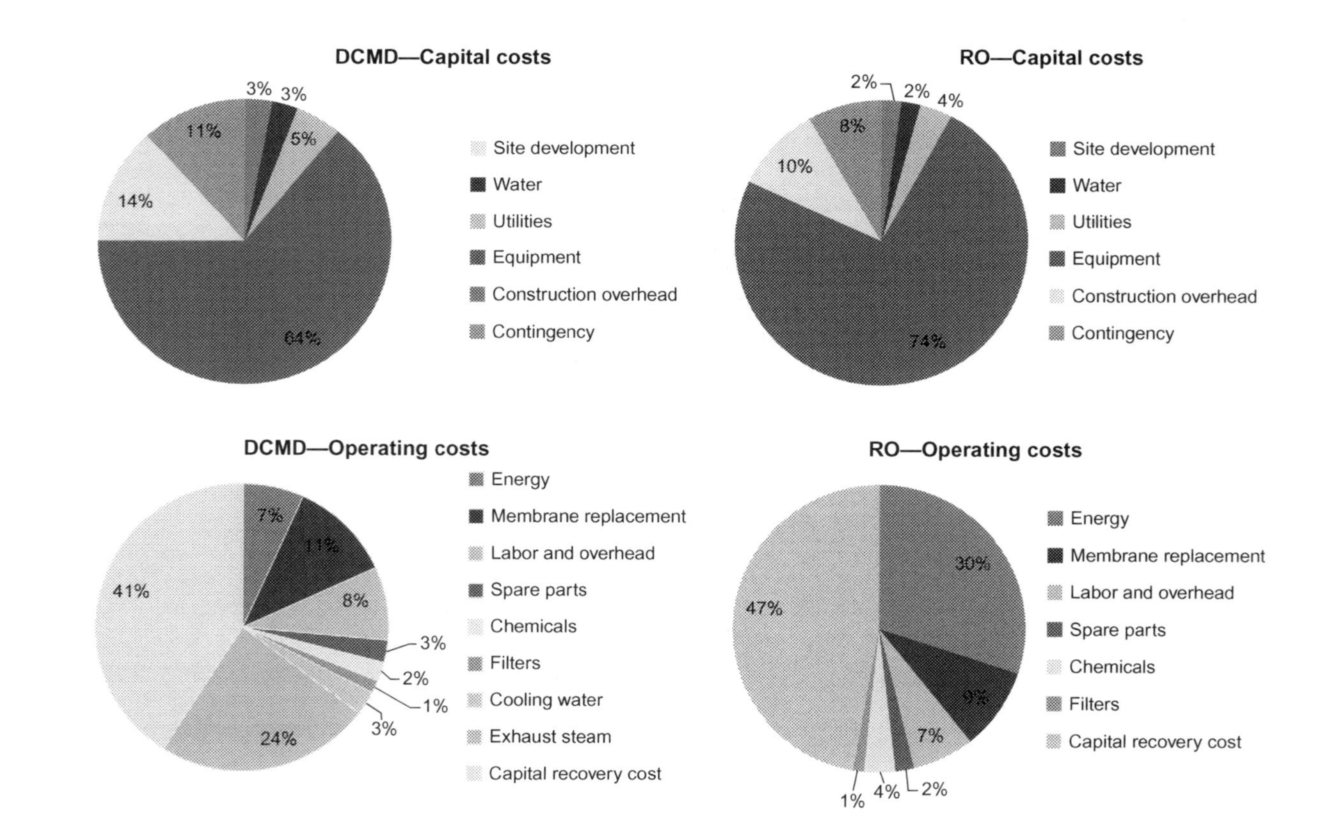

Fig. 5.26 Comparison of capital and operating cost for DCMD and RO [4].

status and variable operational costs which are a function of plant operation conditions. Overall, the fixed operational costs could be included in the capital costs which include the annual costs for providing the investment, service and maintenance, module replacement, personnel and insurance, which are independent of permeate production volume. The variable operational costs depend on the permeate quantity and include thermal energy, electricity, chemical additives and disposal costs, which are related to the operational availably of the plant.

Winter [9] also presented a detailed cost analysis of two MD desalination fields operated by Fraunhofer ISE, one in Pantelleria, Italy in 2010 and another in Gran Canaria, Spain in 2011. Both systems comprised 12 parallel PGMD modules with a total membrane area of $120\,m^2$ and a maximum unit capacity of nearly $10\,m^3$/day. The overall investment costs of these desalination units, excluding the cost for pretreatment, solar energy supply, as well as cooling and lost cost, was estimated to be approximately 60,000 EUR. The detailed breakdown of the investment costs is shown in Fig. 5.27.

Winter compared the investment costs of three common MD configurations including DCMD, PGMD, and AGMD for two different channel lengths ($L = 1$ m, as a high flux module configuration and $L = 10$ as a high heat recovery module configuration) and by considering for reference an MD desalination plant specification for all cost analysis, as summarized in Table 5.5.

In this research and based on the design specifications, the investment cost of all the MD plant configurations varied in the range from 380,000 to 580,000 EUR, depending on the average distillate flux in each setup configuration and the amount of heat recovery accomplished through each module. It was confirmed that a system with higher heat recovery would lead to higher overall cost than a system with higher distillate flux because of the larger membrane surface area required in the former situation. However, for high flux plant designs, the heat exchanger costs will be one of the main costs, which may even exceed the MD module cost.

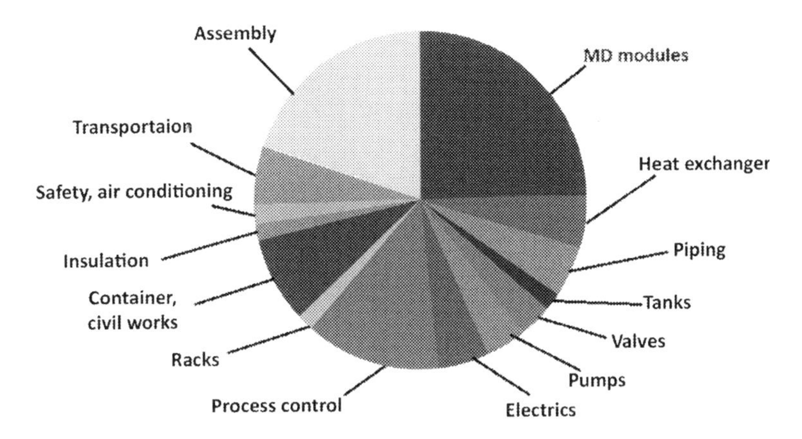

Fig. 5.27 Detailed breakdown investment cost analysis of the desalination MD field test systems in Pantelleria, Italy and in Gran Canaria, Spain [9].

Table 5.5 Reference MD desalination plant used by Winter et al. [9]

Plant		MD modules	Operating conditions
Nominal capacity	200 t/day	$L=1/10\,m$	$T_{Ei}=80°C$
Recovery rate	30%	$H=0.7\,m$	$T_{Ci}=25°C$
Raw water salinity	35 g/kg		$\dot{m}_{Ei}=400\,kg/h$

Winter [9] also conducted a sensitivity analysis on the water unit cost for the defined reference plant in PGMD and vacuum enhanced AGMD (V-AGMD) configurations with the nominal capacity of 200 t/day, as documented in Table 5.5. V-AGMD is established by applying vacuum to the air gap, which reduces the mass transfer resistance through the stagnant air gap layer, so that higher permeate flux may be obtained. In this analysis, the specific thermal energy cost was assumed to be 0.04 EUR/kWh with the specific electrical energy cost of 0.10 EUR/kWh and by assuming a 4% interest rate. Fig. 5.28 provides a detailed breakdown of the water unit cost for the selected factors including capital, MD, and ultrafiltration (UF) module replacement, chemicals, electricity, thermal energy, service and maintenance, insurance,

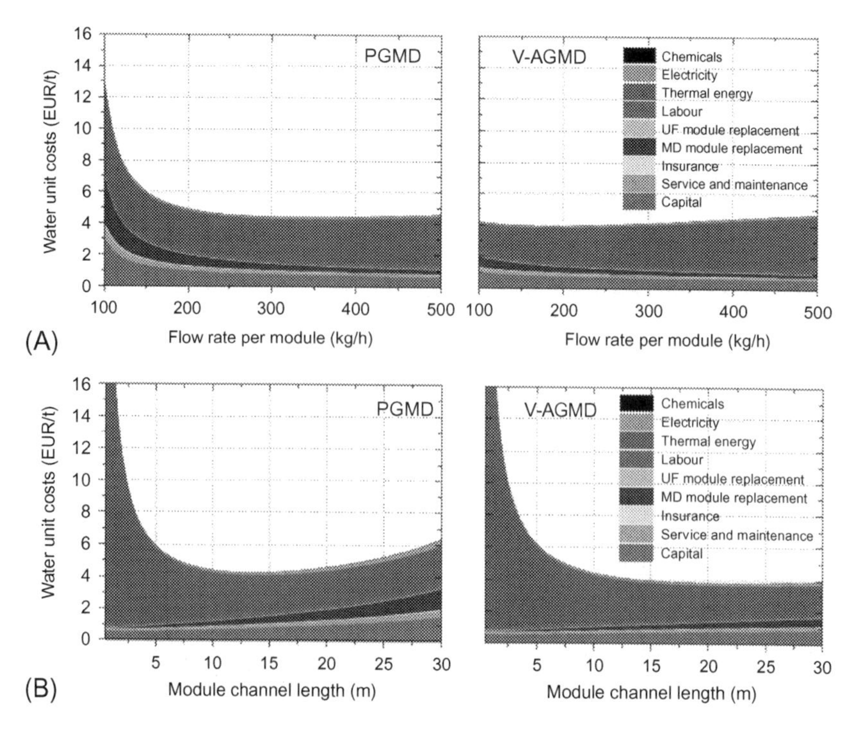

Fig. 5.28 Water unit cost detailed breakdown for PGMD and V-AGMD modules; (A) different flow rates per module and (B) different module channel lengths [9].

and labor. Also, variation of the channel length from 0.5 to 30.0 m and variation of the module flow rate from 100 to 500 kg/h are considered. As is clear from Fig. 5.28, the thermal energy cost is a most important parameter. By increasing the module channel length, the STEC of the system decreased so that the thermal energy cost and overall water unit cost reduced in both configurations. By increasing the module surface area, the MD module replacement cost and the capital cost also necessarily increased. However, for a longer MD module, the thermal energy saving by heat recovery could not compensate for the module cost of the high surface area MD. By analyzing the water unit cost for the PGMD module, the value of 4.25 EUR/t for a channel length of 13.5 m is identified as an optimum value, while for V-AGMD the optimal water unit cost would be 3.81 EUR/t for a channel length of 22 m (as depicted in Fig. 5.28A). The higher average permeate flux in V-AGMD, compared with PGMD, led to a lower increase in water unit cost for V-AGMD by increasing the channel length. Also, by integrating a waste heat source with the MD module, the thermal energy cost could be omitted and the minimum water unit cost will decrease to approximately 1 EUR/t for a shorter module length for both PGMD and V-AGMD modules. Based on this analysis, increasing the module channel length caused higher fixed and variable operational costs. These costs included capital, MD module replacement, and thermal-electrical energy costs.

As depicted in Fig. 5.28B, for the specified reference desalination plant with a nominal capacity of 200 t/day, operating at a low feed flow rate resulted in a lower permeate flux and therefore larger membrane area and higher operational cost. This effect was similar for both PGMD and V-AGMD, but was more influential for PGMD since the system sensitivity to salinity is more significant at a low feed flow rate. On the other hand, by increasing the feed flow rate, lower internal heat recovery would be possible for the system, leading to higher thermal energy cost. In this condition, the optimal value for water unit cost was found to be 4.36 EUR/t for a flow rate of 350 kg/h in the PGMD module, while the optimum feed flow rate for V-AGMD is a lower value, 180 kg/h, because of the lower system sensitivity to salinity. The unit operational cost was investigated by presupposing the availability of a continuous waste thermal power from a 1 MW power station integrated with a PGMD system as shown in Table 5.5. For a module design with GOR and plant capacity, respectively, in the ranges of (5.4–8.2) and (200–300) t/day, the operational cost will change in a range of (1.30–2.30) EUR t/L, which is much lower than that for a system without any waste heat source.

By considering the limited cost analysis study of MD systems, the lack of integrated cost evaluation methods, and the assumption made in terms of different MD desalination system components, it may be concluded that there is uncertainty around the techno-economic feasibility of MD technology as a reliable alternative to conventional desalination technologies. The variations in techno-economic data result not only because of limited experimental data, especially in large-scale plants, but also from the wide diversity in reported data in terms of thermal and electrical energy consumption, distillate flux rate, required component materials of different MD configurations in each scale, and availability of waste and solar heat in different geographical conditions.

5.6 Conclusions and future directions

This chapter introduces PGMD as a newly emerging configuration of MD systems, which has the potential to be developed on a commercial scale and in the future desalination markets. The experimental and numerical results from the existing studies confirm that there is a need for more study in this field. To develop a sustainable PGMD configuration design, from the design point of view, the module length and effective membrane surface area need to be optimized in order to improve the internal heat recovery rate in the system and so reduce the external energy demand, especially in the situation where the external energy source derives from nonrenewable fossil fuels. From the commercial point of view, the economic aspects of the final system need to be considered in order to minimize the overall project cost, including investment and operational cost. Therefore, achieving a reliable compromise between the techno-economic feasibility of MD systems and module design configurations compared to the known developed desalination plants is essential.

The literature survey presented in this chapter focuses on the feasibility of coupling PGMD systems in different configurations, area and plant scale with renewable sources, especially thermal solar and available waste heat sources and shows a high potential for developing a sustainable desalination technology.

Optimistically, it is predicted that by conducting more experimental and project-based activities in MD desalination in terms of MD plant scale, saline or wastewater sources, low cost and available energy sources such as renewable energy and waste heat and by considering appropriate system locations, the current knowledge gap in this field would be overcome in the near future and promising research activities concerning sustainable desalination using MD systems will soon close the current knowledge gap in this field.

References

[1] Charcosset C. A review of membrane processes and renewable energies for desalination. Desalination 2009;245(1):214–31.

[2] Bai F, Akbarzadeh A, Singh R. A novel system of combined power generation and water desalination using solar energy. Solar09, the 47th ANZSES Annual Conference; 2009.

[3] Koschikowski J. Water desalination: when and where will it make sense? 4705—Advancing Science-Serving Society (AAAS) Annual Meeting; 2011.

[4] Camacho L, et al. Advances in membrane distillation for water desalination and purification applications. Water 2013;5(1):94–196.

[5] Meindersma GW, Guijt CM, de Haan AB. Desalination and water recycling by air gap membrane distillation. Desalination 2006;187(1–3):291–301.

[6] Alkhudhiri A, Darwish N, Hilal N. Membrane distillation: a comprehensive review. Desalination 2012;287:2–18.

[7] Khayet M. Solar desalination by membrane distillation: dispersion in energy consumption analysis and water production costs (a review). Desalination 2013;308:89–101.

[8] Gryta M. Long-term performance of membrane distillation process. J Membr Sci 2005;265(1–2):153–9.

[9] Winter D. Membrane distillation: a thermodynamic, technological and economic analysis. Ph.D. Thesis University of Kaiserslautern, Germany: Shaker Verlag Publisher; 2015.

[10] Essalhi M, Khayet M. Fundamentals of membrane distillation, Pervaporation, Vapour Permeation and Membrane Distillation: Principles and Applications. Elsevier; 2015;277–316 (eBook ISBN: 9781782422563).

[11] Alklaibi AM, Lior N. Membrane-distillation desalination: status and potential. Desalination 2005;171(2):111–31.

[12] Khayet M. Membranes and theoretical modeling of membrane distillation: a review. Adv Colloid Interface Sci 2011;164(1–2):56–88.

[13] Sharqawy MH, Lienhard JH, Zubair SM. Thermophysical properties of seawater: a review of existing correlations and data. Desalin Water Treat 2012;16(1–3):354–80.

[14] Emerson WH, Jamieson DT. Some physical properties of sea water in different concentrations. Desalination 1967;3:207–12.

[15] Da Costa AR, Fane AG, Wiley DE. Spacer characterization and pressure drop modelling in spacer-filled channels for ultrafiltration. J Membr Sci 1994;87(1):79–98.

[16] Phattaranawik J, et al. Mass flux enhancement using spacer filled channels in direct contact membrane distillation. J Membr Sci 2001;187(1–2):193–201.

[17] Phattaranawik J, Jiraratananon R, Fane AG. Effects of net-type spacers on heat and mass transfer in direct contact membrane distillation and comparison with ultrafiltration studies. J Membr Sci 2003;217(1–2):193–206.

[18] Xu J, et al. Effect of operating parameters and membrane characteristics on air gap membrane distillation performance for the treatment of highly saline water. J Membr Sci 2016;512:73–82.

[19] Guillén-Burrieza E, et al. Techno-economic assessment of a pilot-scale plant for solar desalination based on existing plate and frame MD technology. Desalination 2015;374:70–80.

[20] Sanmartino JA, Khayet M, García-Payo MC. Desalination by membrane distillation A2. In: Hankins N, Singh R, editors. Emerging membrane technology for sustainable water treatment. Boston: Elsevier; 2016. p. 77–109 [chapter 4].

[21] Dow N, et al. Pilot trial of membrane distillation driven by low grade waste heat: membrane fouling and energy assessment. Desalination 2016;391:30–42.

[22] Winter D, Koschikowski J, Wieghaus M. Desalination using membrane distillation: experimental studies on full scale spiral wound modules. J Membr Sci 2011;375(1–2):104–12.

[23] Cipollina A, et al. Development of a membrane distillation module for solar energy seawater desalination. Chem Eng Res Des 2012;90(12):2101–21.

[24] Winter D, Koschikowski J, Ripperger S. Desalination using membrane distillation: flux enhancement by feed water deaeration on spiral-wound modules. J Membr Sci 2012;423–424:215–24.

[25] Essalhi M. Development of polymer nono-fiber, micro-fiber and hollow-fiber membranes for desalination by membrane distillation. Ph.D. Thesis, Spain: University Complutense of Madrid (UCM); 2014.

[26] Nakoa K, et al. An experimental review on coupling of solar pond with membrane distillation. Sol Energy 2015;119:319–31.

[27] Nakoa K, et al. Sustainable zero liquid discharge desalination (SZLDD). Sol Energy 2016;135:337–47.

[28] Schwantes R, et al. Membrane distillation: Solar and waste heat driven demonstration plants for desalination. Desalination 2013;323:93–106.

[29] Raluy RG, et al. Operational experience of a solar membrane distillation demonstration plant in Pozo Izquierdo-gran Canaria Island (Spain). Desalination 2012;290:1–13.
[30] Shim WG, et al. Solar energy assisted direct contact membrane distillation (DCMD) process for seawater desalination. Sep Purif Technol 2015;143:94–104.
[31] Zaragoza G, Ruiz-Aguirre A, Guillén-Burrieza E. Efficiency in the use of solar thermal energy of small membrane desalination systems for decentralized water production. Appl Energy 2014;130:491–9.
[32] Winter D, et al. Evaluation of MD process performance: effect of backing structures and membrane properties under different operating conditions. Desalination 2013;323:120–33.
[33] Shirazi MMA, Kargari A, Tabatabaei M. Evaluation of commercial PTFE membranes in desalination by direct contact membrane distillation. Chem Eng Process Process Intensif 2014;76:16–25.
[34] Adnan S, et al. Commercial PTFE membranes for membrane distillation application: effect of microstructure and support material. Desalination 2012;284:297–308.
[35] Eykens L, et al. Characterization and performance evaluation of commercially available hydrophobic membranes for direct contact membrane distillation. Desalination 2016;392:63–73.
[36] Swaminathan J, et al. Energy efficiency of permeate gap and novel conductive gap membrane distillation. J Membr Sci 2016;502:171–8.
[37] Warsinger DEM, Swaminathan J, Lienhard JH. In: Effect of module inclination angle on air gap membrane distillation. Proceedings of the 15th International Heat Transfer Conference, IHTC-15; 2014 [IHTC15-9351].
[38] Khalifa A, et al. Experimental and theoretical investigation on water desalination using air gap membrane distillation. Desalination 2015;376:94–108.
[39] Wu HY, Tay M, Field RW. Novel method for the design and assessment of direct contact membrane distillation modules. J Membr Sci 2016;513:260–9.
[40] Sirkar KK, Li B. Novel membrane and device for direct contact membrane distillation-based desalination process: phase III. Denver, CO: Bureau of Reclamation; 2008.

Further reading

[1] Hanemaaijer JH, et al. Memstill membrane distillation—a future desalination technology. Desalination 2006;199(1–3):175–6.

Desalination by pervaporation

Zongli Xie, Na Li†, Qinzhuo Wang*, †, Brian Bolto**
*CSIRO Manufacturing, Clayton South, VIC, Australia, †Xi'an Jiaotong University, Xi'an, China

6.1 Introduction

With the rapid increase in global population and urbanization, water scarcity is becoming one of the major challenges of contemporary society. Among numerous initiatives to improve potable water supply, membrane desalination technology has been widely accepted as an attractive solution for desalination [1–4]. Membrane operation is more competitive than conventional distillation techniques in terms of energy efficiency, separation capacity and selectivity, and capital investments. It also has many advantages such as high operational stability, low chemical costs, and ease of integration and control within industrial process trains [5]. Reverse osmosis (RO) is currently the most mature membrane technology for desalination at a relatively low cost which is employed for approximately 50% of desalination plants in the world [2,3,6]. RO is a process driven by a mechanical pressure which is greater than the osmotic pressure of feed solution. Thus, it is hard to obtain pure water from high concentration feed solution with a reasonable flux. In recent years, emerging membrane technologies like membrane distillation (MD) and pervaporation (PV) have become attractive in treating high total dissolved solids (TDS) source water and in being more resistant to certain types of fouling [7].

Pervaporation, considered as a clean technology, is a potential low energy membrane technology which has been extensively used for separation of mixtures of aqueous-organic or organic liquids. The pervaporation process was first commercialized for ethanol dehydration in the 1980s based on a cross-linked PVA/PAN composite membrane. Since then, both the scope of application and the types of pervaporation membranes have been extensively enlarged [8]. Dehydration of alcohols and other solvents and the removal of small amounts of organic compounds from contaminated waters are now relatively well understood areas of membrane separations and have been successfully utilized in full-scale industrial production [9,10]. For its application in desalination, it has the advantages of near 100% salt rejection due to the non-volatility of salt and the high selectivity of dense polymer membranes or the tunable nanopore of the inorganic membranes [11,12]. Using a partial water vapor pressure gradient as a driving force for the transport of water across the membrane, the energy need for pervaporation is independent of feed salt concentration, and thus it can handle high salt concentrations without losing much of its driving force. For instance, the partial vapor pressure of a 2 M NaCl solution at 75°C is only 8% lower than that of pure water at the same temperature [13]. It has been proved that pervaporation is a feasible

Emerging Technologies for Sustainable Desalination Handbook. https://doi.org/10.1016/B978-0-12-815818-0.00006-0

way of treating produced water from the mineral oil and natural gas extraction industries. The produced water contains a high TDS (1–400 g/L) with sodium chloride being the dominant salt, which makes it difficult for conventional membrane processes like RO to treat as the hydraulic pressures required to overcome the osmotic pressure are extremely high [14]. Pervaporation is capable of concentrating salt solutions up to supersaturated levels, which allows for the recovery of salt by crystallization to sustain zero liquid discharge and minimize pollution of the environment [15].

In general, the main advantages of pervaporation over conventional separation processes are:

- Complete separation (in theory) of ions, macromolecules, colloids, cells, etc. to produce a high-quality permeate.
- Water can be recovered at relatively low temperatures.
- Low-grade heat (solar, industrial waste heat) may be used.
- The feed does not require extensive pretreatment as required for pressure-based membrane treatment processes such as RO.
- Hydrophilic pervaporation membranes are potentially low fouling as compared with hydrophobic membranes used in MD.

Owing to the above advantages of pervaporation in desalination, there are growing interests on studies of pervaporation for desalination. As shown in Fig. 6.1, desalination by pervaporation is a combination of diffusion of water through a membrane and its evaporation into the vapor phase on the other side of the membrane to produce freshwater.

This chapter presents an overview of the principle, transport mechanism, and membrane materials development of the pervaporation process for desalination application. Key factors affecting the pervaporation process, including membrane properties and various operating parameters such as feed temperature and permeate

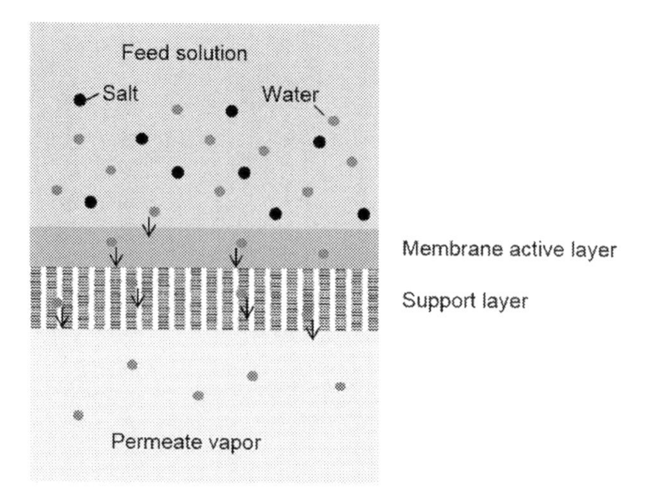

Fig. 6.1 Desalination by pervaporation process where water passes through a pervaporation membrane.

pressure, are reviewed. Some insights into the key operating factors, process design and operation, and engineering aspects for commercialization of pervaporation are also included.

6.2 Principle of pervaporation for desalination

In pervaporation desalination, the heated feed (aqueous salt solution) is brought into contact with a permselective membrane and the permeated product (water) is continuously removed in the form of vapor from the permeate side. The permeability and selectivity of membranes in pervaporation for desalination are generally characterized in terms of the water flux and salt rejection which can be obtained according to the following equations:

$$J = \frac{M}{A \times t} \tag{6.1}$$

$$R = \frac{C_f - C_p}{C_f} \times 100\% \tag{6.2}$$

where M is the amount of permeate (kg), A is the effective membrane area (m^2), and t is the processing time (h). C_f and C_p are the salt concentrations in the feed and permeate, respectively, which are generally determined through conductivity measurement.

Different types of membranes have been used for pervaporation desalination, including dense polymeric membranes, microporous inorganic membranes, and membranes with charge. Transport behavior in pervaporation is membrane-specific due to the intricate interactions among a membrane and the penetrants in the membrane [16]. Hence, the transport and separation mechanism varies with different type of membranes.

For a dense polymeric membrane, the solution-diffusion theory is generally accepted as the mechanism of mass transport [11,17–21]. Analogous to the separation of organic solutions via pervaporation, pervaporation desalination can be regarded as a separation of a pseudo-liquid mixture containing free water molecules and bulkier hydrated salt ions in water [22]. According to this proposal, water molecules are preferentially adsorbed from the liquid mixture onto the membrane surface and then diffuse through the free volume of the membrane matrix and finally evaporate as vapor from the permeate side, while salt is rejected. The driving force for the mass transfer is the chemical potential gradient. Usually, water flux can be written as the form of Fick's law, in which the concentration difference between the feed side and permeate side is taken as the driving force for the mass transport.

$$J_i = -D_i \frac{dc_i}{dx} = \frac{D_i \left(c_{if(m)} - c_{ip(m)} \right)}{l} \tag{6.3}$$

where D is the diffusion coefficient and l is the membrane thickness. The terms $c_{if(m)}$ and $c_{ip(m)}$ denote the concentrations of component i in the membrane at the membrane/feed interface and at the membrane/permeate interface, respectively. $c_{if(m)}$ and $c_{ip(m)}$ can be quantitatively expressed according to the equilibrium of chemical potential of the compound i at the membrane/feed and membrane/permeate interface, which has been derived in detail in the literature [23,24]. As a result, a relationship among flux, concentration of component i in the feed side and the partial pressure of component i in the permeate side can be obtained:

$$J_i = \frac{D_i}{l}\left(K_i c_{if} - K_i^G P_{ip}\right) \tag{6.4}$$

where K_i is the liquid-phase sorption coefficient of component i, which is also called the partition coefficient [16]. It is a characteristic parameter dependent upon the interaction of the species with the membrane. In pervaporation desalination, K_i is related to water uptake which is expressed as the amount of water taken by per unit weight of the membrane. K_i^G is the gas-phase sorption coefficient and P_{ip} is the partial pressure of component i in the permeate side of the membrane.

According to the equations above, the flux through the dense polymeric membrane significantly depends on the physicochemical properties of the membrane, such as thickness (l), affinity for water (K), and the diffusion coefficient of water (D) through the membrane. Many efforts have been devoted to seeking the appropriate membrane materials with maximal D_i and K_i, and thus P_i to achieve improved separation [10,16,25]. Fractional free volume (FFV) is also an important parameter. As reported by Xie et al., water diffusivity increases with the increase of FFV because water can be accommodated and consequently diffused through the membrane at higher FFVs [19].

For charged ion-exchange membranes, the transport may be described by a modified diffusion process as proposed by Korngold et al. [26,27]. In one of these studies, they found a similar water flux through a thicker ion-exchange hydrophilic polyethylene (PE) membrane to that through a commercial microporous hydrophobic PE membrane (Celgard 2500). According to Fick's law, the higher flux of the Celgard membrane is expected than that of a hydrophilic pervaporation membrane because of the thinner thickness and higher diffusion coefficient of vapor in the porous MD membrane. Thus, it is suggested that the ion-exchange membranes present an increased capacity in increasing water flux which is beyond the value calculated from Fick's equation. It was proposed that the diffusion of water in the ion-exchange membrane took place via a continuous pathway of water shells that was formed via the adsorption of water molecules around the sulfonic groups, which is claimed to be faster than the diffusion as clusters of free water in the membrane. The proposal is in agreement with the hypothesis of Cabasso for the mechanism of water diffusion inside an ion-exchange membrane [28].

For porous inorganic membranes, the separation mechanism is based on molecular sieving (or size exclusion) [3,4,12,29–43] and the membrane pores play an important role. The smaller pore sizes generally exclude the permeation of the larger ions Na^+ and Cl^- with hydrated diameters of 0.72 and 0.66 nm, while allowing the preferential diffusion of water molecules with a smaller diameter of 0.26 nm [29,44,45]. For example, Duke et al. [31] pointed out that a methyltriethoxysilane (MTES) membrane with pore diameters of 0.5 nm exhibits a higher flux but lower rejection than a carbonized template molecular sieve silica (CTMSS) membrane with pore diameters of 0.3 nm. Malekpour et al. [33,34] compared the fluxes and rejections of a ZSM-5 zeolite membrane (pore size 0.4 nm) with different single salt aqueous feed solutions containing I^-, MoO_4^{2-}, Cs^+, and Sr^{2+}, respectively. Lower fluxes but higher rejections were obtained for I^- and MoO_4^{2-} because of their larger hydrated radius relative to Cs^+ and Sr^{2+}. Similarly, An et al. [35] carried out pervaporation testing with a nature zeolite membrane (the dimensions of the largest channel were 0.44×0.72 nm) and found that the order of removal efficiency of divalent cations and univalent cations was Mg^{2+} (97.8%) > Ca^{2+} (96.5%) > Na^+ (79.8%) > K^+ (56.9%), which is in agreement with the kinetic sizes of these hydrated ions, that is, Mg^{2+} (0.86 nm) > Ca^{2+} (0.82 nm) > Na^+ (0.72 nm) > K^+ (0.66 nm). This confirms the role of size exclusion of membrane pores in pervaporation separation. Xu et al. [37] developed an ethenylene-bridged organosilica membrane. Owing to the polar C=C and Si—OH groups presenting in the membrane networks, water molecules can initially adsorb onto these hydrophilic portions of the networks via hydrogen bonding and/or dipole-dipole interactions. The adsorbed water molecules then act as a nucleus to attract more water molecules via hydrogen bonds. These water-membrane pore wall interactions and water-water intermolecular interactions play an important role in water transport. Owing to the small pore size of approximately 0.5 nm, water molecules can pass through the membrane while the larger hydrated salt ions would be rejected by size exclusion. Zhou et al. [41] synthesized a zeolite faujasite framework (FAU) membrane with a pore size of 0.74 nm and compared its desalination performance with other reported zeolite membranes with various pore sizes, such as Mordenite Framework Inverted (MFI, pore size: 0.55 nm), Linde Type A (LTA, pore size: 0.4 nm), and Sodalite (SOD, pore size: 0.28 nm). Zeolite SOD membranes with a smaller pore size showed very high ions rejection of over 99.99% but relatively low water flux (0.56 kg/m^2 h at 90°C), while zeolite FAU membranes displayed higher desalination performances with a water flux of 5.64 kg/m^2 h at 90°C and a salt rejection of 99.8% due to its relatively large pore size. Darmawan et al. [42] prepared a kind of nickel oxide silica membranes possessing a dense matrix with lower pore volumes and smaller pore size with which lower water flux and slightly higher salt rejection were obtained. In addition, the phenomenon that zeolite membranes underwent dissolution and thus lost their molecular selective separation function when exposed to a harsh salty environment also demonstrates the important function of size exclusion of membrane pores on mass transport [3].

As for charged porous inorganic membranes, the mechanism of mass transfer and salt rejection can be explained jointly by size exclusion and charge exclusion since the electrostatic interaction between hydrated ions and membrane surface charge cannot be ignored [3,30,33,34]. In terms of charge exclusion, the co-ion cannot move across the membrane channels, so does the counter ion to maintain charge neutrality [34]. Malekpour et al. [33,34] reported that the rejections of I^- and MoO_4^{2-} were higher than those of Cs^+ and Sr^{2+} with a negatively charged ZSM-5 zeolite membrane because of both size exclusion and charge exclusion. In some cases, the charged membranes adsorb certain ions from the feed solution, leading to ion exchange, which is also considered to be one part of the mechanism in pervaporation desalination processes [12,30,32]. Interestingly, the phenomena may lead to a higher flux of the membrane in the presence of seawater than in pure water, which was firstly observed by Duke et al. using a ZSM-5 membrane [12]. The same phenomena were also observed by other researchers with zeolite membranes such as NaA and hydroxyl sodalite in investigating the effect of ion exchange on the flux [30,32]. Zeta potential measurements showed a positively charged surface of NaA zeolite membrane in seawater due to the adsorption of Na^+, whereas they showed a negatively charged surface in pure water due to the adsorption of OH^- [30]. Water molecules have to overcome the electrostatic interaction induced by the surface charge when passing through the membrane. The electrostatic interaction in the Na^+/water pair (radius of hydrated Na^+ 3.58 Å) is smaller than that in the OH^-/water pair (radius of hydrated OH^- 3.0 Å), so water can transport more easily through the Na^+ charged membrane [30,44]. For the same reason, ion exchange also led to an increase in water flux by increasing the NaCl concentration with a hydroxyl sodalite membrane [32]. The sodalite cages are on average occupied by four sodium cations, one hydroxyl anion, and two water molecules. The hydroxyl anion in the membrane can be easily exchanged with Cl^- from a NaCl feed solution. High concentrations of NaCl lead to more ion exchange. Since the electrostatic interaction with the Cl^-/water pair is smaller than that with the OH^-/water pair, water will more easily pass a membrane charged with Cl^-, thus facilitating faster water transport through the membrane.

It has been reported that a small fraction of salts may slowly penetrate/diffuse into and pass through the slightly larger pore of inorganic membranes or voids/gaps in some parts of a polymeric structure [4,12,17,18,30,46]. Salt transport properties can be determined by the kinetic desorption method. In this method, the salt diffusion coefficient D_s in the membrane can be calculated by the initial slopes of sorption curves according to the following Eq. [19,47]:

$$D_s = \frac{\pi \cdot l^2}{16}\left[\frac{d(M_t/M_\infty)}{d(t^{1/2})}\right]^2 \tag{6.5}$$

where M_t is the amount of salt in the water solution at time t (i.e., the mass of salt desorbed from the membrane), M_∞ is the total amount of salt desorbed from the membrane into the solution, and l is the thickness of the membrane. Salt permeability P_s is the product of salt diffusion coefficient D_s and salt solubility K_s. The NaCl solubility K_s is the ratio of the amount NaCl absorbed in the membrane in per unit membrane volume divided by the concentration of NaCl in the original solution. It is related to water uptake based on the hypothesis of hydrated salt being solvated by water in the membrane phase.

In fact, the salt diffusion coefficient and permeability are very low. Though there are trace amounts of salt embedded on the permeate side, as confirmed by SEM images and EDS line-scan profiles, this has little effect on the rejection since the ions (e.g., Na^+ and Cl^-) remained as salts (e.g., NaCl) on the permeate side as they could not evaporate with water [18,21,30]. The rejection, taking into account the concentration of flushed salt from the permeate side, was still high [3,12,34]. There are many factors impacting the salt passage. For example, Lin et al. [4] reported that both the increased salt concentration and temperature of the feed led to a slight increase in salt concentration on the permeate side. This was attributed to the increased mobility of the ions at higher concentration and temperature. However, at a high brine feed concentration of 150 g/L at 75°C, the salt rejection was still above 99% and the salt concentration of the permeate was still below the accepted World Health Organization guidelines of TDS for acceptable drinking water quality. This again reveals the advantage of pervaporation in dealing with high concentration solutions. Another possible factor is that salts embedded in polymer may widen the free volume space of the membrane matrix and further facilitate salt transport across the membrane [46]. Other factors, such as the hydrated size of the dissolved salt, the affinity of the salt to the membrane material, and possible charge interactions between the membrane and the salt, may also play roles in the ability of a given ion to penetrate into the membrane matrix [21]. If scale-forming elements (i.e., Ca^{2+}, Mg^{2+}) are present in the feed, mineral scaling (membrane fouling) may occur and have an impact on membrane performance. The relationship between fouling and the presence of solutes in the feed is relatively well established for pressure-driven membrane processes, but there is a lack of understanding for desalination by pervaporation [21].

Membrane distillation (MD) is another membrane process which has attracted increasing interest as a potential desalination method. It is a thermally driven separation process based on the vapor transport through microporous hydrophobic membranes driven by the vapor pressure gradient across the membrane [48]. There are similarities between MD and pervaporation in which a vapor pressure gradient is established across the membrane by heating the feed on the upstream side and maintaining a lower vapor pressure by applying a vacuum, using a sweeping flow or constructing an air gap on the downstream side. Configurations of both MD and pervaporation have been widely illustrated in the literature [29,49–51]. The main difference between them is the membrane and the role that membrane plays in the separation [52]. Porous and hydrophobic membranes which are used in MD act only as a support medium for the vapor-liquid interface and do not contribute to the separation performance. On the other hand, the dense or molecular sieving hydrophilic membranes which are used in pervaporation contribute to separation due to the attractive interactions between water and the membrane material such as dipole-dipole interactions, hydrogen bonding and ion-dipole interactions [10,25]. Differences in membranes between MD and pervaporation lead to the differences in mass transfer. Mass transfer in MD is governed by the Knudsen diffusion, Poiseuille flow (viscous flow), and molecular diffusion [53]. As mentioned above, mass transport of water in pervaporation consists of solution-diffusion, size exclusion, and charge exclusion. The main differences between MD and pervaporation are summarized in Table 6.1.

Table 6.1 **Differences between membrane distillation and pervaporation**

	Membrane distillation	Pervaporation
Membrane	Porous and hydrophobic	Dense or molecular sieving hydrophilic
Role of membrane	Support medium for the vapor-liquid interface	Selective medium, attractive interactions between water and the membrane
Transport mechanism	Knudsen diffusion, Poiseuille flow (viscous flow), molecular diffusion	Solution-diffusion, size exclusion, charge exclusion
Configurations	Direct contact MD, vacuum MD, sweeping gas MD, air gap MD	Vacuum pervaporation, sweeping gas pervaporation, air gap pervaporation
Challenging problems	Membrane fouling, membrane wetting, long-term stability	Membrane fouling, low flux

6.3 Membrane development

In pervaporation, a wide range of materials including dense metals, zeolites, polymers, ceramics, and biological materials have been used for the manufacture of membranes and polymers are the most widely used material at present. Most pervaporation membranes are nonporous polymeric systems, and membrane selection is focused on polymer design and modification. The low flux of current polymeric membranes is believed to be the main limiting factor for commercial application of the pervaporation process for desalination. The development of an optimum polymer for pervaporation application is a challenge for polymer chemists, which may need entirely new concepts and ideas. Polymer modifications include the incorporation of inorganic additives, cross-linking the polymers, blending a mixture of polymers, and the use of copolymers and polymer graft systems. By combining inorganic nanoparticles with polymers at a molecular level, polymer-inorganic nanocomposite materials can provide novel properties that are not obtained from conventional organic or inorganic materials. Xie et al. found that the introduction of silica nanoparticles in the polymer matrix enhanced both the water flux and salt rejection due to increased diffusion coefficients of water through the membrane [54]. Cross-linking of the polymer tended to form a more compact structure with reduced free volumes, which resulted in an increase in salt rejection [19]. Owing to the incompatibility of different blocks in block copolymers, a spontaneous micro-phase separation occurs to form nanochannels which can provide continuous passages for water to diffuse, thus improving the flux [55]. An overall view of reported materials and performance of pervaporation membranes for desalination is listed in Table 6.2. The earliest materials applied were organic polymers such as PE and cellulose. In the late 2000s, microporous inorganic membranes, mainly zeolites and amorphous silica-based membranes, began to receive

Table 6.2 **List of pervaporation membranes and their performance for desalination**

Membrane	NaCl (g/L)	Conditions in feed side (temp., pressure)	Conditions on permeate side	Membrane thickness (μm)	Flux (kg/m^{2}h)	Salt rejection (%)	Ref.
Sulfonated polyethylene	0–176	25–65°C	Air sweep	100	0.8–3.3	–	[20]
	35	45–65°C	Air sweep	50–180	1.5–3.0	–	[19]
Polyether amide	32	68–70°C solar heated	Cooler tunnel	40	0.56	99.998	[13]
Cotton cellulose	40	40°C	Vacuum	30	4.55–6.7	–	[22]
Cellulose diacetate on PTFE	40	40°C	Vacuum	3.5	4.5–5.1	–	[22]
Cellulose triacetate	100	50°C	Air sweep	10	2.3	99	[21]
Cellulose acetate	40–140	70°C	Vacuum	20–25	5.97–3.45	99.7	[15]
Deacetylated cellulose acetate	120	70°C	Air sweep	22	4.11	99.9	[56]
Polyether ester	3.2–5.2	22–28.7°C	Air sweep	160	0.13–0.16	–	[46]
Polyester	35	20°C	In sand	750	7.1×10^{-3}	99.84	[18]
	100	20°C	Air sweep	20	0.54	99	[21]
Poly(vinyl alcohol)/ polyacrylonitrile composite	5	20°C	Vacuum	0.62	9.04	99.5	[20]
Poly(vinyl alcohol)	30	70°C	Vacuum	0.1	7.4	99.9	[57]
Sulfonated poly(styrene-ethylene/butylene-styrene)	30	63°C	Vacuum	~30	22.8	99.9	[55]
NaA	Seawater	69°C	Vacuum	–	1.9	99.9	[30]
	29	77°C	Vacuum	–	4.4	99.9	[30]
	0.13	25°C	Vacuum	15	0.2	99.4	[34]
	0.1	20°C	Vacuum	10	1.43	99.83	[33]
S-1	3	75°C	Vacuum	6	11.5	99	[3]
ZSM-5	3	75°C	Vacuum	3.3	12.5	99	[3]
	0.13	25°C	Vacuum	10	0.05	99.6	[34]
	38	90°C	Vacuum	–	0.85	99	[12]
ZIF-8	35	50°C	Vacuum	20	8.1	99.8	[43]
FAU	35	90°C	Vacuum	2.3	5.64	>99.8	[41]
Clinoptilolite	0.1	93°C	Vacuum	–	2.5	95.8	[58]

Continued

Table 6.2 **Continued**

Membrane	NaCl (g/L)	Conditions in feed side (temp., pressure)	Conditions on permeate side	Membrane thickness (μm)	Flux (kg/m²h)	Salt rejection (%)	Ref.
Clinoptilolite-phosphate	1.4	95°C	Vacuum	–	15	95	[35]
Hydroxyl sodalite	350	200°C, 2200 kPa	Vacuum	1	3.9	99.99	[32]
Silica from tetraethyl ortho silicate and methyl tri-ethoxy silane	3	25°C, ~700 kPa	Vacuum	–	4.7	93	[31]
Carbon template silica	3	25°C, ~700 kPa	Vacuum	–	2.2	99.9	[31]
	3	20°C	Vacuum	–	3.2	97	[59]
	40	25°C	Vacuum	0.21	2.6	99.9	[60]
	35	20°C	Vacuum	0.5	3.7	98.5	[36]
Silica	35	60°C	Vacuum < 1 Pa	3.3–5	17.8	>99	[40]
Nickel oxide silica	3	25°C	Vacuum		7	99.9	[42]
Microporous silica	35	25°C	Vacuum	5.4–17.3	2.8	99	[38]
Hybrid carbon silica	10	60°C	Vacuum		26.5	99.5	[39]
	150	60°C	Vacuum		9.2	98.6	
Ethenylene-bridged organosilica	20	70°C	Vacuum	0.1 (top layer)	14.2	99.6	[37]
Cobalt oxide silica	10–150	75°C	Vacuum	0.2–0.35	1.8–0.6	99	[4]
Fluoroalkylsilane-Ti	30	40°C	Vacuum	23	5	–	[61]
Graphene oxide/polyacrylonitrile composite	35	90°C	Vacuum	0.1	65.1	99.8	[62]
Graphene oxide framework	35	45–90°C	Vacuum	18	3.4–11.41	99.98	[63]
	100	75°C	Vacuum	18	10.5	99.9	
Graphene oxide/α-Al_2O_3	35	90°C	Vacuum	0.4	48.8	99.7	[64]
Graphene oxide/polyimide	35	90°C	Vacuum	46	36.1	99.9	[65]
Graphene oxide-poly(vinylalcohol)-glutaraldehyde/polyacrylonitrile	35	70°C	Vacuum	0.12	69.1	>99.8	[66]
	35	30°C	Vacuum	0.12	18.3	>99.8	
Poly(vinyl alcohol)/maleic	2	22°C	Vacuum	10	6.9	–	[11]

considerable attention. More recently, novel membranes including hybrid organic-inorganic membranes and thin film composite (TFC) membranes showed attractive promise due to their high water flux [67]. Moreover, graphene oxide (GO), as a new intriguing material with an ultrathin two-dimensional structure and abundant functional groups such as epoxide, carbonyl, and hydroxyl on the surface, has displayed outstanding water permeability and salt rejection. The morphology of the GO film reveals a well-packed structure with a layer-by-layer pattern. It was assumed that the ideal pathway for water molecules to transport through was the tortuous nanocapillaries between the well-stacked GO sheets. The selectivity performance of the membrane was determined by the GO intersheet spacing [62]. By far, the best outcome in the open literature is with hybrid organic-inorganic membranes, followed by ZSM-5, cellulose membranes, silica, ionic polyethylene, and various polyether membranes. A water flux as high as 65.1 kg/m^2 h was reported for a GO/PAN membrane at 90°C with a membrane thickness of 0.1 μm. Recently, studies of GO focus on the enhancement of its mechanical stability [63,64,66].

The design of suitable membranes for desalination by pervaporation is of great importance for its further development. Generally, the membranes used for laboratory-scale trials are always homogeneous and symmetric (Fig. 6.2A) because they are easy to cast and possess the intrinsic separation properties of the polymer.

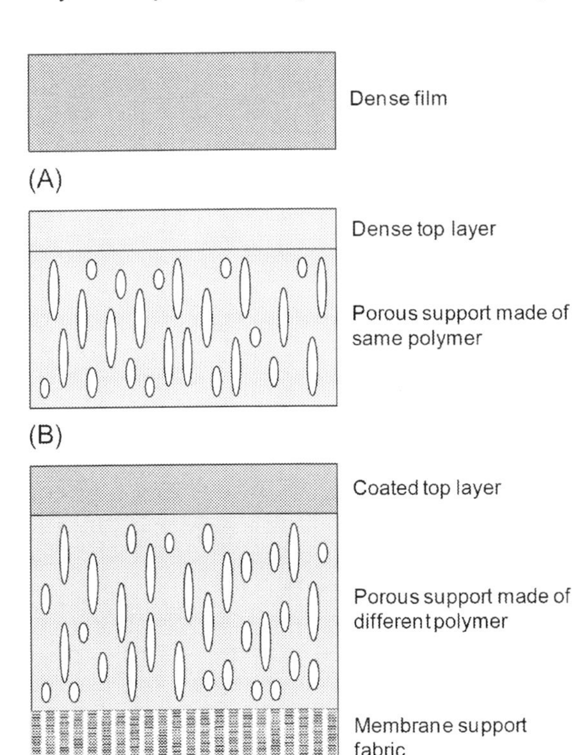

Fig. 6.2 Schematic representation of three different types of membrane morphology including homogeneous membrane (A), asymmetric membrane (B), and composite membrane (C).
Modified from Smitha B, Suhanya D, Sridhar S, Ramakrishna M. Separation of organic–organic mixtures by pervaporation—a review. J Memb Sci 2004;241:1–21.

Table 6.3 **Factors affecting overall mass transport [44]**

Intermolecular interactions	Molecular size and shape Polarity and polarizability Hydrogen bonding Donor-acceptor interaction
State of aggregation of the polymer	Glass transition temperature Ratio of amorphous to crystalline domain
Physical properties	Thickness of selective nonporous layer Porosity of support
Operating conditions	Feed composition Feed temperature Thickness of boundary layers Permeate pressure

However, to attain commercial viability, the membranes successful on the laboratory scale are prepared in an asymmetric or composite form (Fig. 6.2B and C). These two morphologies offer a possibility of making a barrier with a thin effective separation layer, which enables a high flux while maintaining a desirable mechanical strength [68].

Membrane thickness plays an important role in pervaporation performance. The influence of membrane thickness on flux and selectivity in separating different mixtures has been studied by a number of researchers [11,20,21]. In general, the flux decreased inversely with increasing membrane thickness by increasing the membrane resistance. On the other hand, the selectivity was also improved with increasing membrane thickness until the membranes were very thick. The existence of micropores in the nonporous pervaporation membranes would lead to poor selectivity as defects allow the diffusion of molecules through thin membranes. Therefore, in order to obtain a constant selectivity, an optimal membrane thickness with a uniform structure may be required, and the absolute thickness at which this occurs varies between systems.

Membranes with both high permeability and selectivity are desirable. Higher permeability decreases the amount of membrane area requirement, thereby decreasing the capital cost of membrane units. Higher selectivity results in a higher purity product. Table 6.3 summarizes some key factors important for mass transport through a pervaporation membrane. Solubility and diffusivity of low molecular mass solutes in polymers strongly depend on the molecular size and shape of the solute, the polymer/solute interactions, and the chemical and physical properties of the polymer [44].

6.4 Commercial and engineering aspects of pervaporation

6.4.1 Membrane modules

The performance of pervaporation depends not only on membrane properties but also on the module design and process operating conditions. Module design plays an important role in the overall performance of any membrane plant. Key parameters to be taken into consideration include packing density, cost-effective manufacture, easy access for cleaning, reduction in boundary layer effects, and cost-effective membrane replacement. Based on the above, modules can be distinguished as tubular, capillary, hollow fiber, plate-and-frame type, and spiral wound modules [68]. The module configuration has to be carefully optimized in order to reduce the temperature change of the solutions along the membrane module (reduction of heating and cooling costs). One particular issue affecting the pervaporation module design is that the permeate side of the membrane often operates at a vacuum of <100 Torr. The pressure drop required to draw the permeate vapor to the permeate condenser may then be a significant fraction of the permeate pressure. Efficient pervaporation modules must have short, porous permeate channels to minimize this permeate pressure drop [69].

The plate-and-frame type is the dominating module configuration employed in pervaporation since this configuration can provide low resistance channels on both the permeate and feed sides. It also has the advantages of ease of manufacturing and high-temperature operation with efficient interstage heating between stages [16,68]. In plate and frame systems, the feed solution flows through flat, rectangular channels. Packing densities of about 100–400 m^2/m^3 are achievable.

Pervaporation requires volatilization of a portion of liquid feed and the enthalpy of vaporization must be supplied by the feed. Owing to this, a large thermal gradient is established across the membrane with continual heat loss to the permeate, resulting in a reduction in flux. To compensate for this heat loss, interstage heaters within the membrane module are required to reheat the feed. In the case of hollow fiber modules, where surface area-volume ratios are high, hollow fibers may have problems with longitudinal temperature drops and inefficient use of the downstream surface area [68]. In addition, it should be noted that ensuring low transport resistance on the permeate side is a critical consideration in pervaporation module design. This is because the efficient evaporation of the permeate molecules at the downstream face of the membrane needs an extremely low absolute pressure. The presence of resistance in the permeate channel can greatly affect the pervaporation separation process. Owing to this characteristic, the compactness of the membrane module is no longer a preferential consideration for pervaporation modules. It is difficult for a hollow fiber module to be employed in pervaporation unless the fiber length is short, or the fiber diameter is big enough to provide a small temperature gradient along the fiber, for example, 5–25 mm [70]. As a result, tubular membranes seem to be a feasible module configuration for pervaporation, such as tubular zeolite membranes developed by Mitsui Engineering & Shipbuilding Co. However, the manufacturing cost of these modules is high. Capillary modules are generally not used in pervaporation due to their high mass transport resistance when compared to the other modules, due to increased boundary layer thickness.

Spiral wound modules have the advantages of high packing density (>900 m^2/m^3) and a simple design. They are generally considered difficult to develop for pervaporation because of the chemical susceptibility of the adhesive required [68]. However, this may not be the case for desalination applications as there is no solvent in the feed. Therefore, there is potential to develop the spiral wound module for desalination by pervaporation. It should be noted that owing to the thin channels of the module for feed and permeate flow, the rapid drawing of vapor from the long and spiral channel may be difficult and the temperature polarization may be a problem, which leads to the decrease in water flux and the increase in heat lost.

6.4.2 Operating conditions

Membrane performance is also dependent on process operating parameters such as feed salt concentration, feed flow rate, feed temperature, and permeate vapor pressure. To optimize the membrane, it is necessary to study the effect of process parameters on water flux and salt rejection.

For desalination by pervaporation, feed concentration refers to the concentration of salt in the solution. A change of feed concentration directly affects sorption at the liquid/membrane interface. Since diffusion in the membrane is concentration dependent, the water flux generally decreases with increasing salt concentration in the feed. Mass transfer in the liquid feed side may be limited by the extent of concentration polarization. In general, an increase in the feed flow rate increases water flux due to a reduction of transport resistance in the liquid boundary layer and concentration polarization.

Water flux increases as the vacuum increases according to Eq. (6.4), which confirms that when the vacuum degree decreases from 100.7 to 95.7 MPa, the flux drops by nearly 90% [54]. Similar results were also reported by Huth et al. [21], who found that when the vapor pressure gradient increases from 2 to about 12 kPa, the flux of a cellulose triacetate membrane increases from about 0.19 to 0.5 kg/m^2 h and the flux of a polyester membrane increases from 0.4 to 2.3 kg/m^2 h.

Previous works showed that the feed temperature is a crucial parameter; with an increase in feed temperature, an exponential increase of water flux was usually observed [13,20,33,54,62]. As the feed temperature increases, the vapor pressure on the feed side increases exponentially, while the vapor pressure on the permeate side remains constant [54,62]. The raising of vapor pressure leads to an increase in the driving force, thus improving the water flux. Secondly, higher temperature makes it easier for water to permeate through the membrane, which is confirmed by the results that the diffusion coefficient of water increases by four times as the feed temperature is raised from 20°C to 65°C [54]. Moreover, a higher temperature increases the frequency and amplitude of thermal motion of the polymer chain, which can broaden the free volumes of polymer, thus making water transport easier. It should be noted that in pervaporation, temperature polarization may occur and a thermal boundary layer forms via the conduction of sensible heat and the transport of latent heat through the evaporation of water [29]. Owing to the heat transfer, the temperature at the membrane/feed interface is lower than the bulk value, while that of the membrane/permeate interface is higher, which reduces the vapor pressure and hence the flux. A remarkable increase in the flux can be achieved by the creation of turbulence at the interface to reduce the thickness of the temperature polarization layers [21].

The overall effect of temperature on flux, as well as the solubility and diffusivity of component within the membrane, can be reflected by an apparent activation energy $E_{J,i}$ based on the Arrhenius relationship [17,71,72]:

$$J_i = J_0 \exp\left(-\frac{E_{J,i}}{RT}\right) \tag{6.6}$$

where J_i is the water flux of compound i, J_0 is the preexponential factor, R is the gas constant, T is the absolute temperature, and $E_{J,i}$ is the apparent activation energy of compound i. A lower apparent activation energy indicates that there is less of an energy barrier and easier diffusion of water across the membrane.

The benefit of higher feed temperature is attributed to an increase in diffusivity and reduction in flow viscosity that occurs on heating. Hence, expenditure on thermal energy will improve the water flux. Alternative thermal sources that could be exploited are solar, geothermal, and industrial waste heat. Mechanical energy in the form of extra applied pressure or a vacuum can likewise be called upon to enhance the flux.

6.4.3 Techno-economic analysis

The pervaporation process has been used extensively for separation of mixtures of aqueous-organic or organic liquids with significant savings demonstrated in both capital and operating costs when compared with the thermal distillation process. However, there are only few studies on the energy requirement and economics of the pervaporation desalination process. This is mainly due to the low water flux associated with current commercial membranes, as a low water flux increases the amount of membrane area required and consequently increases the capital cost required for the membranes. For good performance of pervaporation, high water flux must be obtained with moderate energy consumption. The energy used in the pervaporation process is primarily heat and electricity. For a typical pervaporation desalination process, as shown in Fig. 6.3, the energy requirement of the pervaporation process can be

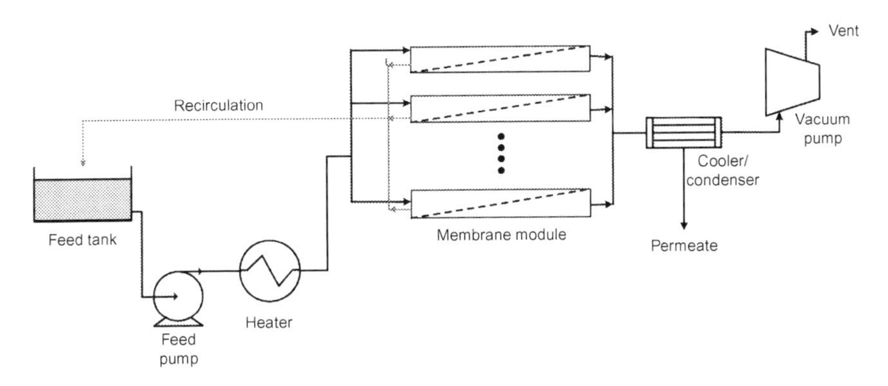

Fig. 6.3 Schematic flow chart of pervaporation process in recirculation mode.

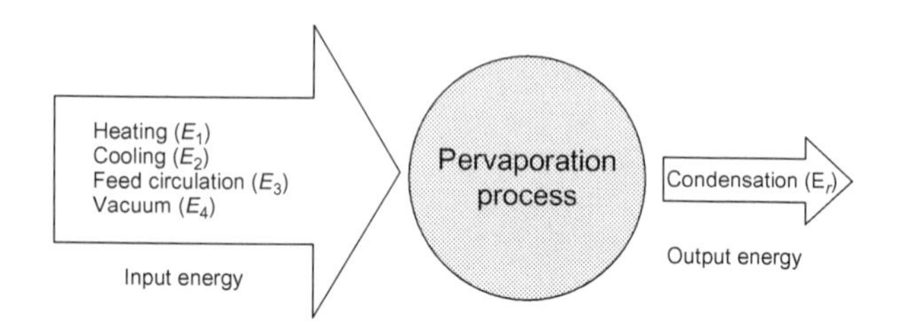

Fig. 6.4 Breakdown of energy requirement for pervaporation.

assessed with the following parameters: the specific energy required for heating, the specific power required for circulating feed and vacuum pump operation, and the specific membrane area. As shown in Fig. 6.4, major energy-consuming components for the process include the feed stream heating, the permeate stream cooling/condensation, the feed pump, and the vacuum pump. Feed heating (Q_h) and permeate cooling energy (Q_c) are classified as the thermal energy requirement, whereas the electrical power associated with feed pump (E_f) and vacuum pump (E_v) is classified as the electrical energy requirement. The latent heat of condensation (E_r) from cooling the water vapor of the outlet permeate stream could be potentially recovered in the process and used for heating the feed.

The overall energy requirement is therefore:

$$E_{\text{Total}} = Q_h + Q_c + E_f + E_v \tag{6.7}$$

If the heat recovery option is considered, depending on the heat recovery efficiency ($x\%$), E_{Total} could be calculated from:

$$E_{\text{Total}} = Q_h + Q_c + E_f + E_v - x\%\ E_v \tag{6.8}$$

Fig. 6.5 shows a breakdown of the thermal energy and electrical power required using the water flux of 11.7 kg/m^2 h and an evaporation efficiency of 90% obtained in a lab recirculation system at a feed inlet temperature of 65°C, feed velocity 0.05 m/s, and vacuum level 6 Torr (800 Pa) [73]. Similar to the MD process, the thermal energy requirement is significant in the pervaporation process. To reduce the process energy required, alternative low-grade, solar, or waste heat can be used to provide the thermal energy for heating the feed. In addition, if the heat recovery option is adopted to recover the latent heat of condensation gained in the condenser, the total energy required for the system could be potentially reduced to a much improved level. This indicates that pervaporation has an advantage against RO at high salt concentration, as its energy needs are essentially independent of the salt concentration in the feed solution. This suggests that pervaporation could be applied in niche markets where RO has limitations, such as RO brine concentration. It should be noted that the value of the low-grade thermal energy becomes higher when the feed temperature is high. It is therefore desirable to improve the membrane permeability as it lowers the required feed temperature and the specific membrane area. Consequently, this reduces the required low-grade thermal energy and the membrane related capital and operating costs.

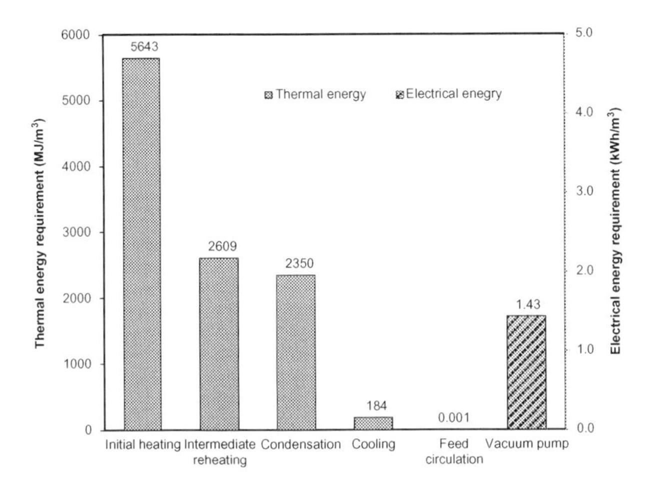

Fig. 6.5 Example of breakdown of thermal and electrical energy requirement for pervaporation process in recirculation mode (feed temperature 65°C, feed velocity 0.05 m/s, vacuum 6 Torr).

6.5 Conclusions and future trends

Pervaporation is widely recognized as a potentially efficient and promising technology for desalination. Desalination by pervaporation is advantageous in coping with a wide feed concentrate range, especially in treating with high concentration saline water. However, despite the pervaporation being extensively used in separation of organic/water and organic/organic compounds, there is no commercial application of pervaporation for desalination. This is mainly due to the low water flux of currently available membranes and the cost of the pervaporation process. It is believed that uptake of the technology will be very fast once there is a breakthrough in membrane materials development since there are already established knowledge and skills in process design and engineering of pervaporation.

Much versatile types of membrane were studied for PV desalination to improve their performance and understand their specific mass transfer and separation mechanisms. Dense polymeric membranes are mostly studied through which mass transport is generally described by solution-diffusion theory. Much effort has been made to improve the sorption and diffusivity of water in the membrane by adjusting the hydrophilicity and free volume of the membrane. The preparation and performance of inorganic membranes such as zeolites and amorphous microporous silica for PV desalination have been studied actively in recent years. The separation mechanism is directly related to the membrane pore structure. The mass transfer and salt rejection of different ions through the charged porous inorganic membranes can be explained jointly by molecular sieving and charge exclusion. For the ion-exchange membranes, a higher capacity in increasing water flux was observed beyond the value calculated from Fick's equation. Also, a higher water flux of the membrane can be observed in the presence of salt water than in pure water, which is attributed to the ion exchange between the charged membranes and the adsorbed certain ions from the feed solution. However, the mechanism of the transport of ions and water in charged membranes for desalination still needs further investigation to guide the design and development of new membranes.

Although some breakthrough on high water flux with some new membrane materials such as GO has been reported, further development is still required to scale up and reduce the cost. Selection of the membrane material is critical as it is related to the inherent permeability. The thickness of the membrane is another vital factor: the active layer should be as thin as possible. To make pervaporation a viable alternative desalination process, it is therefore important to develop a thin pervaporation membrane which can achieve both high water flux and salt rejection. The future growth of the process will still be strongly dependent on the improvement of current membranes or development of novel membrane materials. Furthermore, there is scope to develop novel membrane modules of hollow fiber configuration to replace the dominant plate-and-frame configuration.

Acknowledgments

This work was financially supported by Commonwealth Scientific and Industrial Research Organisation (CSIRO) Manufacturing business unit, and the National Science Foundation of China with the grant No. 21676210.

References

[1] Alghoul MA, Poovanaesvaran P, Sopian K, Sulaiman MY. Review of brackish water reverse osmosis (BWRO) system designs. Renew Sustain Energy Rev 2009;13:2661–7.

[2] Greenlee LF, Lawler DF, Freeman BD, Marrot B, Moulin P. Reverse osmosis desalination: water sources, technology, and today's challenges. Water Res 2009;43:2317–48.

[3] Drobek M, Yacou C, Motuzas J, Julbe A, Ding L, Diniz da Costa JC. Long term pervaporation desalination of tubular MFI zeolite membranes. J Membr Sci 2012;415–416:816–23.

[4] Lin CX, Ding LP, Smart S, da Costa JC. Cobalt oxide silica membranes for desalination. J Colloid Interface Sci 2012;368:70–6.

[5] Drioli E, Stankiewicz AI, Macedonio F. Membrane engineering in process intensification—an overview. J Membr Sci 2011;380:1–8.

[6] García-Rodríguez L. Renewable energy applications in desalination: state of the art. Sol Energy 2003;75:381–93.

[7] Lee S, Boo C, Elimelech M, Hong S. Comparison of fouling behavior in forward osmosis (FO) and reverse osmosis (RO). J Membr Sci 2010;365:34–9.

[8] Jonquieres J, Clemnt R, Lochon P, Neel J, Dresch M, Chrtien B. Industrial state-of-the-art of pervaporation and vapour permeation in the western countries. J Membr Sci 2002;206:87–117.

[9] Feng XS, Huang RYM. Liquid separation by membrane pervaporation a review. Ind Eng Chem Res 1997;36:1048–66.

[10] Chapman PD, Oliveira T, Livingston AG, Li K. Membranes for the dehydration of solvents by pervaporation. J Membr Sci 2008;318:5–37.

[11] Xie Z, Hoang M, Duong T, Ng D, Dao B, Gray S. Sol–gel derived poly(vinyl alcohol)/ maleic acid/silica hybrid membrane for desalination by pervaporation. J Membr Sci 2011;383:96–103.
[12] Duke MC, O'Brien-Abraham J, Milne N, Zhu B, Lin JYS, Diniz da Costa JC. Seawater desalination performance of MFI type membranes made by secondary growth. Sep Purif Technol 2009;68:343–50.
[13] Zwijnenberg H, Koops G, Wessling M. Solar driven membrane pervaporation for desalination processes. J Membr Sci 2005;250:235–46.
[14] Benko KL, Drewes JE. Produced water in the western United States: geographical distribution, occurrence, and composition. Environ Eng Sci 2008;25:239–46.
[15] Naim M, Elewa M, El-Shafei A, Moneer A. Desalination of simulated seawater by purge-air pervaporation using an innovative fabricated membrane, water science and technology: a journal of the international association on water. Pollut Res 2015;72:785–93.
[16] Shao P, Huang RYM. Polymeric membrane pervaporation. J Membr Sci 2007;287:162–79.
[17] Hamouda SB, Boubakri A, Nguyen QT, Amor MB. PEBAX membranes for water desalination by pervaporation process. High Perform Polym 2011;23:170–3.
[18] Sule M, Jiang J, Templeton M, Huth E, Brant J, Bond T. Salt rejection and water flux through a tubular pervaporative polymer membrane designed for irrigation applications. Environ Technol 2013;34:1329–39.
[19] Xie Z, Hoang M, Ng D, Doherty C, Hill A, Gray S. Effect of heat treatment on pervaporation separation of aqueous salt solution using hybrid PVA/MA/TEOS membrane. Sep Purif Technol 2014;127:10–7.
[20] Liang B, Pan K, Li L, Giannelis EP, Cao B. High performance hydrophilic pervaporation composite membranes for water desalination. Desalination 2014;347:199–206.
[21] Huth E, Muthu S, Ruff L. Feasibility assessment of pervaporation for desalinating high-salinity brines. J Water Reuse Desalin 2014;4:109–24.
[22] Kuznetsov YP, Kruchinina EV, Baklagina YG, Khripunov AK, Tulupova OA. Deep desalination of water by evaporation through polymeric membranes. Russ J Appl Chem 2007;80:790–8.
[23] Wijmans JG, Baker RW. The solution-diffusion model a review. J Membr Sci 1995;107:1–21.
[24] Pau DR. Further comments on the relation between hydraulic permeation and diffusion. J Polym Sci Polym Lett Ed 1974;12:1221–30.
[25] Semenova SI, Ohya H, Soontarapa K. Hydrophilic membranes for pervaporation an analytical. Desalination 1997;110:251–86.
[26] Korin E, Ladizhensky I, Korngold E. Hydrophilic hollow fiber membranes for water desalination by the pervaporation method. Chem Eng Process 1996;35:451–7.
[27] Korngold E, Korin E, Ladizhensky I. Water desalination by pervaporation with hollow fiber membranes. Desalination 1996;107:121–9.
[28] Cabasso I, Komgold E, Liu ZZ. On the separation of alcohol water mixtures by polyethylene ion exchange membranes. J Polym Sci Polym Lett Ed 1985;23:577–81.
[29] Elma M, Yacou C, Wang DK, Smart S, Diniz da Costa JC. Microporous silica based membranes for desalination. Water 2012;4:629–49.
[30] Cho CH, Oh KY, Kim SK, Yeo JG, Sharma P. Pervaporative seawater desalination using NaA zeolite membrane: mechanisms of high water flux and high salt rejection. J Membr Sci 2011;371:226–38.
[31] Duke MC, Mee S, da Costa JC. Performance of porous inorganic membranes in non-osmotic desalination. Water Res 2007;41:3998–4004.

[32] Khajavi S, Jansen JC, Kapteijn F. Production of ultra pure water by desalination of seawater using a hydroxy sodalite membrane. J Membr Sci 2010;356:52–7.
[33] Malekpour A, Millani MR, Kheirkhah M. Synthesis and characterization of a NaA zeolite membrane and its applications for desalination of radioactive solutions. Desalination 2008;225:199–208.
[34] Malekpour A, Samadi-Maybodi A, Sadati MR. Desalination of aqueous solutions by LTA and MFI membranes using pervaporation method. Braz J Chem Eng 2011;28: 669–77.
[35] An W, Zhou X, Liu X, Chai PW, Kuznicki T, Kuznicki SM. Natural zeolite clinoptilolite-phosphate composite membranes for water desalination by pervaporation. J Membr Sci 2014;470:431–8.
[36] Ladewig BP, Tan YH, Lin CXC, Ladewig K, Diniz da Costa JC, Smart S. Preparation, characterization and performance of templated silica membranes in non-osmotic desalination. Materials 2011;4:845–56.
[37] Xu R, Lin P, Zhang Q, Zhong J, Tsuru T. Development of Ethenylene-bridged Organosilica membranes for desalination applications. Ind Eng Chem Res 2016;55:2183–90.
[38] Wang S, Wang DK, Smart S, Diniz da Costa JC. Improved stability of ethyl silicate interlayer-free membranes by the rapid thermal processing (RTP) for desalination. Desalination 2017;402:25–32.
[39] Yang H, Elma M, Wang DK, Motuzas J, Diniz da Costa JC. Interlayer-free hybrid carbon-silica membranes for processing brackish to brine salt solutions by pervaporation. J Membr Sci 2017;523:197–204.
[40] Wang S, Wang DK, Motuzas J, Smart S, Diniz da Costa JC. Rapid thermal treatment of interlayer-free ethyl silicate 40 derived membranes for desalination. J Membr Sci 2016;516:94–103.
[41] Zhou C, Zhou J, Huang A. Seeding-free synthesis of zeolite FAU membrane for seawater desalination by pervaporation. Microporous Mesoporous Mater 2016;234:377–83.
[42] Darmawan A, Karlina L, Astuti Y, Sriatun JM, Wang DK, da Costa JCD. Structural evolution of nickel oxide silica sol-gel for the preparation of interlayer-free membranes. J Non Cryst Solids 2016;447:9–15.
[43] Zhu Y, Gupta KM, Liu Q, Jiang J, Caro J, Huang A. Synthesis and seawater desalination of molecular sieving zeolitic imidazolate framework membranes. Desalination 2016;385:75–82.
[44] Staudt-Bickel C, Lichtenthaler RN. Pervaporation thermodynamic properties and selection of membrane polymers, polymer. Science 1994;36:1628–46.
[45] Lin J, Murad S. A computer simulation study of the separation of aqueous solutions using thin zeolite membranes. Mol Phys 2001;99:1175–81.
[46] Quiñones-Bolaños E, Zhou H, Soundararajan R, Otten L. Water and solute transport in pervaporation hydrophilic membranes to reclaim contaminated water for micro-irrigation. J Membr Sci 2005;252:19–28.
[47] Ju H, Sagle AC, Freeman BD, Mardel JI, Hill AJ. Characterization of sodium chloride and water transport in crosslinked poly(ethylene oxide) hydrogels. J Membr Sci 2010;358:131–41.
[48] Wang P, Chung T-S. Recent advances in membrane distillation processes: Membrane development, configuration design and application exploring. J Membr Sci 2015;474:39–56.
[49] Wang Q, Li N, Bolto B, Hoang M, Xie Z. Desalination by pervaporation: a review. Desalination 2016;387:46–60.

[50] El-Bourawi MS, Ding Z, Ma R, Khayet M. A framework for better understanding membrane distillation separation process. J Membr Sci 2006;285:4–29.
[51] Camacho L, Dumée L, Zhang J, Li J-d, Duke M, Gomez J, et al. Advances in membrane distillation for water desalination and purification applications. Water 2013;5:94–196.
[52] Urtiaga AM, Gorri ED, Ruiz G, Ortiz I. Parallelism and differences of pervaporation and vacuum. Sep Purif Technol 2001;22–23:327–37.
[53] Alkhudhiri A, Darwish N, Hilal N. Membrane distillation: a comprehensive review. Desalination 2012;287:2–18.
[54] Xie Z, Ng D, Hoang M, Duong T, Gray S. Separation of aqueous salt solution by pervaporation through hybrid organic–inorganic membrane: effect of operating conditions. Desalination 2011;273:220–5.
[55] Wang Q, Lu Y, Li N. Preparation, characterization and performance of sulfonated poly(styrene-ethylene/butylene-styrene) block copolymer membranes for water desalination by pervaporation. Desalination 2016;390:33–46.
[56] Elewa MM, El-Shafei AA, Moneer AA, Naim MM. Effect of cell hydrodynamics in desalination of saline water by sweeping air pervaporation technique using innovated membrane. Desalin Water Treat 2016;57:23293–307.
[57] Chaudhri SG, Rajai BH, Singh PS. Preparation of ultra-thin poly(vinyl alcohol) membranes supported on polysulfone hollow fiber and their application for production of pure water from seawater. Desalination 2015;367:272–84.
[58] Swenson P, Tanchuk B, Gupta A, An W, Kuznicki SM. Pervaporative desalination of water using natural zeolite membranes. Desalination 2012;285:68–72.
[59] Wijaya S, Duke MC, Diniz da Costa JC. Carbonised template silica membranes for desalination. Desalination 2009;236:291–8.
[60] Singh PS, Chaudhri SG, Kansara AM. Cetyltrimethylammonium bromide–silica membrane for seawater desalination through pervaporation. Bull Mater Sci 2015;38:565–72.
[61] Kujawski W, Krajewska S, Kujawski M, Gazagnes L, Larbot A, Persin M. Pervaporation properties of fluoroalkylsilane (FAS) grafted ceramic membranes. Desalination 2007;205:75–86.
[62] Liang B, Zhan W, Qi G, Lin S, Nan Q, Liu Y, et al. High performance graphene oxide/polyacrylonitrile composite pervaporation membranes for desalination applications. J Mater Chem A 2015;3:5140–7.
[63] Feng B, Xu K, Huang A. Covalent synthesis of three-dimensional graphene oxide framework (GOF) membrane for seawater desalination. Desalination 2016;394:123–30.
[64] Xu K, Feng B, Zhou C, Huang A. Synthesis of highly stable graphene oxide membranes on polydopamine functionalized supports for seawater desalination. Chem Eng Sci 2016;146:159–65.
[65] Feng B, Xu K, Huang A. Synthesis of graphene oxide/polyimide mixed matrix membranes for desalination. RSC Adv 2017;7:2211–7.
[66] Cheng C, Shen L, Yu X, Yang Y, Li X, Wang X. Robust construction of a graphene oxide barrier layer on a nanofibrous substrate assisted by the flexible poly(vinylalcohol) for efficient pervaporation desalination. J Mater Chem A 2017;5:3558–68.
[67] Lau WJ, Ismail AF, Misdan N, Kassim MA. A recent progress in thin film composite membrane: a review. Desalination 2012;287:190–9.
[68] Smitha B, Suhanya D, Sridhar S, Ramakrishna M. Separation of organic–organic mixtures by pervaporation—a review. J Membr Sci 2004;241:1–21.
[69] Baker RW. Membrane technology and applications. Hoboken, NJ: John Wiley & Sons Ltd; 2004:355–92.

[70] Bowen TC, Kalipcilar H, Falconer JL, Noble RD. Pervaporation of organic/water mixtures through B-ZSM-5 zeolite membranes on monolith supports. J Membr Sci 2003;215:235–47.
[71] Feng XS, Huang RYM. Estimation of activation energy for permeation in pervaparation process. J Membr Sci 1996;118:127–31.
[72] Peng F, Lu L, Sun H, Jiang Z. Analysis of annealing effect on pervaporation properties of PVA-GPTMS hybrid membranes through PALS. J Membr Sci 2006;281:600–8.
[73] Xie Z. Hybrid organic-inorganic pervaporation membranes for desalination. In: School of Engineering & Science. Melbourne: Victoria University; 2012.

Humidification-dehumidification desalination cycle

7

Saeed Dehghani, Abhijit Date, Aliakbar Akbarzadeh
RMIT University, Melbourne, VIC, Australia

Nomenclature

Symbol

T	temperature (°C)
$\dot{m}$	flow rate (kg/s)
$\dot{Q}$	heat rate (kW)
$\dot{H}$	enthalpy rate (kW)
mr	water to air mass flow ratio
h	specific enthalpy (kJ/kg); heat transfer coefficient (W/m^2 K)
RR	recovery ratio
GOR	gain output ratio
HCR	modified heat capacity rate ratio for HME devices
SWP	specific water production (kg/m^2 day)
c_p	specific heat capacity at constant pressure (J/kg K)
q	heat transfer (kW)
NTU	number of transferred units
C	heat capacity at constant pressure (W/K); heat capacity ratio
A	area (m^2)
Me	Merkel number
R_{wall}	thermal resistance of wall (K/W)
D	diameter (m)
L	length (m)
R	thermal resistance (K/W)
R''	fouling factor
q	heat transfer rate (W)
I	modified Bessel function
x	length (m)

Greek letters

ω	absolute humidity of dry air or humidity ratio (kg$_w$/kg$_a$)
φ	relative humidity
ε	effectiveness
Δ	difference or change
η	efficiency

Emerging Technologies for Sustainable Desalination Handbook. https://doi.org/10.1016/B978-0-12-815818-0.00007-2

Subscripts

a	air
da	dry air
db	dry bulb
dw	distilled water
b	bottom
t	top
sw	seawater
max	maximum
min	minimum
m	middle
br	brine
h	humidifier
d	dehumidifier
i	inside
o	outside
s	surface
f	fin
v	vapor
in	inlet
sc	solar collector

Acronyms

CAOW	closed-air open-water system
CWOA	closed-water open-air
HDH	humidification-dehumidification
HME	heat and mass exchanger

7.1 Introduction

The availability of potable water is essential requirement for sustaining human life on Earth. Although there is an abundant amount of freshwater in nature, the lack of potable water in some regions of the world is still a serious issue. For some of these areas, seawater or brackish water desalination can be a beneficial solution. There are two major desalination processes by which impurities are removed from water: (i) evaporation-based methods and (ii) semipermeable membrane-based methods [1].

The International Desalination Association (IDA) estimated in 2010 that there were over 15,000 contracted desalination plants with a capacity of 71.7 billion liters of drinkable water per day [2]. The online capacity of these plants was over 65.2 billion liters per day, indicating that an additional 6.5 billion liters per day of future capacity was under construction. Humidification-dehumidification (HDH) desalination technology has the potential to dramatically increase the availability of potable water.

This chapter presents a detailed account of the HDH desalination process. HDH system components including the humidifier, dehumidifier, and heat source are separately analyzed in the context of thermodynamic laws as well as heat and mass transfer analysis. This chapter also explains performance parameters for the HDH system which are beneficial for evaluating system performance.

7.2 HDH system

HDH desalination distillation, a well-developed desalination technology for small-scale applications, has attracted researchers' attention for further development. The main characteristics of HDH desalination, including its ability to supply water to remote areas, its small-scale production rate, and the ease with which it may be coupled with solar energy make it a possible alternative to other desalination technologies. For instance, people living in remote and arid areas of the Middle East or on small islands elsewhere lack sufficient access to potable water, making it more economical to produce drinkable water on-site than to import it. These regions also possess substantial solar radiation and a significant amount of seawater or underground saline water [3].

HDH desalination technology can be considered a viable alternative when large-scale thermal desalination systems such as multistage flash (MSF) and multieffect desalination (MED) are inappropriate choices due to cost and size constraints, or where there is inadequate electric power supply to operate reverse osmosis (RO).

It should be noted here that the predecessor of the single-stage HDH cycle is a simple solar still, whose cost is very high. Even with very good insulation, the solar still produces distilled water inefficiently depending on the operating conditions. The still's major problem, its low efficiency, is mainly attributable to its glass cover, where water condensation results in high loss of energy in the form of latent heat. The need to eliminate this heat loss leads to consideration of more efficient methods, such as the HDH cycle. HDH desalination is a more efficient modified version of a solar still [4]. The HDH desalination system imitates the natural rain phenomenon to generate distilled water. In nature's system, vapor is produced from seawater when sun shines on it; next, this vapor is carried by air in cloud form until its temperature drops, causing it to condense into rain. A humidifier evaporates seawater to produce moist air, and a dehumidifier condenses that air's moisture.

This process separates pure water from a mixture of impure seawater. A humidifier transfers water vapor into an air stream, while concentrated seawater is collected at the bottom and rejected as brine. The vapor inside the air stream from the humidifier is condensed out of the air stream and extracted from the system. Water can be heated after preheating in a dehumidifier or before spraying it in a humidifier. Air can be heated before entering the humidifier. Circulating air and water through the system requires electric work which is negligible in comparison to the thermal energy consumed by the cycle.

It is essential to realize the relative technical advantages of each of these cycles and select the one that is most beneficial with respect to cycle efficiency and cost of water production. Fig. 7.1 presents different types of HDH desalination cycles.

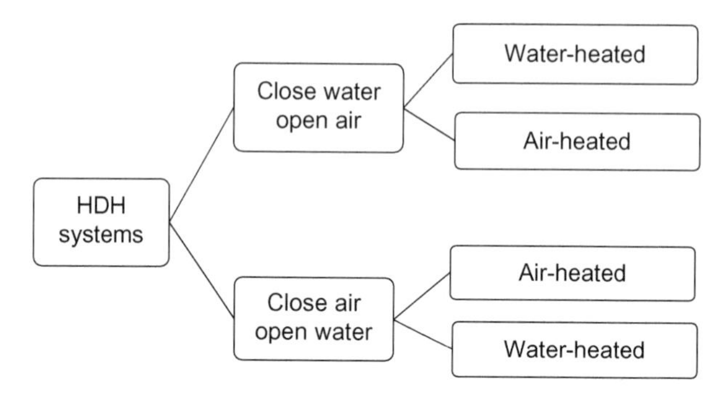

Fig. 7.1 Classification of the HDH systems types.

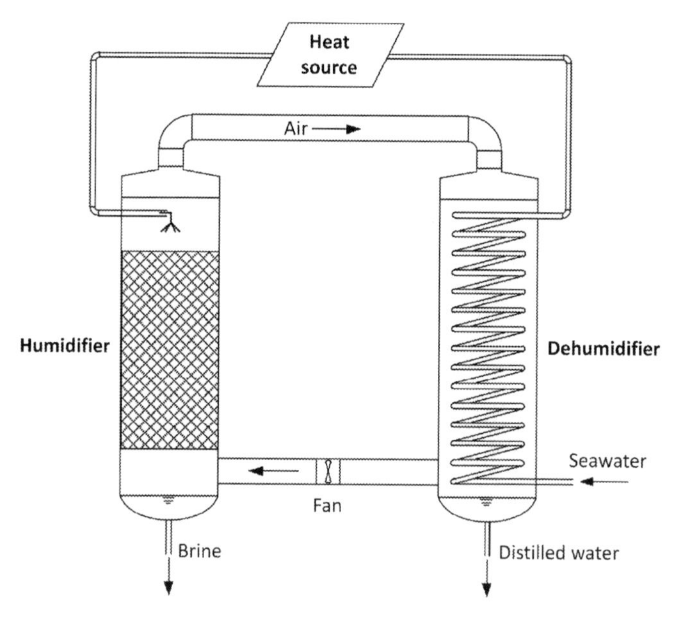

Fig. 7.2 Overview of the CAOW water-heated humidification-dehumidification desalination cycle.

HDH cycles are classified according to the configuration with which they heat the stream, and whether that stream is open or closed. For example, a closed-air open-water (CAOW) water-heated cycle is demonstrated in Fig. 7.2. Other types such as the CAOW air-heated cycle, closed-water open-air (CWOA) water-heated cycle, CWOA cycle, and the air-heated cycle can be used for different applications [5]. In closed air/water cycles, air/water is circulated in a closed loop between a humidifier and dehumidifier while the water/air is in an open loop. The air in these systems can move by either natural or forced circulation by adding a fan or blower [6]. These cycles will be discussed in detail in the following sections.

7.2.1 CAOW water-heated cycles

Fig. 7.2 shows the HDH cycle. The air stream coming to the humidifier is moisturized by spraying hot seawater, which increases the air stream's temperature, then, the hot, moist air goes to another device called a dehumidifier, which decreases the air stream's temperature using cold seawater, condensing the air stream vapor and producing distilled water. The cycle consists of two main components: a humidifier and dehumidifier. A humidifier is a direct-contact, counter-flow, heat-and-mass exchange device employed to humidify the air stream. Dehumidifiers, indirect-contact, counter-flow, heat-and-mass exchange devices, condense liquid water from hot, moist air. The heat source required to heat the air or water stream can be an electric or gas heater, solar collector, etc. A heat-and-mass exchanger (HME) is a device which simultaneously transfers heat and mass between streams (in this case, air and seawater).

7.2.2 Multieffect CAOW water-heated cycle

One way to improve the HDH cycle's efficiency is to use a system with multiple extractions (multiextraction) of air from a humidifier. Fig. 7.3 shows a schematic diagram of the multiextraction HDH cycle in which hot, humid air can be extracted from different points of a humidifier and forwarded correspondingly to the dehumidifier. However, this system is more complex in design and fabrication; it can yield higher heat recovery in a dehumidifier. This method leads to a small temperature gap between the humid air coming from the humidifier and seawater passing through

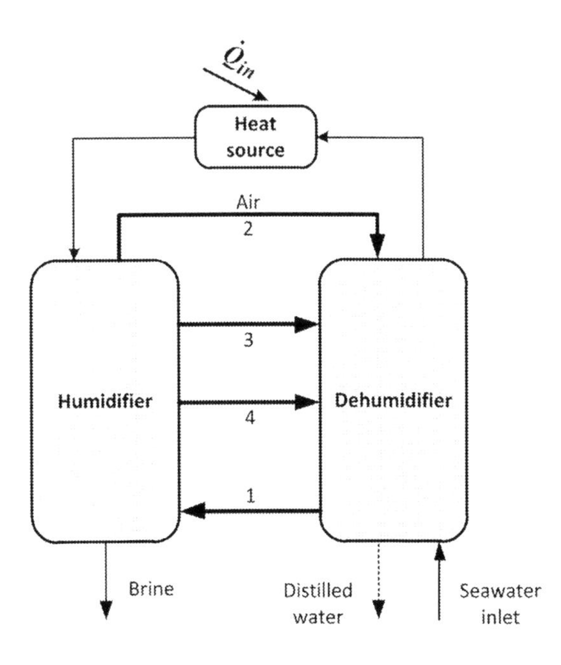

Fig. 7.3 Schematic diagram of the multieffect CAOW water-heated HDH cycle.

the dehumidifier coil to keep the process running. A commercial water management company called Tinox GmbH utilizes this type of HDH cycle. This is perhaps the first case in which the HDH cycle concept has been commercialized.

7.2.3 CAOW air-heated cycles

The next type of HDH cycle is the CAOW air-heated cycle, and a schematic diagram is shown in Fig. 7.4. Air circulates through the cycle and heats up before the humidifier. The air solar collector can be used to increase the air temperature to 80–90°C. The multiextraction concept can be applied in this configuration as well. Humidifying the air in the humidifier simultaneously saturates the air and causes its temperature to drop. Then, this humid, warm air moves to a dehumidifier for further processing. The main drawback of this method is that air's lower heat capacity prevents it from being moisturized at high temperature when leaving the humidifier. Air cannot effectively carry vapor after humidification, which results in less water distillation in the dehumidifier.

To solve this problem, a multistage heating and humidification cycle can be implemented. At each stage, the air is heated, then enters separate humidifiers to be sufficiently saturated [7]. After ward, this hot, moist air is ready for dehumidification. Higher air temperature obviously improves cycle efficiency; however, it creates a significant disadvantage for solar air collectors. It should be noted that a multistage method simultaneously results in higher water production and higher thermal energy consumption.

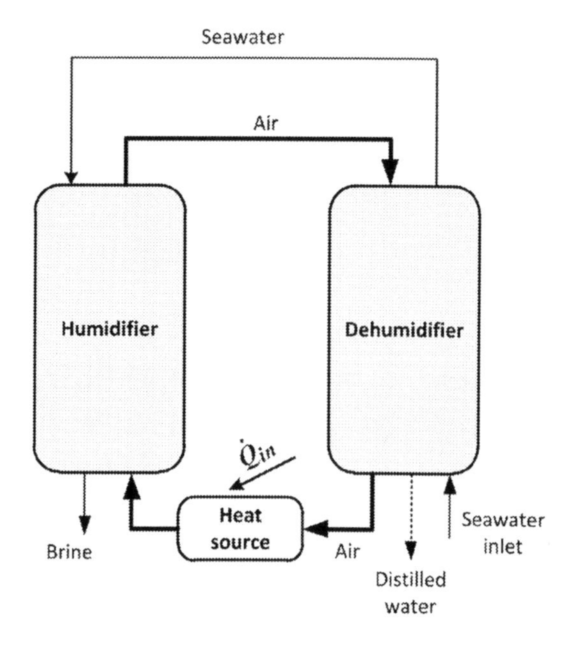

Fig. 7.4 Schematic diagram of the CWOA air-heated HDH cycle.

7.2.4 CWOA water-heated cycles

A schematic diagram of the CWOA water-heated system is shown in Fig. 7.5. In this cycle configuration, ambient air is introduced through the humidifier and moisturized by spraying hot seawater. After preheating in a dehumidifier, this seawater is heated utilizing an external heat source (such as a solar collector) and a water heat recovery system, and is run through a closed loop. To compensate for the seawater's concentration after the humidifier's makeup, more seawater is applied. After dehumidification, the air will be discharged to the ambient atmosphere.

The drawback of this cycle configuration is that when the ambient air temperature and humidity at the humidifier's inlet are not low enough, the outlet temperature of the seawater in the humidifier will rise. As a result, the moist air in the dehumidifier cannot be sufficiently cooled to distill more freshwater than in open cycles. But using an efficient humidifier to reduce the outlet seawater's temperature to a level close to the ambient temperature can considerably improve the cycle's efficiency, bringing it close to that of an open-water system.

7.2.5 CWOA air-heated cycles

Another type of the CWOA HDH cycle heats the air instead of the water (Fig. 7.6). In this configuration, ambient air is introduced to a heat source (such as a solar air heater), then moisturized by a humidifier. Humidification considerably reduces the

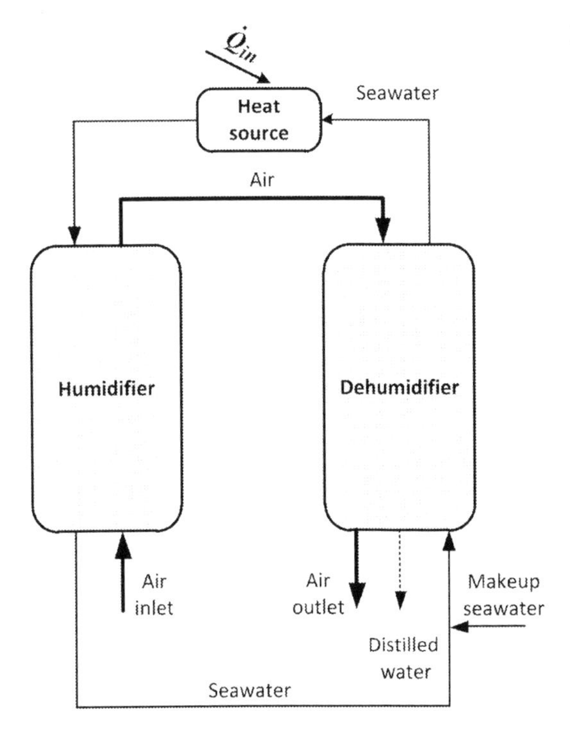

Fig. 7.5 Schematic diagram of the CAOW water-heated HDH cycle.

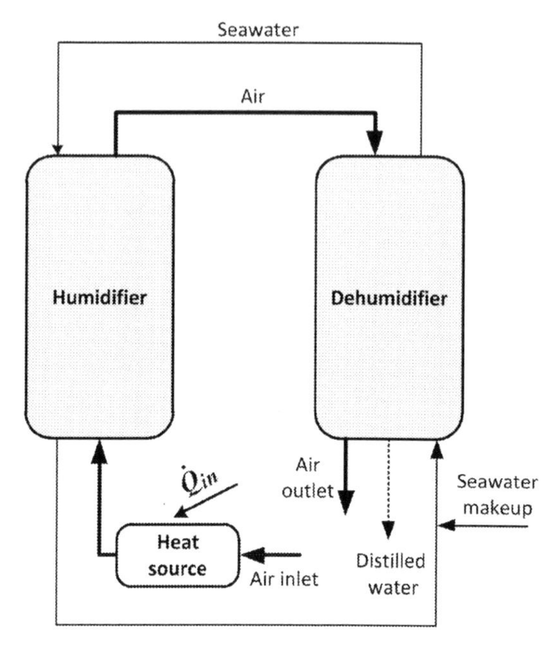

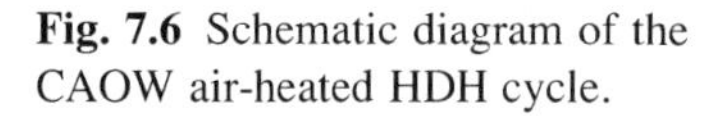
Fig. 7.6 Schematic diagram of the CAOW air-heated HDH cycle.

air temperature, which substantially affects cycle efficiency. The drawback of this cycle is that air's low specific heat capacity in comparison to water) enables its air temperature to rise very well at a humidifier. This lowers the vapor content of the humid air carried to the dehumidifier, decreasing the water production rate. A multistage heating and humidifying process can improve this. A more efficient humidifier would help lower the brine temperature at the humidifier outlet when the ambient air is too hot, improving cycle efficiency.

Air-heated systems have higher thermal energy consumption than water-heated systems. This is because air heats the water in the humidifier, and this energy is not subsequently recovered from the water; in the water-heated cycle, the water stream is cooled in the humidifier, and part of the thermal energy (which is carried by the air) is recovered in the dehumidifier, too.

7.2.6 Cycle potential

The HDH process works because air can be mixed with large quantities of water vapor. Air's ability to carry water vapor rises with its temperature. For example, 1 kg of dry air can carry 0.5 kg of vapor and about 2800 kJ when its temperature rises from 30°C to 80°C. When this circulating air is in contact with seawater, a certain quantity of vapor will be extracted by the air, which causes cooling. On the other hand, cooling the air through contact with a cooler surface condenses its vapor content, producing distilled water. Fig. 7.7 shows the relationship between dry bulb temperature and the absolute humidity of air at different humidity ratios.

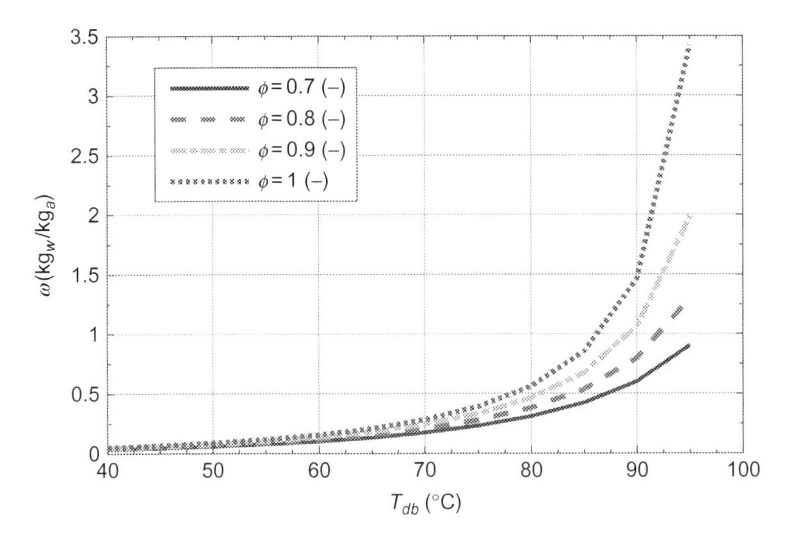

Fig. 7.7 Effect of dry bulb temperature on the vapor content of humid air (ω) at various relative humidities (ϕ).

It can be seen that for a temperature higher than 90°C, the air's vapor content will substantially increase. In other words, a cycle configuration that can increase moist air temperature at the humidifier outlet has more potential for water production.

7.3 Humidifier

Humidification is a process in which seawater is brought into contact with air in order to increase the air's moisture content. Water diffuses into the air as vapor and increases of the air's humidity. The force driving humidification is the difference in concentration between the water-air interface and the water vapor in the air. This concentration difference depends on the vapor pressure at the gas-liquid interface and the partial pressure of water vapor in the air. Various devices can be used as humidifiers, such as packing-filled towers, spray towers, bubble columns, and wetted-wall towers [8].

Spray towers consist of an empty cylindrical vessel made of steel or plastic, and nozzles that spray liquid into the vessel. The inlet gas stream usually enters at the bottom of the tower and moves upward, while the liquid is sprayed downward from one or more levels. This type of spray tower is simple in design, with a low-pressure drop on the gas side, low maintenance cost, and fairly inexpensive capital cost. The drawbacks of these devices include a substantial pressure drop on the water side caused by the spray nozzles. For better performance, drift eliminators must be installed before humid air exits the vessel [9]. The diameter-to-length ratio is a significant variable in spray tower design. When this ratio is high, air will be completely mixed with the sprayed water, while a low ratio means that sprayed water rapidly touches the tower walls, creating a liquid film on the walls and reducing system the effectiveness. Direct heat and mass transfer principles must be implemented to design spray towers. Empirical correlations and design procedures are provided in Ref. [9].

Another type of humidifier is the bubble column. In a bubble column, a vessel is filled with hot seawater and air is injected through the vessel using a sparger located at the vessel's bottom. The sparger injects a turbulent stream of air directly into the water, enabling vapor diffusion into the bubbles, moisturizing the air. Bubble towers are simple in design, and numerous forms of bubble tower construction exist. The water can be in parallel flow or a countercurrent. The diffusion of water into the air bubbles depends on various parameters, such as air and water temperature, heat and mass transfer coefficient, the ratio of bubble-to-water volume, and bubble characteristics such as size and velocity.

To improve air humidification efficiency using seawater, the humidifier's heat and mass transfer area should be improved. This will substantially increase water droplet travel timing. In chemical engineering, devices that use this technique are called packed beds. A packed bed is a hollow tube, pipe, or other vessel filled with a packing material. The packing can be randomly filled with small objects like ranching rings, or it can be a specifically designed, structured packing. The seawater sprays at the top of column while air blows from the bottom and passes through the packing materials, resulting in humidification. In industry, wet cooling towers are well-known packed-bed devices. Packed-bed towers' higher effectiveness make them quite popular for air-to-water systems. Different packing materials may be used, including ceramic ranching rings, wooden-slat packing, honeycomb paper, corrugated cellulose material, and plastic packing.

In selecting a packing material, different factors should be considered, including volumetric surface area (heat-and-mass-transfer surface area), maximum temperature tolerance, durability, pressure drop, and cost. Film fills are a packing which provides high thermal performance along with high surface area per volume, demonstrating lower pressure drop as it increases. Various film fill materials' performance is presented in Ref. [10]. The main disadvantage of choosing a film fill is scaling, which decreases the effectiveness of the humidification process.

As shown in Fig. 7.2, a humidifier operation is like a counterflow, wet-cooling tower in which one fluid stream of hot water sprays over a packing material with a high surface area and another stream is a mixture of air and water vapor. These two streams of seawater and air are in direct contact with each other and a simultaneous transfer of heat and mass occurs throughout the chamber from water to the air stream. Any packing fill with high surface area per volume and tolerance for high temperatures over 80°C (like PVC packing already used in wet cooling towers or wooden packing used in evaporative coolers) can be used for a humidifier.

To evaluate humidifier performance and study the effect of related parameters on humidifier effectiveness, first considering a humidifier a black box. The energy and mass balance is applied in order to thermodynamically investigate the relationship between input and output variables. Then, using a ε-NTU model, heat and mass transfer analysis through the humidifier is performed.

7.3.1 Mass and energy balance

The following assumptions have been made to create the HME devices' theoretical modeling:

- Cycles operate at a steady state and in steady flow conditions.
- All components are adiabatic, and there is no heat loss from any of the cycle components to their surroundings.
- Pumping and fan powers are negligible in comparison to the total thermal energy input of the system.
- Kinetic and potential energy terms are excluded in the energy balance.

Considering a control volume around the humidifier, as shown in Fig. 7.8, and applying mass and energy balance, the governing equations can be obtained as follows:

$$\dot{m}_{sw} + \dot{m}_{da}\omega_{a,b} = \dot{m}_{br} + \dot{m}_{da}\omega_{a,t} \tag{7.1}$$

$$\dot{m}_{sw}h_{sw,t} + \dot{m}_{da}h_{a,b} = \dot{m}_{br}h_{br} + \dot{m}_{da}h_{a,t} \tag{7.2}$$

It should be noted that the dry mass flow of air is constant through the humidifier. Also, enthalpy of humid air is considered a binary mixture of dry air and water vapor, or, in other words: $h_a = h_{da} + \omega h_v$.

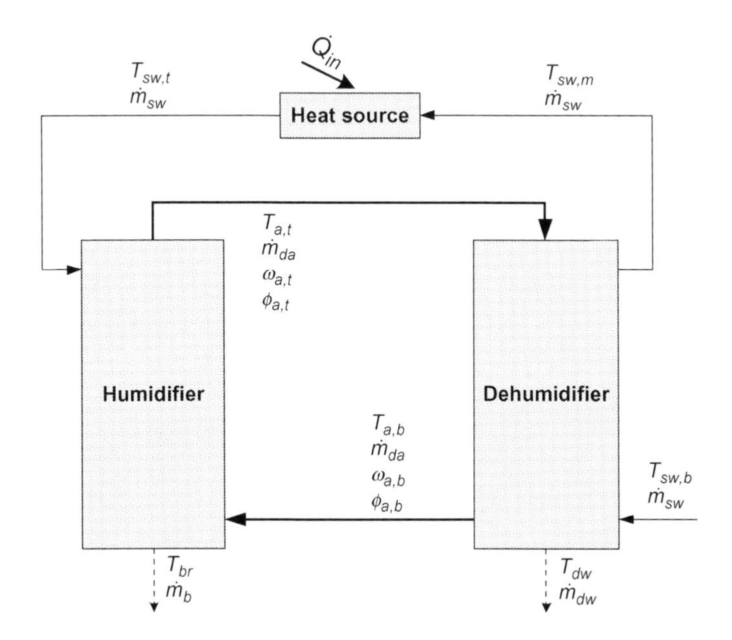

Fig. 7.8 Schematic diagram of the CAOW water-heated HDH cycle.

The thermodynamic properties of moist air and water are provided in the ASHRAE handbook [11]. In addition, the Engineering Equation Solver (EES) software [12] calculates the moist air properties by the formulation presented by Hyland and Wexler [13], as well as water properties using the formulation of IAPWS (International Association for Properties of Water and Steam) [14], which are accurate equations of state to model the properties of moist air and water.

The thermophysical properties of seawater including density, specific heat capacity, thermal conductivity, boiling point elevation, latent heat of vaporization, specific enthalpy, and specific entropy at different salinity, temperature, and pressure can be obtained in Refs. [15,16].

7.3.2 Humidifier effectiveness

To realize outlet stream conditions in the humidifier, a supplementary effectiveness equation needs to be considered for thermodynamic investigations. Effectiveness compares the actual versus ideal thermal energy transferred from each stream and is defined as actual enthalpy variation to the maximum possible enthalpy variation, or in other words, $\varepsilon = \Delta\dot{H}/\Delta\dot{H}_{\max}$ [17]. Therefore, the effectiveness of the humidifier would be expressed as follows:

$$\varepsilon_h = \max\left(\frac{\dot{H}_{a,t} - \dot{H}_{a,b}}{\dot{H}^{\text{ideal}}_{a,t} - \dot{H}_{a,b}}, \frac{\dot{H}_{sw,t} - \dot{H}_{br}}{\dot{H}_{sw,t} - \dot{H}^{\text{ideal}}_{br}}\right) \tag{7.3}$$

In a humidifier, the ideal outlet air enthalpy occurs when the outlet air is fully saturated at the water inlet temperature, and the ideal outlet seawater enthalpy is when its temperature is equivalent to the inlet air dry-bulb temperature.

7.3.3 Effectiveness-NTU relation of the humidifier

Conceptually, effectiveness is the ratio of the actual heat transfer to the maximum possible heat transfer which can be applied for both the heat exchanger and HME device, and the general form for heat exchangers is as follows:

$$\varepsilon = \frac{q}{q_{\max}} = \frac{\Delta\dot{H}}{\Delta\dot{H}_{\max}} \tag{7.4}$$

where q is the total rate of heat transfer between the hot and cold fluids $q = \Delta\dot{H}_c = \Delta\dot{H}_h = \Delta\dot{H}$. In addition, $q_{\max}$ is the maximum transferred heat or maximum enthalpy rate change, and can be as follows:

$$\Delta\dot{H}_{\max} = \min\left(\Delta\dot{H}_h^{\text{ideal}}, \Delta\dot{H}_c^{\text{ideal}}\right) \tag{7.5}$$

The ideal enthalpy rate change of hot fluid happens when the hot fluid outlet temperature is equal to the cold fluid inlet temperature. Also, the ideal enthalpy rate change of cold fluid happens when the cold fluid outlet temperature is equal to the hot fluid inlet temperature.

The maximum heat transfer is limited by irreversibility of the system due to the second law of thermodynamics. In heat exchangers, it is straightforward to determine the relationship between effectiveness and number of transferred units (NTUs) by applying energy balance to the differential element in the hot and cold fluids and considering some approximations such as constant fluid specific heats (c_p) as well as the constant overall heat transfer coefficient (UA). A differential element is shown in Fig. 7.9 for a counterflow heat exchanger for ε-NTU equation extraction for better understanding [18,19].

The same approach can be applied for determining ε-NTU relation in the HME devices. Nonetheless, in HMEs, the capacity rate ratio (HCR) is not constant, as it strongly depends on the amount of mass transferred from one stream to another. Moreover, it is difficult to determine the overall mass transfer coefficient between the stream exchanging heat and mass simultaneously. Then, the Lewis factor which relates the mass transfer coefficient to the heat transfer coefficient can be used. Usually, the Lewis factor can be experimentally determined. An empirical relation of the Lewis factor for air-water systems was developed by Bosnjakovic [20].

Analytical expressions for the effectiveness of an HME as a function of the parameters like heat capacities and heat and mass transfer characteristics are generally not available unless significant approximations are considered. Using the numerical simulation technique is a convenient way for solving the governing equations of the simultaneous HME.

Merkel used the differential control volume method for cooling towers and extracted the governing equations in the form of three differential equations [21]. Following that, Jaber and Webb [19], by considering some approximations, solved the equations and presented the analytical relation for effectiveness by analogy to a counterflow heat exchanger. The effectiveness relation has been used for cooling towers calculations. Finally, Narayan et al. [17] modified the ε-NTU relation presented by Jaber and Webb, and they showed that a better-correlated numerical result for energy effectiveness can be

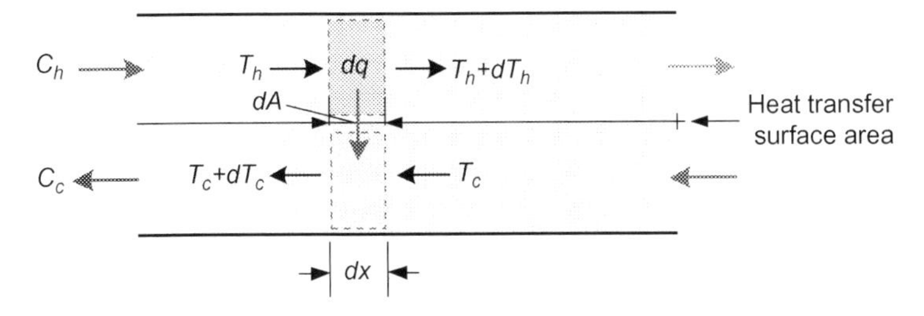

Fig. 7.9 Differential element in a counterflow heat exchanger.

obtained by comparing their results with Jaber and Webb. Their modified ε-NTU relation that can be used for humidifier effectiveness is as follows:

$$\varepsilon_h = \frac{1 - \exp(-\mathrm{NTU}(1 - \mathrm{HCR}))}{1 - \mathrm{HCR} \cdot \exp(-\mathrm{NTU}(1 - \mathrm{HCR}))} \tag{7.6}$$

where NTU and HCR are the number of transferred units and heat capacity rate ratio, respectively. They are defined as follows:

$$\mathrm{HCR} = \min\left(\frac{\Delta\dot{H}\mathrm{max}_{,sw}}{\Delta\dot{H}\mathrm{max}_{,a}}, \frac{\Delta\dot{H}\mathrm{max}_{,a}}{\Delta\dot{H}\mathrm{max}_{,sw}}\right) \tag{7.7}$$

$$\begin{cases} \mathrm{NTU} = Me \ \text{ if } \Delta\dot{H}\mathrm{max}_{,sw} > \Delta\dot{H}\mathrm{max}_{,a} \\ \mathrm{NTU} = Me{,}mr \ \text{ if } \Delta\dot{H}\mathrm{max}_{,a} > \Delta\dot{H}\mathrm{max}_{,sw} \end{cases} \tag{7.8}$$

In a humidifier, the maximum enthalpy change of air is when the air is in its ideal outlet condition; in addition, maximum enthalpy change of seawater is when the outlet seawater is in its ideal outlet condition.

Also, Me is the Merkel number, which is a function of the nature of the packing and flow characteristics. Packing fill manufacturers mainly provide the Merkel number within product specification, and it is a function of packing height and mass flow rate ratio of liquid to gas (mr).

7.4 Dehumidifier

There are wide ranges of heat exchangers that can be used in the HDH cycle as dehumidifiers, for example, flat-plate heat exchangers made of polypropylene (double-webbed slabs) [22]. Other most commonly used types include finned-tube heat exchangers. There have been various types of dehumidifiers used by researchers. For instance, Farid et al. [23] designed different condensers for the HDH cycle in the pilot plant scale. The dehumidifier was made of a long copper-galvanized steel tube (3 m length, 170 mm diameter) with 10 longitudinal fins of 50 mm height on the outer tube surface and 9 fins on the inner side. Also, in another project, they utilized a simplified stack of flat condensers made of 2 m × 1 m galvanized steel plates with long copper tubes mounted on each side of the plate for supplying a large heat-and-mass transfer surface area. They made fairly large condensers, as the heat transfer coefficients on both the air and water sides were small. The heat-and-mass transfer coefficients mainly depend on low air velocity as well as low water flow rates. In another design for dehumidifiers, they used a copper pipe 27 m in length and 10 mm in OD (outer diameter), which was mechanically bent and shaped to form a helical coil 4 m in length and fixed in the PVC pipe [24].

In another study, two types of dehumidifiers were presented in which they used galvanized steel plates for both bench and pilot proposed modules [25]. A copper tube with 11 mm OD and 18 m length was welded to a galvanized plate in a helical shape for fabrication of a pilot unit. Also, the tube's outside diameter of 8 mm and length of

3 m were specified for making the bench unit. Then, these condensers were connected vertically and in a series by locating them in a duct. In one module, the condenser was basically a cylinder 170 mm in diameter and made of galvanized steel plates. Ten longitudinal fins were soldered to the outer surface of the cylinder, and nine similar fins were soldered to the inner surface with a height of 50 mm for both the inside and outside fins. The cylinder was made of a plate with a thickness of 1.0 mm. A copper tube with a 9.5-mm inside diameter was soldered to the surface of the cylinder. The condenser was fixed vertically in the PVC pipe 316 mm in diameter and connected to the humidifier section using two short horizontal pipes.

The humidifier module proposed by Orfi et al. [26] contains two rows of long cylinders made of copper in which the feed water flows. Longitudinal fins were soldered to the outer surface of the cylinders. The condenser is characterized by a heat-transfer surface area of 1.5 m^2 by having 28 m as the total length of the coil.

Moreover, direct-contact heat exchangers can be used for dehumidifying purposes, for example, bubble column and packing fill dehumidifiers. Additionally, direct-contact dehumidifiers in combination with a shell-and-tube heat exchangers can be used to provide improved condensation and better heat recovery for the cycle. However, these types of systems will increase both the cost and complexity of the system.

A dehumidifier (see Fig. 7.2) is a counterflow condenser in which cold seawater comes from the bottom and goes into a coil, making the coil surface cold; on the other hand, hot moist air comes from the top and interfaces with a cold surface, reaching its dew point. And as a result, vapor will be condensed on the cold surface of the coil and produce some distilled water. Any type of coil which can resist against seawater corrosion with high heat transfer surface area can be used for a dehumidifier.

As in a humidifier, the air and water streams in a dehumidifier can be in direct contact with each other. Therefore, there is another configuration of dehumidifier which is called a "direct-contact dehumidifier" (DC dehumidifier), such as bubble column and packing fill dehumidifiers.

In a DC dehumidifier, humid hot air is in direct contact with cold fresh water. By using a DC dehumidifier, the size and cost of a dehumidifier can be decreased. One type, the bubble column dehumidifier, has been shown to have high heat recovery [27–31]. Fig. 7.10 demonstrates a schematic diagram of the bubble column dehumidifier. In a bubble column dehumidifier, hot, moist air coming from a humidifier is bubbled though a column of freshwater cooled by indirect heat exchange with seawater. The concentration gradient from the hot bubble center to the cool bubble surface drives condensation on the surface of the bubble. The presence of noncondensable gas leads to a low condensation heat-transfer coefficient. Nonetheless, the key advantage of the bubble column dehumidifier lies in moving the condensation process off an expensive solid surface and onto the surface of a swarm of bubbles. The large interfacial area leads to very low resistance, and the heat leaving the bubbles is transferred to the cooling seawater at a high heat-transfer coefficient through a coil with a relatively small surface area. Finally, cold dry air is collected and returned to the humidifier [28,32].

The second type of DC dehumidifier is known as the packing fill dehumidifier. Fig. 7.11 shows the schematic diagram of the packing fill dehumidifier. In this type

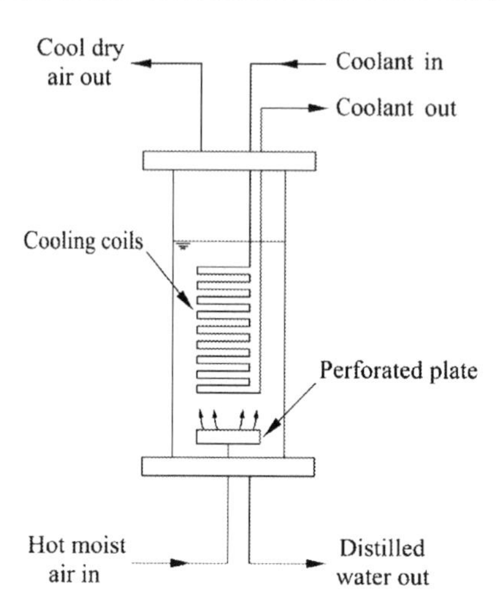

Fig. 7.10 Schematic diagram bubble column dehumidifier [32].

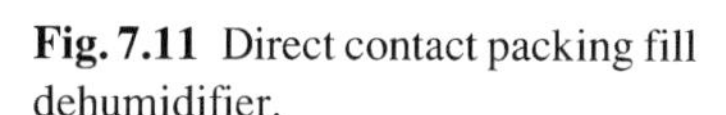
Fig. 7.11 Direct contact packing fill dehumidifier.

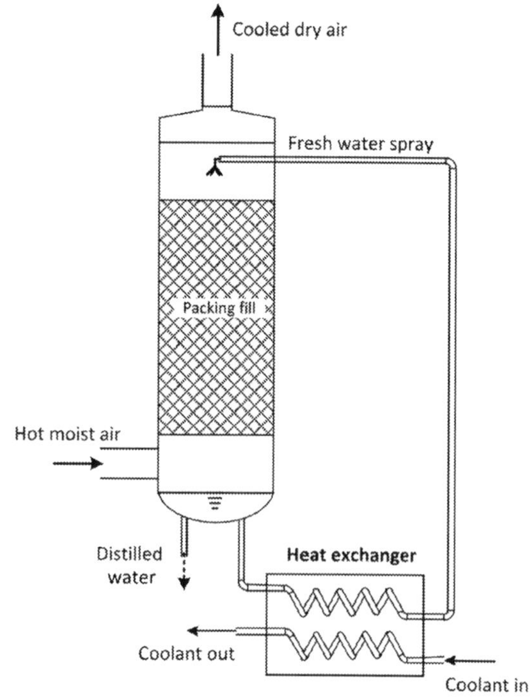

of dehumidifier, hot, moisturized air coming from a humidifier enters the chamber from the bottom. Meanwhile, cold freshwater sprays from the top over a packing fill and cools down the hot, moist air. As a result, by decreasing the hot, moist air's temperature, distilled water is produced. The packing material is used to increase the heat-and-mass transfer area between the air and freshwater streams. The cooled dry air leaves the chamber and is transferred to the humidifier again for moisturizing. The clean sprayed water is collected at the bottom and transferred to a heat exchanger to be cooled down using a coolant. The cold seawater can be considered a coolant which not only cools down the freshwater but is also preheated before heating in the main heat source. Next, the cooled water is circulated through the system and sprayed again. The distilled water will be collected from the dehumidifier by adjusting the freshwater flow rate. The packing-filled dehumidifier is more economical and offers simplicity of design, which makes it an attractive option.

7.4.1 Mass and energy balance

Like a humidifier, a control volume is drawn based on (Fig. 7.8) the illustration of an HDH cycle around the dehumidifier. By applying mass and energy balance on a dehumidifier, the following equations can be extracted.

$$\dot{m}_{sw} + \dot{m}_{da}\omega_{a,t} = \dot{m}_{sw} + \dot{m}_{da}\omega_{a,t} + \dot{m}_{dw} \quad (7.9)$$

$$\dot{m}_{sw}h_{sw,b} + \dot{m}_{da}h_{a,t} = \dot{m}_{sw}h_{sw,m} + \dot{m}_{da}h_{a,b} + \dot{m}_{dw}h_{dw} \quad (7.10)$$

It should be noted that the same assumptions as humidifier modeling have been considered in dehumidifier modeling. In addition, distilled water temperature is assumed to be the average value of the air temperature at the inlet and outlet of the dehumidifier, or, in other words, $T_{dw} = (T_{a,t} + T_{a,b})/2$ [6].

The thermodynamic properties of moist air and water as well as thermophysical properties of seawater are provided in Section 7.3.1.

It should be mentioned that for the DC dehumidifier, the same mass and energy balance approach which is used for a dehumidifier can be applied for extracting the governing equations.

7.4.2 Effectiveness of a dehumidifier

To calculate outlet streams conditions in the dehumidifier, supplementary effectiveness equations are required for conducting thermodynamic analysis. Based on thermodynamic definitions, an effectiveness equation is considered an actual enthalpy variation to the maximum possible enthalpy variation; in other words, $\varepsilon = \Delta\dot{H}/\Delta\dot{H}_{\max}$ [17]. Consequently, humidifier effectiveness would be stated as

$$\varepsilon_d = \max\left(\frac{\dot{H}_{a,t} - \dot{H}_{a,b} + \dot{H}_{dw}}{\dot{H}_{a,t} - \dot{H}^{\text{ideal}}_{a,b} + \dot{H}_{dw}}, \frac{\dot{H}_{sw,m} - \dot{H}_{sw,b}}{\dot{H}^{\text{ideal}}_{sw,m} - \dot{H}_{sw,b}}\right) \quad (7.11)$$

In a dehumidifier, the ideal outlet air enthalpy happens when the outlet air is fully saturated at the water inlet temperature, and the ideal outlet seawater enthalpy is when its temperature is equivalent to the inlet air dry-bulb temperature.

Furthermore, distilled water temperature can be assumed to be the average temperature of the humid air at the inlet and outlet of the dehumidifier, or, in other words, $T_{dw} = (T_{a,t} + T_{a,b})/2$ [6].

7.4.3 Effectiveness-NTU relation of a dehumidifier

Examining the effectiveness relation with the NTUs is one of the commonly used methods for heat exchanger analysis. This method is applied to determine the required heat transfer surface area for a specified effectiveness and mass flow rate ratio.

A normal dehumidifier in which fluids remain unmixed can be considered a kind of heat exchanger, and ε-NTU relation can be applied for it with some approximation. Fig. 7.12 shows a multipass, cross-flow heat exchanger with unmixed fluids. By assuming a dry-surface condition as well as neglecting the latent heat transferred by condensation of water vapor of the moist air, the ε-NTU relation of the heat exchangers can be applied. These approximations will be attuned by calculating the overall heat-transfer coefficient that accounts for the effect of condensation. Hence, for designing a heat exchanger with a considered effectiveness and given mass flow rate ratios, the heat exchanger sizing (heat transfer area) can be performed by using the following set of equations [33,34]:

$$\varepsilon_D = 1 - e^{-\mathrm{NTU}(1-C)}\left[I_0\left(2\mathrm{NTU}\sqrt{C}\right) + \sqrt{C}I_1\left(2\mathrm{NTU}\sqrt{C}\right) - \frac{1-C}{C}\sum_{n=2}^{\infty} C^{n/2} I_n\left(2\mathrm{NTU}\sqrt{C}\right)\right] \tag{7.12}$$

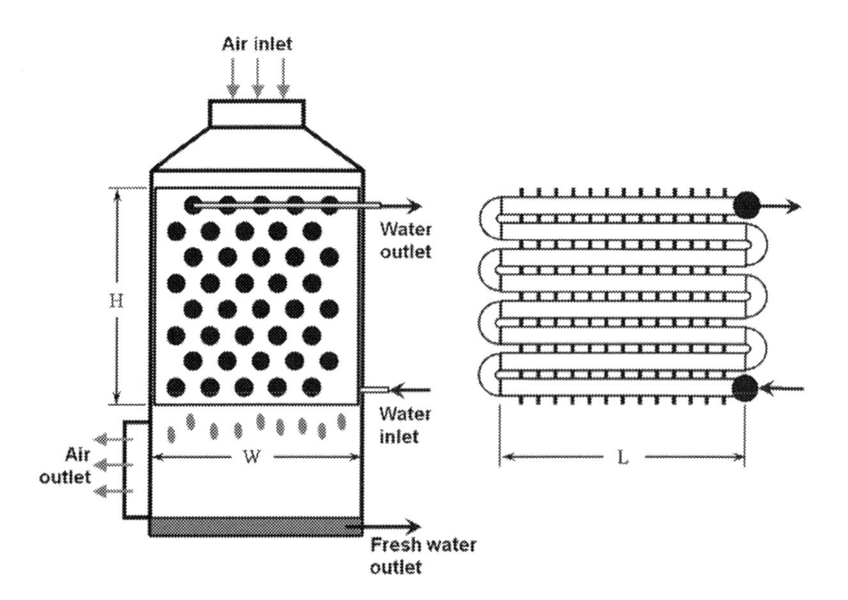

Fig. 7.12 Schematic of a multipass cross-flow finned-tube dehumidifier [33].

I_0, I_1, and I_n indicate the zeroth, first, and nth orders of the modified Bessel functions, respectively. C is the minimum value of the heat-capacity ratio of the air and water streams and NTU is the number of transfer units, given as follows:

$$\mathrm{NTU} = \frac{\mathrm{UA}}{C_{\min}} \tag{7.13}$$

In which $C_{\min}$ is the minimum heat capacity rate ratio of the air and water streams. It is related to the mass flow rate ratio of seawater to dry air of the HDH cycle and is defined as follows:

$$C_{\min} = \min\left(\frac{mr \times c_{sw}}{c_{da}}, \frac{mr \times c_{da}}{c_{sw}}\right) \tag{7.14}$$

where mr is the mass flow rate ratio of seawater to dry air; in addition, c_{sw} and c_{da} are the specific heat capacity of seawater and dry air, respectively.

In order to find the specified effectiveness of a dehumidifier, the heat transfer surface area and overall heat transfer coefficient should be known. Moreover, the overall heat transfer coefficient is based on details of various parameters including tube and fin geometry, their materials, as well as air and water mass flow rate. The overall heat transfer coefficient is presented as follows [18]:

$$\frac{1}{\mathrm{UA}} = \frac{1}{h_i A_i} + \frac{R''_{f,i}}{A_i} + R_{\mathrm{wall}} + \frac{R''_{f,o}}{\eta_s A_o} + \frac{1}{\eta_s h_o A_o} \tag{7.15}$$

where A is the total (fin plus exposed base) surface area and h_i and h_o are the convection heat transfer coefficient on the inside and outside of the pipe, respectively. A_i and A_o refer to the inner (cold side) and outer (hot side) tube surfaces, respectively. Also, $R''_{f,i}$ and $R''_{f,o}$ represent fouling factors at inside and outside, respectively. R_{wall} is the thermal resistance of the wall which is $\ln(D_o/D_i)/2\pi \mathrm{k\,L}$ for radial conduction in a cylindrical wall. However, the thermal resistance of the tubes wall can be neglected in most cases if metal tubes are used.

In this equation, assuming that fins are machined as an integral part of the wall from which they extend, there is no contact resistance at their base. However, fins may be manufactured separately and attached to the wall by a metallurgical or adhesive joint. In such cases, there is a thermal contact resistance, and the last term in Eq. (7.15) turns to $1/\eta_{s(c)} h_o A_o$ where $\eta_{s(c)}$ is the corresponding overall surface efficiency. More details about the overall surface efficiency can be found in heat and mass transfer books [18]. The fouling and contact resistances can have a significant effect, and thus, should not be ignored. Values of these resistances can be found in heat exchangers design text books [18,35].

For calculating the convection heat transfer coefficient of the water side (h_i), the Dittus and Boelter correlation for flow inside the tubes can be used [35]. Also, the convection heat-transfer coefficient of air side (h_o) can be calculated from the correlations provided by Wang et al. [36,37] for wet plain fins.

The η_s is the overall surface efficiency of the finned surface, and here just the outside of the pipe is finned. Therefore, the overall surface efficiency can be expressed as

$$\eta_s = 1 - \frac{A_f}{A}\left(1 - \eta_f\right) \tag{7.16}$$

where A_f is the entire fin surface area and η_f is the efficiency of a single fin. The efficiency of the various fin types is presented in heat transfer textbooks [18,38].

Mostly, the fin efficiency for the design of a dry finned surface without condensation on it is given as a function of the fin parameter (m) and the fin length. Because simultaneous heat and mass transfer happens in a dehumidifier, there is condensation on the fin which changes the finned surface's performance; consequently, the fin efficiency is different. So, a modified relation should be applied. A correction factor method is used [39,40] for calculating the fin efficiency for the case of simultaneous heat and mass transfer on a fully wet surface.

7.5 Performance parameters

In order to understand the HDH desalination cycle performance in terms of thermal efficiency, water production potential, and each component's effectiveness, performance parameters are defined. Basically, these nondimensional parameters are used as cycle operational metrics to evaluate system behavior. Also, this parameter provides a useful tool for comparing an HDH cycle with other desalination technologies.

Gain output ratio (GOR): GOR is the ratio of the latent heat of evaporation of the distillate water produced to the total heat input to the cycle from the heat source. It is a nondimensional parameter and represents the amount of heat recovery in the cycle. In an HDH system, if a GOR of at least 8 (which corresponds to energy consumption rates of about 300 kJ/kg) can be achieved, then it will have thermal performance comparable to MSF distillation or multiple-effect distillation (MED).

$$\text{GOR} = \frac{\dot{m}_{dw} h_{fg}}{\dot{Q}_{in}} \tag{7.17}$$

Recovery ratio (RR): Recovery ratio is the amount of distilled water over inlet impure water to the cycle, which is a criterion for the water production efficiency of the cycle. RR is a nondimensional parameter and is presented in percentage. It should be noted that, for low recovery ratios, the brine disposal process is not required.

$$\text{RR} = \frac{\dot{m}_{dw}}{\dot{m}_{sw}} \times 100 \tag{7.18}$$

Modified heat capacity ratio (HCR): This parameter is the ratio of the maximum enthalpy change of cold stream to maximum enthalpy change of hot stream in a HME device of the system, which is used to evaluate performance of cycle components like the humidifier and dehumidifier based on the second law of thermodynamics. In other words, when a heat and mass exchange device is working at the HCR = 1 the amount of entropy generation is minimized [41–44]. It is a nondimensional parameter. The idea is similar to that of a heat exchanger when the temperature difference in terminals is minimized.

$$\mathrm{HCR_{component}} = \frac{\Delta\dot{H}\mathrm{max}_{,\mathrm{cold}}}{\Delta\dot{H}\mathrm{max}_{,\mathrm{hot}}} \tag{7.19}$$

$$\mathrm{HCR}_h = \frac{\dot{H}_{a,t}^{\mathrm{ideal}} - \dot{H}_{a,b}}{\dot{H}_{sw,t} - \dot{H}_{sw,b}^{\mathrm{ideal}}} \tag{7.20}$$

$$\mathrm{HCR}_d = \frac{\dot{H}^{\mathrm{ideal}}_{sw,m} - \dot{H}_{sw,b}}{\dot{H}_{a,t} - \dot{H}^{\mathrm{ideal}}_{a,b}} \tag{7.21}$$

Mass flow rate ratio: One of the key parameters in HDH desalination which mainly appears in relation with other performance parameters of the cycle is the mass flow rate ratio (mr), which is defined as the ratio of mass flow rate liquid to gas as follows:

$$mr = \frac{\dot{m}_{sw}}{\dot{m}_a} \tag{7.22}$$

For a humidifier and dehumidifier of a fixed size, there is an optimum mass flow rate ratio in which the effectiveness of these components is maximum. Also, at a certain working condition of the HDH cycle in which temperatures of saline water are known, as well as a fixed size of humidifier and dehumidifier, there is a particular value of mass flow rate ratio at which the overall HDH system efficiency is optimal, and this was observed by several other researchers [6,24,42,45,46].

Specific water production (SWP): This parameter is basically applicable for a solar-driven HDH cycle. It is the amount of water produced per square meter of solar collector area per day. This parameter indicates the solar energy efficiency of the HDH cycle as well as capital cost of solar collectors. The cost of the HDH system for a solar collector is 40%–45% of the total capital cost for air-heated systems [7] and 20%–35% of the total capital cost for water-heated systems [47].

$$\mathrm{SWP} = \frac{\dot{m}_{dw}}{A_{sc}} \ \left(\mathrm{kg/m^2\,day}\right) \tag{7.23}$$

7.6 Process control

Natural or forced convection can be considered for circulation of the air stream through the cycle. The remarkable advantage of the natural convection air loop inside the HDH cycle is related to the fact that it is self-adjusting to the optimum operating state based on the present parameters. While using a forced circulation, based on cycle input parameters like saline water inlet temperature, the available thermal energy optimum working condition should be adjusted. This can be performed by controlling the mass flow rate of saline water and volumetric flow rate of air based on the theoretical simulations. Therefore, by monitoring the available input parameters' condition and adjusting controllable parameters, optimum working conditions can be achieved.

For a situation where thermal energy input is not constantly provided for the cycle, like heat provided by solar thermal collectors, the supplementation of the system by a high-temperature thermal storage tank makes it possible to keep the humidifier inlet temperature constant and on a 24-hour, continuous operation.

7.7 Heat sources

There are various types of low-grade heat sources that can be utilized for supplying the thermal energy for the HDH cycle. Industrial waste heat is one of the examples in which, by using a heat exchanger, it is possible to transfer heat to the air/water stream in cycles. All sorts of industrial waste heat like exhaust hot gases from furnaces, boilers, and so on, can be considered a heat source for providing heat. Finned-tube heat exchangers are one of the technologies which can be used for extracting heat and delivering it to an air or water stream of an HDH system.

Solar energy is another option that can be used for this purpose. Mainly, there are two kinds of solar collectors: nonconcentrating, or stationary and concentrating. A nonconcentrating collector has the same area for intercepting and for absorbing solar radiation, while a sun-tracking concentrating solar collector generally includes concave reflecting surfaces to intercept and focus the sun's beam radiation to a smaller receiving area, which increases the radiation flux. A large number of solar collectors are available on the market [48]. Stationary solar collectors such as flat plate and evacuated solar tubes are among the most viable options in terms of cost and temperature ranges for integration with the HDH system as a heat source. They have been used in several experimental studies [49–54]. Evacuated solar collectors and flat plate solar collectors are among the low-grade heat sources that can be options for application in HDH desalination systems [48].

Additionally, geothermal energy can be used for providing heat for the HDH system. The geothermal industry mainly uses the standard heat exchangers usually utilized in the chemical industry, namely, plate and/or shell-and-tube type exchangers such as are commercially readily available.

7.8 Economics of the system

For supplying freshwater to remote areas, different responsibilities which have different influences on the system should be taken into consideration. These mainly include water extraction, treatment, storage, and distribution. For desalination systems, maintaining the balance between investment and operational costs is a major consideration in making primary decisions between two possible approaches toward designing a system for low lifetime water costs. Therefore, decision makers first need to determine their approach between low investment and low operational cost. For medium-scale installation (water production rate of 5000–10,000 m^3/day), the investment share on lifetime water costs is relatively low in comparison to small-scale installations. On the basis of the currently available data, it is reported in the range of 0.2–0.5 €/m^3, depending on the local conditions and the method employed [47,55–57]. The costs of water distribution and costs due to leakages in the distribution grid are rarely monitored. These costs usually vary between 0.1 and 0.8 €/m^3, based on the local conditions.

Operational costs related to energy consumption, chemical additives dosing, system maintenance, and spare parts are in the range of 0.2–0.7 €/m^3. All of this adds up to lifetime water costs for midrange installations of between 0.5 and 2 €/m^3. These are the costs for installations in medium size towns and for larger holiday resorts, including a number of hotels and island installations.

For small-scale applications from the size of 500 to 5000 m^3/day water production rates, water costs are reported in the range of 0.7–3.1 €/m^3. For a small-scale application which includes capacities between 5 and 100 m^3/day, the main implemented technology is RO installations. Because of local conditions, plants within this range are mostly driven by brackish water with salinities in the range of 2000 and 8000 ppm TDS (total dissolved solid). RO is highly sensitive to salt content because of the limitations imposed by membrane retaining factor and pressure, so the investment costs are highly reliant on the raw water source. For the case of seawater desalination, the investment cost allocation for the desalination system is in the range of 0.5–1.4 €/m^3. Operation and maintenance costs are closely related to installations of this size. For seawater RO, based on the maintenance companies' reports, the maintenance and energy costs are in the range of 0.9–2.8 €/m^3, counting labor costs. Therefore, the overall water production cost is in the range of 1.4–4.2 €/m^3. The key feature of the HDH desalination is its low maintenance demand and no need for chemical pretreatment. Moreover, because the main energy demanded for the cycle is low-grade thermal energy, various heat sources such as a generator's waste heat or relatively inexpensive solar heat can be implemented, where adequate and stable energy cannot be supplied from the main grid, while for electric-powered desalination systems like RO units, an extra power supply unit has to be taken into account, for example, from photovoltaic cells (PV) or a small diesel generator.

By assuming specific energy consumption for a remote-area RO system at a value of 7 kWh_{el}/m^3, the extra costs per m^3 are about 2.8 €/m^3 for the integrated PV-RO, as well as 1.4 €/m^3 for RO by means of generator power. By comparing those costs with the anticipated cost of water production of solar-driven, multiple-effect humidification dehumidification (MEHDH) desalination, this system has its relative cost advantages for some locations as compared to electric-powered RO units, where its application is economically feasible. Decentralized resorts, weekend houses, military stations, remote villages, and small marinas are some examples. In the case of saving thermal energy waste from the mechanical power production units located in remote areas, an MEHDH system can completely compete with an RO system. This can be

Table 7.1 **Cost comparison for small-scale remote area water supply methods [47]**

Costs in €/m^3 operation/total	Heat source	Electricity source	1 m^2/day	5 m^2/day	10 m^2/day
MEHDH waste heat	CHP	CHP	4.56/6.20 €	2.86/3.94 €	2.40/3.34 €
MEHDH solar thermal	Solar coll.	Grid	7.22/8.87 €	5.17/6.25 €	4.77/5.71 €
MEHDH autonomous	Solar coll.	Photovoltaic	8.8/10.2 €	5.94/6.78 €	5.13/5.73 €
RO grid connected	–	Grid	–	0.90–2.80 €/1.40–4.20 €	
RO-Genset	–	Generator	–	1.00–2.70 €/2.80–5.60 €	
RO-PV	–	Photovoltaic	–	0.70–2.6 €/4.20–7 €	

enforced by use of waste heat from small diesel or gas electrical generators in a combined heat and power (CHP) production system [47].

The cost comparison for different decentralized water production technologies is presented in Table 7.1. It can be seen that reverse osmosis has light cost advantages also in small-scale applications over 5 m^3/day in installations where a standard electrical-grid connection is available and electricity prices are at or below 0.15 €/kWh. In all other cases, the use of thermal desalination systems as the MEHDH system is more beneficial when considering the costs and ease of operation.

7.9 Summary

Material selection is a key consideration that not only will benefit the system's life time, but can also reduce the labor costs related to system maintenance. Therefore, all parts which are in contact with brine can be made of polypropylene-based materials or high-alloyed stainless steel. Muller et al. [47] used a complete casing which is vapor-tight and lined with welded stainless steel. Heat and mass exchange surfaces are made from special temperature-treated and heat-conduction-enhanced polypropylene. To enhance the effectiveness of the condenser plates, they aggregated them applying a time-optimized extrusion-welding method to stacks of condensation units. They designed their commercial HDH system for long-term durability based on improvement of their pilot setup for 20 years of life time, ensuring reasonable water prices and low maintenance demand during operation.

The HDH technology already is under development to be more optimized in terms of material cost, and most of the fabricated setups are designed for running experiments and proof of concept. Therefore, all of these setups' components, such as the humidifier and dehumidifier, are not specifically designed for satisfying an integrated setup module, which results in ineffectiveness as well as higher cost of the system. Since the humidifier design and performance are fairly close to the cooling towers which already have been studied very well, the components of a humidifier such as nozzle sprayers and packing fillers are totally compatible with the humidifier design requirements. On the other hand, humidifier chamber materials which hold and support the packing filler and nozzle spray system must be specifically designed. The material needs to be corrosion resistant and endure in either fed saline water or seawater. Both polypropylene-based material and stainless steel are good options for this purpose. Providing a cheaper material that can be mounted in humidifier shape (with square or circle cross-section area) that can tolerate a maximum temperature of 100°C is still a challenge. Moreover, a lack of a well-designed heat and mass exchange devices for dehumidifying purposes can be widely observed in the body of knowledge. There is no almost specified designed dehumidifier for an HDH system, and current setups mainly use their own unsophisticated, designed dehumidifier [3,26,42,51,58,59] for experimental purposes. Therefore, design and fabrication of an affordable dehumidifier with a high heat-transfer area can remarkably improve the cycle's performance. The dehumidifier should be corrosion resistant from the inside in which saline water passes, while on the tubes' outside area, where they are in contact with moist air, they are less sensitive to corrosion. Also, to increase the heat transfer area, dehumidifier tubes can be finned. Kabeel et al. used a cylindrical shell galvanized steel with a finned copper tube [53,60], and Prakash Narayan et al. [42] used polypropylene plate-and-tube dehumidifiers. Both were used for experimental purposes with a lack of consideration of cost and durability in long-term applications.

References

[1] Giwa A, Akther N, Housani AA, Haris S, Hasan SW. Recent advances in humidification dehumidification (HDH) desalination processes: improved designs and productivity. Renew Sust Energ Rev 2016;57:929–44.

[2] Pankratz T. IDA desalination yearbook 2010-2011. Section 1, p. 1. IDA desalination yearbook: Media Analytics Ltd, 2010.

[3] Kabeel AE, Hamed MH, Omara ZM, Sharshir SW. Experimental study of a humidification-dehumidification solar technique by natural and forced air circulation. Energy 2014;68:218–28.

[4] Alhazmy MM. Minimum work requirement for water production in humidification—dehumidification desalination cycle. Desalination 2007;214(1):102–11.

[5] Narayan GP, Sharqawy MH, Summers EK, Lienhard JH, Zubair SM, Antar MA. The potential of solar-driven humidification–dehumidification desalination for small-scale decentralized water production. Renew Sust Energ Rev 2010;14(4):1187–201.

[6] Narayan GP, Sharqawy MH, Lienhard VJH, Zubair SM. Thermodynamic analysis of humidification dehumidification desalination cycles. Desalin Water Treat 2010;16 (1–3):339–53.

[7] Chafik E. A new type of seawater desalination plants using solar energy. Desalination 2003;156(1):333–48.

[8] Treybal RE. Mass transfer operations. New York: McGraw-Hill; 1980.

[9] Kreith F, Boehm RF. Direct-contact heat transfer. Hemisphere Pub. Corp.: Washington; 1988.
[10] Aull RJ, Krell T. Design features of cross-fluted film fill and their effect on thermal performance. CTI J 2000;21(2):12–33.
[11] ASHRAE. ASHRAE handbook—fundamentals (SI edition). Atlanta, GA: American Society of Heating, Refrigerating and Air-Conditioning Engineers, Inc.; 2009.
[12] Klein SA, Alvarado FL. Engineering equation solver Version 8.400, http://www.fchart.com/ees/.
[13] Hyland RW, Wexler A. Formulations for the thermodynamic properties of the saturated phases of H2O from 173.15 K to 473.15 K. ASHRAE Trans 1983;(Part 2A).
[14] Pruss A, Wagner W. The IAPWS formulation 1995 for the thermodynamic properties of ordinary water substance for general and scientific use. J Phys Chem Ref Data Monogr 2002;31(2):387–535.
[15] Nayar KG, Sharqawy MH, Banchik LD, Lienhard VJH. Thermophysical properties of seawater: a review and new correlations that include pressure dependence. Desalination 2016;390:1–24.
[16] Sharqawy MH, Lienhard JH, Zubair SM. Thermophysical properties of seawater: a review of existing correlations and data. Desalin Water Treat 2010;16(1–3):354–80.
[17] Narayan GP, Mistry KH, Sharqawy MH, Zubair SM, Lienhard JH. Energy effectiveness of simultaneous heat and mass exchange devices. Front Heat Mass Transf 2010;1(2).
[18] Bergman TL, Lavine AS, Incropera FP, DeWitt DP. Fundamentals of heat and mass transfer. New York: John Wiley; 2011.
[19] Jaber H, Webb RL. Design of cooling towers by the effectiveness-NTU method. J Heat Transf 1989;111(4):837–43.
[20] Bosnjakovic F. Technical thermodynamics. New York: Holt, Rinehart and Winston; 1965.
[21] Merkel F. "Verdunstungskühlung," VDI Forschungsarbeiten, 275. Berlin. 1925.
[22] Müller-Holst H, Engelhardt M, Herve M, Schölkopf W. Solarthermal seawater desalination systems for decentralised use. Renew Energy 1998;14(1):311–8.
[23] Farid MM, Parekh S, Selman JR, Al-Hallaj S. Solar desalination with a humidification-dehumidification cycle: mathematical modeling of the unit. Desalination 2003;151(2):153–64.
[24] Al-Hallaj S, Farid MM, Rahman Tamimi A. Solar desalination with a humidification-dehumidification cycle: performance of the unit. Desalination 1998;120(3):273–80.
[25] Nawayseh NK, Farid MM, Al-Hallaj S, Al-Timimi AR. Solar desalination based on humidification process—I. Evaluating the heat and mass transfer coefficients. Energy Convers Manag 1999;40(13):1423–39.
[26] Orfi J, Laplante M, Marmouch H, Galanis N, Benhamou B, Nasrallah SB, et al. Experimental and theoretical study of a humidification-dehumidification water desalination system using solar energy. Desalination 2004;168:151–9.
[27] Deleted in review.
[28] Tow EW, Lienhard VJH. Experiments and modeling of bubble column dehumidifier performance. Int J Therm Sci 2014;80:65–75.
[29] Liu H, Sharqawy MH. Experimental performance of bubble column humidifier and dehumidifier under varying pressure. Int J Heat Mass Transf 2016;93:934–44.
[30] Rajaseenivasan T, Shanmugam RK, Hareesh VM, Srithar K. Combined probation of bubble column humidification dehumidification desalination system using solar collectors. Energy 2016;116:459–69.
[31] Srithar K, Rajaseenivasan T. Performance analysis on a solar bubble column humidification dehumidification desalination system. Process Saf Environ Prot 2017;105:41–50.

[32] Sharqawy MH, Liu H. The effect of pressure on the performance of bubble column dehumidifier. Int J Heat Mass Transf 2015;87:212–21.

[33] Sharqawy MH, Antar MA, Zubair SM, Elbashir AM. Optimum thermal design of humidification dehumidification desalination systems. Desalination 2014;349:10–21.

[34] Baclic BS. A simplified formula for cross-flow heat exchanger effectiveness. J Heat Transf 1978;100(4):746–7.

[35] Mujumdar AS. Heat exchanger design handbook T. Kuppan Marcel Dekker Inc., New York 2000, 1118 pages. Dry Technol 2000;18(9):2167–8.

[36] Chen H-T, Wang H-C. Estimation of heat-transfer characteristics on a fin under wet conditions. Int J Heat Mass Transf 2008;51(9–10):2123–38.

[37] Wang C-C, Lin Y-T, Lee C-J. An airside correlation for plain fin-and-tube heat exchangers in wet conditions. Int J Heat Mass Transf 2000;43(10):1869–72.

[38] Cengel YA. Heat transfer: a practical approach with EES CD. New York: McGraw-hill Higher Education; 2002.

[39] Sharqawy MH, Zubair SM. Efficiency and optimization of straight fins with combined heat and mass transfer—an analytical solution. Appl Therm Eng 2008;28(17–18):2279–88.

[40] Sharqawy MH, Zubair SM. Efficiency and optimization of an annular fin with combined heat and mass transfer—an analytical solution. Int J Refrig 2007;30(5):751–7.

[41] Prakash Narayan G, Lienhard VJH, Zubair SM. Entropy generation minimization of combined heat and mass transfer devices. Int J Therm Sci 2010;49(10):2057–66.

[42] Prakash Narayan G, St. John MG, Zubair SM, Lienhard V JH. Thermal design of the humidification dehumidification desalination system: an experimental investigation. Int J Heat Mass Transf 2013;58(1–2):740–8.

[43] Narayan GP, Chehayeb KM, McGovern RK, Thiel GP, Zubair SM, Lienhard VJH. Thermodynamic balancing of the humidification dehumidification desalination system by mass extraction and injection. Int J Heat Mass Transf 2013;57(2):756–70.

[44] Bejan A. Entropy generation minimization: the method of thermodynamic optimization of finite-size systems and finite-time processes. Boca Raton, FL: Taylor & Francis; 1995.

[45] Farid M, Al-Hajaj AW. Solar desalination with a humidification-dehumidification cycle. Desalination 1996;106(1):427–9.

[46] Narayan GP, McGovern RK, Zubair SM, Lienhard JH. High-temperature-steam-driven, varied-pressure, humidification-dehumidification system coupled with reverse osmosis for energy-efficient seawater desalination. Energy 2012;37(1):482–93.

[47] Müller-Holst H. Solar thermal desalination using the multiple effect humidification (MEH)-method. In: Rizzuti L, Ettouney HM, Cipollina A, editors. Solar desalination for the 21st century: a review of modern technologies and researches on desalination coupled to renewable energies. Dordrecht: Springer Netherlands; 2007. p. 215–25.

[48] Kalogirou SA. Solar thermal collectors and applications. Prog Energy Combust Sci 2004;30(3):231–95.

[49] Yamalı C, Solmus İ. A solar desalination system using humidification–dehumidification process: experimental study and comparison with the theoretical results. Desalination 2008;220(1):538–51.

[50] Abdel Dayem AM, Fatouh M. Experimental and numerical investigation of humidification/dehumidification solar water desalination systems. Desalination 2009;247 (1):594–609.

[51] Elminshawy NAS, Siddiqui FR, Addas MF. Experimental and analytical study on productivity augmentation of a novel solar humidification–dehumidification (HDH) system. Desalination 2015;365:36–45.

[52] Hamed MH, Kabeel AE, Omara ZM, Sharshir SW. Mathematical and experimental investigation of a solar humidification–dehumidification desalination unit. Desalination 2015;358:9–17.

[53] Kabeel AE, El-Said EMSA. Hybrid solar desalination system of air humidification–dehumidification and water flashing evaporation: part I. A numerical investigation. Desalination 2013;320:56–72.

[54] Yuan G, Wang Z, Li H, Li X. Experimental study of a solar desalination system based on humidification–dehumidification process. Desalination 2011;277(1):92–8.

[55] The desalination plants market in Europe, the Middle East and North Africa. Marketing Study by Frost and Sullivan; 2004.

[56] Eslamimanesh A, Hatamipour MS. Economical study of a small-scale direct contact humidification–dehumidification desalination plant. Desalination 2010;250(1):203–7.

[57] Zamen M, Amidpour M, Soufari SM. Cost optimization of a solar humidification–dehumidification desalination unit using mathematical programming. Desalination 2009;239(1):92–9.

[58] Farshchi Tabrizi F, Khosravi M, Shirzaei Sani I. Experimental study of a cascade solar still coupled with a humidification–dehumidification system. Energy Convers Manag 2016;115:80–8.

[59] Abdel Dayem AM. Efficient solar desalination system using humidification/dehumidification process. J Sol Energ Eng 2014;136(4):041014–9.

[60] Kabeel AE, El-Said EMS. A hybrid solar desalination system of air humidification, dehumidification and water flashing evaporation: part II. Experimental investigation. Desalination 2014;341:50–60.

A spray assisted low-temperature desalination technology

Qian Chen, Kum J. M, Kian J. Chua
National University of Singapore, Singapore, Singapore

Nomenclature

A	area
BPE	boiling point elevation
C	constant
c_p	specific heat
d	droplet diameter
D	diffusion coefficient
$\dot{D}$	production rate
h	heat transfer coefficient/reactor height
h_{fg}	latent heat of evaporation/condensation
h_m	mass transfer coefficient
k	thermal conductivity
LMTD	log mean temperature difference
$\dot{m}$	mass flowrate
N_{total}	total number of stages
Nu	Nusselt number
p	pressure
Pr	Prandtl number
Q	heat flux
r	radial axis
Re	Reynold number
S	salinity
SEC	specific energy consumption
Sh	Sherwood number
Sc	Schmidt number
T	temperature
t	residence time
TBT	top brine temperature
u	droplet velocity
U	overall heat transfer coefficient
z	vertical distance

Emerging Technologies for Sustainable Desalination Handbook. https://doi.org/10.1016/B978-0-12-815818-0.00008-4

Greek letters

Δ	difference
ε	effectiveness
θ	dimensionless temperature difference
μ	dynamic viscosity
ν	kinetic viscosity
ρ	density

Subscripts

b	brine
c	condenser
cl	liquid in condenser
cv	vapor in condenser
cw	cooling water
e	evaporator
el	liquid in evaporator
ev	vapor in evaporator
f	feed
h	heating
i	ith stage, inside
in	inlet
l	liquid
loss	temperature loss in demister
o	outlet, outside
r	heat recovery
sw	sea water
v	vapor

8.1 Introduction

Freshwater is one of the key essential resources for the continuity of human lives and all socioeconomic activities. With the growth of global population and development of the world economy, world water demand is increasing rapidly and world water scarcity is becoming progressively pervasive. In order to address the global water deficit, desalination technologies have been widely used in the past few decades. >16,000 desalination plants have been installed in around 150 countries [1], and global online desalination capacity has reached 83.09 million m^3/day in 2014 [2]. However, current desalination capacity is still lower than the world's water deficit, and desalination technologies have huge potential to contribute to fresh water supply.

One of the challenges that all desalination technologies are facing is the high energy consumption. The specific energy consumption of major thermal desalination technologies, that is, multieffect distillation (MED) and multistage flash distillation (MSF), spans 250–330 kJ/kg-water [3]. Membrane-based desalination technologies have lower energy consumption because no phase change is involved in the separation process. The most advanced membrane process, namely, the reverse osmosis (RO) system, has achieved a specific energy consumption of 8.3 kJ/kg-water [4]. However, RO systems can only be electrically driven. Owing to high costs of primary energy and greater concerns on environmental issues, new interests have been reported on employing renewable energy sources for desalination.

Attempts have been made to integrate various renewable energy sources into existing desalination technologies, including RO/PV (photovoltaic) plants [5], solar thermal/MED plants [6], MED plants driven by geothermal heat [7,8], combined CSP (concentrating solar power) and desalination (MED or MSF) plant for cogeneration of electricity and freshwater [9,10], etc. New technologies are also being developed, including improved solar distillation [11], humidification-dehumidification (HDH) desalination systems powered by solar heat [12,13] and industry waste heat [14], and solar driven membrane distillation [15].

Spray assisted low-temperature desalination is one of the novel technologies that potentially enable the employment of renewable energy sources. It applies direct contact heat and mass transfer mechanism and eliminates the need to use metallic surfaces inside the chambers. In this technology, seawater is first preheated outside and then injected into the evaporation chamber at a lower pressure. The superheated seawater undergoes extreme flash evaporation and a portion of it turns into vapor. Freshwater is consequently obtained by condensing the vapor. Compared with traditional thermal desalination technologies, the spray assisted low-temperature desalination possesses the key merits of high heat and mass transfer rates, simplicity of system design, lower scaling and fouling potential, and lower initial cost.

The technological merits for the spray assisted low-temperature desalination technology have been demonstrated in a pilot plant located in Egypt [16]. Several experimental studies have been carried out on spray flash evaporation to evaluate its thermal performance [17,18], enhance the evaporation rate [19], investigate the impact of design and operating parameters (direction of injection [20]; initial liquid temperature, spray velocity, and degree of superheat [21–23]), and evaluate its application in renewable energy utilization [24] and waste heat recovery [25,26]. Analytical studies have also been carried out to model the heat and mass transfer processes in the spray evaporator [27–29] as well as to investigate the thermal performance of the whole system [16,30]. The conclusions of the previous studies are summarized in Table 8.1.

This chapter discusses the spray assisted low-temperature desalination process both in depth and breadth. The flashing and shattering phenomena in the spray evaporator are first introduced, followed by detailed mathematical modeling on the spray evaporator which enables precise performance prediction. Then a simple spray desalination system is presented, and the major sources of energy consumption are elucidated and quantified. Next, a multistage spray assisted low-temperature desalination system is introduced. A detailed thermodynamic model is then developed to highlight the transport phenomena within the multistage system. Key input parameters affecting the system performance are taken into consideration during the model formulation. Applying this model, both the production rate and thermal efficiency of the multistage spray assisted low-temperature desalination system are analyzed and optimized under

Table 8.1 Previous studies on the spray assisted low-temperature desalination technology

Configuration	Method	Objective	Conclusion	Reference
Single stage	Experimental study	Effect of temperature, degree of superheat and flow velocity	Evaporation rate is higher than pool boiling in MSF; the temperature profile of the spray undergoes a two-exponential decay	[17,18]
Single stage	Experimental study	Effect of droplet diameter	Microbubble injection reduces the droplet diameter and promotes evaporation rate	[19]
Single stage	Experimental study	Effect of direction of the spray	The evaporation process completed with a shorter distance in the upward spray	[20]
Single stage	Experimental study	Effect of degree of superheat, temperature	Production rate is increased under high degree of superheat and initial temperature	[21–23]
Single stage	Experimental study	Desalination driven by solar thermal energy	The production rate and the thermal efficiency are strongly affected by the inlet temperature. Maximum daily production rate is 9 L/m^2	[24]
Single stage	Mathematical modeling	Desalination by industrial waste heat	Integration of the desalination increases the exergy efficiency of the power plant by 28% and is economically beneficial	[25,26]

Table 8.1 Continued

Configuration	Method	Objective	Conclusion	Reference
Single stage	Mathematical modeling	Effect of degree of droplet diameter, flow velocity	Evaporation rate is promoted by smaller droplet diameter and lower flow velocity	[27–29]
Multistage	Mathematical modeling	Integration of the desalination system with a CSP plant	Cogeneration of 2.2 MW electricity in the CSP plant and 520 m^3/day fresh water in the desalination system	[16]
Multistage	Mathematical modeling	Effect of top brine temperature, operating stages, cooling water flowrate	Production rate and energy efficiency are promoted by higher operating temperatures and operating stages, while there is an optimal value for cooling water flowrate	[30]

different operating parameters, namely, heat source temperature, total number of stages, and cooling water flowrate.

8.2 Spray evaporator

8.2.1 Flashing and shattering phenomena

A liquid stream, when injected into a low-pressure zone, shatters into fine droplets when its temperature is slightly higher than the environmental saturation temperature. Such a process is referred to as "flash atomization" [31], as schematically shown in Fig. 8.1. Small bubbles are formed due to nucleation when the superheated liquid approaches the nozzle exit. The bubbles expand rapidly under a low-pressure environment. When the liquid-bubble mixture ejects from the nozzle, the bubbles achieve a burst phenomenon, which shatters the liquid into fine droplets. The droplet diameter is several orders of magnitude smaller than the nozzle diameter. For example, a droplet diameter of several hundred micrometers is available via a circular nozzle having a

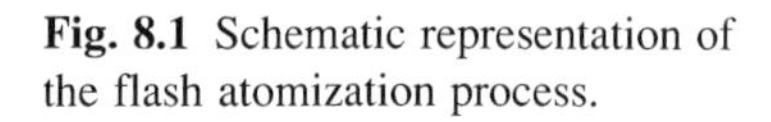

Fig. 8.1 Schematic representation of the flash atomization process.

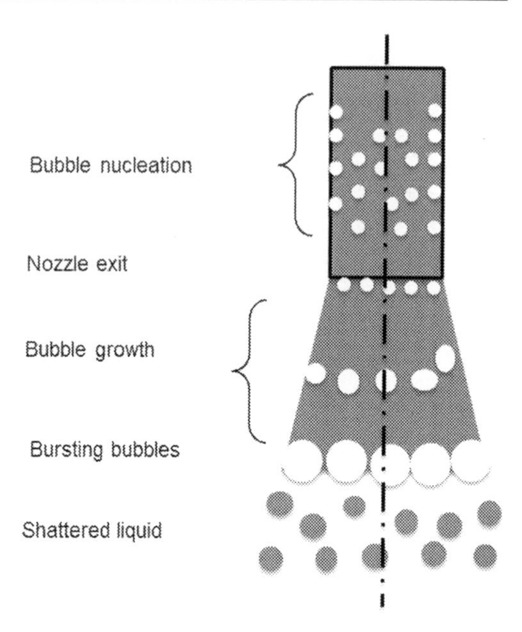

diameter of several millimeters. Under the vacuum environment, the liquid at the droplet surface flashes into vapor due to the great partial pressure difference between the liquid surface and the surrounding vapor. The energy consumed during the flash evaporation process comes from the sensible heat of the liquid, which reduces the liquid temperature. The flash evaporation stops when the droplets are cooled to the environment temperature.

The small droplet formed in the flash atomization process markedly increases the specific surface area for evaporation and leads to a high evaporation rate, while the application of a circular nozzle not only reduces the pumping power but also minimizes the chances of potential scaling and fouling. Spray evaporation is therefore a promising method for thermal desalination.

8.2.2 Heat and mass transfer modeling in the spray evaporator

The schematic representation of the spray evaporator is shown in Fig. 8.2. Superheated seawater at a temperature of T_f is injected into the evaporator which is at a saturation temperature of T_{ev}. The seawater stream shatters into droplets and partially evaporates. The rest is drained off at the bottom of the evaporator. The resultant vapor is directed to the condenser so that the vapor pressure in the evaporator is kept constant. To evaluate the evaporation process in the spray evaporator, the following assumptions are made:

(1) the spray completely breaks into small sphere droplets after a certain vertical distance from the nozzle exit;

(2) heat conduction inside the droplet is one-dimensional, particularly, along the radial direction;

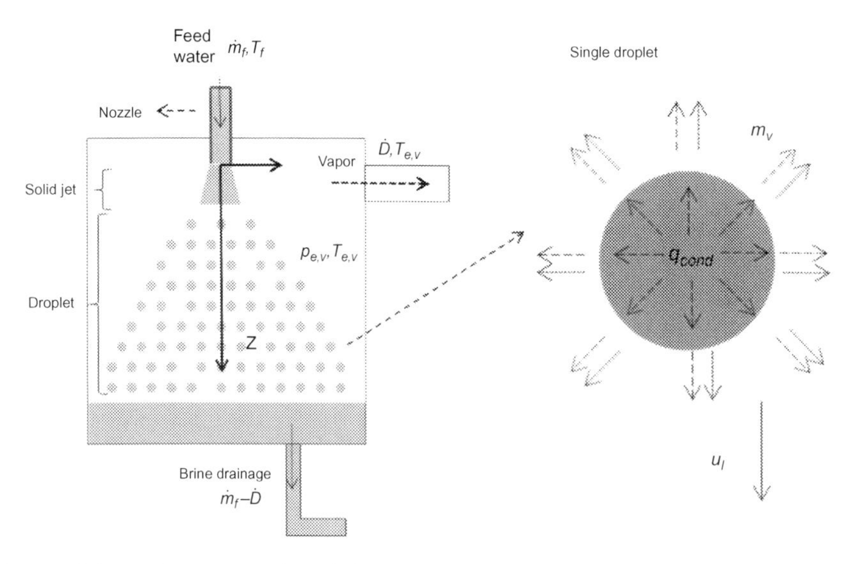

Fig. 8.2 Schematic representation of the spray evaporator.

(3) the influence of seawater droplets on each other is negligible, and the motion and heat transfer of each droplet are independent; and

(4) the change in droplet diameter due to evaporation is negligible (according to energy conservation, $<3\%$ of water evaporates for a degree of superheat lower than 15°C).

8.2.2.1 Flash evaporation process of the droplet

The heat conduction inside the liquid droplet is described by the conduction function in the spherical coordinate as

$$\rho_{sw} c_{p,sw} \frac{\partial T_l}{\partial t} = k_{sw} \frac{1}{r^2} \frac{\partial}{\partial r} \left(r^2 \frac{\partial T_{sw}}{\partial r} \right) \tag{8.1}$$

where ρ_{sw}, $c_{p,sw}$, and k_{sw} are the density, specific heat, and thermal conductivity of seawater, respectively.

All the droplets are assumed to have the same temperature as the feedwater in the feed pipeline, and the initial condition of the droplet is

$$T_{sw}(0, r) = T_f \tag{8.2}$$

Since heat conduction is symmetric within the droplet, the boundary condition in the droplet center is written as

$$k_{sw} \frac{\partial T_{sw}}{\partial r} |_{r=0} = 0 \tag{8.3}$$

In the liquid-vapor interface, the heat conduction is balanced by the sensible heat of convection and the latent heat of vaporization. Therefore, the boundary condition at the droplet surface is described by

$$k_{sw}\frac{\partial T_{sw}}{\partial r}\Big|_{r=d/2} = -h(T_{sw} - T_{ev}) - h_m(\rho_{sw} - \rho_v)h_{fg} \tag{8.4}$$

where T_{ev} is the environmental vapor temperature; d is the droplet diameter; ρ_{sw} and ρ_v are the densities of vapor at the droplet surface and the atmosphere, respectively; T_{ev} is regarded as the saturation temperature that corresponds to the vapor pressure in the chamber (p_v), and ρ_v is the density of saturated vapor at the droplet surface temperature; h and h_m are the respective heat and mass transfer coefficients obtained from correlations developed by Ranz and Marshall [32]:

$$Nu = \frac{hd}{k_v} = 2 + 0.495{Re_v}^{0.55}{Pr_v}^{0.33} \tag{8.5}$$

$$Sh = \frac{h_m d}{D_v} = 2 + 0.495{Re_v}^{0.55}{Sc_v}^{0.33} \tag{8.6}$$

In the above equations, k_v is the thermal conductivity of water vapor, D_v is the diffusion coefficient, Pr_v and Sc_v are the respective Prandtl number and Schmidt number of the vapor, while Re_v, the droplet Reynolds number, is defined as

$$Re_v = \frac{u_{sw}d}{\nu_v} \tag{8.7}$$

where u_{sw} and ν_v are the droplet velocity and dynamic viscosity of vapor, respectively.

Droplet velocity changes over time due to the effect of gravity, buoyancy, and drag force. It can be described by the equation:

$$\frac{du_{sw}}{dt} = \left(1 - \frac{\rho_v}{\rho_{sw}}\right)g - \frac{3}{16}\frac{C_D\rho_v}{d\rho_{sw}}{u_{sw}}^2 \tag{8.8}$$

where g is the gravitational constant, and C_D is the drag coefficient obtained from the Lapple-Shepherd correlation [33] given by

$$C_D = \frac{24}{Re_v}\left(1 + 0.125{Re_v}^{0.72}\right) \tag{8.9}$$

and the residence time for a given axial distance is calculated via

$$t_z = \int_0^H \frac{dz}{u_{sw}(z)} \tag{8.10}$$

where H is the height of the evaporator.

Accordingly, the brine temperature at the bottom of the evaporator is obtained as

$$T_b = \frac{1}{{}^4/_3\pi(d/2)^3}\int_0^{d/2} T_{sw}(t_z, r)4\pi r^2 dr \tag{8.11}$$

To simplify the expression, the dimensionless temperature difference θ_e, the ratio of the temperature difference between the brine and the surrounding vapor to the initial degree of superheat, is defined as

$$\theta_e = \frac{T_b - T_{ev}}{T_f - T_{ev}} = \frac{T_b - T_{ev}}{\Delta T} \tag{8.12}$$

where ΔT is the initial degree of superheat, namely, the temperature difference between the feed seawater and the surrounding vapor. The dimensionless temperature difference θ_e is an indicator of the completeness of evaporation and the degree of thermal utilization. When θ_e equals 0, all the available heat source has been used for evaporation and the amount of evaporated water reaches the maximum value. The curvature of θ_e denotes the rate of evaporation, with a larger curvature indicating a higher evaporation rate.

Viewing from a physical perspective, the spray contains droplets with varying diameters. This effect can be accounted for by using the Sauter Mean Diameter [34] as the average diameter of the droplets.

8.2.2.2 Solution method

The partial differential equations for heat conduction in the droplet are solved using PDEPE, the solver for initial-boundary value problems for systems of parabolic and elliptic partial differential equations in MATLAB [35]. The PDEPE solver converts the PDEs to ordinary differential equations (ODEs) using a second-order accurate spatial discretization [36] and then solves ODEs using the explicit Runge-Kutta method. MATLAB offers a full code package to achieve the computational solution.

8.2.3 Performance analysis of the spray evaporator

The major parameters that affect the evaporation process include: (1) droplet diameter, (2) droplet velocity, (3) degree of superheat, and (4) feedwater temperature. This section investigates the effect of these parameters on the spray evaporation process. The reference condition is designated as $d=400\ \mu m$, $u=5\,m/s$, $\Delta T=4°C$, and $T_f=40°C$.

The effect of droplet mean diameter on the dimensionless temperature difference θ, the ratio of the temperature difference between the brine and the surrounding vapor to the initial degree of superheat, is plotted in Fig. 8.3. It is apparent from the figure that for a long vertical distance ($z=0.9\,m$), droplets with diameters ranging from 150 to 600 μm can reach θ values that are smaller than 0.1. In contrast, at shorter vertical distances, θ values increase dramatically with an increase in the diameter, and only

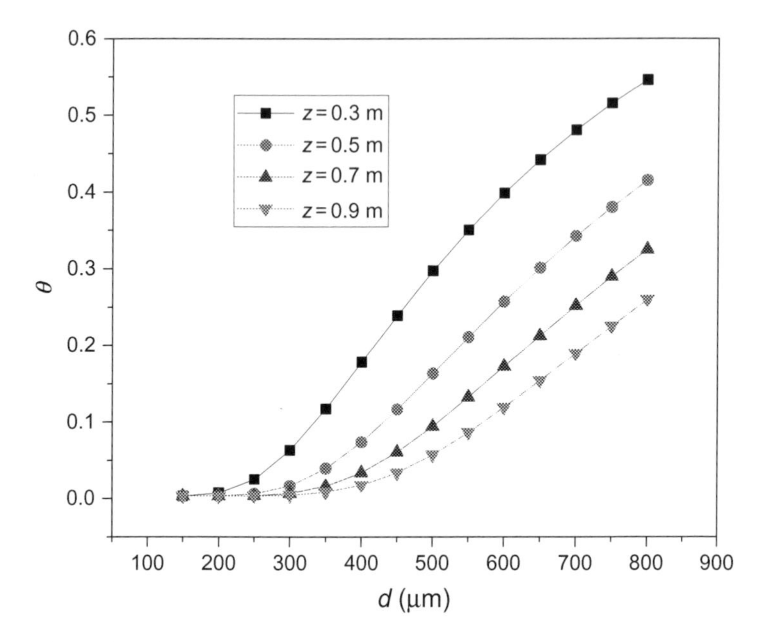

Fig. 8.3 Effect of droplet mean diameter on dimensionless temperature difference.

smaller droplets are able to reach high completeness of evaporation. This is because larger droplets have smaller specific surface areas and higher thermal conduction resistance. Such droplets evaporate slowly and require longer residence time to complete the evaporation process. Consequently, reducing the droplet diameter is key to increasing the evaporation rate and minimizing the evaporator size.

Fig. 8.4 shows the effect of varying initial droplet velocity on the dimensionless temperature difference θ, the ratio of the temperature difference between the brine and the surrounding vapor to the initial degree of superheat. It is apparent from Fig. 8.4 that the θ value is lower for droplets with a lower initial velocity. Droplets with an initial velocity of 2 m/s reach θ values of 0.1 within 0.2 m, while the values are 0.29, 0.41, 0.51, and 0.56 for droplets with initial velocities of 5, 8, 12, and 16 m/s, respectively. The droplet velocity affects residence time as well as impacts heat and mass transfer coefficients. Residence time decreases dramatically with higher initial droplet velocities, while a higher velocity only has marginal impact on heat and mass transfer coefficients. Thus, the increase of the θ value with droplet velocity is significant.

Degree of superheat is the driving force for spray evaporation. Fig. 8.5 compares the dimensionless temperature difference under different degrees of superheat. It is apparent from Fig. 8.5 that the θ values are almost the same for different degrees of superheat. Although a higher degree of superheat provides a larger driving force, it also translates to a longer evaporation process. These two effects offset each other, resulting in marginal change in θ values at different degrees of superheat.

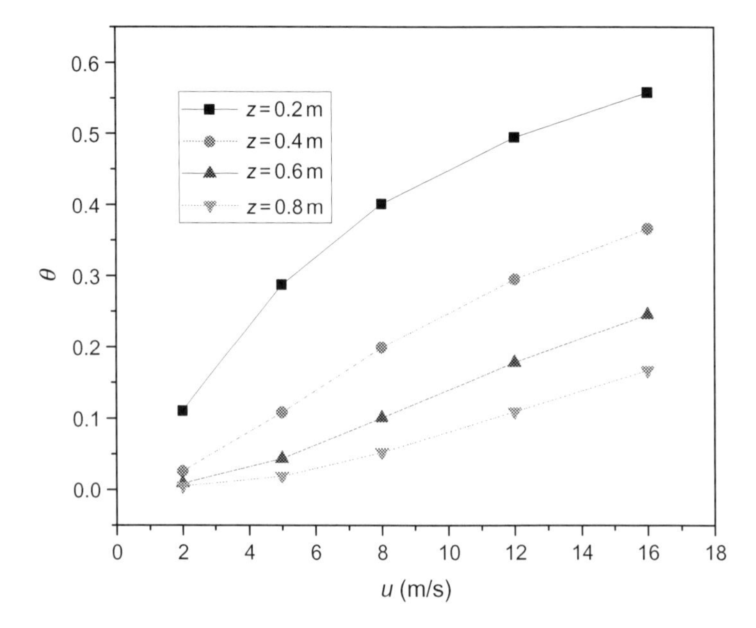

Fig. 8.4 Effect of droplet initial velocity on dimensionless temperature difference.

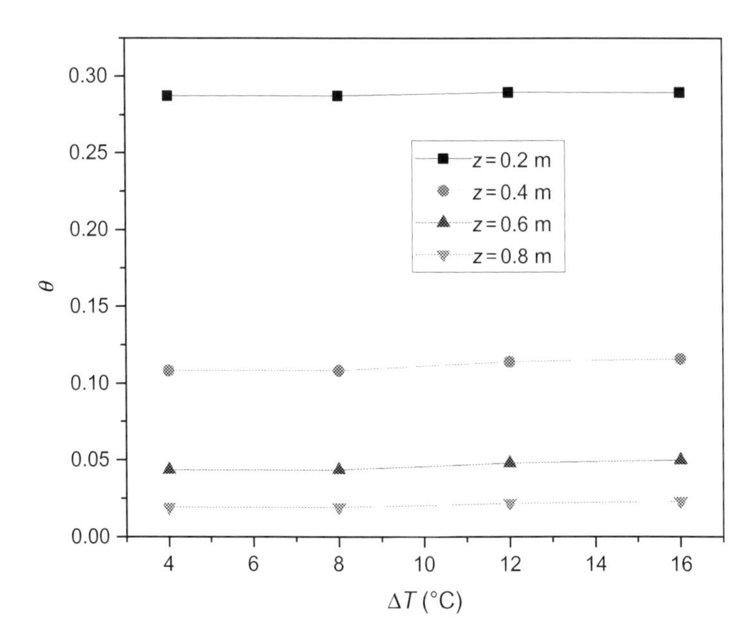

Fig. 8.5 Effect of the degrees of superheat on dimensionless temperature difference.

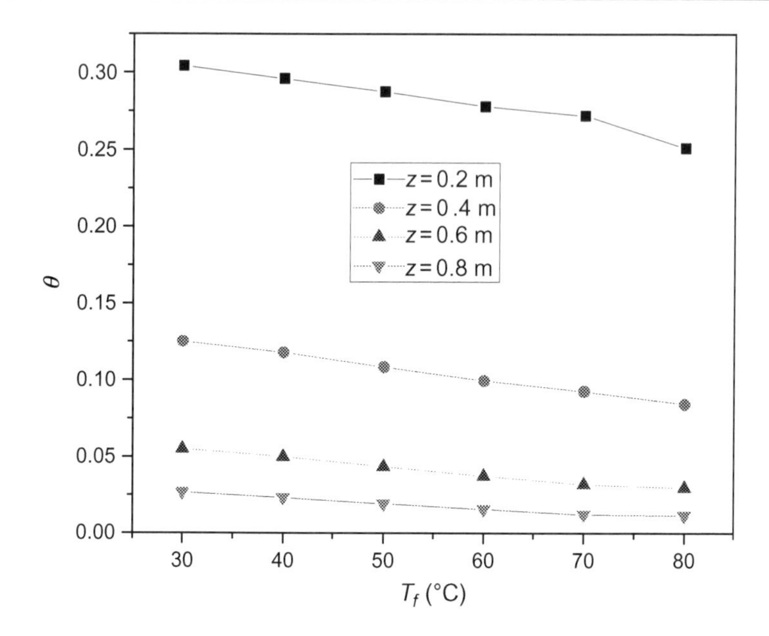

Fig. 8.6 Effect of the initial temperature on dimensionless temperature difference.

Fig. 8.6 shows the effect of changing the initial feedwater temperature on the evaporation process. As illustrated in the figure, θ values are progressively lower at higher initial temperatures. This is attributed to different vapor pressure gradients at different temperatures. At higher temperatures, the increase of vapor pressure is more pronounced. Consequently, the partial pressure difference between the droplet surface and the environment is higher at the same degree of superheat, leading to a greater driving force and a faster evaporation process.

8.3 Single-stage system

A simple spray desalination system evolves when combining a spray evaporator with a coil condenser, as schematically shown in Fig. 8.7. In the desalination system, feed seawater at the temperature of T_f is injected into the evaporator. Part of the seawater ($\dot{D}$) flashes into vapor and the rest ($m_f - \dot{D}$) is drained off. The vapor is directed to the condenser, which is separated from the evaporator with a wire mesh demister to prevent carryover of mist. Vapor is condensed at the outer surface of the condensation coil, while the cold stream ($m_f + m_{cw}$) is circulated inside the coil. The released condensation heat increases the seawater temperature from $T_{sw,in}$ to $T_{sw,r}$. Then a portion of the cold stream (m_{cw}) is rejected, while the rest (m_f) is further heated to T_f using the external heat source at the temperature of T_h. The condensate is collected at the bottom of the condenser.

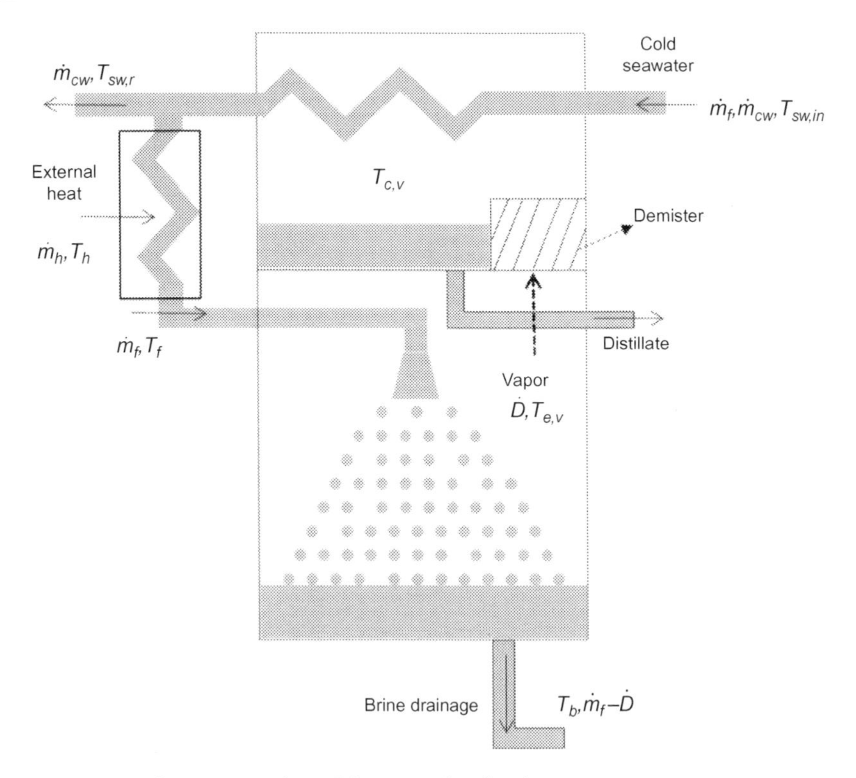

Fig. 8.7 Schematic representation of the spray desalination system.

To evaluate the thermal performance of the system, the following assumptions are made:

(1) heat losses to the ambient are negligible;
(2) the vapor is salt free, and there is no entrainment of mist by the vapor; and
(3) there is no subcooling of condensate leaving the system.

8.3.1 Mathematical modeling

8.3.1.1 Heat and mass balances

The evaporation rate is obtained from the energy conservation in the evaporator:

$$\dot{D}_v = \frac{\dot{m}_f c_{p,sw}(T_f - T_b)}{h_{fg,sw}} = \frac{\dot{m}_f c_{p,sw} \Delta T}{h_{fg,sw}}(1 - \theta_e) \tag{8.13}$$

The heat transfer and energy conservation in the condenser is modeled as

$$\dot{Q}_c = \dot{D}_v h_{fg,sw} = (\dot{m}_f + \dot{m}_{cw}) c_{p,sw}(T_{sw,r} - T_{sw,in}) = (UA)_c (LMTD)_c \tag{8.14}$$

where U is the overall heat transfer coefficient, A is the condenser surface area, and $LMTD$ is the log-mean temperature difference in the condenser defined as

$$(LMTD)_c = \frac{(T_{cv} - T_{sw,r}) - (T_{cv} - T_{sw,in})}{\log[(T_{cv} - T_{sw,r})/(T_{cv} - T_{sw,in})]} \quad (8.15)$$

The condensation temperature T_{cv} is lower than the evaporation temperature attributed to (1) the boiling point elevation due to the dissolved salt and (2) the saturation temperature depressions associated with the pressure loss along the demisters:

$$T_{c,v} = T_{e,v} - BPE - \Delta T_{\text{demis}} \quad (8.16)$$

where BPE is the boiling point elevation and ΔT_{demis} is the temperature loss caused by the demisters.

Heat transfer and energy balance during the heat input process are modeled as

$$\dot{Q}_h = \dot{m}_h h_{fg,sw} = \dot{m}_f c_{p,sw}(T_f - T_{sw,r}) = (UA)_h (LMTD)_h \quad (8.17)$$

where m_h is the mass flowrate of the heating steam, U is the overall heat transfer coefficient, A is the surface area of the heat exchanger, and $LMTD$ is the log-mean temperature difference defined as

$$(LMTD)_h = \frac{(T_h - T_f) - (T_h - T_{sw,r})}{\log[(T_h - T_f)/(T_h - T_{sw,r})]} \quad (8.18)$$

8.3.1.2 Performance indicator

The performance of a thermal desalination system is normally evaluated by the specific energy consumption, the energy consumption for producing a unit mass of freshwater

$$SEC = \dot{Q}_h / \dot{D} \quad (8.19)$$

8.3.2 Performance analysis of the single-stage system

To analyze the performance of a single-stage system, we adopt a spray evaporator operating at a seawater flowrate of 10 kg/s, a cooling water flowrate of 10 kg/s, a seawater temperature of 25°C, and a heating steam temperature of 45°C. The heat transfer areas in the condenser and the heater are designed to sustain a terminal temperature difference of 5°C. The sum of the BPE and the demister temperature loss is assumed to be 0.5°C, and the θ value is designated as 0.2.

Applying the developed model, the evaporation temperature is computed to be 33.21°C; the condensation temperature is 32.71°C while the seawater temperature

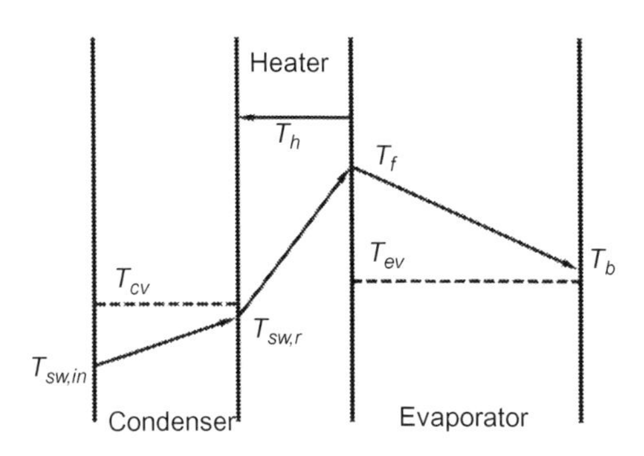

Fig. 8.8 Schematic representation of the temperature profile.

leaving the condenser is 27.71°C. The distillate production rate is 0.095 kg/s and the specific energy consumption of the system is estimated to be 5227 kJ/kg.

The temperature profile for seawater is schematically depicted in Fig. 8.8. The heat recovery approach reduces the heat input, since seawater is heated up from $T_{sw,r}$ to T_f, instead of $T_{sw,in}$ to T_f. However, the heat input does not completely translate into freshwater production. This is because the brine temperature T_b is higher than the seawater temperature entering the heater $T_{sw,r}$. Such a temperature difference is attributed to (a) incomplete evaporation, which makes the brine temperature T_b higher than the evaporation temperature $T_{e,v}$; (b) boiling temperature elevation and demister temperature loss, which makes the condensation temperature $T_{c,v}$ lower than the evaporation temperature $T_{e,v}$; and (c) insufficient heat recovery, which makes the seawater temperature leaving the condenser $T_{sw,r}$ lower than the condensation temperature $T_{c,v}$.

The percentages of various causes for heat input are charted in Fig. 8.9. As can be seen, the thermal performance of the system is influenced by the degree of heat recovery and the completeness of evaporation, which account for 52% of the heat input.

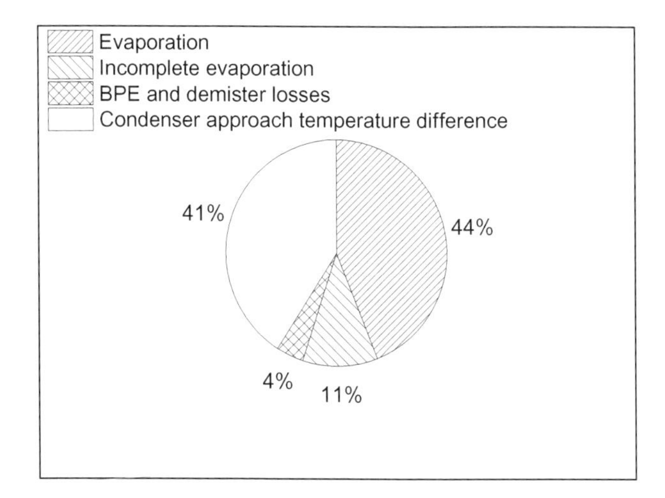

Fig. 8.9 Sources of heat input requirement.

Thus, higher degrees of heat recovery and higher completeness of evaporation promote system thermal performance. Also, improvement of the production rate reduces the weight of the heat losses and consequently enhances the energy efficiency. This is achieved by either increasing the heating source temperature or the cooling water flowrate. The effects of these approaches on production rate and specific energy consumption are plotted in Fig. 8.10. It is apparent that production rate increases significantly with smaller θ, lower ΔT_c, higher heating source temperatures, and higher cooling water flowrates, while the corresponding *SEC* is lower.

It is worth noting that the improvement of heat recovery is associated with a larger heat transfer area, which leads to a higher initial cost. The heating steam temperature is determined by the temperature of the available heating medium. A higher degree of evaporation completeness can be achieved through proper evaporator design and operation, that is, providing sufficient evaporator height, evolving smaller droplets, and providing lower spray velocities, as readily inferred from the previous section.

It is impossible for the *SEC* of a single-stage desalination system to be lower than the latent heat of vaporization (h_{fg}). This is because the condensation heat is reused only once. This limitation hinders the single-stage system from large-scale applications. In order to improve the thermal performance, the system should operate in a multistage mode.

8.4 Multistage system

A multistage system has a lower *SEC* simply due to the multiple reuses of condensation heat. Additionally, spray condensers, which have a similar configuration with spray evaporators, can be applied for condensation. The high completeness of condensation in spray condensers enables a more efficient manner to recover the condensation heat, while the simplicity of the component configuration further simplifies the system design.

8.4.1 Process description

The multistage system employs a similar process flow as the MSF system, while the temperature and pressure dynamics are similar to the MED system [16]. The system includes several process stages as well as two sets of heat exchangers for heat recovery and heat input. Each stage comprises a spray evaporator and a spray condenser. The evaporator and condenser are connected to facilitate vapor flow, and a wire mesh demister is placed in between to prevent the entrainment of seawater droplets.

Fig. 8.11 shows a schematic diagram of a proposed multistage system. The feed seawater at the top brine temperature *TBT* is injected into the evaporator of the first stage (E_1). In the evaporator, the seawater partially flashes into vapor. The remaining seawater at a lower temperature $T_{el,2}$ is injected into the second evaporator (E_2). E_2 is subjected to a lower pressure condition, and consequently the seawater reaches superheated conditions and evaporates again. Such a process is repeated in the downstream evaporators at progressively lower pressure and temperature conditions. The vapor

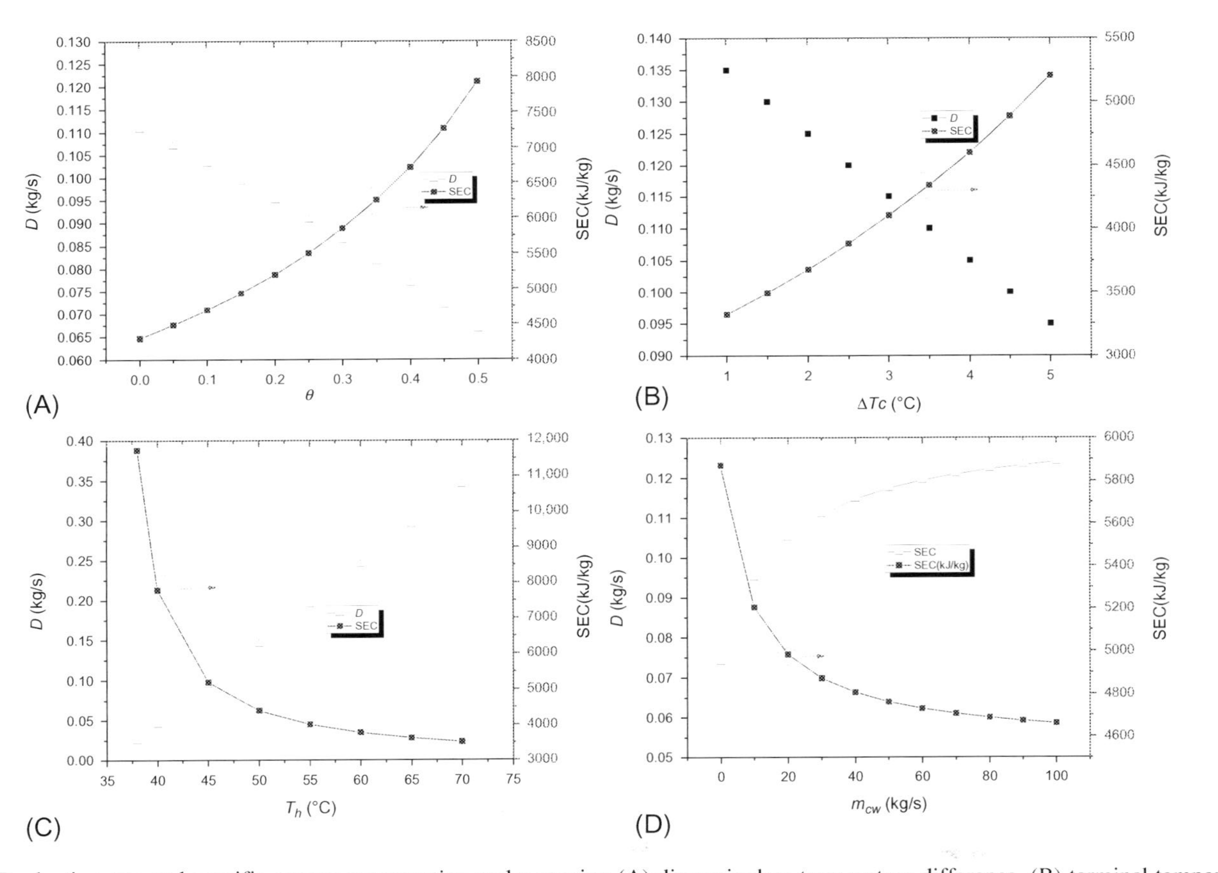

Fig. 8.10 Production rate and specific energy consumption under varying (A) dimensionless temperature difference, (B) terminal temperature difference of the condenser (C) heat source temperature, and (D) cooling water flowrate.

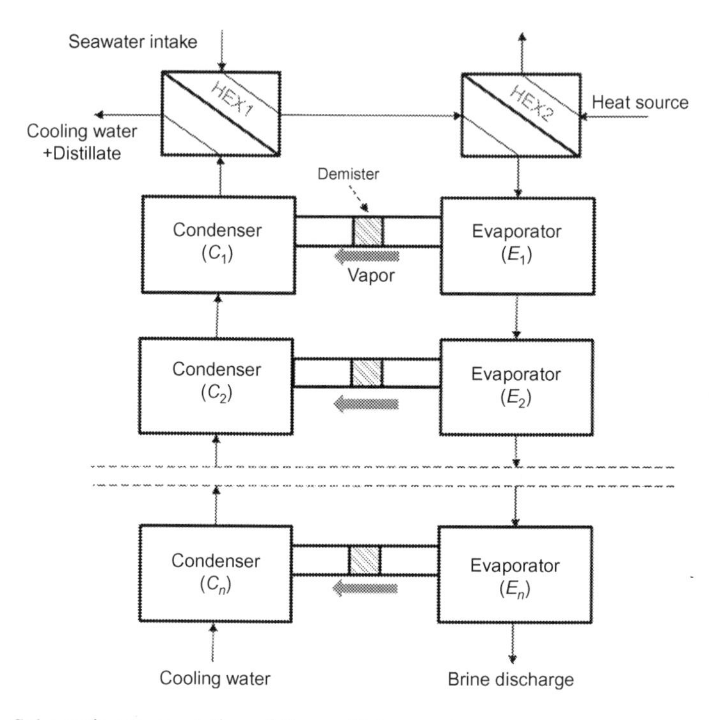

Fig. 8.11 Schematic representation of the multistage system.

produced in each evaporator is directed to the condensers in the corresponding stages. Cooling water at the ambient temperature is injected from the last stage (C_n) to condense the vapor from E_n. The released condensation heat increases the temperature of the cooling water, which is then injected into the previous condenser at higher pressure and temperature. Accordingly, the cooling water becomes subcooled again and the condensation process is repeated. Finally, the cooling water exits the condenser at the first stage (C_1), carrying with it the desired distilled water.

The temperature of the cooling water leaving the system ($T_{cl,1}$) is much higher than the feed seawater temperature ($T_{sw,in}$). Thus, it is employed to heat up the feed seawater in a heat recovery manner via the heat exchanger *HEX1*. The heat recovery process minimizes the heat input and reduces the cooling water temperature. The cooling water is reused after the produced distilled water is extracted. Seawater can be further heated to the required temperature by potentially employing a low-grade heat source, such as solar energy, geothermal energy, or waste heat.

8.4.2 Mathematical modeling

An analytical model is developed to describe the spray-assisted low-temperature desalination process. The model is based on analyzing the evaporation and condensation processes in each stage, as well as the principles of heat balance, mass balance, and

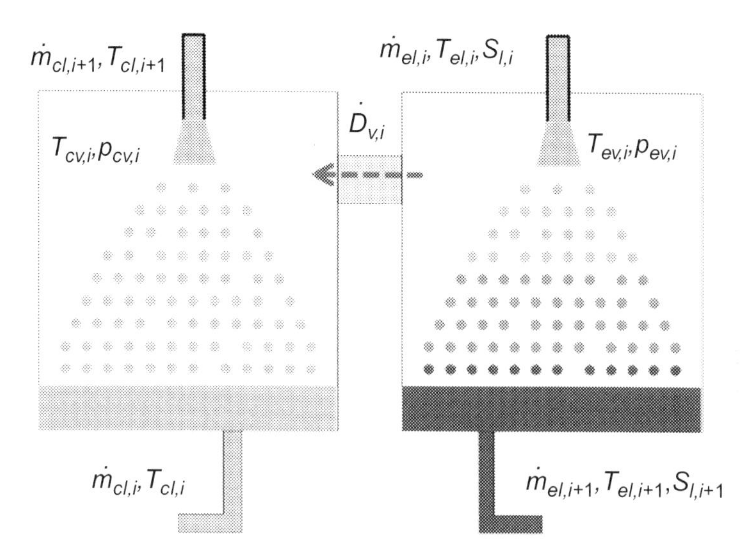

Fig. 8.12 Schematic diagram for single stage.

salt balance in each stage. The schematic representation of an intermediate stage is shown in Fig. 8.12. The following key assumptions are made:

(1) the system is adiabatic, and heat losses to the ambient are negligible;
(2) seawater properties are considered to be constant in each stage, and are determined by the mean temperature and salinity in the corresponding stage; and
(3) the vapor from any stage is salt free, and there is no entrainment of mist by the vapor.

8.4.2.1 Flash evaporation and spray condensation

The spray evaporation model derived in the previous section is judiciously applied to obtain the dimensionless temperature differences $\theta_{e,i}$ in each evaporator. The same process can be applied for calculating the dimensionless temperature differences $\theta_{c,i}$ in the condensers which are defined as

$$\theta_{c,i} = \frac{T_{cv,i} - T_{cl,i}}{T_{cv,i} - T_{cl,i+1}} = \frac{T_{cv,i} - T_{cl,i}}{\Delta T_{c,i}} \tag{8.20}$$

The only difference is that the convection thermal resistance is marginal, and the boundary condition at droplet surface is written as

$$T_{cl}\big|_{r=d/2} = T_{cv,i} \tag{8.21}$$

8.4.2.2 Heat and mass balance

The variation of temperatures for seawater and cooling water in different stages can be expressed as

$$T_{el,i+1} = T_{ev,i} + (T_{el,i} - T_{ev,i})\theta_{e,i} \tag{8.22}$$

$$T_{cl,i} = T_{cv,i} - (T_{cv,i} - T_{cl,i+1})\theta_{c,i} \tag{8.23}$$

Similar to the single-stage system, the condensation temperatures in the condensers are lower than the boiling temperatures in the evaporators of corresponding stages. The differences are attributed to the boiling point elevation due to the dissolved salt and the saturation temperature depressions associated with the pressure loss along the demisters. The condensation temperature is expressed as

$$T_{cv,i} = T_{ev,i} - BPE_i - \Delta T_{\text{demis},i} \tag{8.24}$$

The seawater temperatures at the inlet of the first evaporator and the outlet of the last evaporator are the top brine temperature and the lower brine temperature written as

$$T_{el,1} = TBT \tag{8.25}$$

$$T_{el,n+1} = T_{sw,o} \tag{8.26}$$

Similarly, the cooling water temperatures at the inlet of the last condenser and the outlet of the first condenser correspond to the cooling water inlet and outlet temperatures are expressed as

$$T_{cl,n+1} = T_{cw,in} \tag{8.27}$$

$$T_{cl,1} = T_{cw,o} \tag{8.28}$$

Under an adiabatic condition, the evaporation and condensation rates in each stage can be obtained from the energy balance for each evaporator and condenser. They are derived as

$$\dot{D}_{v,i} = \dot{m}_{el,i}\frac{c_{pel,i}(T_{el,i} - T_{el,i+1})}{h_{fge,i}} = \dot{m}_{el,i}\frac{c_{pel,i}(T_{el,i} - T_{ev,i})(1-\theta_{e,i})}{h_{fge,i}} \tag{8.29}$$

$$\dot{D}_{v,i} = \dot{m}_{cl,i}\frac{c_{pcl,i}(T_{cl,i} - T_{cl,i+1})}{h_{fgc,i}} = \dot{m}_{cl,i}\frac{c_{pcl,i}(T_{cv,i} - T_{cl,i})(1-\theta_{c,i})}{h_{fgc,i}} \tag{8.30}$$

In the evaporators, the brine from the bottom of each stage is injected into the next stage. Therefore, the mass balance for each evaporator is written as

$$\dot{m}_{el,i+1} = \dot{m}_{el,i} - \dot{D}_{v,i} \tag{8.31}$$

Similarly, the mass balance in the condenser side is written as

$$\dot{m}_{cl,i} = \dot{m}_{cl,i+1} + \dot{D}_{v,i} \tag{8.32}$$

Since no salt is carried over by the vapor, all the dissolved salts stay in the evaporators. Thus, the salt balance in the evaporator is written as

$$\dot{m}_{el,i+1}S_{l,i+1} = \dot{m}_{el,i}S_{l,i} \tag{8.33}$$

8.4.2.3 Heat input and heat recovery

Fig. 8.13 shows the schematic representation of the heat recovery and heat input processes. Heat recovery is achieved via a liquid-liquid heat exchanger and the process is expressed as

$$\dot{Q}_R = \dot{m}_{sw,in}c_{p,sw}(T_{sw,r} - T_{sw,in}) = \dot{m}_{cw,o}c_{p,cw}(T_{cl,1} - T_{cl,r}) \tag{8.34}$$

In the heat input process, hot steam is employed as the heat source and the process is written as

$$\dot{Q}_H = \dot{m}_{sw,in}c_{p,sw}(TBT - T_{sw,r}) = \dot{m}_{h,v}h_{fg} \tag{8.35}$$

8.4.2.4 Performance indicator

Similar to the single-stage system, the thermal performance of the multistage is also evaluated by the specific energy consumption which is expressed as

$$SEC = \dot{Q}_h/\dot{D} \tag{8.36}$$

where the overall production rate, $\dot{D}_v$, is the sum of production rate in each stage and is written as

$$\dot{D}_v = \sum \dot{D}_{v,i} \tag{8.37}$$

8.4.2.5 Solution method

As evidenced from the previous section, the dimensionless temperature difference values change marginally with the degree of superheat. Consequently, they are

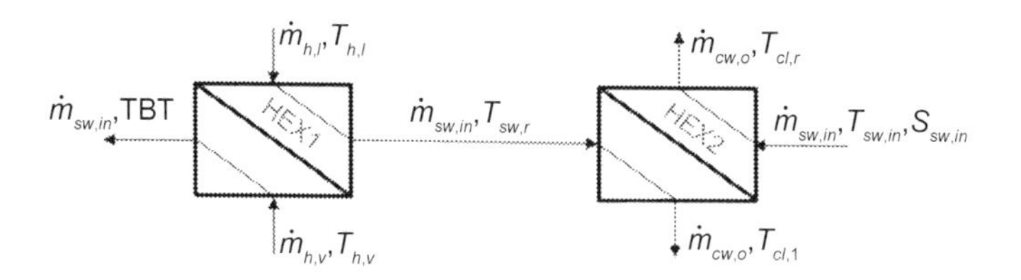

Fig. 8.13 Schematic diagram of heat recovery and heat input process.

calculated using preassigned values of degrees of superheat. The heat and mass balance equations are then solved by the successive iteration method to get the temperatures, salinity values, and production rates in each stage. The overall production rate and the specific energy consumption are subsequently obtained.

8.4.3 Performance analysis of the multistage system

The performance of the multistage system varies under different design and operating conditions. Table 8.2 lists the parameters adopted for simulations. The chamber height is designed to provide sufficient residence time to complete evaporation and condensation, and the top brine temperature spans 55–90°C for potential utilization of renewable heat sources.

Other considerations and simplifications involved in the calculation process include:

(1) The vapor temperature drop $\Delta T_{\text{demis},i}$ across the demister is affected by the geometry and size of the demister. Taking reference from published operation data on MSF plants, it is designated as 0.15°C in each stage [37].
(2) The terminal temperature differences in each heat exchanger are kept fixed. An approach temperature difference of 3°C is set for the liquid-liquid heat exchanger in the heat recovery process, while a temperature difference of 5°C is applied to the liquid-vapor heat exchanger in the heat input process [38].

8.4.3.1 Temperature and production rate profiles

Fig. 8.14A portrays the temperature distributions of both feedwater and cooling water. Seawater at the temperature of 70°C enters the first evaporator. The temperature depreciates almost linearly along stages due to evaporation, and finally seawater leaves the last evaporator at the temperature of 34.2°C. On the condenser side, the cooling water enters the last condenser at the temperature of 30°C. It absorbs the condensation heat and finally leaves the first condenser at 65.8°C. In *HEX1*, the cooling

Table 8.2 Parameter set employed for simulations

Parameter		Base case	Range
Design parameter	Droplet diameter in evaporator	600 μm	–
	Droplet diameter in condenser	600 μm	–
	Height of evaporator	1.1 m	–
	Height of condenser	1.1 m	–
Operation parameter	Seawater temperature	30°C	–
	Seawater salinity	0.035 kg/kg	–
	Cooling water temperature	30°C	–
	Top brine temperature	70°C	55–90°C
	Operating stages	10	3–14
	Feedwater flowrate	10 kg/s	–
	Cooling water flowrate	9.5 kg/s	7–15 kg/s

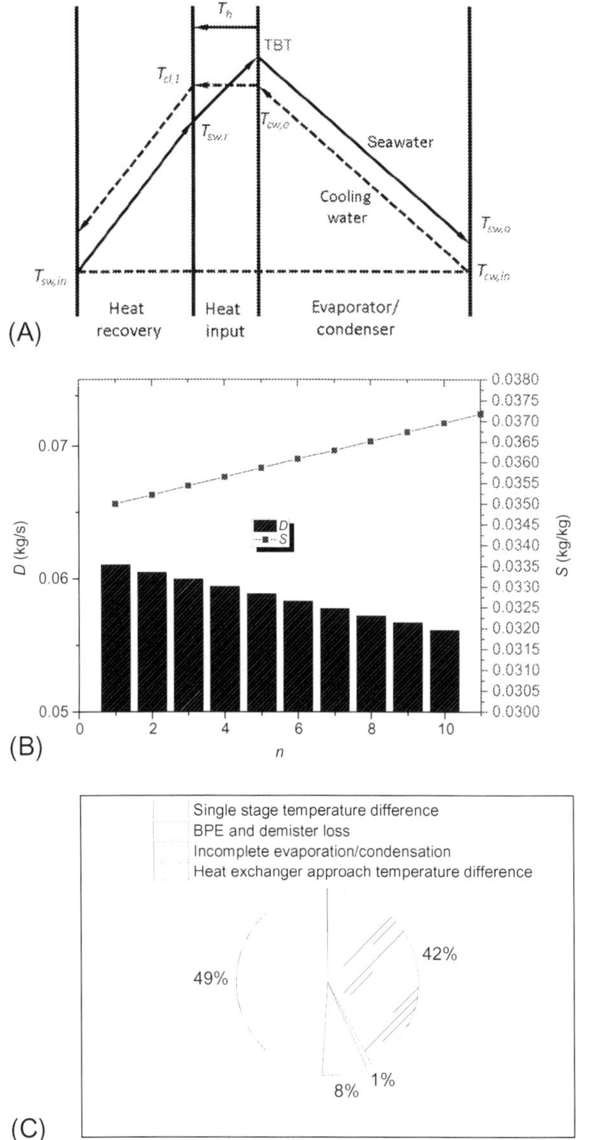

Fig. 8.14 (A) Profiles of feedwater temperatures and cooling water temperatures, (B) production rate and salinity for each stage, and (C) sources for heat input under TBT = 70°C, $T_{cw,in}$ = 30°C, $S_{sw,in}$ = 0.035 kg/kg, and N_{total} = 10.

water heats up the seawater to 62.8°C. The seawater is further heated up to 70°C by taking advantage of the external heat source in *HEX2*. The overall production rate is 0.6 kg/s and the corresponding specific energy consumption is 504.4 kJ/kg. The production rate is several times higher than that achieved in the single-stage system, while the specific energy consumption is much lower.

The production rate is highest in the first stage and decreases marginally in subsequent stages, as evidenced in Fig. 8.14B. This observation is attributed to the reduction of both feedwater flowrate and cooling water flowrate along the stages which are caused by the feedwater evaporation and the accumulation of condensate in the cooling water. Such flowrate changes render previous stages higher capacities than those in later stages. Fig. 8.14B also illustrates the increase of seawater salinity in the evaporators due to evaporation. However, the increase in salinity is considered to be insignificant due to small evaporation rates.

Fig. 8.14C illustrates the sources for heat input requirement. Nearly half of the heat input is dominated by the operating temperature difference of the first stage. The second rank is the approach temperature difference of the heat exchanger, which accounts for 42% of the heat input. The heat input requirement caused by incomplete evaporation and condensation is marginal, because the droplet diameters are small, which results in high process completeness.

8.4.3.2 Effect of design and operating parameters

The effect of dimensionless temperature differences on the system performance is plotted in Fig. 8.15. It is apparent that smaller θ values in the evaporators and condensers significantly promote both production rate and energy efficiency. As defined, the value of θ quantifies the degree of complete evaporation/condensation process. Smaller θ implies a higher degree of evaporation/condensation completion level. Consequently, a higher degree of thermal utilization, that is, higher D and lower SEC, ensues.

As evidenced previously, nearly half of the heat input is dominated by the operating temperature difference of the first stage. The single-stage temperature difference is almost inversely proportional to the total number of operating stages. Thus, higher numbers of operating stages promote system performance, as shown in Fig. 8.16A. This is because a smaller single-stage temperature difference lowers the brine temperature in the last stage ($T_{sw,o}$) while increasing the cooling water outlet temperature in the first stage ($T_{cw,o}$), as depicted in Fig. 8.16B. A lower $T_{sw,o}$ indicates a more

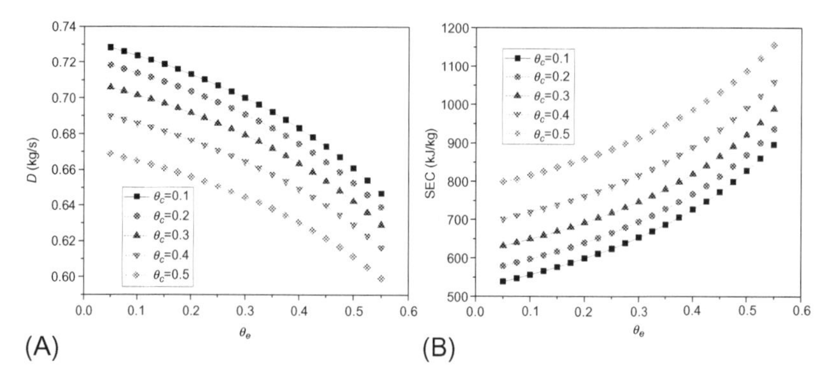

Fig. 8.15 Effect of dimensionless temperature difference on (A) production rate and (B) specific energy consumption for TBT = 70°C, $T_{cw,in}$ = 30°C, S = 0.035 kg/kg, and N_{total} = 10.

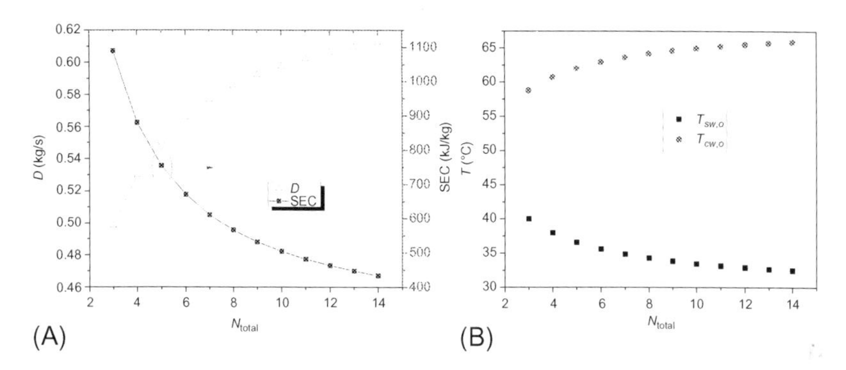

Fig. 8.16 Effect of operating stages on (A) production rate and specific energy consumption and (B) lower brine temperature and cooling water outlet temperature for TBT = 70°C, $T_{cw,in}$ = 30°C, and S = 0.035 kg/kg.

complete utilization of the available thermal energy in the feedwater and leads to a higher production rate. A higher $T_{cw,o}$ enhances heat recovery and minimizes the heat input, which renders a lower specific energy consumption.

Fig. 8.17 shows the production rate and specific energy consumption under varying top brine temperatures. It is apparent that both production rate and energy efficiency are higher under higher *TBT*. A higher top brine temperature provides more heat source for evaporation and, as a result, increases the production rate proportionally. The *SEC* is affected by both the production rate and heat input. The required heat input

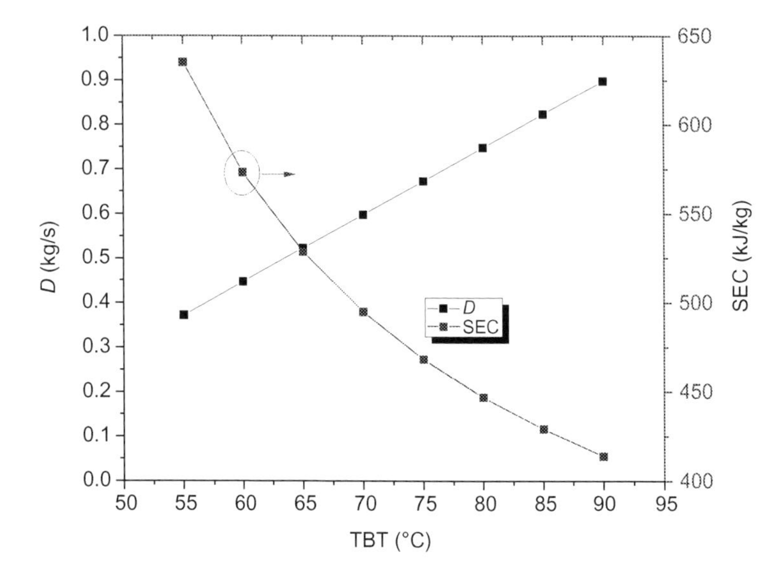

Fig. 8.17 Effect of top brine temperature on production rate and specific energy consumption for $T_{cw,in}$ = 30°C, S = 0.035 kg/kg, and N_{total} = 10.

is caused by the single-stage temperature difference, the temperature losses due to incomplete evaporation/condensation, demister loss, and *BPE*. While the single-stage temperature difference increases proportionally with *TBT*, the temperature losses due to incomplete evaporation, demister loss, and *BPE* remain unchanged. Consequently, the increase in heat input is less significant than the improvement of production rate, and a lower *SEC* is observed.

Fig. 8.18 shows the effect of varying cooling water flowrate on production rate and specific energy consumption. In computing the heat transfer rate, the effectiveness-heat transfer unit (ε-NTU) method is employed, and the heat transfer area is kept constant (set as the heat transfer area when the cooling water flowrate equals the seawater flowrate, namely, 10kg/s, and the approach temperature difference is 3°C). It is apparent from the figure that the production rate markedly increases with higher cooling water flowrate when the flowrate is small, and the trend gradually stops when the cooling water flowrate exceeds 12kg/s. The key reason is that a higher cooling water flowrate improves the condensation rate, which subsequently improves the evaporation rate and consequently improves the production rate. The upward trend becomes saturated at higher cooling water flowrates because the disposed seawater approaches the temperature of the incoming cooling water. In contrast to the production rate, the *SEC* has a minimal value when the cooling water flowrate is around 9.8kg/s. The production rate is low under a small cooling water flowrate. Although a higher cooling water flowrate improves

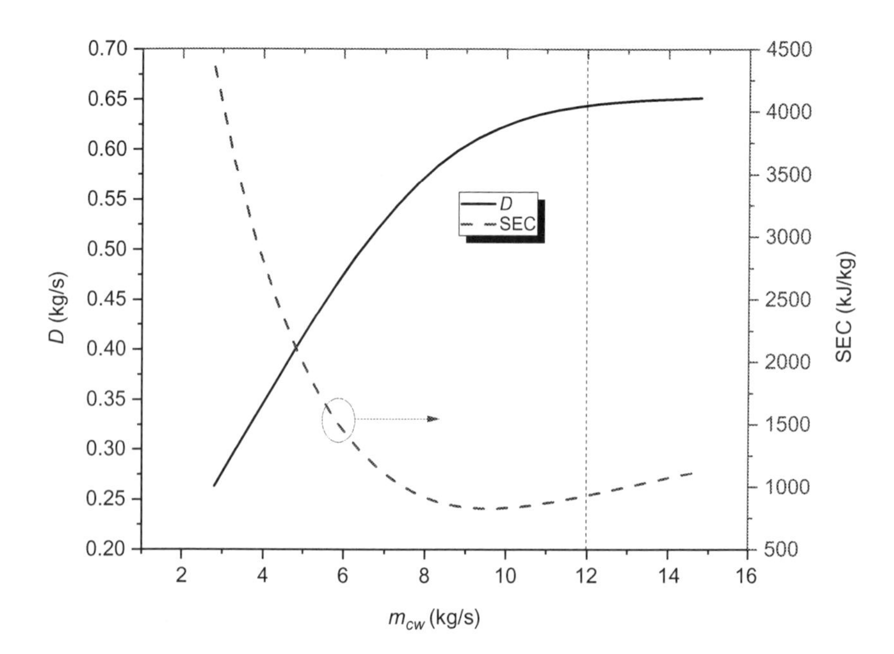

Fig. 8.18 Effect of cooling water flowrate on production rate and specific energy consumption, $m_{sw,in} = 10$ kg/s.

the production rate, the cooling water outlet temperature is also lower, and the condensation heat collected by the cooling water cannot be fully reused. In this case, more external energy is required to heat up the seawater to the top brine temperature, and a higher *SEC* ensues. A cooling water flowrate that is close to the feedwater flowrate ensures optimal energy efficiency, since it renders a high production rate and a high cooling water outlet temperature for heat recovery.

8.4.3.3 Comparison with other thermal desalination technologies

Fig. 8.19 compares the specific energy consumption of MED, MSF, and the spray-assisted low-temperature desalination system under varying operating stages. Data for MED and MSF plants are adopted from the literature [39]. The top brine temperatures are 90°C for both MSF and the current system, while the value is 70°C for the MED plant to prevent the contamination of $CaSO_4$. It is clear from the figure that the *SEC* of the current system is lower than that of the MSF plants. Even though the MED plants have lower *SEC* under the same operating stages, the spray-assisted low-temperature desalination system is able to operate under more stages due to its higher operating temperature. Therefore, the *SEC* of the spray-assisted low-temperature desalination system is comparable with that of the MED plants.

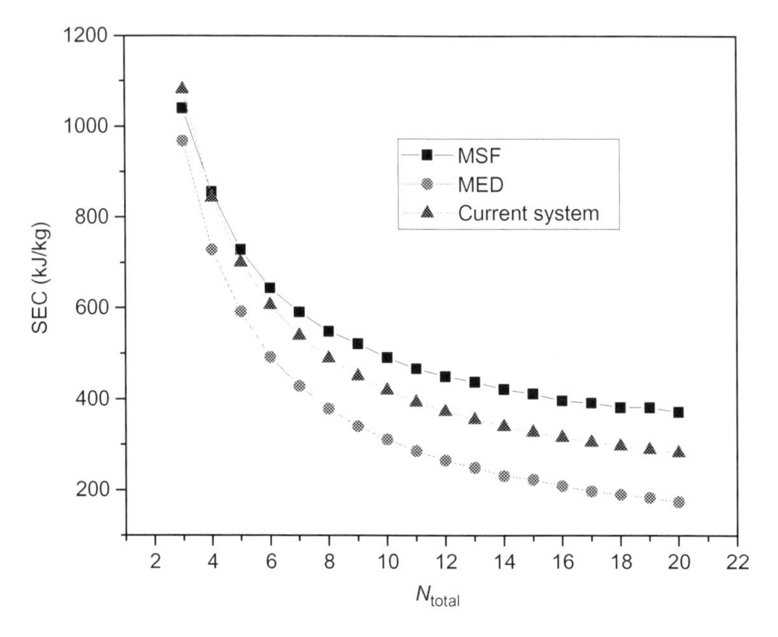

Fig. 8.19 Comparison of specific energy consumption among various desalination technologies.

8.5 Summary

This chapter introduces a novel technology: a spray assisted low-temperature desalination technology. The working principle of the spray evaporator is first discussed, followed by the investigation of a single-stage spray desalination system. In the later section, a multistage system is presented. It recovers condensation heat more effectively and yields a higher thermal efficiency. The analysis on the system leads to the following key findings:

(1) The flashing and shattering phenomena enable the spray evaporators to conduct high intensity of heat and mass transfer. The completeness of the evaporation process is promoted by smaller droplet diameters and lower flow velocities, while the effects of the degree of superheat and the process temperature are marginal. A mean droplet diameter of 600 μm along with a flow velocity of 5 m/s allows the evaporation process to be completed in a highly compact evaporator.

(2) A simple spray desalination system evolves when combining a spray evaporator with a coil condenser. The performance of such a simple system improves by higher degrees of heat recovery, high completeness of evaporation, higher heat source temperatures, and higher cooling water flowrates.

(3) Operating the system in a multistage mode further improves the system performance. A greater number of operating stages and higher top brine temperatures markedly improve the production rate and energy efficiency, while a cooling water flowrate close to that of the feedwater flowrate renders the lowest specific energy consumption with a sufficient level of production rate. A specific energy consumption of 353.85 kJ/kg can be achieved for a 14-stage system with a top brine temperature of 90°C under the optimal cooling water flowrate.

Moving forward in attempting to evolve technologies that are more competitive and appealing for water treatment, future fundamental research efforts can focus on the following:

(1) Developing a dynamic system model that is capable of investigating the unsteady state performance subject to varying operating conditions.

(2) Optimal integration of the spray desalination system with renewable sources, for example, solar energy, geothermal energy, and industry waste heat, to evolve more sustainable "green" systems.

(3) Conducting detailed entropy and exergy analysis to identify the major irreversibilities and to optimize the system performance based on the second law of thermodynamics efficiency.

References

[1] Henthorne L, Pankratz T, Murphy S. Desalination: sustainable solutions for a thirsty planet, the state of desalination 2011. IDA World congress on desalination and water reuse.

[2] International Desalination Association. IDA desalination yearbook 2015–2016 http://idayearbook.desalination.com/?issueID=1&pageID=85.

[3] Latteman S. Development of an environmental impact assessment and decision support system for seawater desalination plants. Boca Raton, FL: CRC Press; 2010.

[4] Gude VG. Energy consumption and recovery in reverse osmosis. Desalin Water Treat 2011;36(1–3):239–60.
[5] Garcí L. Renewable energy applications in desalination: state of the art. Sol Energy 2003;75(5):381–93.
[6] Fiorenza G, Sharma VK, Braccio G. Techno-economic evaluation of a solar powered water desalination plant. Energ Conver Manage 2003;44(14):2217–40.
[7] Christ A, Regenauer-Lieb K, Chua HT. Thermodynamic optimisation of multi effect distillation driven by sensible heat sources. Desalination 2014;336:160–7.
[8] Christ A, Regenauer-Lieb K, Chua HT. Application of the boosted MED process for low-grade heat sources—a pilot plant. Desalination 2015;366:47–58.
[9] Darwish M, Mohtar R, Elgendy Y, Chmeissani M. Desalting seawater in Qatar by renewable energy: a feasibility study. Desalin Water Treat 2012;47(1–3):279–94.
[10] Olwig R, Hirsch T, Sattler C, Glade H, Schmeken L, Will S, et al. Techno-economic analysis of combined concentrating solar power and desalination plant configurations in Israel and Jordan. Desalin Water Treat 2012;41(1–3):9–25.
[11] Arunkumar T, Jayaprakash R, Ahsan A, Denkenberger D, Okundamiya MS. Effect of water and air flow on concentric tubular solar water desalting system. Appl Energy 2013;103:109–15.
[12] Deniz E, Çınar S. Energy, exergy, economic and environmental (4E) analysis of a solar desalination system with humidification-dehumidification. Energ Conver Manage 2016;126:12–9.
[13] Elminshawy NA, Siddiqui FR, Addas MF. Development of an active solar humidification-dehumidification (HDH) desalination system integrated with geothermal energy. Energ Conver Manage 2016;126:608–21.
[14] He WF, Han D, Yue C, Pu WH. A parametric study of a humidification dehumidification (HDH) desalination system using low grade heat sources. Energ Conver Manage 2015;105:929–37.
[15] Zaragoza G, Ruiz-Aguirre A, Guillén-Burrieza E. Efficiency in the use of solar thermal energy of small membrane desalination systems for decentralized water production. Appl Energy 2014;130:491–9.
[16] Wellmann J, Neuhäuser K, Behrendt F, Lehmann M. Modeling an innovative low-temperature desalination system with integrated cogeneration in a concentrating solar power plant. Desalin Water Treat 2015;55(12):3163–71.
[17] Miyatake O, Tomimura T, Ide Y, Fujii T. An experimental study of spray flash evaporation. Desalination 1981;36(2):113–28.
[18] Miyatake O, Tomimura T, Ide Y, Yuda M, Fujii T. Effect of liquid temperature on spray flash evaporation. Desalination 1981;37(3):351–66.
[19] Miyatake O, Tomimura T, Ide Y. Enhancement of spray flash evaporation by means of the injection of bubble nuclei. J Sol Energy Eng 1985;107(2):176–82.
[20] Ikegami Y, Sasaki H, Gouda T, Uehara H. Experimental study on a spray flash desalination (influence of the direction of injection). Desalination 2006;194(1):81–9.
[21] Mutair S, Ikegami Y. Experimental study on flash evaporation from superheated water jets: influencing factors and formulation of correlation. Int J Heat Mass Transf 2009;52(23):5643–51.
[22] Mutair S, Ikegami Y. Experimental investigation on the characteristics of flash evaporation from superheated water jets for desalination. Desalination 2010;251(1):103–11.
[23] El-Fiqi AK, Ali NH, El-Dessouky HT, Fath HS, El-Hefni MA. Flash evaporation in a superheated water liquid jet. Desalination 2007;206(1):311–21.
[24] El-Agouz SA, El-Aziz GA, Awad AM. Solar desalination system using spray evaporation. Energy 2014;76:276–83.

[25] Araghi AH, Khiadani M, Lucas G, Hooman K. Performance analysis of a low pressure discharge thermal energy combined desalination unit. Appl Therm Eng 2015;76:116–22.
[26] Araghi AH, Khiadani M, Hooman K. A novel vacuum discharge thermal energy combined desalination and power generation system utilizing R290/R600a. Energy 2016;98:215–24.
[27] Hwang TH, Moallemi MK. Heat transfer of evaporating droplets in low pressure systems. Int Commun Heat Mass Transf 1988;15(5):635–44.
[28] Muthunayagam AE, Ramamurthi K, Paden JR. Modelling and experiments on vaporization of saline water at low temperatures and reduced pressures. Appl Therm Eng 2005;25 (5):941–52.
[29] Chen Q, Thu K, Bui TD, Li Y, Ng KC, Chua KJ. Development of a model for spray evaporation based on droplet analysis. Desalination 2016;399:69–77.
[30] Chen Q, Li Y, Chua KJ. On the thermodynamic analysis of a novel low-grade heat driven desalination system. Energ Conver Manage 2016;128:145–59.
[31] Sher E, Bar-Kohany T, Rashkovan A. Flash-boiling atomization. Prog Energy Combust Sci 2008;34(4):417–39.
[32] Ranz WE, Marshall WR. Evaporation from drops. Chem Eng Progr 1952;48(3):141446.
[33] Scala F, editor. Fluidized bed technologies for near-zero emission combustion and gasification. Sawston, Cambridge, UK: Elsevier; 2013 p. 42–76 [p. 45]https://doi.org/10.1533/9780857098801.1.42.
[34] Lapple CE, Shepherd CB. Calculation of particle trajectories. Ind Eng Chem 1940;32 (5):605–17.
[35] pdepe–MathWorks; 2015 May 05. Retrieved from: http://www.mathworks.com/help/matlab/math/partial-differential-equations.html.
[36] Skeel RD, Berzins M. A method for the spatial discretization of parabolic equations in one space variable. SIAM J Sci Statist Comput 1990;11(1):1–32.
[37] Darwish MA, Alsairafi A. Technical comparison between TVC/MEB and MSF. Desalination 2004;170(3):223–39.
[38] Rahimi B, Christ A, Regenauer-Lieb K, Chua HT. A novel process for low grade heat driven desalination. Desalination 2014;351:202–12.
[39] Kucera J, editor. Desalination: water from water. Hoboken, NJ: John Wiley & Sons; 2014.

Nanocomposite membranes

Mohammad Amin Alaei Shahmirzadi, Ali Kargari
Amirkabir University of Technology (Tehran Polytechnic), Tehran, Iran

9.1 Introduction

In the 21st century, the most important problem affecting humane societies around the world is the increasing demand for water and its global scarcity. The rapid growth of world population and industries has resulted in substantial increase in the water consumption, leading to a serious water shortage particularly in arid regions [1,2]. Based on the United Nations reports, around 1.2 billion people live in the region of physical scarcity, another 500 million people are approaching this condition, and 1.6 billion people (one-quarter of the world's population) are facing economic water shortage [3]. Humans have long searched for cost-effective technologies to supply fresh water. Hence, desalination, the idea of converting salt water into fresh water, has been expanded and used during recent decades in >120 countries in the world [3,4]. Over the past few years, different types of desalination technologies have been developed [e.g., multieffect distillation, reverse osmosis (RO), and electrodialysis] [5].

Membrane-based processes are expected to play an increasingly substantial role in the sea and brackish water desalination, because they offer ease of operation, does not comprise phase changes or chemical additives, require lower footprint, and can be made modular for facile scale-up [6]. Among the membrane technologies employed in the desalination, the market for nanofiltration (NF), RO, and forward osmosis (FO) will be increased in the next few years [1].

Polymers are the most widely used materials for desalination due to their prominent properties. However, desalination by polymeric membranes is still restricted by several challenges such as low permeability, relatively high-energy consumption, low resistance to fouling, and short lifetime. The development of membranes with desired properties, energy efficient, and cost effective is much needed for desalination. Incorporation of nanomaterials into polymer matrix is advanced to overcome the challenges of the existing polymeric membranes. Graphene, zeolites, carbon nanotubes (CNTs), silica, silver, titanium dioxide, and other metal oxides are the predominantly tested nanomaterials in the recent researches [7]. Nanocomposite membranes could be used for a variety of applications such as gas-gas, liquid-liquid, and liquid-solid separations. The nanocomposite membranes have gained considerable attention for desalination due to overcoming trade-off between permeability and solute rejection as well as reduction of fouling propensity and is known as the cutting edge of manufacturing the next generation of high-performance membrane for desalination [8].

Despite the outstanding benefits of nanomaterials, several limitations have been met in the fabrication and application of these newly emerging membrane materials,

Emerging Technologies for Sustainable Desalination Handbook. https://doi.org/10.1016/B978-0-12-815818-0.00009-6

but still there exist some challenges for the industrial development. This chapter focuses on the recent scientific and technological developments related to the nanocomposite membrane fabrication and its applications in desalination. The concept of nanocomposite membrane, history and development membrane technology for desalination, classification of nanocomposite membrane, the methods of incorporation of nanomaterials into the polymer matrix, limitations and challenges of preparation and application of the nanocomposite membrane, economic and feasibility of nanocomposite membrane in large-scale application, and future prospectives will be discussed.

9.2 Nanostructured or nano-enhanced membranes

In membrane technology, there is a clear difference between the two terms nano-enhanced and nanostructured membranes. The term "nano-enhanced" is referred to the membranes which are functionalized with nanoparticles, nanotubes, nanosheets, or nanofibers, and the term "nanostructured" refers only to the pore size or internal structure of the membrane, or size of the solutes (about 1 nm) that are separated by the membrane. The general term for all membranes including nanotechnology, which is defined by the Network of Excellence NanoMem-Pro (EU FP6 project), is "nanoscale-based membrane technologies" [9,10].

In spite of the significant success and widespread applications of nanostructured membranes (NF membranes), some limitations remain, such as fouling, limited flux and trade-off between flux and selectivity. To overcome these limitations, nanomaterials can be incorporated into membrane matrix to improve membrane characteristics such as better antifouling and antimicrobial properties, increased selectivity, and increased flux (through altering the membrane hydrophilicity). The embedding of nanomaterials in the membrane can also decrease the energy consumption, the use of chemicals for membrane cleaning, and total cost of operation [9,10]. The technology development level and risk of application of nano-enhanced and nanostructured membranes differ significantly. The main producers in the field of membrane technology are still not active in large-scale production of nano-enhanced membranes [11]. In the following sections, the recent developments in the synthesis, characterization, and production of nanocomposite (nano-enhanced) membranes are presented.

9.3 History and development of membrane technology for desalination

9.3.1 Conventional membranes

During the recent years of widespread research on membrane preparation, there are two major techniques for membrane fabrication including the phase-inversion process and interfacial polymerization (IP) that are widely applied for the production of the

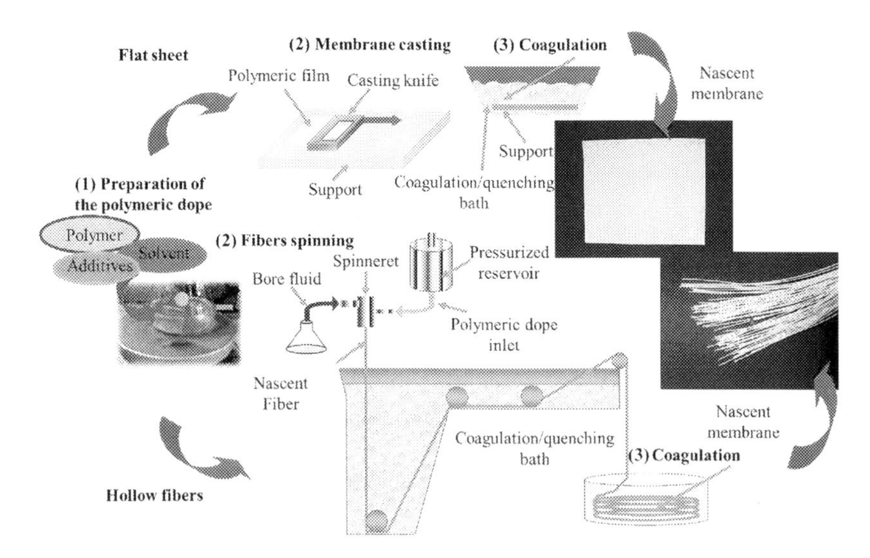

Fig. 9.1 Flat sheet and hollow fiber membranes preparation via phase inversion [12].

commercial membranes. The fabrication of conventional membranes is performed via phase inversion technique. An initially homogenous solution (dope solution) made up by dissolving the specified amounts of polymer and additive (nanomaterial) in the respective solvent and then immersed in a coagulation bath containing nonsolvent(s) [12] is shown in Fig. 9.1. The nonsolvent diffuses into the dope solution and the solvent escapes from the dope solution to the coagulation bath, known as the demixing process, which will give rise to a polymer-rich and a polymer-lean phase during phase-inversion process. The polymer-rich phase forms the solid membrane matrix, while the polymer-lean creates membrane pores [13].

In the late 1950s, Reid and Breton prepared a symmetric membrane by cellulose acetate, achieving 98% salt rejection, but the water flux was very low. Next, the Loeb-Sourirajan cellulose acetate membrane was the first made RO membrane in practice. Then Loeb-Sourirajan fabricated a cellulose acetate asymmetric membrane with a 200 nm thin and dense active layer over a microporous support. This membrane indicates a water flux of at least an order of magnitude higher than the initial symmetric membrane [14]. Fig. 9.2 summarizes the major and early developments of asymmetric membranes up to the 1980s. In the following years, researchers focused on the improvement of cellulose acetate materials such as cellulose triacetate or cellulose diacetate to obtain a membrane with higher permeability and simplify the manufacturing to bring the technology to the industrial application [14]. For example, a blend of cellulose triacetate and cellulose diacetate finally displayed higher permeability and selectivity than cellulose acetate membranes, as well as provide greater mechanical resistance to compaction [15]. Acetate-based materials have some limitations including the susceptibility of the acetate group to hydrolysis in both acidic and alkaline media and sensitivity to microbial impurity [16]. Until the 1960s, the

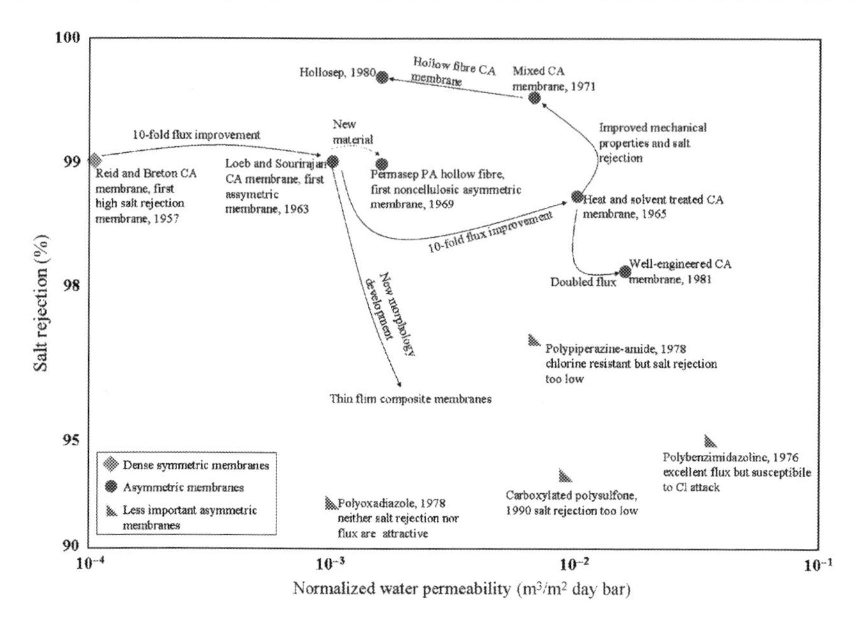

Fig. 9.2 The development of asymmetric RO membrane [14].

synthesis of alternative polymer materials was insignificant. The first significant and noncellulosic membrane materials were aromatic polyamide (PA) used for fabricating asymmetric hollow-fiber membrane and then commercialized by Du Pont company (B-9 Permasep) [14]. Due to the susceptibility of PAs to chlorine attack, polypiperazine-amides have subsequently been synthesized [17]. This type of membrane was not commercialized because of its relatively low salt rejection. After that, sulfonated polysulfone was developed to augment permeate flux and the mechanical, chemical, and biological stabilities. Although, the rejection of solutes was lower than the acceptable level required for commercialization. Carboxylated polysulfone, polybenzimidazoline, and polyoxadiazole are also used as membrane materials but are not commercially attractive for desalination purpose [14].

9.3.2 Thin film composite (TFC) membrane

The water permeability of the membrane is a function of not only intrinsic properties of membrane materials, but also of membrane thickness. In practice, there is a tradeoff between membrane thickness and mechanical resistance. A very thin membrane can be practical if it is located on a thick and porous substrate. As mentioned above, the earliest commercial RO membranes were integrally asymmetric and derived from cellulose acetate with dense skin layer which is supported on a more open substructure [11]. Only a few appropriate polymers can be used for preparing integrally asymmetric structure in one-step casting method, and a few of them are commercially attractive in the case of high water flux and solute rejection. Moreover, the collapse of middle transition layer of the integrally asymmetric membranes takes place under high operating pressure [14]. In order to overcome the limitations of integrally asymmetric membranes, TFC membranes are known to be the most effective for desalination. These types of membranes improve both solute rejection and permeability with lower energy consumption compared to typical integrally asymmetric membranes [18]. The TFC membrane fabrication is a two-step method that provided optimization of the materials used for preparing the substrate and active layer, for mechanical stability, and the acceptable salt rejection and permeability, respectively. In addition, a wide diversity of polymeric materials can be employed for the active and support layer, separately [14]. The real breakthrough in the fabrication of TFC membrane was started with the finding of Cadotte, published in 1978: a very thin PA active layer which is created on a support using interfacial polycondensation [18]. The thickness of the active layer is $<0.2\,\mu m$ with the interstitial void size of $<0.5\,nm$. The most common reactive monomers are aliphatic/aromatic diamine such as piperazine (PIP), m-phenylenediamine (MPD), and p-phenylenediamine (PPD) and acid chloride monomers such as trimesoyl chloride (TMC), isophthaloyl chloride (IPC) and 5-isocyanatoisophthaloyl chloride (ICIC) [19,20]. Fig. 9.3 demonstrates the structure and morphology of the active PA layer on the porous support layer. According to a general approach for preparing TFC membrane, an aqueous solution containing a reactive prepolymer is deposited in the voids of a microporous substrate (polysulfone). The amine impregnated substrate

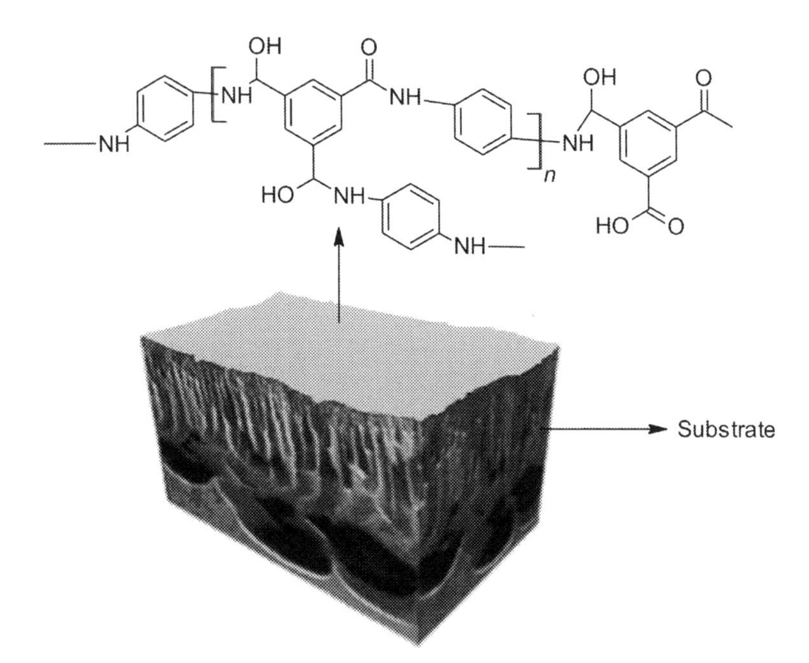

Fig. 9.3 Schematic representation of TFC membrane with PA active layer established on top of the microporous substrate [18].

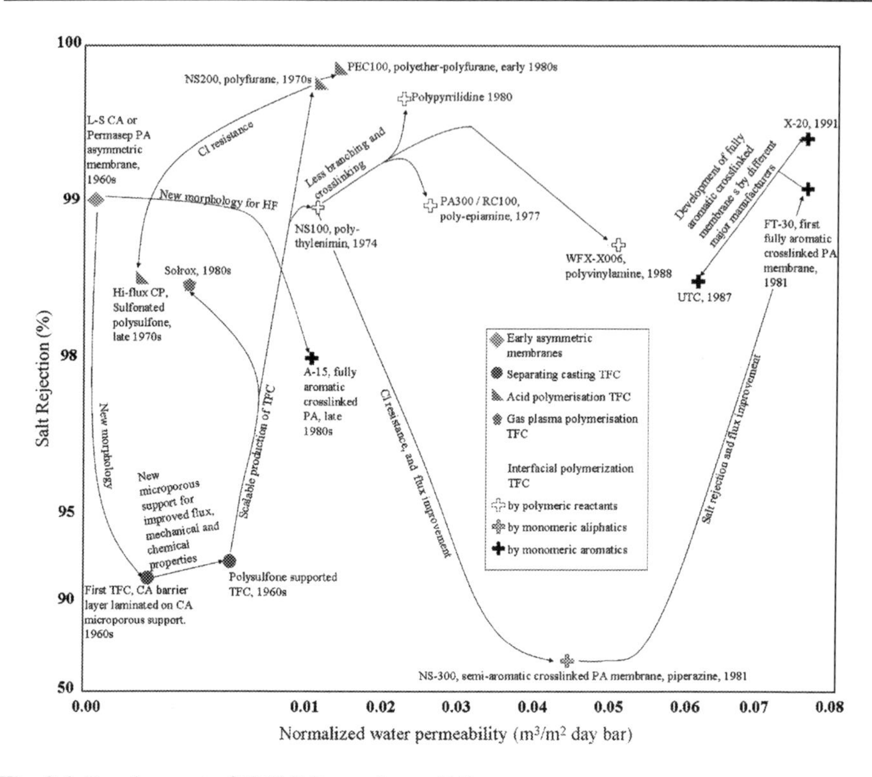

Fig. 9.4 Development of TFC RO membrane [14].

is then soaked in a water-soluble solvent containing a reactant, such as a diacid chloride in hexane. The amine and acid chloride react at the interface of the low immiscible solutions to create a thin and dense active layer [11,18].

The recent development of TFC RO membrane is illustrated in Fig. 9.4. The first TFC membrane was prepared by using a cellulose acetate as top and substrate layer. An ultrathin layer was formed on the water surface by float-casting method followed by annealing and lamination onto a microporous substrate [21]. The membrane prepared with this method never produced commercially because its flux was lower than integrally asymmetric cellulose acetate membrane for higher manufacturing costs [14]. After extensive research studies, polysulfone was selected as the best polymer for substrate due to its mechanical and thermal stability, a proper amount of permeate flux, and its stability in acidic media which provides further improvement of TFC membrane by acid polycondensation and IP [14,22]. In order to solve the scaling-up problems in float-casting technique, a dip-coating method including acid polycondensation was proposed. The first membrane based on dip-coating was NS-200 which showed superior solute rejection but suffered from irreversible swelling and hydrolysis of the sulfate linkage [14]. After that, Toray Company introduced PEC-1000 TFC RO membrane prepared by acid polycondensation method with 1,3,5-tris (hydroxyethyl)isocyanuric acid instead of polyoxyethylene [23]. Sulfonated polysulfones, PAs via polyethylenimine, polypyrrolidine, polypiperazine-amide, and cross-linked fully aromatic PA are the most notable TFC RO membrane materials [24]. After the significant success in the synthesis of TFC membrane, by introducing of cross-linked fully aromatic PA as membrane material of active layer, research and development for the synthesis of new materials for TFC RO membranes has reduced significantly [22]. The current TFC membranes in the market are still prepared according to the original chemistry discovered during the 1980s. Based on this chemistry, DOW FILMTECH currently sells products with the trade name of FT-30; membranes produced by the Toray company are based on UTC-70; Hydranautics supplies membranes based on NCM1, which is identical to CPA2; and Trisep membranes are based on X-20 [14]. Further studies were performed to optimize and control the IP synthesis which is discussed comprehensively in the literature [14,20,22,25]. In Fig. 9.5, the development of high-performance series by DOW FILMTEC and Toray is illustrated.

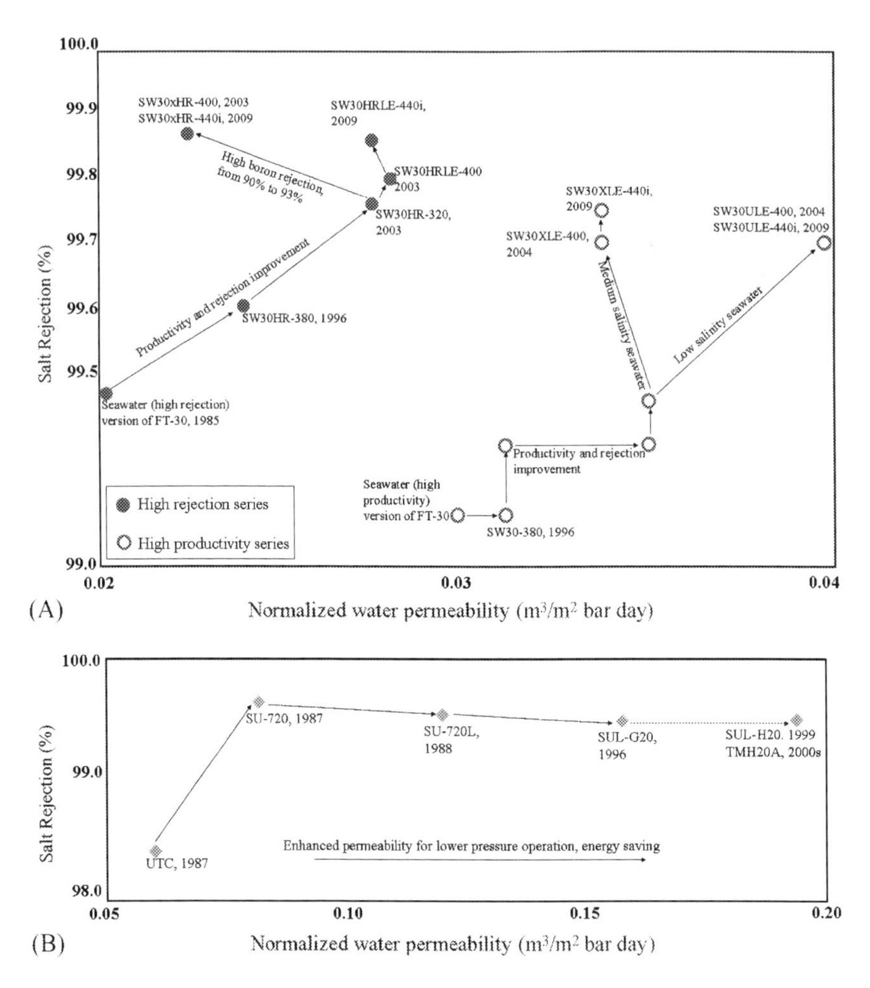

Fig. 9.5 Development of RO membrane by reaction optimization and postsynthesis surface modifications: (A) Dow Filmtec seawater series and (B) Toray brackish water series [14].

The development of FT-30 and UTC-70 are the outcome of surface modification of membranes, optimization of IP reactions, and design of efficient modules. In order to achieve a comprehensive overview of the research studies on TFC membranes, peruse reviews published on the topic [11,14]. Fig. 9.4 presents the recent patent and research activity.

In 2011, GE Osmonics Company introduced a posttreatment method to enhance salt rejection and antifouling properties. The unreacted amine-based materials of the formed PA react with aryl sulfonyl groups, which leads to the modification of the membrane without changing the reactants or fabrication method and condition used to form the basic membrane [26]. The Woongjin Chemical Company also uses a same postmodification strategy to modify RO membrane [27].

Different types of surface modification methods such as free radical, photochemical, radiation, redox, and plasma-induced grafting are applied to attach some monomers onto the membrane surface. Hayashi et al. reported applying an atmospheric pressure plasma technique to treat the membrane surface which has superior chemical stability against the routine or intermittent chlorine sterilization [22].

The addition of some additive to amine solution such as water-soluble polymers or sulfur-containing compounds can enhance permeate flux without considerable change in salt rejection. Additive (phosphate-containing compounds) can also be added to acyl chloride solution to minimize concurrent hydrolysis of acyl chloride [11,18].

9.3.3 Thin film nanocomposite

The TFN membrane is an emerging class of TFC membranes which are prepared by incorporating nanomaterials within the active dense layer with the purpose of modifying the properties of the surface thin layer. For instance, to enhance the hydrophilicity and/or surface charge density, without decreasing the solute rejection [25]. The most prevalent method for TFN membrane fabrication is done by the in situ IP process between an aqueous phase (MPD) and an organic phase (TMC) as exhibited in Fig. 9.6. The nanomaterials can be dispersed either in aqueous or in organic phase [8]. The term "Thin film nanocomposite (TFN) membrane" was presented by Hoek et al. for the first time in their research in early 2007 [28]. In their study, they recommended a new approach for the preparation of RO membrane via the IP process by incorporating zeolite NaA nanoparticles into the PA active layer. A significant permeate flux enhancement was obtained due to zeolite NaA nanoparticle embedding, with remaining salt rejection compared to TFC virgin membrane. It was proposed that water molecules tend to transport preferentially through the super-hydrophilic and molecular sieve nanomaterials pores or channels. In the starting of 2017, when this chapter was prepared, >100 relevant articles have been published, reporting on the employment of nanomaterials in fabricating PA TFN membranes for various application. Although the idea of the TFN membrane is still relatively imperfect in comparison to the TFC membrane, the development of TFN membranes has gained growing interest among scientists.

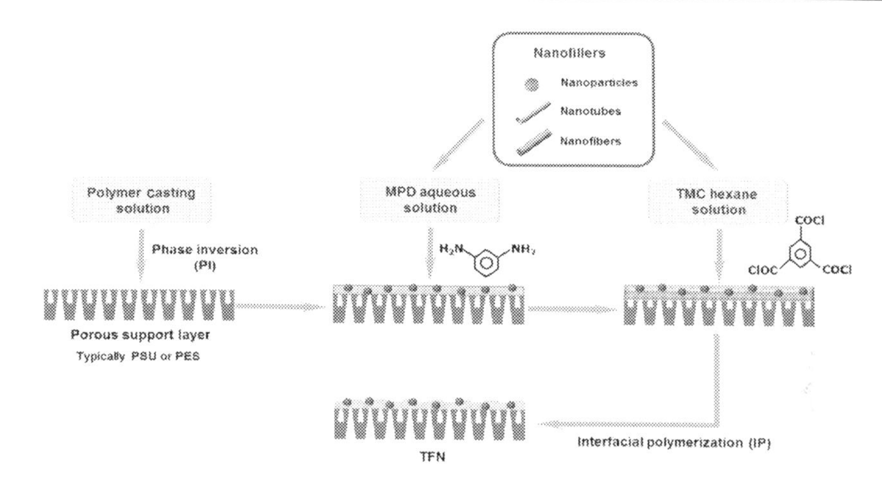

Fig. 9.6 Fabrication of TFN membranes through the IP process [8].

9.4 Classification of nanocomposite membrane based on the fabrication method and structure

Based on the fabrication method and location of nanomaterials, nanocomposite membrane with different structure and performance can be obtained. Generally, the nanocomposite membranes are divided into four categories: (1) conventional (integrally asymmetric) nanocomposite; (2) TFN; (3) TFC with nanocomposite substrate; and (4) surface located nanocomposite [8]. The typical structure and the publication numbers of each type of the nanocomposite membrane are illustrated in Fig. 9.7. As depicted in this figure, in recent years, the number of publication in the field of nanocomposite membrane have increased. In the fabrication of nanocomposite membrane, different types of nanomaterials consisting of nanoparticles, nanofibers, nanotubes, and nanosheets can be used which is obtained from inorganic, (bio) organic, and hybrid origins.

9.4.1 Methods of incorporation of nanomaterials into polymer matrix

The effect of nanomaterials on properties and performance of nanocomposite membrane strongly depends on the type of nanomaterial, size, and the amount of nanomaterials and the way they were embedded into the polymeric matrix [29]. Typically, there are two common methods used for incorporation of the nanomaterials into polymer matrix: (1) blending nanomaterials with casting solution (dope solution) before phase-inversion process and (2) deposition of nanomaterials on the surface of the presynthesized polymeric membrane. Studies reveal that the deposition technique is a more effective although the entrapping the nanomaterials into polymer matrix makes it more stable [13,18].

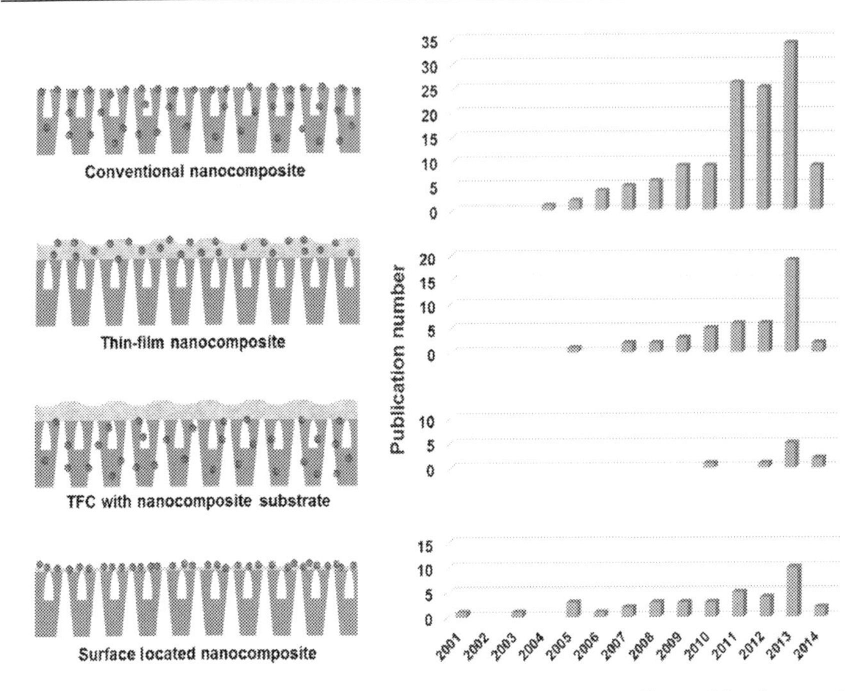

Fig. 9.7 Typical types of nanocomposite membranes and corresponding publication numbers related to water treatment applications [8].

Synthesis of nanomaterials in order to use for fabrication of nanocomposite membrane can be performed by two methods: (1) presynthesizing nanoparticles and then direct physical mixing (ex situ synthesis) and (2) in situ synthesis of nanoparticles in the polymeric matrix (Fig.9.8) [29]. Presynthesizing nanoparticles and then direct physical mixing approach have been widely used to prepare nanocomposite

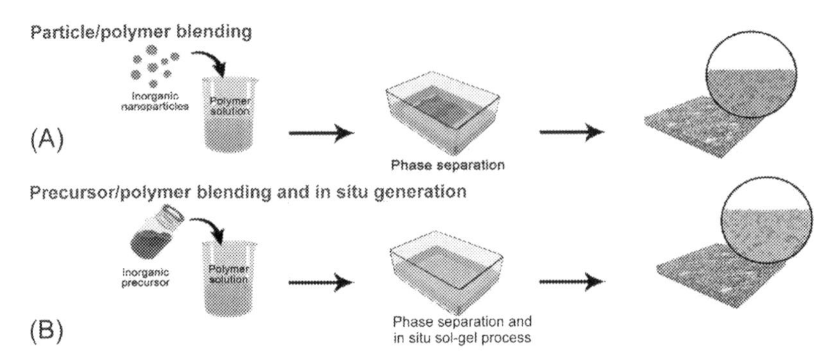

Fig. 9.8 Schematic diagram of different fabrication processes for nanocomposite membranes: (A) particle/polymer blending method; (B) precursor/polymer blending followed by in situ generation [30].

membrane mainly because of its ease of operation, relatively low cost, and propriety for large-scale manufacturing of membranes. Exclusive of the types of nanomaterial, the inorganic additive and dope solution are prepared separately and then blended together through solution, emulsion, fusion, or mechanical forces. Direct physical mixing of nanomaterials and polymer has been limitedly successful for most systems due to the tendency of nanomaterials to form larger aggregates during mixing. Various surface treatments of nanomaterials and addition of proper dispersing agents or compatibilizers are employed to alter the particle dispersion and compatibility between the nanomaterials and the polymer matrix [31].

The in situ synthesis method is also used to fabricate nanocomposite membrane, and many transition metal sulfide or halide nanomaterials can be easily preloaded into a polymer matrix using this synthesis approach [31]. Based on nanomaterial type, several methods such as sol-gel, ion exchange and reduction, and chemical reduction are proposed for in situ nanomaterials synthesis [29].

Depending on the different precursors (starting materials) and fabrication methods, in situ synthesis can be commonly categorized into three approaches, as exhibited in Fig. 9.9. Metal ions are preloaded into a membrane matrix to be used as starting materials (nanomaterials precursors) where the ions are anticipated to disperse uniformly into the polymer matrix. Then, the precursors are subjected to the corresponding liquid and gas for the in situ synthesis of the nanomaterials (Fig. 9.9A). Another in situ synthesis method utilized the monomers of the base polymer and the target nanomaterials as the starting materials. Generally, the nanomaterials are first

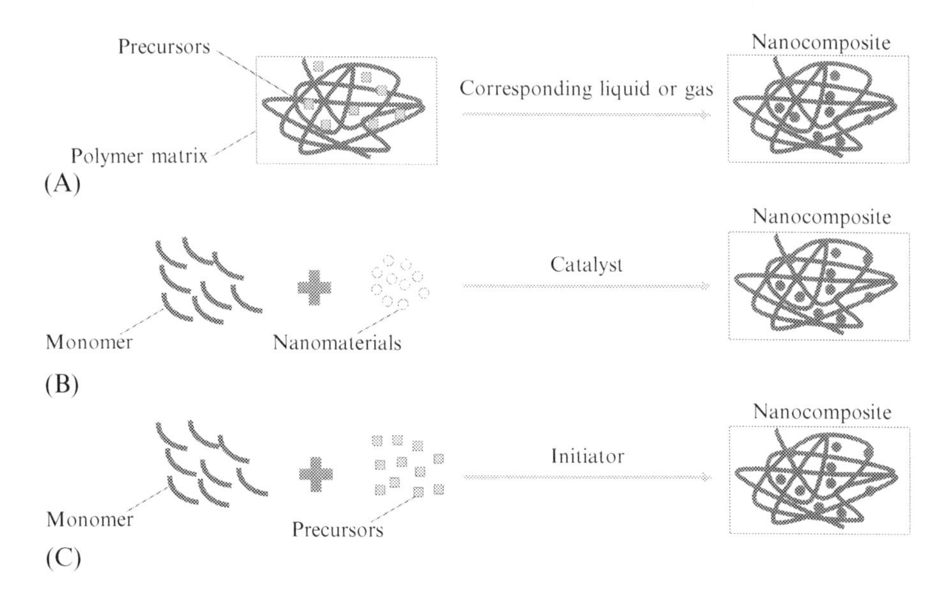

Fig. 9.9 Illustration of in situ synthesis process of mixed-matrix nanocomposite membranes: (A) metal ions are preloaded within polymer matrix to serve as nanoparticle precursor; (B) monomers of polymer matrix and the nanofillers as the starting materials; and (C) blending of the nanoparticle precursors and monomers of polymers in solvent.

dispersed into the monomers of polymer matrix followed by in situ polymerization reaction in the presence of appropriate catalyst under desirable condition (Fig. 9.9B). This approach allows the fabrication of nanocomposites membrane with favorable physical properties and provides a suitable and direct distribution of the nanomaterials into the liquid monomers or precursors which reduce the tendency of nanomaterials aggregation in the polymer matrix and consequently improves the interfacial interactions between organic and inorganic phases. In the third approach of in situ synthesis, nanomaterials and host polymers can be made altogether by mixing the precursors of nanomaterials and the monomers of polymers as starting materials with an initiator in the appropriate solvent (Fig. 9.9C) [31].

9.4.1.1 Blending of nanomaterials with dope solution

In the preparation of membrane with phase-inversion process, additives (organic or inorganic) are added to the dope solution to modify the morphology and performance of the membrane. Additive in dope solution exhibited several functions. In order to predict the final morphology of the membrane and filtration performance, the influence of the additive on kinetic and thermodynamic properties of the dope solution should also be investigated [29]. Effect of nanomaterials on the final morphology and performance of the prepared membranes were rationalized by evaluating the trade-off between thermodynamic enhancement and kinetic hindrance [32]. The incorporation of nanomaterials augments the phase separation and shows deterrent effects. From a thermodynamic point of view, the addition of nanomaterials into dope solution decreases the solvent power of the solution and works as a nonsolvent agent [33]. Therefore, less nonsolvent is needed for the phase separation of dope solution. This phenomenon leads to instantaneous demixing and results in a membrane with porous structure (Fig. 9.10) [29]. The nature of most nanomaterials used to fabricate nanocomposite membrane for desalination is hydrophilic. The hydrophilic nature of nanomaterials seems to be the main reason for the reduction in the miscibility (compatibility) of the polymer/solvent system through increasing the diffusional exchange rate between solvent and nonsolvent, accelerating the penetration rate of nonsolvent into polymeric film and speeding up the demixing process, which results in porous structure [29].

In many other studies and from the rheological point of view, the addition of nanomaterials causes an increase in the viscosity of dope solution [29]. Consequently, nanomaterials reduce the diffusion factor of the dope solution, decrease the diffusional exchange rate between solvent and nonsolvent which induce the delayed demixing and create a dense structure [34].

As reported previously, nanomaterials simultaneously have two opposite effects (thermodynamic enhancement and kinetic hindrance) during the phase inversion process. Thermodynamic enhancement is usually predominant at low-nanomaterial loading, and nanomaterials enhance the demixing rate and cause a porous structure. However, the rheological hindrance is important at high-nanomaterials loading and results in membrane with dense structure. Therefore, there is usually an exclusive threshold of nanomaterial loading for any specific dope solution system

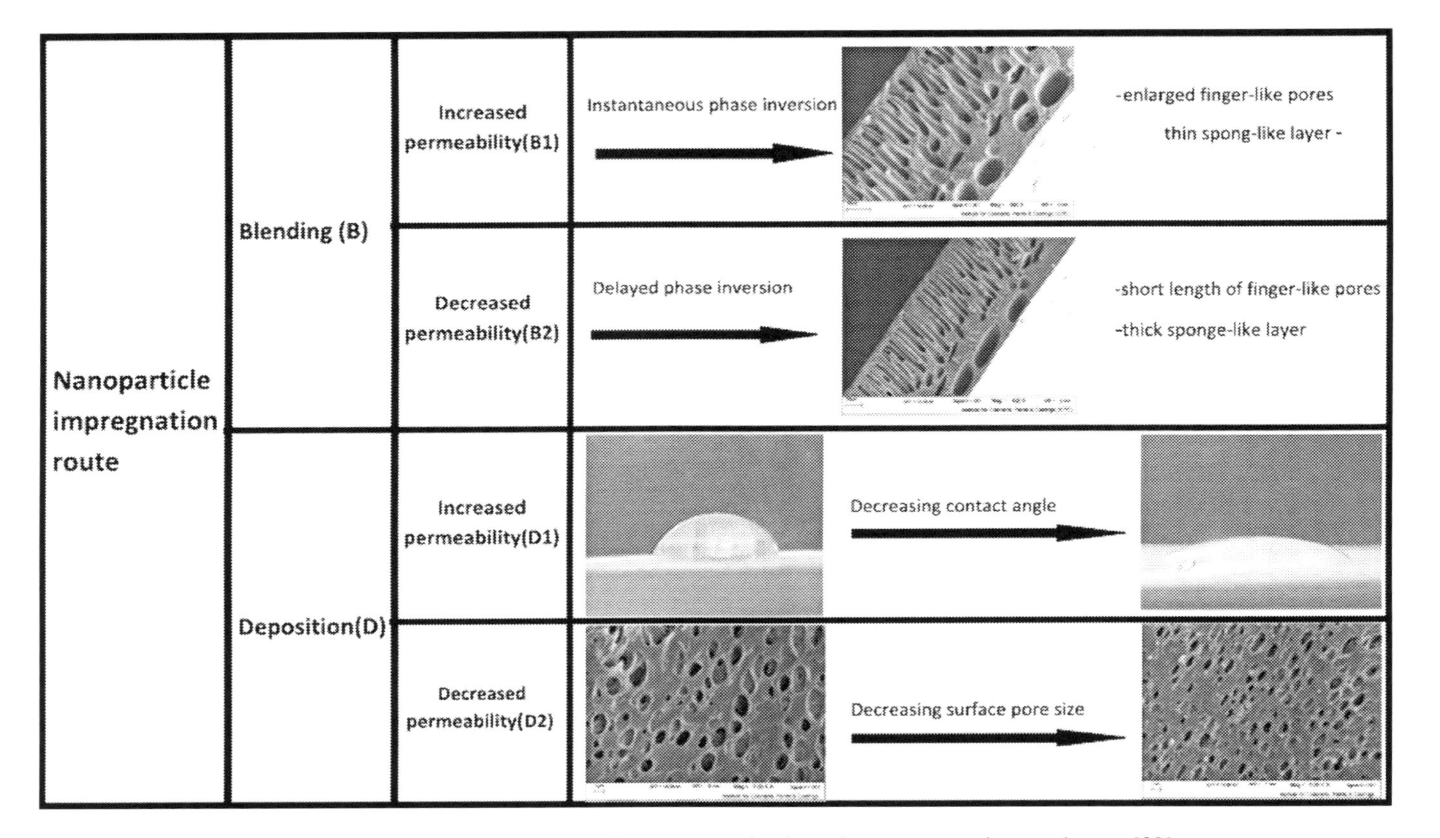

Fig. 9.10 Effect of the nanoparticle impregnation method on the structure of polymeric nanocomposite membranes [29].

(polymer/solvent/nanomaterial) which can be determined by studying the morphology and performance of membrane by changing the additive concentration in constant polymer and solvent concentration. Generally, thermodynamic enhancement is important at concentrations below the threshold, while rheological hindrance is predominant at concentrations above the threshold [29].

9.4.1.2 Deposition of nanomaterials on the surface of polymeric membranes

The majority of nanocomposite membranes are fabricated by adding nanomaterials to the polymer matrix but in recent years, deposition of nanomaterials on the membrane surface has been extensively used to prepare nanocomposite membranes [30]. The deposition of nanomaterials on the surface of the prepared membrane may have two opposite effects. Nanomaterials, on the one hand, may increase the water permeability of membrane due to the improvement of hydrophilicity and surface roughness and on the other hand, may decrease the water permeability and enhance the selectivity through reducing the surface pore size [29].

In situ synthesis of nanomaterials on membrane surface is used in the fabrication of the nanocomposite membrane. Self-assembly of the coated layer has been reported in some cases. Madaeni and Ghaemi [35] studied the effects of the coating of membrane surface with titanium dioxide particles. This approach not only creates photocatalytic functionality but also enhances surface hydrophilicity. The coating was performed by self-assembly of titanium dioxide particles through coordinance bonds with OH functional groups of the top layer (poly vinyl alcohol). The permeability of the modified membrane with titanium dioxide particles was altered remarkably compared to an unmodified membrane. This result reveals that the self-cleaning property has been generated by coating the nanoparticles on membrane surface. Lee et al. [36] also evaluated the effect of colloidal solution concentration on membrane performance. As the colloidal solution concentration increased, the water flux reduced and the rejection of magnesium sulfate increases due to the decrement in the surface pore size. This trend continues until the concentration of titanium dioxide reaches a critical concentration (5 wt%). At above 5 wt% concentration, hydrophilicity and surface roughness of membrane improved resulting in an increment in the permeability and a decrease in the salt rejection and mechanical stability of membrane. In another research, Andrade et al. [37] fabricated a membrane with polysulfone as the base polymer and silver nanoparticles as the additive by the wet phase-inversion process. They incorporated nanomaterials into the polymer matrix both by presynthesis (ex situ) and by in situ synthesis of nanomaterials. Based on ex situ methodology, 45 nm average size silver nanoparticles were uniformly dispersed in the internal structure of the membranes. However, by the in situ synthesis method, the silver nanoparticles were uniformly and preferentially dispersed on both the top and bottom surfaces of the membrane. The release of nanomaterials during filtration process is a main challenge in the nanocomposite membranes. The release of silver nanoparticles during filtration was measured using inductively coupled plasma mass spectrometry, which indicated a silver leaching of about 2 μg/L. The nanocomposite NF membranes prepared by the in situ method displayed a better antibacterial activity compared to other method, and also a reduction in 90% *Escherichia coli* adhered cells in comparison to the bare polysulfone membrane. The researchers have shown that the in situ synthesis can be considered a feasible, simple, and reproducible technique to uniformly disperse the silver nanoparticles into polymer matrix.

9.4.2 Conventional nanocomposite

9.4.2.1 Performance and properties of conventional nanocomposite membranes

An efficient membrane for water desalination should have the appropriate structure (e.g., porosity and pore size) and surface properties (e.g., hydrophilicity, surface charge density). Incorporating nanomaterials with different origins and geometries into the polymer matrix, not only alter the morphology and physical, chemical, and mechanical properties of membranes, but also create special characteristics such antibacterial, photocatalytic, and magnetic properties into the membrane. Some reported conventional nanocomposite membranes for water desalination are summarized in Table 9.1. Graphene, zeolites, CNTs, silica, silver, and titanium dioxide are the most common tested nanomaterials in recent studies. Membranes containing graphene, zeolites, and CNTs, all have been shown to improve the membrane water permeability. Silica and hydrophilic nanomaterials have been illustrated to display high affinity for water and control the membrane fouling characteristics through improvement in surface hydrophilicity and roughness. The silver and titanium dioxide result in antibacterial characteristics, zirconium dioxide and Fe provide catalytic properties and can be used to develop nanocomposite membranes that are less susceptible to biofouling. Moreover, embedding the nanomaterials into polymer matrix

Table 9.1 Conventional nano-composite membrane for desalination and water treatment

Filler	Polymer	Enhanced performance	Ref.
Zeolite	PSf, PVDF, PU, Ultem	Molecular sieving, hydrophilicity	[38–41]
CNTs	PES, PEI, PAN, PSf	Electrical conductivity, potential water channel, mechanical property, hydrophilicity after modification	[42], [43–46]
GO	PSf, PES	Hydrophilicity, negative surface charge	[47,48]
Clay	PSf, PVDF, PES	Mechanical property, hydrophilicity	[49–51]
TiO_2	PSf, PVDF, PA, PES, CA	Hydrophilicity	[52–56]
Al_2O_3	PVDF, PES	Hydrophilicity	[57,58]
Ag	CA, PI, PSf, PES, PVDF	Antimicrobial functionality	[39], [59–62]

leads to stronger chemical, thermal, and mechanical stabilities which specify the membrane durability and life cycle under various conditions [42].

The metals/metal oxides are one of the main class of inorganic nanomaterials that have been widely studied in the literature. The most commonly used metal oxides are based on silver, iron, zirconium, silica, aluminum, titanium, and magnesium. Each of these nanomaterials can be incorporated with most of the polymers to produce nanocomposite membrane with desirable properties which are resulted from synergetic characteristic between the polymeric materials and nanoparticles [61]. The modification of polyvinyl chloride membranes with zinc oxide (ZnO) nanoparticle as the additive was taken into consideration. The results indicated that the pure water flux of the nanocomposite membranes has increased up to 3 wt% (optimized concentration) with ZnO addition and at this concentration, the nanocomposite membranes were less susceptible to be fouled. The addition of nanoparticles also alter the morphology of modified membranes and they became more porous with higher interconnectivity between the pores [62]. Arsuaga et al. [63] investigated the modification of nanocomposite membrane by dispersing nanoparticles of TiO_2, Al_2O_3, and ZrO_2 in a polyethersulfone solution. Entrapped metal oxides altered the membrane morphology to more porous with open internal structure. Results showed that there was an abrupt increase in the particle size of metal oxides as a result of dispersion into the polymeric matrix, implicating that the metal-oxide particles became larger compared to their primary nanosize. The contact angle and permeability data showed that the presence of metal-oxide nanoparticles led to a higher affinity of nanocomposite membranes and lower fouling resistance as a result of water affinity and pore formation properties. These physicochemical properties show good agreement with the distribution of nanoparticles in the membrane structure. The rejection potential of nanocomposite membrane was hardly affected by the addition of the metal oxides, this being similar to the pristine membrane. In another research [64], integrally asymmetric poly(phenylene ether-ether sulfone)-based NF membrane has been prepared by TiO_2 nanoparticles as the additive for water desalination application. Data reveals that addition of the TiO_2 nanoparticles into the polymer matrix led to a sharp increment of water flux from 15.72 to 133.85 L/m^2 h. The results also showed that solute rejection was decreased initially with the increase in TiO_2 nanoparticle concentration up to 0.05 wt% in the membrane matrix, but with more nanoparticle content, it began rising.

An emerging class of carbon-based nanomaterials, particularly graphene-based materials, and CNTs, has been widely studied for its extraordinary water transport and sieving properties. The CNTs present excellent properties such as high flexibility, low mass density, and the efficient π-π stacking interaction between CNTs and aromatic compounds. The recent developments in the graphene-based membrane and CNT-based membranes are illustrated in Fig. 9.11. As exhibited in Fig. 9.11, graphene-based membranes can be categorized into single-layer and composite membranes while CNT-based membranes are typically classified into vertically and randomly aligned CNT membranes. From the fabrication point of view, the composite graphene membranes can be divided into surface modified, stacked, and mixed-matrix membranes. Likewise, randomly aligned CNT membranes are generally obtained via surface modification and mixed-matrix approaches [65]. Although there is a great

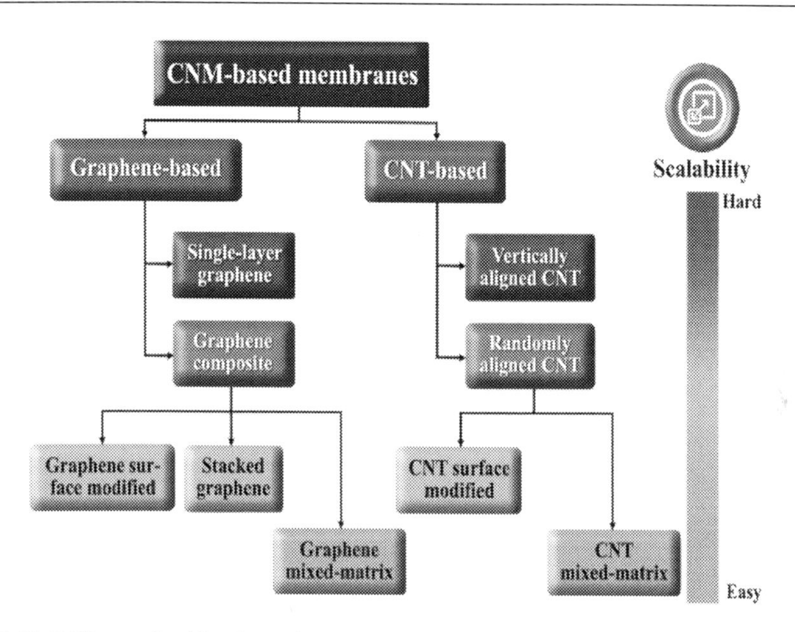

Fig. 9.11 Different classifications of carbon nanomaterial-based membranes and their relative levels of scalability [65].

number of advantages of CNTs for fabrication of nanocomposite membranes, CNTs still suffer from some issues for dispersion and dissolution in most organic solvents and polymers as well as weak interaction of the CNTs and the base polymer of the membrane. Several methods are proposed in the literature to obviate these problems such as attachment of hydrophilic functional groups on CNTs surface, functionalization by chemical agents, and introducing polar groups to CNT sidewalls [61]. Functionalization of CNTs can introduce positive ($—NH_3^+$), negative ($—COO^-$, sulfonic acids), and hydrophobic (aromatic rings) groups on CNT surfaces. These modifications make CNT-containing membrane selective for solutes rejection and increase water permeability as well as mechanical and thermal stabilities, fouling resistance, pollutant degradation, and self-cleaning functions. Functionalization of the CNTs is performed by two methods: (1) tip functionalizing and (2) core functionalizing. Tip functionalized CNT-containing membranes have selective functional agent on the nanotube aperture and the core functionalized CNT have functionalities at the sidewall or interior of the nanotube (Fig. 9.12). Both modification approaches demonstrates improved water permeability and selective rejection of solutes. Functionalization reduces energy demand through increased permeability and physical adjustability. The CNTs can also be treated with various nanoparticles including Cu, Ag, Au, Pt, Pd, TiO_2, polymers, and biomolecules which create noteworthy membrane properties and thus make them attractive in water desalination [66]. Vatanpour et al. [34,67] have used a surface modification of CNTs with chemical agents and acid treatment to improve their dispersibility in the polymer matrix. They have fabricated nanocomposite membranes from polyether sulfone as the base polymer and acid-treated functionalized and

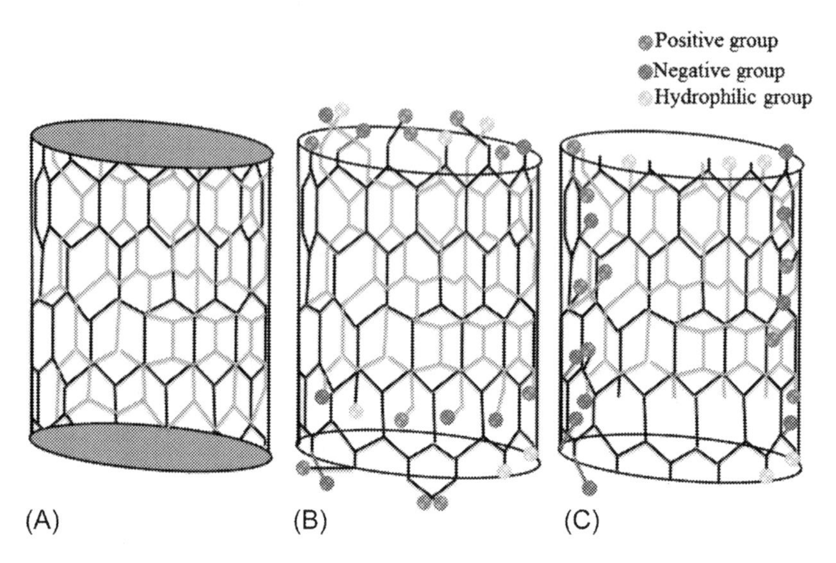

Fig. 9.12 Functionalization of CNTs (A) Pristine CNT, (B) Tip functionalized CNT, and (C) core functionalized CNT [66].

amine-functionalized multiwall carbon nanotubes (MWCNTs) as the inorganic additive. The modified MWCNTs showed good compatibility with polyether sulfone and caused an enhancement in hydrophilicity and water permeability as well as influenced pore size and porosity of the prepared membrane. The salt rejection of the nanocomposite membranes exhibited that the mechanism of solute rejection is governed by Donnan effect, in which membrane surface was negatively charged. Vatanpour et al. [44] studied acid oxidized MWCNTs which were also coated by titanium dioxide (TiO_2) nanoparticles via the deposition of $TiCl_4$ as the precursor on the sidewall of MWCNTs and used in the fabrication of NF membranes (Fig. 9.13). Coating of the TiO_2 nanoparticles on the surface of oxidized MWCNTs has augmented the hydrophilicity of the nanocomposite membrane and enhanced the induced effects of photocatalytic and antibiofouling properties due to the synergistic effect of coupling nanomaterials compared to the pristine TiO_2. Acid-treated MWCNTs display high negative Zeta potential in neutral media, causing a negative surface charge of the nanocomposite membrane and improving the salt rejection.

Graphene oxide (GO) is attaining much more interest in the field of nanocomposite membrane due to its two-dimensional structure, high surface area, mechanical stability, superior electron transport, hydrophilic nature, promote negative surface charges, and better miscibility with host polymer [68,69]. The selection of GO as an additive is attributed to its hydrophilicity and pH-sensitive behavior. Also, it has been investigated and demonstrated that GO can induce surface negative charge. Because of the different hydrophilic functional groups located on the GO surface, it can adsorb water very lightly and it has also been well established that water transport enhances as the degree of oxidation increases [69]. Ganesh et al. [68] oxidized the graphite to GO using $KMnO_4$ to introduce hydrophilic functional groups onto the additive surface and then the prepared GO was incorporated into polysulfone matrix to prepare

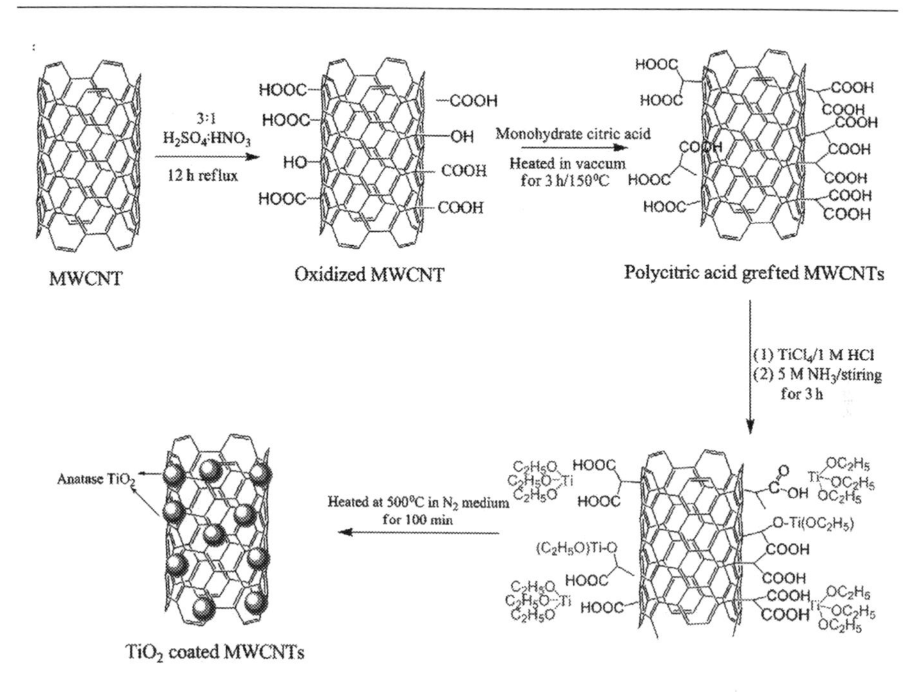

Fig. 9.13 Schematic representation of the preparation of TiO_2-coated MWCNTs [44].

nanocomposite membrane. The GO/PSf membrane resulted in enhanced hydrophilicity, permeate flux, and salt rejection of the membrane. At 2000 ppm GO loading, the membrane showed maximum salt rejection of 72% for 1000 ppm Na_2SO_4 solution at 4 bar transmembrane pressure. It has also been displayed that increasing pH of feed solution influenced salt rejection performance due to the negative surface charge of membrane. This negative surface charge is caused by ionizable functional groups on GO surface [25]. The oxygen-containing functional groups on GO surface can be mostly removed to form reduced GO (rGO) which could exhibit higher thermal stability due to the better graphitization and improved van der Waals forces between layers through the deoxygenation process [70]. The rGO-based nanocomposite membrane is prepared by eco-friendly enzymatic reaction-based method. The synthesis method is illustrated in Fig. 9.14. The rGO/PANI was produced via oxidative polymerization to use as nanomaterials in polysulfone nanocomposite membrane for water desalination application. The incorporation of nanofiller into the polymer matrix led to higher porosity and enhanced macro-void formation in the substrate. Because of the hydrophilic nature of the nanofiller, higher permeability can be obtained and the nanocomposite membrane showed a maximum of 82% NaCl rejection [70].

9.4.2.2 New functionalities introduced by nanomaterials

Embedding of nanomaterials in polymer matrix provided not only filter the solute in aqueous phase but also synergetic functionalities including adsorbing, photocatalysis, and antimicrobial activity. Functionalization of the membrane are performed with the specific application [8].

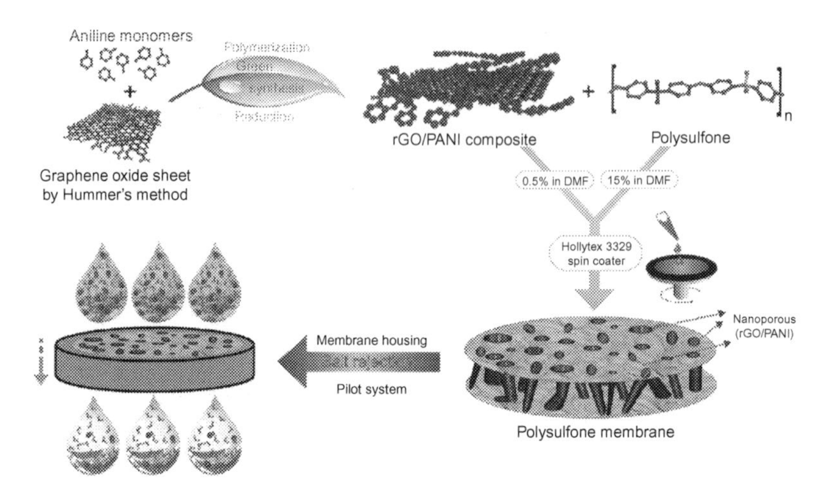

Fig. 9.14 Schematic diagram for the preparation of the PSf-rGO/PANI composite membrane [70].

Adsorption

Incorporation of porous nanomaterials such as CNTs into the base polymer are done to enhance the selectivity of the nanocomposite membranes which is attributed to synergic influences of molecular sieving, selective adsorption, and difference in diffusion rates of species. Some polymer matrixes such as sodium alginate, chitosan, etc., also improve the selective adsorption of metal ions due to the presence of some functional groups in their backbones which makes their interaction feasible with low-molecular weight solute via complexation. Chitosan displays cationicity in acidic solution due to the presence of amino groups in its backbone. This feature provides chelation of chitosan with transition metal in near-neutral solution [31]. Tetala and Stamatialis [71] prepared a mixed matrix membrane to adsorb copper ions from aqueous solutions. The fabricated flat sheet membrane consists of chitosan beads introduced in ethylene vinyl alcohol (EVAL) polymer as a base polymer. Dynamic adsorption of copper ions from aqueous media with mixed matrix membrane is similar to static adsorption but is achieved much faster (15 vs 60 min). Copper ions desorbed successfully from the mixed matrix membrane without loss of chitosan beads from the membrane structure and make it possible to reuse. Many studies confirm that preparation of nanocomposite membrane containing porous adsorbents for adsorption of metal ions from aqueous solution is possible. However, the short contact time and low adsorption capacity may limit their practical applications.

Photocatalytic

Photocatalytic is a major functionality of some nanomaterials such as titanium dioxide. Titanium dioxide is widely used for water treatment due to its unique properties such as photodegradation, stability, commercial availability, and ease of preparation. Introducing titanium dioxide into polymer matrix provides membrane with photocatalytic activities [8]. Shaban et al. [72] studied the fabrication, characterization, and performance evaluation of polyether sulfone/titanium dioxide nanotubes nanocomposite membranes. The permeate flux, salt rejection, and hydrophilicity were improved by embedding the titanium dioxide nanotubes.

Antimicrobial activity

Biofouling is one of the main drawbacks of membrane technology for water and wastewater treatment. Microbial growth and biofilm formation decreases membrane flux and augments the energy demand. Preparation of antimicrobial membranes improves efficiency and application duration considerably. The most studied nanomaterial for creating antimicrobial activity is silver due to its outstanding biocidal properties. One problem with the silver nanoparticles within the polymer matrix is its weak resistance to washing, so the antimicrobial capability of membranes would vanish during filtration. Many researches have been performed to overcome this problem by immobilizing the silver nanoparticle into the dense top layer or grafting on membrane surface. Other nanofillers such as copper and selenium also have potential for fabrication of antimicrobial membrane [8].

9.4.3 TFN

In the following sections, an overview on the properties and performance of TFN membranes consisting of various types of inorganic nanofillers will be provided. We focus on how the incorporation of nanofiller with different geometries, surface properties, and functionality could change the properties of the active layer of the TFN membrane and enhance solutes separation performance.

9.4.3.1 Performance and properties of TFN

In TFC, active layer (typically PA) determines the membrane properties and performance in terms of permeate flux, solute rejection, and fouling propensity. The introduction of nanofiller into active layer could alter the physicochemical features of the surface and/or bulk of membrane such as contact angle, surface charge density, porosity, and cross-linking or may supply water channel which may improve the permeate flux at remained selectivity. An enhancement in membrane hydrophilicity facilitates water transport through the membrane and improves the mass transfer and water flux in TFC membrane based on solution-diffusion mechanism. Nanomaterials used for fabricating TFN membrane should have proper size, internal structure, surface functionality, and appropriate interfacial interactions with the polymer matrix. In order to modify the dispersibility of nanomaterials and its interaction with polymer chains, it is usual to alter the nanomaterials surface prior to the incorporation. Also, due to the ultrathin thickness of active layer, the nanomaterials should not be too large to destroy the barrier [8]. Table 9.2 presents some scientific reports on TFN membrane for desalination.

The embedding of the hydrophilic nanomaterials in active layer leads to improvement in hydrophilicity. For example, by adding oxidized MWNTs into PA layer, the

Table 9.2 **TFN membrane for desalination and water treatment**

Filler	Polymer	Application	Enhanced performance	Ref.
Oxidized MWNTs	PVA	UF	Permeability increased, solute rejection decreased	[73]
Carboxylic MWNTs	Polyester	NF	Permeability increased, solute rejection increased	[74]
MWNTs	PA	RO	Chlorine resistance increased	[75]
Functionalized MWNTs	PA	NF	Permeability increased, no change in solute rejection	
Aluminosilicate SWNTs	PVA	NF	Hydrophilicity increased, permeability increased, solute rejection increased	[76]
Zwitterion functionalized CNTs	PA	RO	Permeability increased, solute rejection increased	[77]
PMMA modified MWNTs	PA	NF	Permeability increased, solute rejection increased	[78]
CNTs	PA	FO	Permeability increased	[79]
Sulfonated MWNTs	PA	NF	Permeability increased, no change in solute rejection, improved antifouling	[80]
Zeolite (NaA)	PA	RO	Permeability increased, solute rejection increased	[81]
Zeolite (AgA)	PA	RO	Hydrophilicity increased, permeability increased, no change in solute rejection, improved antibacterial properties	[82]
Zeolite (LTA)	PA	RO	Permeability increased, defects and molecular-sieving was dominant mechanism	[83]
Zeolite (NaY)	PA	FO	Permeability increased, solute rejection decreased	[84]
Zeolite (Silicalite-1)	PA	RO	Hydrophilicity increased, permeability increased	[85]
Zeolite (NaA)	PA	RO	Permeability increased, solute rejection increased	[86]
Silica (LUDOXs HS-40)	PA	–	Permeability increased, solute rejection decreased	[87]
Functionalized Silica	PA	RO	Permeability increased, solute rejection decreased, improved thermal stability	[88]

Table 9.2 **Continued**

Filler	Polymer	Application	Enhanced performance	Ref.
Mesoporous silica (MCM-41) and nonporous silica	PA	RO	Hydrophilicity increased, permeability increased, no change in solute rejection	[89]
Aminated hyper branched silica	PA	RO	Chlorine resistance increased, permeability increased	[90]
MCM-48 mesoporous silica	PA	RO	Permeability increased, no change in solute rejection	[91]
Aminopropyl-functionalized MCM-41	PA	RO	Hydrophilicity increased, permeability increased	[92]
Hollow mesoporous silica	PA	–	Hydrophilicity increased, permeability increased, no change in solute rejection	[93]
Ag nanoparticles	PA	NF	Hydrophilicity increased, permeability increased, no change in solute rejection, improved antifouling	[94]
GO	PA	NF	Hydrophilicity increased, permeability increased, solute rejection increased, improved antifouling	[95]
Zeolite imidazolate framework/ graphene oxide	PA	NF	Improved antimicrobial activity	[96]
Metal alkoxide (TTIP, BTESE, PhTES)	PA	RO/NF	Permeability increased, no change in solute rejection	[97]
Alumina nanoparticles	PA	NF	Hydrophilicity increased, permeability increased, no change in solute rejection	[98]
Organoclay (Cloisite 15Aand30B)	Chitosan	NF	Solute rejection increased	[99]

contact angle decreased from 70 to 25 degrees with increasing MWNTs concentration from 0% to 0.2% (w/v) in the aqueous phase [97]. In some studies, nanomaterials were added to organic phase. For example, Yin et al. [86] investigated the incorporation of mesoporous silica nanoparticles into PA layer. The contact angle was decreased from 57 to 28 degrees with increasing additive concentration from 0% to 0.1% (w/w) in the organic phase. In all studies, nanocomposite membrane displayed an enhanced permeate flux with increasing hydrophilic nanofiller concentration. The embedding of the nanomaterials into active layer decreases the contact angle by two reasons: (1) nanomaterials may be hydrated and release heat when added to MPD aqueous solution and may influence the IP process and subsequently the chemical nature of the PA layer. If more acyl chloride groups in TMC remained on the surface without reacting with amine groups, the hydrolysis of acyl chloride could generate carboxylic acid functional groups; thus, surface hydrophilicity would increase [8,98], and (2) incorporated hydrophilic nanofillers can be located on membrane surface, creating more hydrophilic functional groups on membrane surface.

Another effect of adding nanomaterials on permeate flux and salt rejection is controlling the degree of cross-linking and thickness of top layer. Embedding nanomaterials in the active layer could reduce the degree of the cross-linking, leading to thinner top layer and higher permeate flux. This phenomenon is caused by disturbing the chemical reaction between amine groups and acyl chloride groups or creating nanovoids near the interfaces between nanomaterials and PA chain [8].

The introduction of nanomaterials may also provide water channel facilitating the transport of water molecules but not solutes across the membrane. The addition of zeolite-A nanofiller into PA matrix, which resulted in higher permeability with constant salt rejection due to preferential flow channel for water molecules, was studied. After blocking the zeolite cages (pores), data indicates that permeability enhancement reduced, although was still higher than virgin TFC membrane. Yin et al. used two types of silica (nonporous and mesoporous silica nanoparticles) to indicate the effect of hexagonal pores of mesoporous silica on water permeability. Permeate flux nanocomposite membrane with nonporous silica is higher than virgin TFC membrane but it is less than that containing mesoporous silica. This outcome is attributed to the hexagonal internal pores of nanoparticles, which contributed remarkably to increased water flux.

Similar to conventional nanocomposite, the incorporation of hydrophilic nanomaterials into active layer enhances the hydrophilicity and decreases the fouling propensity. Zhao et al. [99] concluded that the embedding of carboxyl-functionalized MWCNTs into active layer structure could improve membrane antifouling property. In comparison to virgin TFC membrane, the TFN membrane incorporated with 0.1 wt% functionalized MWCNTs demonstrated smaller flux decline during the filtration process. These results may be attributed to the better antifouling property of TFN membrane as well as the greater negative surface charge upon addition of nanomaterials.

The antibacterial materials (e.g., Ag) could also be used for TFN membrane fabrication to reinforce the antibacterial resistance during the filtration process as well as minimizing membrane biofouling [25]. Research [91] has revealed that TFN membranes with Ag nanoparticles in the active layer displayed higher permeate flux and antibacterial effects on the growth of *Pseudomonas aeruginosa* PAO1. The improved permeability by the addition of Ag nanoparticles is interesting because the Ag nanoparticles do not have any water pathway. Yin et al. [100] grafted the Ag nanoparticles on the TFC PA surface in order to decrease the release of Ag

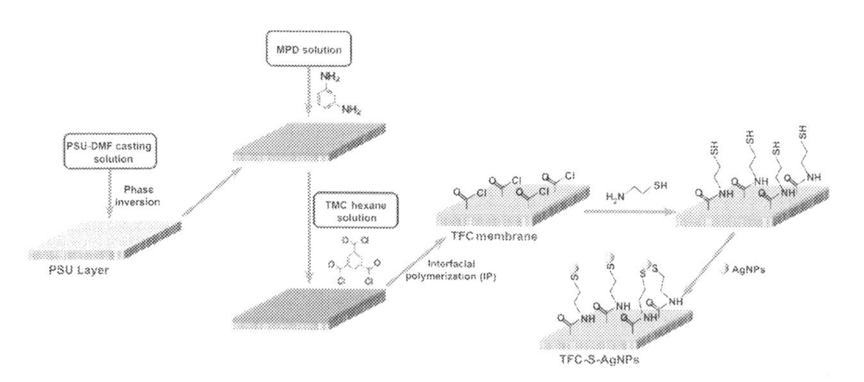

Fig. 9.15 Schematic diagram of immobilization of Ag nanoparticles onto the surface of PA TFC membrane [100].

nanoparticles and ions to the aqueous phase. This approach was performed by chemical immobilization of Ag nanoparticles on the TFC membrane as illustrated in Fig. 9.15. Besides exhibiting increased permeate flux with slightly lower salt rejection (compared with TFC membrane), the resultant TFN membrane also displayed supreme antibacterial property. The grafting method provides distinct benefits over physically embedding approaches as the virgin active layer structure is left with only a few change in top layer thickness and is likely to decrease Ag releasing during preparation and filtration process.

The amide linkage in PA-based TFC membranes is sensitive to attack by strong oxidants especially chlorine, leading to the unwanted degradation that could reduce the membrane efficiency. To overcome this problem, researches focused on coating PA top layer with chlorine-resistant materials and creating specific functional groups or nanomaterials to the amide structure [8]. Park et al. [101] employed TFN membrane modified with MWCNTs. The authors concluded that the effective interaction between the carboxylic group of nanotubes and the amide bond in the PA active layer was the main reason for enhanced resistance of TFN membrane against chlorine attack. In another research [87], a TFN membrane with improved chlorine resistance was fabricated using hyperbranched aromatic PA-grafted silica (HBP-g-silica) nanoparticles. In addition to improvement in permeate flux, the TFN membrane was also showed to defend the PA top layer from chlorine attack.

9.4.3.2 Challenges of TFN membrane fabrication

Besides the attractive features of TFN membrane for water treatment and desalination, there are several significant challenges during TFN membrane fabrication. Rectifying these challenges is the key factor for industrial application of TFN membranes.

The most important challenge during TFN membrane preparation is the agglomeration of nanoparticles into the top layer, which is likely to decrease the active surface area of nanomaterials and the formation of defects (holes) at the active layer structure. This could be attributed the low dispersibility of nanomaterials in both the aqueous and organic solutions used in the IP process. The agglomeration of nanofillers, particularly in the nonpolar organic solvent, leads to aggregation. As a result, some parts of the active layer contains no nanomaterials at all [25].

As reported previously, utilization hydrophilic nanotubes could provide a preferential channel for the transport of water molecules, resulting in an increased permeability. Nanotubes typically have a dimension between 10 and 50 mm (equivalent to 10,000 and 50,000 nm). They cannot be incorporated within the thin active layer of 100–500 nm unless all of the nanotubes are horizontally aligned, which is highly improbable for IP and it is still inconsistent with the results of many researchers that water transports through vertically aligned channels. Additional research is required to show the water and solute transport mechanism across these types of membranes [25].

Other problems encountered during preparation and application of TFN membranes are releasing and leaching out of nanomaterials during IP reaction and/or filtration process, making the fabricated TFN membrane less useful. The future research must answer following questions: (1) Is the physical interaction between nanomaterials and polymer matrix sufficient to maintain the fillers in the active layer particularly under the high-pressure condition? (2) What will happen if the functional groups of nanomaterials are chemically bounded with the polymer chains of active layer? (3) Will the chemical interaction be efficient for enhancing TFN durability for long-term filtration? [25].

9.4.3.3 Possible approaches to overcome TFN fabrication problems

Surface modification of nanomaterials

This method is still known as the effective method to modify the dispersibility of nanomaterials in a nonpolar organic solvent. The modified nanomaterials usually contain specific functional groups to homogenously disperse in the nonpolar organic solution during IP process to decrease the particle aggregation in the active layer or to attach to active layer (PA) network with chemical binding [25]. For example, in order to facilitate uniform dispersion of UZM-5 zeolite nanofiller in hexane as organic solvent by functionalizing the UZM-5 using amino silane coupling agent-3-aminopropyldiethoxymethylsilane (APDEMS). The result indicated that the grafting of the amino functional group (—NH_2) on the surface of nanofiller not only assist uniform dispersion of nanofiller in the organic solvent but also enables a covalent bond with TMC molecules, increasing the chemical interaction between nanofiller with the PA chains [102].

Use of metal alkoxides

As discussed in the literature, good dispersion of solid hydrophilic nanoparticles in the nonpolar solvent is difficult to obtain. To overcome this limitation, Kong et al. [94] employed metal alkoxides-tetraisopropoxide (TTIP) instead of TiO_2 nanoparticles. The metal alkoxides used in this study were titanium tetraisopropoxide, bis (triethoxysilyl) ethane, and phenyltriethoxysilane. Metal alkoxides could be also hydrolyzed to yield inorganic nanofiller with the smaller size. Data reveals that the permeability of the nanocomposite membranes was augmented notably at higher metal-alkoxide (phenyltriethoxysilane) loadings at constant rejection [25,94]. The advantage of employing metal alkoxides in TFN membrane preparation is still not very obvious because there are no open literature [25].

Modified/novel interfacial polymerization techniques

Many efforts have been made continuously by the researchers for the fabrication of TFC membranes by modifying IP processes to obtain membranes with better interfacial properties. For example, use of a secondary amine for neutralization of unreacted acyl chloride groups is an effective method which could be easily done by placing the substrate membrane in an aqueous solution for a determined time or introduction of intermediate organic solvent between the aqueous amine solution and the organic acid chloride solution to reduce possible pinhole formation on PA layer [25].

9.4.4 TFC with nanocomposite substrate

Substrate remarkably influences the properties and performance of TFC membrane. Among all polymers, polysulfone is a commercially available and the most frequently used polymer for the fabrication of the substrate [19]. The widespread use of polysulfone is due to its excellent mechanical, chemical, and thermal stabilities, and antibacterial resistance compared to other polymeric materials. However, polysulfone is hydrophobic and needs to be modified [19]. In addition to polysulfone, polyethersulfone and polyacrylonitrile are used as the substrate in researches [8,103]. Embedding of nanomaterials into substrate alters the hydrophilicity and compaction behavior [8,18]. The hydrophilic substrate is desirable for FO membrane since it increases the force to draw water into the membrane substrate and enhances the permeability. However, high hydrophilicity is not the only important factor for the selection of support's polymer and filler, because too high substrate hydrophilicity debilitates the adhesion between the top layer and the support [18]. Silica, zeolite, CNTs, and titanium dioxide are the most commonly used nanomaterials reported in the literature [8]. Representative publications on the development and employment of nanomaterials into the substrate of TFC membranes are summarized in Table 9.3. Pendergast et al. [118] studied the incorporation of silica or zeolite nanofillers into polysulfone substrate to prepare TFC membranes with IP for RO application. The modified membranes displayed higher initial permeate flux and less flux decline during the compaction. The addition of nanomaterials into polymer matrix provided necessary mechanical support to reduce the collapse of the microporous substrate upon physical compaction as exhibited in Fig. 9.16. In addition, this modified TFN membrane exhibited great potential against irreversible fouling propensity because of greater resistance to physical compaction. Emadzadeh et al. [119] investigated the incorporation of TiO_2 nanoparticles (Degussa P25) into the porous substrate of the TFC membrane for FO application. The impact of hydrophilic substrate is much more important in FO application compared to RO and NF as both the active layer and the porous substrate are simultaneously exposed to aqueous solutions during filtration process. In the case of osmotic-drive membrane processes, there are two types of concentration polarization phenomena

Table 9.3 **TFC membranes with nanocomposite substrate**

Filler	Polymer	Application	Enhanced performance	Ref.
SiO_2 zeolite	PSf	RO	Resistance to physical compaction	[104]
MWNTs	PSf	NF/RO	Hydrophilicity, water flux	[94]
Zeolite A	PSf	RO	Water flux, salt rejection, resistance to physical compaction	[105]
TiO_2	PSf	FO	Hydrophilicity, water flux	[106]
SiO_2	PSf	FO	Hydrophilicity, water flux, resistance to internal concentration polarization	[107]
GO	PSf	FO	Hydrophilicity, water flux, porosity	[108]
Zeolite	PSf	FO	Hydrophilicity, water flux, surface porosity	[109]
LDH nanoparticles	PSf	FO	Hydrophilicity, porosity, mechanical and thermal stability	[110]
HNTs	PSf	FO	Hydrophilicity, water flux	[111]
$CaCO_3$	PSf	FO	Porosity, water flux	[112]
CNTs	PES	FO	Hydrophilicity, resistance to internal concentration polarization	[113]
CN/rGO	PES	FO	Hydrophilicity, water flux	[114]
CNTs	PEI	FO	Hydrophilicity, water flux, porosity, mechanical strength	[115]
SiO_2	PEI	FO	Hydrophilicity, water flux, porosity, pore size	[116]
Functionalized MWCTNs GO	PSF	FO	Hydrophilicity, water flux, solute rejection	[117]

[25]: external concentration polarization which occurs at membrane active layer surface due to the build-up of rejected solute and internal concentration polarization which is created by the formation of the polarized layer within the porous substrate of TFC membrane. Experimental data revealed that the best performance of FO membrane was obtained when 0.5 wt% TiO_2 nanoparticles were embedded into the microporous substrate. The increase in permeability was supported by the formation of long finger-like pore extended from the top to the bottom of the substrate [119]. The effect of incorporating zeolite NaY nanoparticles into the substrate of TFN-FO membrane was also studied by Ma et al. [106], where the 0.5 wt% was the best nanofiller concentration for nanocomposite substrate in order to obtain a good performance. This desirable results are attributed to higher porosity, better hydrophilicity, and additional water pathways through porous nanoparticles.

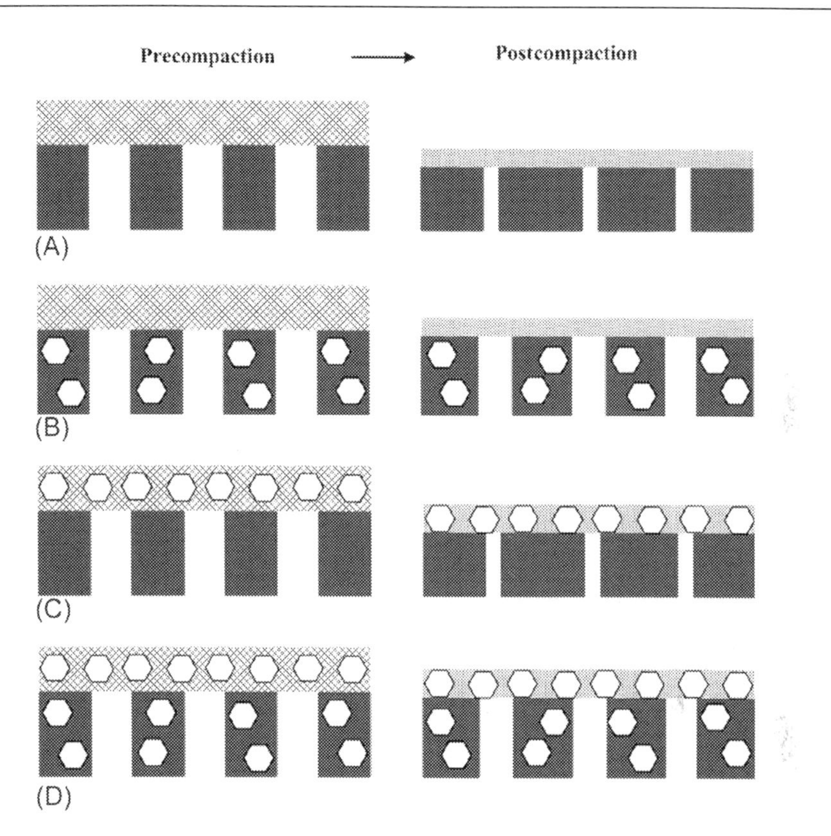

Fig. 9.16 Schematic drawings showing the proposed physical changes to support and thin film structure during compaction in (A) TFC, (B) TFC with nanocomposite substrate, (C) TFN, and (D) TFN with nanocomposite substrate [104].

9.4.5 Surface located nanocomposite

Modification of membrane surface could considerably alter its efficiency and properties such as hydrophilicity, pore size, surface roughness, and charge density. This method of membrane preparation has great potentials to use for improving the surface properties of commercially available membranes because of minimal effect on membrane bulk properties. Self-assembly, coating/deposition, and chemical grafting are the main approaches implemented in individually or be involved simultaneously in preparing surface located nanocomposite membrane [8].

Self-assembly

The most straightforward method for exploiting nanomaterials is to locate nanomaterials to the surface of an available membrane by self-assembly. This method is based on the soaking of the whole or top layer of the membrane in a dilute solution containing nanomaterials. Then after the membrane drying and solvent evaporation,

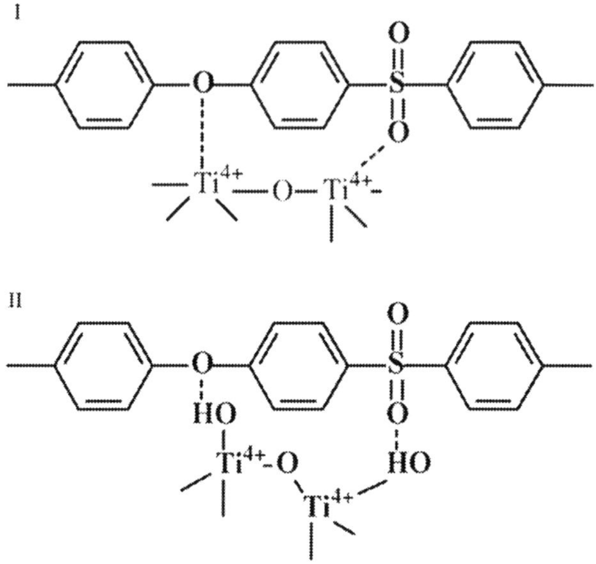

Fig. 9.17 Mechanism of self-assembly of TiO_2 nanoparticles: (I) by a coordination of sulfone group and ether bond to Ti^{4+}; (II) by an H-bonding between sulfone group and ether bond and surface hydroxyl group of TiO_2 [116].

the nanomaterials settled on both or top surface of the membrane as a self-adhering thin layer. The thickness of formed layer varies with the concentration of nanomaterials in the colloidal solution. No reaction other than spontaneous association of nanomaterials with the functional group of polymeric materials occurs which is possible only for nanomaterials for which such interactions are possible such as TiO_2 [31]. Self-assembly and attachment of TiO_2 nanomaterials on membrane surface are possible for specific type of polymers containing —COOH, —SO_3^- H^+, and sulfone groups through coordination and H-bonding interactions. Membrane materials without these functional groups could be treated to create such specific group prior to the self-assembly process [8]. For the first time, Bae and Tak [115] introduced the self-assembly method for the formation of a thin layer of TiO_2 to a polyethersulfone and sulfonated polyethersulfone substrate. Fig. 9.17 exhibits the formation of the TiO_2 nanoparticles layer and its interaction with the functional group of polyethersulfone. The membranes containing —COOH groups have a similar mechanism (Fig. 9.18). This procedure can be used to polyacrylonitrile and PVDF membranes and a blend of poly(styrene-*alt*-maleic anhydride) and poly(vinylidene) difluoride [5,12]. It is considered that the nanomaterials make the membrane more hydrophilic, which would enhance the permeate flux and antifouling property [31]. Consequently, increased permeability is not always observed, but another functionality of TiO_2 nanoparticles, that is, its photocatalytic activity can be considered. This is an important factor for the implementation self-assembly method, because in this method the nanoparticles remain on top layer, therefore, easily accessible for light sources [31].

Thin-film layer of TFC RO membrane

Self-assembly of TiO_2 nanoparticles onto thin-film layer

I. By a bidendate coordination of carboxylate to Ti^{4+}.

II. By a H-bond between carbonyl group and surface hydroxyl group of TiO_2.

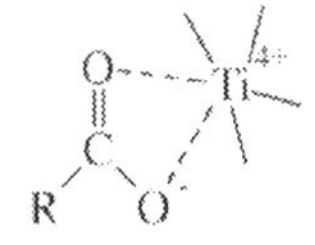

Fig. 9.18 Self-assembly of TiO_2 nanoparticles on the membrane surface. (I) Bidentate coordination of carboxylate to Ti4+ and (II) H-bonding between a carbonyl group and a surface hydroxyl group of TiO_2 [117].

Coating/deposition

Coating/deposition is known as very simple procedures to fabricate surface located nanocomposite membrane through dipping method or filtration-deposition method to create a layer(s) of nanomaterials on the membrane surface. For example, Ahmed et al. [120] investigated the effects of deposition of single-walled CNT on antimicrobial activity of nitrocellulose membranes. Membranes coated with nanotubes displayed considerable antimicrobial property and virus separation of ~2.5 logs. Bae and Tak [115] also deposited TiO_2 nanoparticles onto the different substrates (PSU, PVDF, and PAN) in order to compare their efficiency with conventional nanocomposite membrane for MBR application. Data reveals that TiO_2 deposited membranes have a greater fouling mitigation effect. The deposition of nanoparticles on membrane surface may have some drawbacks such as releasing the nanomaterials to aqueous solution during filtration process due to the weak attachment of nanomaterials to membrane surface, which hindered its applications [5,12].

Layer-by-layer (LBL) assembly

LBL assembly method, which can conduct the formation of the film in nanodimension on the substrate by alternate deposition of cationic and anionic nanostructures via different interaction mechanisms, provides a tunable, easy, robust, reproducible, flexible, and environment-friendly way to modify the separation efficiency and properties of the membrane. Conventional surface modifications techniques lead to an additional hydraulic-resistant layer on the membrane surface and consequently lower permeate flux, but the layer created by the LBL assembly is in highly ordered nanostructured with controlled thickness. The LBL assembled layers can also create more binding sites for functional groups of nanomaterials to introduce multifunctionalization to membrane. Although this method is step by step, but these steps are always frequented including sequential immersion and washing procedures without chemical bonding. The main benefit of the utilization of LBL assembly is that the nanomaterials used are highly versatile (Fig. 9.19). The possible interactions which may take place in LBL assembly consist of electrostatic interactions, hydrogen bonding, charge-transfer interactions, host-guest interactions, biologically specific interactions, coordination interactions, covalent bonding, stereo-complexation, and

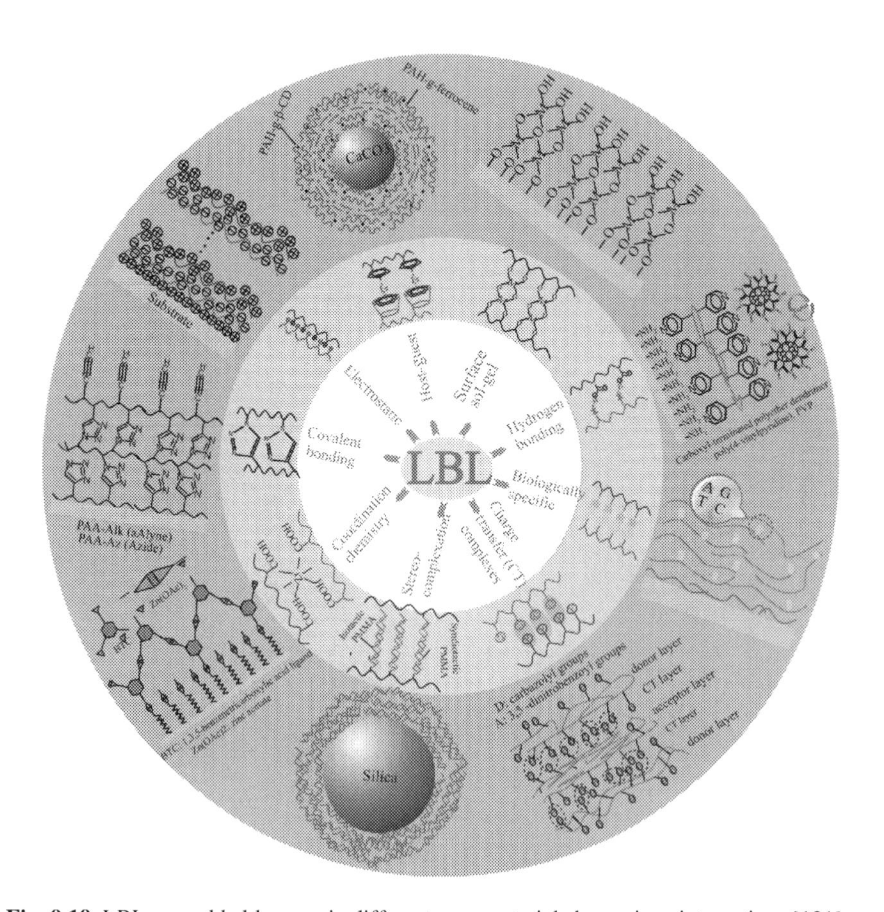

Fig. 9.19 LBL assembled layers via different nanomaterials by various interactions [121].

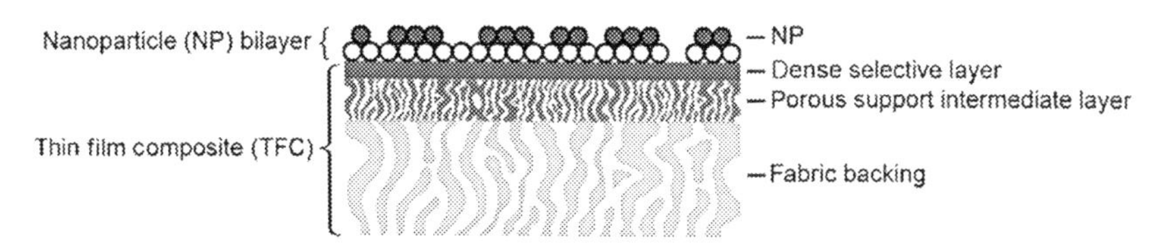

Fig. 9.20 Schematic illustration of the novel TFC membranes with surface modified by inorganic SiO_2 nanoparticles [122].

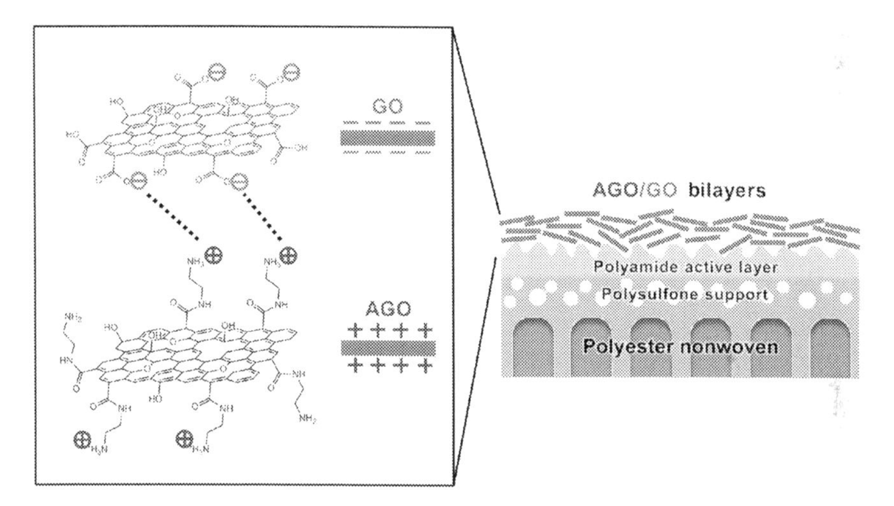

Fig. 9.21 Schematic illustration of a multilayered graphene oxide (GO) coating on a PA thin-film composite membrane surface via LbL deposition of oppositely charged GO and aminated-GO (AGO) nanosheets [123].

surface sol-gel process [121]. For instance, Chan et al. [122] used electrostatic LBL deposition of inorganic nanoparticles to improve the permselectivity of a commercially available NF membrane. Colloidal anionic and cationic SiO_2 nanoparticles were alternately deposited on the surface of Dow FILMTEC NF270 membrane, as illustrated in Fig. 9.20. They found that a single layer of nanoparticles was sufficient to enhance the permselectivity of the membrane by about 50%, compared to the bare TFC membrane. Choi et al. [123] have prepared GO nanosheets on the surface of virgin TFC membranes by sequential deposition of the oppositely charged GO nanosheets via LBL assembly method (Fig. 9.21). Results exhibit that the conformal GO coating layer can modify the surface hydrophilicity and antifouling property as well as reduce the surface roughness. The chemically inert nature of GO nanosheets also acts as chlorine resistance barrier for the underlying PA layer.

Chemical grafting

One of the drawbacks related to surface located nanocomposite membrane is releasing of the deposited nanomaterials to the solution in contact with membrane during the filtration process for those that have weak interaction with the membrane surface. This phenomenon will gradually exhaust the favorable functionality of the nanocomposite membranes and cause secondary pollution. Immobilization of nanomaterials on the membrane surface by chemical bonding is a critical issue solved with chemical grafting method. For instance, Yin et al. [100] studied the effective attachment of silver nanoparticles to the surface of PA-TFC membrane via covalent bonding. The surface of virgin TFC membrane was the first thiol derivatized by reacting with NH_2-$(CH_2)_2$-SH in ethanol solution, and then the freshly synthesized silver nanoparticles were connected onto the PA layer via the Ag—S chemical bonding. The results exhibited good durability of immobilized silver nanoparticles and superior antibacterial properties with desirable permeability and salt rejection.

9.5 Applications of nanocomposite membrane for desalination

9.5.1 RO membrane

RO is the most commonly used technology for the desalination and water treatment and has overtaken conventional technologies such as multistage flash and multieffect distillation [14,124,125]. In recent years, significant developments have been made in the utilization and preparation of RO membranes. These developments are carried out in the synthesis of new membrane materials, module design, process optimization, pretreatment retrofitting, and reduction in energy consumption [14]. In recent years, the improved mechanical, biological, and chemical stabilities as well as minimized fouling and concentration polarization of RO membranes have reduced the membrane replacement, energy consumption, and the water cost [11,14,126]. Due to attractive features of RO technology, the development of the membranes have been greatly studied with the purpose of attaining higher permeate flux, lower fouling, and flux decline, while also maintaining or even enhancing the salt rejection. The most effective and commercially available RO membranes in the market are TFC membrane. RO membranes incorporated with various types of nanomaterials have been investigated at the research level to obtain attractive permeability and many scientists expect nanomaterials to make a milestone in desalination industry [11].

The water transport across RO membranes is driven by applying hydraulic pressure [Fig. 9.22A] and controlled by solution-diffusion mechanism, and no open pathways exist for pore flow mechanism [4]. In the solution-diffusion model, water transport through the dense layer occurs in three separate steps: adsorption of water onto the membrane surface, diffusion of water molecules across the membrane, and finally desorption of water at the permeate side [4].

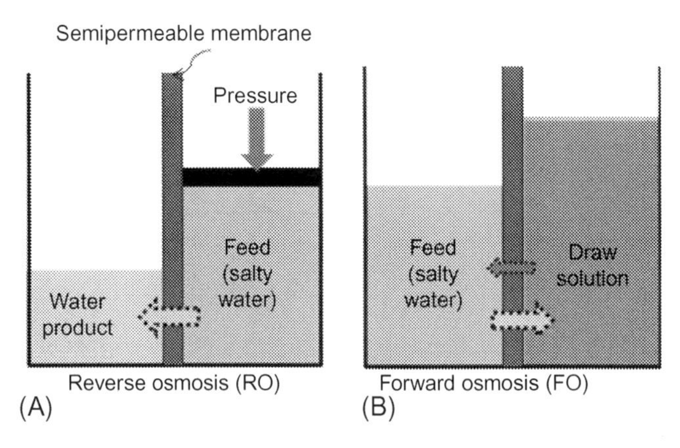

Fig. 9.22 Schematic diagram illustrating the working principles of (A) RO and (B) FO processes. Water flows in different directions in RO and FO (indicated by *larger arrows*); *smaller arrow* shows reverse salt diffusion from draw solution to feed in FO. Hydraulic pressure is used as driving force in RO, while osmotic pressure differential between feed and draw solution serves as driving force in FO [7].

9.5.2 NF membrane

The NF membranes are relatively a newer type of the pressure-driven membranes that lie between that of RO and ultrafiltration (UF) membranes [31]. The term nanofiltration was originally coined at Filmtec Company to describe the "loose RO" or "tight UF" membranes with pores diameters greater than about 1 nm. The NF membrane definition is based on some approximate characteristics [127]: (1) pore diameters <2 nm, (2) passage of sensible amount of monovalent ions ($>30\%$) across the membrane, (3) significant rejection of multivalent ions ($>90\%$), (3) molecular weight cutoff (MWCO) for neutral species is in the 150–2000 Da range, and (4) rejection of neutrals and positive species according to their size and shape. The superior performance of the NF membrane, such as higher permeability, lower energy consumption, and lower capital cost, make it attractive in various industries, such as water softening, sea and brackish water desalination, pretreatment of desalination plants, wastewater treatment, food, beverage, and pharmaceutical industry [31]. The NF membrane is a noteworthy candidate for sea and brackish water desalination where very high salt rejection is not only essential or even acceptable. The NF membranes salt rejection is about 10%–90%, compared with that of up to 99.5% by RO [31]. The governing mechanisms in the NF membranes are a complex combination of solution-diffusion, Donnan effect, dielectric exclusion, and electromigration [13]. The NF membranes in contact with aqueous media with different pH also possess a surface charged due to the breakdown of surface functional groups or adsorption of charge species from an aqueous solution which leads to electrostatic charge repulsion between membrane surface and ionic species (e.g., polymeric contain carboxylic groups and sulfonic acid groups) [128].

9.5.3 FO membrane

The FO is considered as emerging membrane technology with widespread applications such as desalination, power generation, food industry, drug release, and wastewater treatment and reuse [31,129]. The FO desalination processes possess unique features including low-energy consumption and infrastructure requirements as well as membrane fouling propensity [7]. Generally, FO includes two major steps: the first is an osmotic dilution of the draw solution and the second is a generation of fresh water from the diluted draw solution [129]. Unlike RO that requires hydraulic pressure to operate, FO is a naturally occurring process and driven by osmotic pressure gradient across a semipermeable membrane to separate the water from solutes (Fig. 9.22B) [129]. Among various applications, FO has indicated noteworthy potential as a pretreatment system for RO stages, for example, to dilute feed water (sea or brackish water) before RO to decrease osmotic pressure and subsequently energy required for RO function [7]. The FO hybrid system with RO provides the desalination of water with high salinity which is not feasible using stand-alone RO process [129].

The most significant parts of FO relies on two critical things: (1) a membrane with desired properties including high porosity, great hydrophilicity to decrease fouling propensity and enhance permeate flux, and extra-thin thickness, and (2) an efficient draw solutes with high osmotic pressure and easy separation [31]. Despite the researches in the development of FO membrane, only a few membranes have been commercialized due to their better solutes rejection and mechanical stability, even though FO has low permeate flux. For example, Hydration Technologies, Inc. (HTI) produced the first commercial FO membrane including a thin polyester mesh as the support and cellulose triacetate the active layer. In recent years, development of nanostructured materials has led to the fabrication of novel FO membranes that are incorporated with nanofiller [12,13].

9.6 Economic and feasibility of nanocomposite membranes

The economic aspects of nanocomposite membranes are concentrated on higher efficiency and lower energy demand. The energy consumption is considered to have greater impacts on the process economics than the benefit of combined processing. It is assumed that ~1% of total energy consumption (2900 TWh in 2008) is spent on water treatment and desalination, approximately 5% of these electricity costs are attributed to the filtration processes, and that electricity costs are at 0.08€ for 1 kWh. It can be calculated that approximately €116 M is spent annually on electricity for water and wastewater filtration within EU27. In comparison the energy consumption of the whole water and wastewater process, which is around 4% of total national energy consumption, is a total of €9.3 bn; the majority of this energy consumption is related to pumping stations. The estimated Compound Annual Growth Rate (CAGR) for the related electricity demand in the United State was around 0.8% in 2015 and thus displays a relatively steady market. The above economic data are credible for conventional water treatment processes. Potable water supply by seawater desalination technologies consumes at least six times more energy than that supplied from surface or ground water. Thus, there are major economic impacts expected from making seawater desalination less energy intensive, especially in the areas with limited access to fresh water [9].

In 2010, a continuous manufacturing line of TFN membranes was started-up and the fabricated membranes rolled into conventional TFC spiral-wound modules known commercially as QuantumFlux (Qfx). Based on standard element testing, the resulting Qfx membrane modules with 8 in. diameter (37.2 m^2 membrane area) displayed flux of 51.9 m^3/day with salt rejection of 99.8%. Recent modification on Qfx membrane provided membranes with salt rejection of 99.85%, which produce 25% better water quality compared to 99.8%. The modified membrane is suitable for warm high-salinity waters and hybrid designs [130].

Hofs et al. [131] compared the performance of a recently developed TFN membrane with inorganic nanomaterials incorporated into the top layer, and standard TFC spiral wound seawater desalination RO membranes. The following membranes were employed: Qfx SW 75ES RO membranes from Nano-H_2O (El Segundo, California, United States), which are TFN membranes with inorganic nanoparticles, and Filmtec SW30HRLE 4040 units via Lenntech (Delft, The Netherlands), which are standard TFC membranes. The TFN membrane showed a higher flux than the TFCs (SW30HRLE and aged SW30XHR), at similar salt rejections (slightly lower salt rejection found for SW30HRLE). In the case of industrial application of TFN membrane, Nano-H_2O Company successfully conducted a field test of a TFN membrane at Port Hueneme United States Navy Facility in September 2008. They reported that water flux of the TFN membrane was approximately twice the flux of the TFC membrane with NaCl rejection maintained at >99.7% [71,95]. Two years later in September 2010, Nano-H_2O Company installed its first commercial pilot seawater desalination using TFN membrane modules under the Qfx brand followed by official launching into the RO membrane desalination market in April 2011 [25]. >400 desalination plants in over 50 countries have installed LG Nano-H_2O s RO elements. A pilot with Qfx elements tested by EMALSA (Empresa Municipal de Aguas de Las Palmas, SA) as a second-stage system was used to increase recovery by desalinating the brine from an SWRO system. The plant capacity increased by 50% and desalted water with higher quality and lower energy consumption was produced. Energy consumption decreased by approximately 4%. In another desalination plant at Cayman Brac in the Caribbean, replacing the conventional TFC membranes with TFN membrane led to an energy saving of 28% [130].

9.7 Conclusions and future direction

The development of nanocomposite membrane is the major research application in MF, UF, NF, RO, and FO technologies for water/wastewater treatment and desalination and has attracted wide attention in recent years. Compared to the typical membrane, it has been widely reported that incorporation of nanomaterials significantly modifies physicochemical properties of the membrane (hydrophilicity, porosity, charge density, membrane durability, thermal, chemical, and mechanical stability), and has great potential in resolving the problem of the trade-off between permeability and selectivity. Despite the significantly greater achievements, there are some challenges encountered during nanocomposite membrane fabrication which still need to be addressed for practical applications at a large scale. These challenges consist of poor dispersibility of nanomaterials in solvent and polymer, aggregation of nanomaterials within the polymer matrix, weak chemical interaction between nanomaterials and host polymer, lack of compatibility of nanomaterials with the polymer, alignment control of nanotubes in the active layer, and nanomaterial leakage and its environmental toxicity. In order to overcome challenges, several innovative approaches have been offered, for example, modification of nanomaterials surfaces or optimizing the embedding process, development of novel nanomaterials with specific pore structure/charge properties, and optimization of loading concentration and durability of nanocomposite membranes. In terms of characterization, the techniques that can give a high-resolution observation in the angstrom scale to understand the physical and the chemical interaction between the polymer matrix and nanomaterials as well as transport mechanism of solute species in the nanocomposite membranes. The selection and employment of appropriate nanomaterials for nanocomposite membrane fabrication should depend on the specifications of feed (such as type of ions, bacteria, chlorine, etc.) to be treated, because there exists no nanocomposite membrane that is universally applicable for any types of applications (water/wastewater treatment or sea/brackish water desalination).

The industrial implementation of nanocomposite membranes for water treatment and desalination is still in progress. There are many research activities in laboratory and pilot scale on the application of nanocomposite membranes, but very few reports exist on the large-scale production and practical application. More research studies are required to develop the industrial production and application of nanocomposite membrane including the synthesis and supply of suitable nanomaterials, effective approaches for nanomaterials incorporation, and evaluation of the long-term stability of fabricated membranes under practical conditions.

References

[1] Ong CS, Goh PS, Lau WJ, Misdan N, Ismail AF. Nanomaterials for biofouling and scaling mitigation of thin film composite membrane: a review. Desalination 2016;393:2–15.

[2] Shahmirzadi MAA, Hosseini SS, Tan NR. Enhancing removal and recovery of magnesium from aqueous solutions by using modified zeolite and bentonite and process optimization. Korean J Chem Eng 2016;33:3529–40.

[3] Ang WL, Mohammad AW, Hilal N, Leo CP. A review on the applicability of integrated/hybrid membrane processes in water treatment and desalination plants. Desalination 2015;363:2–18.

[4] Greenlee LF, Lawler DF, Freeman BD, Marrot B, Moulin P. Reverse osmosis desalination: water sources, technology, and today's challenges. Water Res 2009;43:2317–48.

[5] Mezher T, Fath H, Abbas Z, Khaled A. Techno-economic assessment and environmental impacts of desalination technologies. Desalination 2011;266:263–73.

[6] Hosseini SS, Bringas E, Tan NR, Ortiz I, Ghahramani M, Shahmirzadi MAA. Recent progress in development of high performance polymeric membranes and materials for metal plating wastewater treatment: a review. J Water Process Eng 2016;9:78–110.
[7] Li D, Yan Y, Wang H. Recent advances in polymer and polymer composite membranes for reverse and forward osmosis processes. Prog Polym Sci 2016;61:104–55.
[8] Yin J, Deng B. Polymer-matrix nanocomposite membranes for water treatment. J Membr Sci 2015;479:256–75.
[9] http://nanopinion.archiv.zsi.at/sites/default/files/briefing_no.16_nanoenhanced_membranes_for_water_treatment.pdf.
[10] http://nanopinion.archiv.zsi.at/sites/default/files/observatorynano_briefing_no.13_nanostructured_membranes_for_water_treatment.pdf.
[11] Buonomenna MG. Nano-enhanced reverse osmosis membranes. Desalination 2013;314:73–88.
[12] Figoli A, Marino T, Simone S, Di Nicolò E, Li X-M, He T, et al. Towards non-toxic solvents for membrane preparation: a review. Green Chem 2014;16:4034.
[13] Alaei Shahmirzadi MA, Hosseini SS, Ruan G, Tan NR. Tailoring PES nanofiltration membranes through systematic investigations of prominent design, fabrication and operational parameters. RSC Adv 2015;5:49080–97.
[14] Lee KP, Arnot TC, Mattia D. A review of reverse osmosis membrane materials for desalination—development to date and future potential. J Membr Sci 2011;370:1–22.
[15] King WM, Cantor PA, Schoellenback LW, Cannon CR. High-retention reverse-osmosis desalination membranes from cellulose acetate, membranes from cellulose derivatives. Interscience Publisher: New York; 2001.
[16] Edgar KJ, Buchanan CM, Debenham JS, Rundquist PA, Seiler BD, Shelton MC, et al. Advances in cellulose ester performance and application. Prog Polym Sci 2001;26:1605–88.
[17] L. Credali, G. Baruzzi, V. Guidotti, Reverse osmosis anisotropic membranes based on polypiperazine amides, (1978).
[18] Ismail AF, Padaki M, Hilal N, Matsuura T, Lau WJ. Thin film composite membrane—recent development and future potential. Desalination 2015;356:140–8.
[19] Misdan N, Lau WJ, Ismail AF. Seawater reverse osmosis (SWRO) desalination by thin-film composite membrane-current development, challenges and future prospects. Desalination 2012;287:228–37.
[20] Lau WJ, Ismail AF, Misdan N, Kassim MA. A recent progress in thin film composite membrane: a review. Desalination 2012;287:190–9.
[21] Francis PS. Fabrication and evaluation of new ultrathin reverse osmosis membranes. Available from Natl. Tech. Inf. Serv. Springf. VA 22161 as PB-177 083, Price Codes A 04 Pap. Copy, A 01 Microfich. OSW Res. Dev. Prog. Rep; 1966.
[22] Li D, Wang H. Recent developments in reverse osmosis desalination membranes. J Mater Chem 2010;20:4551.
[23] Kurihara M, Kanamaru N, Harumiya N, Yoshimura K, Hagiwara S. Spiral-wound new thin film composite membrane for a single-stage seawater desalination by reverse osmosis. Desalination 1980;32:13–23.
[24] Saeid S, Bringas E, Tan NR, Ortiz I. Recent progress in development of high performance polymeric membranes and materials for metal plating wastewater treatment: a review. J Water Process Eng 2016;9:78–110.
[25] Lau WJ, Gray S, Matsuura T, Emadzadeh D, Paul Chen J, Ismail AF. A review on polyamide thin film nanocomposite (TFN) membranes: history, applications, challenges and approaches. Water Res 2015;80:306–24.

[26] C.J. Kurth, I.K. Iverson, S.D. Kloos, L.T. Hodgins, Modified polyamide matrices and methods for their preparation (2011).

[27] J. Koo, S.P. Hong, J.H. Lee, K.Y. Ryu, Selective membrane having a high fouling resistance (2011).

[28] Jeong B-H, Hoek EMV, Yan Y, Subramani A, Huang X, Hurwitz G, et al. Interfacial polymerization of thin film nanocomposites: a new concept for reverse osmosis membranes. J Membr Sci 2007;294:1–7.

[29] Homayoonfal M, Mehrnia MR, Mojtahedi YM, Ismail AF. Effect of metal and metal oxide nanoparticle impregnation route on structure and liquid filtration performance of polymeric nanocomposite membranes: a comprehensive review. Desalin Water Treat 2013;51:3295–316.

[30] Yang H-C, Hou J, Chen V, Xu Z-K. Surface and interface engineering for organic–inorganic composite membranes. J Mater Chem A 2016;4:9716–29.

[31] Hilal N, Ismail AF, Wright C. Membrane fabrication. Boca Raton, FL: CRC Press; 2015.

[32] Sadrzadeh M, Bhattacharjee S. Rational design of phase inversion membranes by tailoring thermodynamics and kinetics of casting solution using polymer additives. J Membr Sci 2013;441:31–44.

[33] Lee K-W, Seo B-K, Nam S-T, Han M-J. Trade-off between thermodynamic enhancement and kinetic\rhindrance during phase inversion in the preparation of\rpolysulfone membranes. Desalination 2003;159:289–96.

[34] Vatanpour V, Madaeni SS, Moradian R, Zinadini S, Astinchap B. Fabrication and characterization of novel antifouling nanofiltration membrane prepared from oxidized multiwalled carbon nanotube/polyethersulfone nanocomposite. J Membr Sci 2011;375:284–94.

[35] Madaeni SS, Ghaemi N. Characterization of self-cleaning RO membranes coated with TiO_2 particles under UV irradiation. J Membr Sci 2007;303:221–33.

[36] Lee HS, Im SJ, Kim JH, Kim HJ, Kim JP, Min BR. Polyamide thin-film nanofiltration membranes containing TiO_2 nanoparticles. Desalination 2008;219:48–56.

[37] Andrade PF, de Faria AF, Oliveira SR, Arruda MAZ, Gonçalves MC. Improved antibacterial activity of nanofiltration polysulfone membranes modified with silver nanoparticles. Water Res 2015;81:333–42.

[38] Ciobanu G, Carja G, Ciobanu O. Preparation and characterization of polymer–zeolite nanocomposite membranes. Mater Sci Eng C 2007;27:1138–40.

[39] Yin J, Zhu G, Deng B. Multi-walled carbon nanotubes (MWNTs)/polysulfone (PSU) mixed matrix hollow fiber membranes for enhanced water treatment. J Membr Sci 2013;437:237–48.

[40] Majeed S, Fierro D, Buhr K, Wind J, Du B, Boschetti-de-Fierro A, et al. Multi-walled carbon nanotubes (MWCNTs) mixed polyacrylonitrile (PAN) ultrafiltration membranes. J Membr Sci 2012;403:101–9.

[41] Wu H, Tang B, Wu P. Novel ultrafiltration membranes prepared from a multi-walled carbon nanotubes/polymer composite. J Membr Sci 2010;362:374–83.

[42] Giwa A, Akther N, Dufour V, Hasan SW. A critical review on recent polymeric and nano-enhanced membranes for reverse osmosis. RSC Adv 2016;6:8134–63.

[43] Mago G, Kalyon DM, Fisher FT. Membranes of polyvinylidene fluoride and PVDF nanocomposites with carbon nanotubes via immersion precipitation. J Nanomater 2008;2008:17.

[44] Vatanpour V, Madaeni SS, Moradian R, Zinadini S, Astinchap B. Novel antibifouling nanofiltration polyethersulfone membrane fabricated from embedding TiO_2 coated multiwalled carbon nanotubes. Sep Purif Technol 2012;90:69–82.

[45] Zinadini S, Zinatizadeh AA, Rahimi M, Vatanpour V, Zangeneh H. Preparation of a novel antifouling mixed matrix PES membrane by embedding graphene oxide nanoplates. J Membr Sci 2014;453:292–301.
[46] Wu H, Tang B, Wu P. Development of novel SiO 2–GO nanohybrid/polysulfone membrane with enhanced performance. J Membr Sci 2014;451:94–102.
[47] Ghaemi N, Madaeni SS, Alizadeh A, Rajabi H, Daraei P. Preparation, characterization and performance of polyethersulfone/organically modified montmorillonite nanocomposite membranes in removal of pesticides. J Membr Sci 2011;382:135–47.
[48] Lai CY, Groth A, Gray S, Duke M. Investigation of the dispersion of nanoclays into PVDF for enhancement of physical membrane properties. Desalin Water Treat 2011;34:251–6.
[49] Monticelli O, Bottino A, Scandale I, Capannelli G, Russo S. Preparation and properties of polysulfone-clay composite membranes. J Appl Polym Sci 2007;103:3637–44.
[50] Abedini R, Mousavi SM, Aminzadeh R. A novel cellulose acetate (CA) membrane using TiO_2 nanoparticles: preparation, characterization and permeation study. Desalination 2011;277:40–5.
[51] Cao X, Ma J, Shi X, Ren Z. Effect of TiO_2 nanoparticle size on the performance of PVDF membrane. Appl Surf Sci 2006;253:2003–10.
[52] Xiao Y, Wang KY, Chung T-S, Tan J. Evolution of nano-particle distribution during the fabrication of mixed matrix TiO_2-polyimide hollow fiber membranes. Chem Eng Sci 2006;61:6228–33.
[53] Wu G, Gan S, Cui L, Xu Y. Preparation and characterization of PES/TiO_2 composite membranes. Appl Surf Sci 2008;254:7080–6.
[54] Yang Y, Wang P, Zheng Q. Preparation and properties of polysulfone/TiO_2 composite ultrafiltration membranes. J Polym Sci Part B Polym Phys 2006;44:879–87.
[55] Maximous N, Nakhla G, Wong K, Wan W. Optimization of Al2O3/PES membranes for wastewater filtration. Sep Purif Technol 2010;73:294–301.
[56] Yan L, Hong S, Li ML, Li YS. Application of the Al2O3–PVDF nanocomposite tubular ultrafiltration (UF) membrane for oily wastewater treatment and its antifouling research. Sep Purif Technol 2009;66:347–52.
[57] Basri H, Ismail AF, Aziz M. Polyethersulfone (PES)–silver composite UF membrane: Effect of silver loading and PVP molecular weight on membrane morphology and antibacterial activity. Desalination 2011;273:72–80.
[58] Chou W, Yu D, Yang M. The preparation and characterization of silver-loading cellulose acetate hollow fiber membrane for water treatment. Polym Adv Technol 2005;16:600–7.
[59] Deng Y, Dang G, Zhou H, Rao X, Chen C. Preparation and characterization of polyimide membranes containing Ag nanoparticles in pores distributing on one side. Mater Lett 2008;62:1143–6.
[60] Zodrow K, Brunet L, Mahendra S, Li D, Zhang A, Li Q, et al. Polysulfone ultrafiltration membranes impregnated with silver nanoparticles show improved biofouling resistance and virus removal. Water Res 2009;43:715–23.
[61] Vatanpour V, Safarpour M, Khataee A, Visakh PM, Nazarenko O. Mixed matrix membranes for nanofiltration application. In: Nanostructured polym. membr. appl. vol. 2. Beverly, MA: Wiley; 2016. p. 441–71.
[62] Rabiee H, Vatanpour V, Farahani MHDA, Zarrabi H. Improvement in flux and antifouling properties of PVC ultrafiltration membranes by incorporation of zinc oxide (ZnO) nanoparticles. Sep Purif Technol 2015;156:299–310.
[63] María Arsuaga J, Sotto A, del Rosario G, Martínez A, Molina S, Teli SB, et al. Influence of the type, size, and distribution of metal oxide particles on the properties of nanocomposite ultrafiltration membranes. J Membr Sci 2013;428:131–41.

[64] Mobarakabad P, Moghadassi AR, Hosseini SM. Fabrication and characterization of poly(phenylene ether-ether sulfone) based nanofiltration membranes modified by titanium dioxide nanoparticles for water desalination. Desalination 2015;365:227–33.
[65] Goh K, Karahan HE, Wei L, Bae TH, Fane AG, Wang R, et al. Carbon nanomaterials for advancing separation membranes: a strategic perspective. Carbon NY 2016;109:694–710.
[66] Das R, Ali ME, Hamid SBA, Ramakrishna S, Chowdhury ZZ. Carbon nanotube membranes for water purification: a bright future in water desalination. Desalination 2014;336:97–109.
[67] Vatanpour V, Esmaeili M, Farahani MHDA. Fouling reduction and retention increment of polyethersulfone nanofiltration membranes embedded by amine-functionalized multi-walled carbon nanotubes. J Membr Sci 2014;466:70–81.
[68] Ganesh BM, Isloor AM, Ismail AF. Enhanced hydrophilicity and salt rejection study of graphene oxide-polysulfone mixed matrix membrane. Desalination 2013;313:199–207.
[69] Aditya Kiran S, Lukka Thuyavan Y, Arthanareeswaran G, Matsuura T, Ismail AF. Impact of graphene oxide embedded polyethersulfone membranes for the effective treatment of distillery effluent. Chem Eng J 2016;286:528–37.
[70] Akin I, Zor E, Bingol H, Ersoz M. Green synthesis of reduced graphene oxide/polyaniline composite and its application for salt rejection by polysulfone-based composite membranes. J Phys Chem B 2014;118:5707–16.
[71] Tetala KKR, Stamatialis DF. Mixed matrix membranes for efficient adsorption of copper ions from aqueous solutions. Sep Purif Technol 2013;104:214–20.
[72] Shaban M, AbdAllah H, Said L, Hamdy HS, Abdel Khalek A. Titanium dioxide nanotubes embedded mixed matrix PES membranes characterization and membrane performance. Chem Eng Res Des 2015;95:307–16. https://doi.org/10.1016/j.cherd.2014.11.008.
[73] Baroña GNB, Choi M, Jung B. High permeate flux of PVA/PSf thin film composite nanofiltration membrane with aluminosilicate single-walled nanotubes. J Colloid Interface Sci 2012;386:189–97.
[74] Chan W-F, Chen H, Surapathi A, Taylor MG, Shao X, Marand E, et al. Zwitterion functionalized carbon nanotube/polyamide nanocomposite membranes for water desalination. ACS Nano 2013;7:5308–19.
[75] nan Shen J, chao Yu C, min Ruan H, jie Gao C, Van der Bruggen B. Preparation and characterization of thin-film nanocomposite membranes embedded with poly (methyl methacrylate) hydrophobic modified multiwalled carbon nanotubes by interfacial polymerization. J Membr Sci 2013;442:18–26.
[76] Son M, Novotny V, Choi H. Thin-film nanocomposite membrane with vertically embedded carbon nanotube for forward osmosis. Desalin Water Treat 2016;57:26670–9.
[77] Zheng J, Li M, Yu K, Hu J, Zhang X, Wang L. Sulfonated multiwall carbon nanotubes assisted thin-film nanocomposite membrane with enhanced water flux and anti-fouling property. J Membr Sci 2017;524:344–53.
[78] Huang H, Qu X, Dong H, Zhang L, Chen H. Role of NaA zeolites in the interfacial polymerization process towards a polyamide nanocomposite reverse osmosis membrane. RSC Adv 2013;3:8203–7.
[79] Lind ML, Jeong B-H, Subramani A, Huang X, Hoek EMV. Effect of mobile cation on zeolite-polyamide thin film nanocomposite membranes. J Mater Res 2009;24:1624–31.
[80] Lind ML, Eumine Suk D, Nguyen T-V, Hoek EMV. Tailoring the structure of thin film nanocomposite membranes to achieve seawater RO membrane performance. Environ Sci Technol 2010;44:8230–5.
[81] Ma N, Wei J, Liao R, Tang CY. Zeolite-polyamide thin film nanocomposite membranes: towards enhanced performance for forward osmosis. J Membr Sci 2012;405:149–57.

[82] Huang H, Qu X, Ji X, Gao X, Zhang L, Chen H, et al. Acid and multivalent ion resistance of thin film nanocomposite RO membranes loaded with silicalite-1 nanozeolites. J Mater Chem A 2013;1:11343–9.
[83] Cay-Durgun P, McCloskey C, Konecny J, Khosravi A, Lind ML. Evaluation of thin film nanocomposite reverse osmosis membranes for long-term brackish water desalination performance. Desalination 2017;404:304–12.
[84] Singh PS, Aswal VK. Characterization of physical structure of silica nanoparticles encapsulated in polymeric structure of polyamide films. J Colloid Interface Sci 2008;326:176–85.
[85] Jadav GL, Aswal VK, Singh PS. SANS study to probe nanoparticle dispersion in nanocomposite membranes of aromatic polyamide and functionalized silica nanoparticles. J Colloid Interface Sci 2010;351:304–14.
[86] Yin J, Kim E-S, Yang J, Deng B. Fabrication of a novel thin-film nanocomposite (TFN) membrane containing MCM-41 silica nanoparticles (NPs) for water purification. J Membr Sci 2012;423:238–46.
[87] Kim SG, Chun JH, Chun B-H, Kim SH. Preparation, characterization and performance of poly (aylene ether sulfone)/modified silica nanocomposite reverse osmosis membrane for seawater desalination. Desalination 2013;325:76–83.
[88] Liu L, Zhu G, Liu Z, Gao C. Effect of MCM-48 nanoparticles on the performance of thin film nanocomposite membranes for reverse osmosis application. Desalination 2016;394:72–82.
[89] Zhu G, Bao M, Liu Z, Gao C. Preparation of spherical mesoporous aminopropyl-functionalized MCM-41 and its application in polyamide thin film nanocomposite reverse osmosis membranes. Desalin Water Treat 2016;1–10.
[90] Zargar M, Hartanto Y, Jin B, Dai S. Hollow mesoporous silica nanoparticles: A peculiar structure for thin film nanocomposite membranes. J Membr Sci 2016;519:1–10.
[91] Kim E-S, Hwang G, El-Din MG, Liu Y. Development of nanosilver and multi-walled carbon nanotubes thin-film nanocomposite membrane for enhanced water treatment. J Membr Sci 2012;394:37–48.
[92] Lai GS, Lau WJ, Goh PS, Ismail AF, Yusof N, Tan YH. Graphene oxide incorporated thin film nanocomposite nanofiltration membrane for enhanced salt removal performance. Desalination 2016;387:14–24.
[93] Wang J, Wang Y, Zhang Y, Uliana A, Zhu J, Liu J, et al. Zeolitic Imidazolate framework/Graphene oxide hybrid Nanosheets functionalized thin film Nanocomposite membrane for enhanced antimicrobial performance. ACS Appl Mater Interfaces 2016;8:25508–19.
[94] Kong C, Kamada T, Shintani T, Kanezashi M, Yoshioka T, Tsuru T. Enhanced performance of inorganic-polyamide nanocomposite membranes prepared by metal-alkoxide-assisted interfacial polymerization. J Membr Sci 2011;366:382–8.
[95] Saleh TA, Gupta VK. Synthesis and characterization of alumina nano-particles polyamide membrane with enhanced flux rejection performance. Sep Purif Technol 2012;89:245–51.
[96] Daraei P, Madaeni SS, Salehi E, Ghaemi N, Ghari HS, Khadivi MA, et al. Novel thin film composite membrane fabricated by mixed matrix nanoclay/chitosan on PVDF microfiltration support: Preparation, characterization and performance in dye removal. J Membr Sci 2013;436:97–108.
[97] Zhang L, Shi G-Z, Qiu S, Cheng L-H, Chen H-L. Preparation of high-flux thin film nanocomposite reverse osmosis membranes by incorporating functionalized multi-walled carbon nanotubes. Desalin Water Treat 2011;34:19–24.
[98] Kim CK, Kim JH, Roh IJ, Kim JJ. The changes of membrane performance with polyamide molecular structure in the reverse osmosis process. J Membr Sci 2000;165:189–99.

[99] Zhao H, Qiu S, Wu L, Zhang L, Chen H, Gao C. Improving the performance of polyamide reverse osmosis membrane by incorporation of modified multi-walled carbon nanotubes. J Membr Sci 2014;450:249–56.
[100] Yin J, Yang Y, Hu Z, Deng B. Attachment of silver nanoparticles (AgNPs) onto thin-film composite (TFC) membranes through covalent bonding to reduce membrane biofouling. J Membr Sci 2013;441:73–82.
[101] Park J, Choi W, Kim SH, Chun BH, Bang J, Lee KB. Enhancement of chlorine resistance in carbon nanotube based nanocomposite reverse osmosis membranes. Desalin Water Treat 2010;15:198–204.
[102] Namvar-Mahboub M, Pakizeh M, Davari S. Preparation and characterization of UZM-5/polyamide thin film nanocomposite membrane for dewaxing solvent recovery. J Membr Sci 2014;459:22–32.
[103] Lu P, Liang S, Zhou T, Mei X, Zhang Y, Zhang C, et al. Typical thin-film composite (TFC) membranes modified with inorganic nanomaterials for forward osmosis: a review. Nanosci Nanotechnol Lett 2016;8:906–16.
[104] Pendergast MM, Ghosh AK, Hoek EMV. Separation performance and interfacial properties of nanocomposite reverse osmosis membranes. Desalination 2013;308:180–5.
[105] Park MJ, Phuntsho S, He T, Nisola GM, Tijing LD, Li X-M, et al. Graphene oxide incorporated polysulfone substrate for the fabrication of flat-sheet thin-film composite forward osmosis membranes. J Membr Sci 2015;493:496–507.
[106] Ma N, Wei J, Qi S, Zhao Y, Gao Y, Tang CY. Nanocomposite substrates for controlling internal concentration polarization in forward osmosis membranes. J Membr Sci 2013;441:54–62.
[107] Lu P, Liang S, Qiu L, Gao Y, Wang Q. Thin film nanocomposite forward osmosis membranes based on layered double hydroxide nanoparticles blended substrates. J Membr Sci 2016;504:196–205.
[108] Ghanbari M, Emadzadeh D, Lau WJ, Riazi H, Almasi D, Ismail AF. Minimizing structural parameter of thin film composite forward osmosis membranes using polysulfone/halloysite nanotubes as membrane substrates. Desalination 2016;377:152–62.
[109] Liu Z, Yu H, Kang G, Jie X, Jin Y, Cao Y. Investigation of internal concentration polarization reduction in forward osmosis membrane using nano-CaCO3 particles as sacrificial component. J Membr Sci 2016;497:485–93.
[110] Wang Y, Ou R, Ge Q, Wang H, Xu T. Preparation of polyethersulfone/carbon nanotube substrate for high-performance forward osmosis membrane. Desalination 2013;330:70–8.
[111] Wang Y, Ou R, Wang H, Xu T. Graphene oxide modified graphitic carbon nitride as a modifier for thin film composite forward osmosis membrane. J Membr Sci 2015;475:281–9.
[112] Tian M, Wang Y-N, Wang R. Synthesis and characterization of novel high-performance thin film nanocomposite (TFN) FO membranes with nanofibrous substrate reinforced by functionalized carbon nanotubes. Desalination 2015;370:79–86.
[113] Tian M, Wang Y-N, Wang R, Fane AG. Synthesis and characterization of thin film nanocomposite forward osmosis membranes supported by silica nanoparticle incorporated nanofibrous substrate. Desalination 2016;401:142–50.
[114] Morales-Torres S, Esteves CMP, Figueiredo JL, Silva AMT. Thin-film composite forward osmosis membranes based on polysulfone supports blended with nanostructured carbon materials. J Membr Sci 2016;520:326–36.
[115] Bae T-H, Tak T-M. Effect of TiO_2 nanoparticles on fouling mitigation of ultrafiltration membranes for activated sludge filtration. J Membr Sci 2005;249:1–8.

[116] Luo M-L, Zhao J-Q, Tang W, Pu C-S. Hydrophilic modification of poly (ether sulfone) ultrafiltration membrane surface by self-assembly of TiO_2 nanoparticles. Appl Surf Sci 2005;249:76–84.
[117] Kim SH, Kwak S-Y, Sohn B-H, Park TH. Design of TiO_2 nanoparticle self-assembled aromatic polyamide thin-film-composite (TFC) membrane as an approach to solve biofouling problem. J Membr Sci 2003;211:157–65.
[118] Pendergast MTM, Nygaard JM, Ghosh AK, Hoek EMV. Using nanocomposite materials technology to understand and control reverse osmosis membrane compaction. Desalination 2010;261:255–63.
[119] Emadzadeh D, Lau WJ, Matsuura T, Rahbari-Sisakht M, Ismail AF. A novel thin film composite forward osmosis membrane prepared from PSf-TiO_2 nanocomposite substrate for water desalination. Chem Eng J 2014;237:70–80.
[120] Ahmed F, Santos CM, Mangadlao J, Advincula R, Rodrigues DF. Antimicrobial PVK: SWNT nanocomposite coated membrane for water purification: performance and toxicity testing. Water Res 2013;47:3966–75.
[121] Xu G-R, Wang S-H, Zhao H-L, Wu S-B, Xu J-M, Li L, et al. Layer-by-layer (LBL) assembly technology as promising strategy for tailoring pressure-driven desalination membranes. J Membr Sci 2015;493:428–43.
[122] Chan EP, Mulhearn WD, Huang Y-R, Lee J-H, Lee D, Stafford CM. Tailoring the permselectivity of water desalination membranes via nanoparticle assembly. Langmuir 2014;30:611–6.
[123] Choi W, Choi J, Bang J, Lee J-H. Layer-by-layer assembly of graphene oxide nanosheets on polyamide membranes for durable reverse-osmosis applications. ACS Appl Mater Interfaces 2013;5:12510–9.
[124] Khazaali F, Kargari A. Treatment of phenolic wastewaters by a domestic low-pressure reverse osmosis system. J Membr Sci Res 2017;3:22–8.
[125] Khazaali F, Kargari A, Rokhsaran M. Application of low-pressure reverse osmosis for effective recovery of bisphenol a from aqueous wastes. Desalin Water Treat 2014;52:7543–51.
[126] Kargari A, Khazaali F. Effect of operating parameters on 2-chlorophenol removal from wastewaters by a low-pressure reverse osmosis system. Desalin Water Treat 2015;55:114–24.
[127] Paul M, Jons SD. Chemistry and fabrication of polymeric nanofiltration membranes: a review. Polymer (United Kingdom) 2016;103:417–56.
[128] Mohammad AW, Teow YH, Ang WL, Chung YT, Oatley-Radcliffe DL, Hilal N. Nanofiltration membranes review: recent advances and future prospects. Desalination 2015;356:226–54.
[129] Akther N, Sodiq A, Giwa A, Daer S, Arafat HA, Hasan SW. Recent advancements in forward osmosis desalination: a review. Chem Eng J 2015;281:502–22.
[130] Reisner DE, Pradeep T. Aquananotechnology: global prospects. London: CRC Press; 2014.
[131] Hofs B, Schurer R, Harmsen DJH, Ceccarelli C, Beerendonk EF, Cornelissen ER. Characterization and performance of a commercial thin film nanocomposite seawater reverse osmosis membrane and comparison with a thin film composite. J Membr Sci 2013;446:68–78.

Further reading

[1] Husain S, Koros WJ. Macrovoids in hybrid organic/inorganic hollow fiber membranes. Ind Eng Chem Res 2009;48:2372–9.
[2] Liao C, Yu P, Zhao J, Wang L, Luo Y. Preparation and characterization of NaY/PVDF hybrid ultrafiltration membranes containing silver ions as antibacterial materials. Desalination 2011;272:59–65.
[3] Hoek EMV, Ghosh AK, Huang X, Liong M, Zink JI. Physical–chemical properties, separation performance, and fouling resistance of mixed-matrix ultrafiltration membranes. Desalination 2011;283:89–99.
[4] Ma H, Yoon K, Rong L, Shokralla M, Kopot A, Wang X, et al. Thin-film nanofibrous composite ultrafiltration membranes based on polyvinyl alcohol barrier layer containing directional water channels. Ind Eng Chem Res 2010;49:11978–84.
[5] Wu H, Tang B, Wu P. MWNTs/polyester thin film nanocomposite membrane: an approach to overcome the trade-off effect between permeability and selectivity. J Phys Chem C 2010;114:16395–400.
[6] Emadzadeh D, Lau WJ, Ismail AF. Synthesis of thin film nanocomposite forward osmosis membrane with enhancement in water flux without sacrificing salt rejection. Desalination 2013;330:90–9.
[7] Liu X, Ng HY. Fabrication of layered silica–polysulfone mixed matrix substrate membrane for enhancing performance of thin-film composite forward osmosis membrane. J Membr Sci 2015;481:148–63.
[8] Kurth CJ, Burk R, Green J. Utilizing nanotechnology to enhance RO membrane performance for seawater desalination, In: Proceeding of International Desalination Association World Congress, Perth, West Australia; 2011.

Part Two

Recent trends and applications

Electrochemically active carbon nanotube (CNT) membrane filter for desalination and water purification

10

Zaira Z. Chowdhury, Kaushik Pal†, Suresh Sagadevan‡, Wageeh A. Yehye*, Rafie B. Johan*, Syed T. Shah*, Abimola Adebesi*, Md. E. Ali*, Md. S. Islam§, Rahman F. Rafique¶*
*Nanotechnology and Catalysis Research Center (NANOCAT), Kuala Lumpur, Malaysia, †Wuhan University, Wuchang, PR China, ‡AMET University, Chennai, India, §Royal Melbourne Institute of Technology (RMIT), Melbourne, VIC, Australia, ¶The State University of New Jersey, New Brunswick, NJ, United States

10.1 Introduction

The escalating population growth together with rapid industrialization and upsurge in energy demands has led to serious challenges of providing freshwater sources [1–3]. This has steered numerous research activities into new water treatment technologies which will ensure and develop innovative water supply strategies [4]. The techniques should be highly energy efficient as well as environmentally friendly [5]. Difficulties with seawater are its salinity which makes it unsuitable for consumption. The salinity for seawater is mainly caused by the presence of NaCl [6]. The elemental mass of Cl is 16 times higher than Mg, 48 times higher than Br and K, and 22 times higher than Mg. The alkali metal Na is 9 times higher than the alkaline earth metal Mg, 12 times higher than the nonmetal of S, 17 times higher than K, and 180 times higher than Br. Therefore, the usage of seawater for domestic purposes needs energy-efficient techniques whereby less input energy with minimum amount of chemicals should be used to minimize the adverse effect toward the environment [7,8]. The membrane-based separation method has been considered ahead of the other traditional methods of wastewater treatment in view of its numerous advantages, including lower energy demand unlike distillation or electrolysis and less space requirement compared to sorption methods. Furthermore, it has higher separation selectivity and can be operated continuously [9,10]. However, to maintain a robust strength mechanically so as to provide maximum permeation, and to be selective through its favorable pore size distribution which must be narrow and maintain a high chemical inertness, the separation membrane should be thin [11–14].

To make a selection between the available polymeric material- and inorganic material-based membrane, factors such as flux, selectivity, fabrication cost, and

Emerging Technologies for Sustainable Desalination Handbook. https://doi.org/10.1016/B978-0-12-815818-0.00010-2

stability under usage are usually considered. While polymeric membranes provide high selectivity and high flux rates, they often exhibit high intolerance toward high temperatures, corrosive surroundings, and organic solvents. Inorganic membranes on the other hand enjoy high thermal, structural, and mechanical strengths coupled with high selectivity but limited by extremely low flux and high cost of fabrication [15–17]. In order to resolve the limitations of these membrane types, the use of well-defined nanostructured materials such as zeolites [18], organic frameworks made of metals [19], and carbon-based materials [20,21] have been considered for the development of membranes. Among these materials, carbon nanotubes (CNTs) have received the highest consideration due to their one-dimensional hollow structure and high mechanical strength [22,23].

Water treatment is considered as one of the challenging issues which should be taken under consideration to lessen release of greenhouse gas and global warming: To scale-up wastewater treatment and seawater desalination plants require high energy and capital investment [1,2]. Careful process design and optimization with efficient recovery of energy and a substantial upgrading in the performance of existing polymeric membranes have gradually reduced the energy constraint of the membrane-initiated water treatment plants (Fig. 10.1) [2]. Proper thermodynamic calculation of a process can additionally reduce the entire energy requirements for filtration processes. However, further development can only be made by incorporating novel membrane materials which should be able to increase the water flux of a membrane for specified pressure changes. Although this would not lessen the extent of energy required to obtain 1 m^3 of water (kWh/m^3), a higher flux can be ensured from smaller-scale plant, lesser emissions of carbon with overall reduced capital costs [24]. The CNTs exhibited ultrahigh water flow rates in the early 2000s. Thus, it has been hoped that CNTs might be the impeccable candidate to fabricate membranes with higher flux values than the

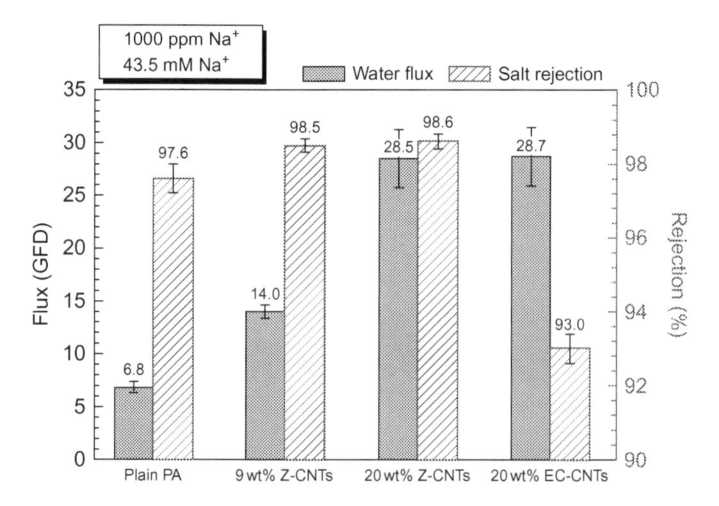

Fig. 10.1 Rate of water flux and salt rejection in terms of CNT concentration in TFC polyamide membrane [2].

commercial polymeric ones [25,26]. Water flux was enhanced up to fourfold compared to a normal thin film composite (TFC) polyamide membrane when CNTs were incorporated randomly with the active layer of the membrane [27]. However, incorporation of CNTs could not alter the salt rejection rate as this is still dependent on the polymer itself. Basically, the tubes are completely implanted inside the active layer and they perform as the interior fast lanes for water molecules to pass after the water has already infiltrated through the surface of the polymer.

Membranes or filter bed based on micro- or nanostructured carbonaceous materials provide a convenient means to overcome these difficulties. Based on geometrical arrangement, CNTs can be of three different types, namely: armchair, zigzag, and chiral [e.g., zigzag (n, 0); armchair (n, n); and chiral (n, m)] (Fig. 10.2) [28]. This structural difference is based on the rolling up of graphene sheets during preparation process [28]. The nanoscale cylindrical-shaped graphene is considered as CNTs. The outer radius of single-walled CNTs (SWNTs) can be 0.5–1.5 nm having inner radius of 0.2–1.2 nm (Fig. 10.2). The radius of multiwalled CNTs (MWNTs) can vary from ~1 nm (double-walled nanotubes) up to ~50 nm with tens of walls. Three types of geometrical shapes are shown by CNTs (Fig. 10.3).

Recently, electrochemically active CNT membrane filters are extensively used for desalination and purification of aqueous streams [29,30]. The CNT-based filters were used earlier for separation of organic compound [31], proteins [32], virus [33], azo dyes [34], pharmaceuticals [35], fluorine-based chemicals [36], and mono- and multisubstituted phenolic compounds [37]. Electrochemically active CNT membrane can reduce the fouling of filtration medium by biological inactivation [31,38]. It can ensure optimum permeability of the membrane [39]. The modified ultrafiltration (UF) membrane containing CNT poly-vinylidene fluoride-based cathode can reduce organic fouling to a greater extent [40]. Even it has demonstrated its high potential for reducing energy consumption up to twofold than the unmodified ultrafiltration

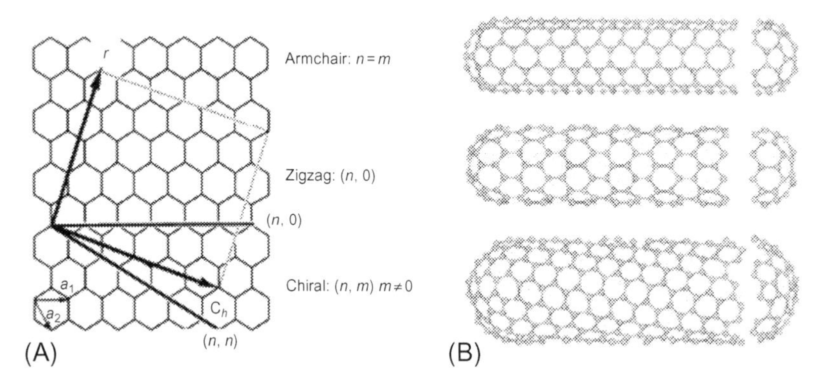

Fig. 10.2 Graphical representation of (A) formation of SWCNTs by rolling up of a graphene sheet along lattice vector, (B) structural orientation of armchair, zigzag, and chiral tubes [28].

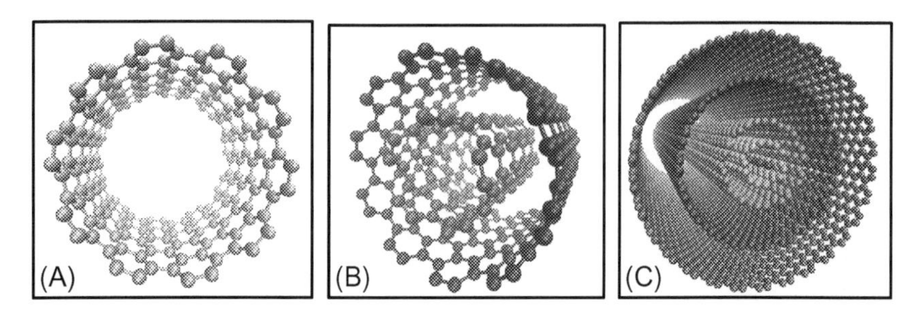

Fig. 10.3 Different types of CNTs: (A) single walled, (B) double walled, and (C) multiwalled CNTs [28].

membrane (UUF). The CNT-based electrode is more efficient than glass electrode due to its less overpotential and can be used effectively for micro-pollutant sensors [41]. However, the CNTs have their challenges such as high cost of fabrication together with problems associated with vertical alignment of CNT arrays with very high density thereby limiting its usage to theoretical studies only [42].

In spite of using carbonaceous nanomaterials for desalination and purification techniques, several challenges still exist. The major focus of this chapter is to provide a comprehensive knowledge about the utilization of CNT-based separation process used for desalination as well as purification of specific contaminants from waste streams. The chapter highlighted the advantages and disadvantages of different types of membrane technologies. Based on this perspective, the feasibility of incorporating CNT-based materials in membrane technology, the existing obstacles, and imminent challenges including its toxicological effects are reviewed. A brief summary of existing literature on CNT filters will help to develop an energy-efficient and cost-effective technique for water treatment technologies.

10.2 Filtration mechanism for desalination and purification

10.2.1 Mechanism for desalination

Desalination process efficiently separates the salt ions from water using different mechanism and can be classified as follows.

10.2.1.1 Mechanisms involving phase change

To separate a mixture of salt and water would normally involve a phase change from liquid to either a vapor or ice stage. This has been a popular mechanism used in desalination. While multi-effect distillation (MED) involves evaporation of brine that may be available on the surfaces of pipes supplying hot air. Multistage flash distillation (MSF), on the other hand, do take place through flashing of salt-containing water in a low-pressure chamber thereby leading to evaporation. Distillation processes such as humidification or dehumidification, solar distillation, and/or membrane distillation all make use of evaporation technique to operate [43]. Another popular technique usually adopted in desalination is freezing which is considered to be more advantageous than vaporization. Some of its advantages include lower energy consumption and less scaling problems [44]. However, it has its limitation in the plant design which is considered complex. There is possible risk of secondary contamination which may occur during the process of freezing [45,46].

10.2.1.2 Mechanisms relating short-distance ($<\sim 1$ nm) interactions

Another method that can be used for desalination is the short-distance interactions between H_2O molecules or ions and other materials. This may include a combination of two or more of the following interactions: dispersion, steric hindrance, electrostatic, and dipole. During these interactions, desalination is allowed to occur through preferential transportation of the H_2O molecules, anions, or cations through the selected substrate or by absorption or adsorption of ions onto the chelating substances. The trace metallic cations can be eliminated by chelation or adsorption methods [47,48] while transportation through selective materials is more suitable for desalination. Selective materials can either be the flexible type or rigid type. The sorption process starts with adsorption and then followed by absorption of the molecules [47,48].

10.2.1.3 Mechanisms relating distant ($>\sim 1$ nm) electrostatic interfaces

The use of distant electrostatic fields arises due to availability of positive or negative charges on ions and nonavailability of net charge on water which enables the fields to utilize the ions selectively. Electrical fields have the ability to deplete, accumulate, or even transport ions. The subject that deals with this study is known as electrokinetics [49]. Thus, it is possible for the surfaces with charge to attract ions with opposite charges but repel ions with same charges. This phenomenon results in the formation of electric double layer (EDL). The EDL thickness depends on the concentration of the salt which usually ranges from 1 μm to 1 nm for deionized water and seawater, respectively. It is therefore possible to be adsorbed preferentially on the surfaces with specific charges or transported selectively through the pores of the membrane due to long-range electrostatic interactions.

Overall the classification of desalination process is illustrated in Fig. 10.4.

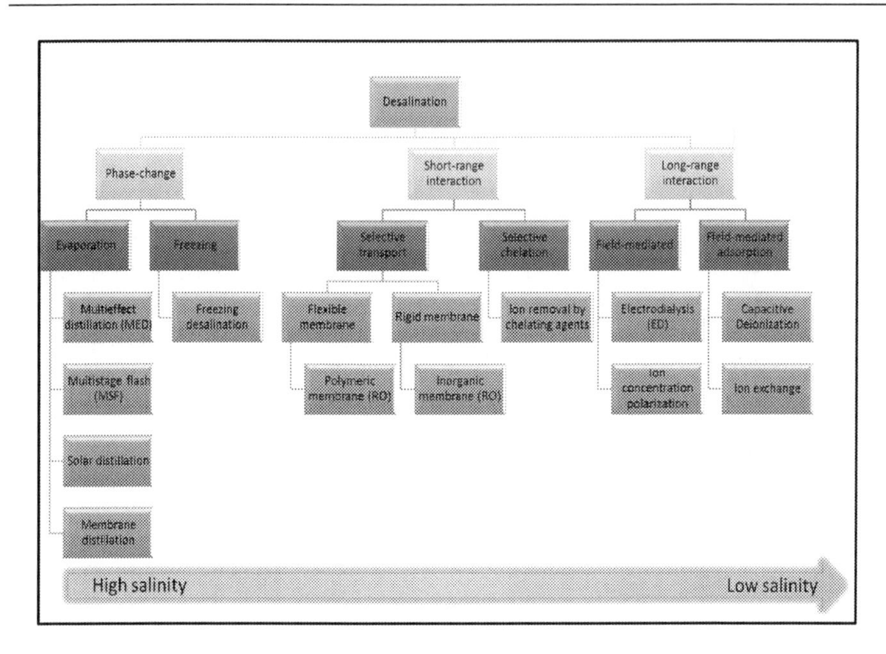

Fig. 10.4 Classification of desalination techniques.

10.2.2 Mechanism for purification

10.2.2.1 Electrochemically activated electrode

The filter bed containing CNT as purification medium has at least tens of hundreds of different tubes that are entangled with each other through van der Waals forces [50]. Thus, the resulting surface area becomes big and can vary from 30 to 500 m^2/g that can be utilized for chemical and biological contaminants. The configuration of a representative electrochemical CNT filter is shown in Fig. 10.5 [34]. Previously, a commercial polycarbonate-based filtration medium was prepared by using stainless steel or titanium cathode and a titanium anode ring inside which multiwalled nanotube (MWNT) anode can be used for electrooxidation [34]. Nevertheless, electrooxidation of organic contaminants using CNT-based filter medium is less time consuming than the conventional biological wastewater treatment systems [51]. Even some contaminants which are difficult to be adsorbed can be oxidized electrochemically easily [52]. The CNT-based electrochemically active filtration medium has following three advantages [34]:

1. the mass transfer is increased hydrodynamically,
2. ensure physical or chemical sorption/desorption which is temperature dependent, and
3. transfer of electron which is dependent on overall voltage.

Numerous chemical contaminants are removed by CNT-based filtration medium and the research showed that the breakthrough point reached within few hours using column-based adsorption techniques [53–55].

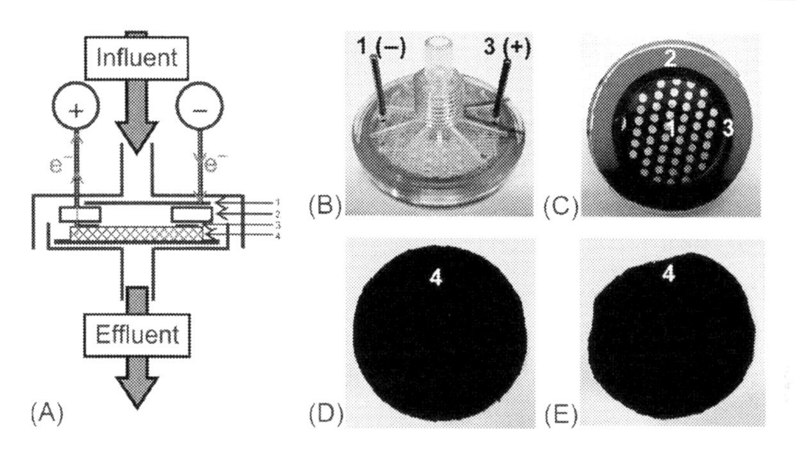

Fig. 10.5 (A) Basic layout of electrochemically active CNT filter: (1) a perforated stainless steel cathode, (2) an insulating silicone rubber separator and seal, (3) a titanium anodic ring that is pressed into the CNT anodic filter, and (4) the MWNT anodic filter supported by a PTFE membrane. (B) and (C) modified filter casing, (D) MWCNT filter bed before filtration, and (E) MWCNT filter bed after electrochemical filtration, respectively [34].

10.2.2.2 Electrochemically active Fenton process

Previously, CNT was used in electro-Fenton system where H_2O_2 was used as Fenton agent to improve the treatment efficiency [56]. Overall, the system contains two cathodic medium: one is purely made up of CNT cathode and the other layer of CNT-COOFe^{n+} cathode. A PTFE separator was used after the cathode layer to separate it from CNT anode. The schematic representation of this process illustrated earlier in literature using Fig. 10.5 [56].

In this system, first reduction of oxygen will produce hydrogen per oxide at cathode which will later on react with Fe^{2+} to produce H_2O_2 first, and H_2O_2 further reacts with Fe^{2+} to produce hydroxyl radical (OH·). These radicles will further oxidize the contaminants according to the following equations:

$$H_2O_2 + Fe^{2+} + H^+ \rightarrow Fe^{3+} + H_2O + OH^{-1} \tag{10.1}$$

$$Fe^{3+} + e \rightarrow Fe^{2+} \tag{10.2}$$

The reaction will generate Fe^{3+} in the cathode and Fe^{2+} is reused in the subsequent reactions. The residual intermediates are oxidized in the CNT anode. In this method, H_2O_2 is produced on the spot, the expenditure and hazard related to handling of H_2O_2 can be reduced. The system does not require addition of acid or bases as the process is carried out under neutral pH. Due to electro-regeneration of Fe^2, there is no need for the addition of Fe in the reaction medium and thus the process is free from sludge formation (Fig. 10.6).

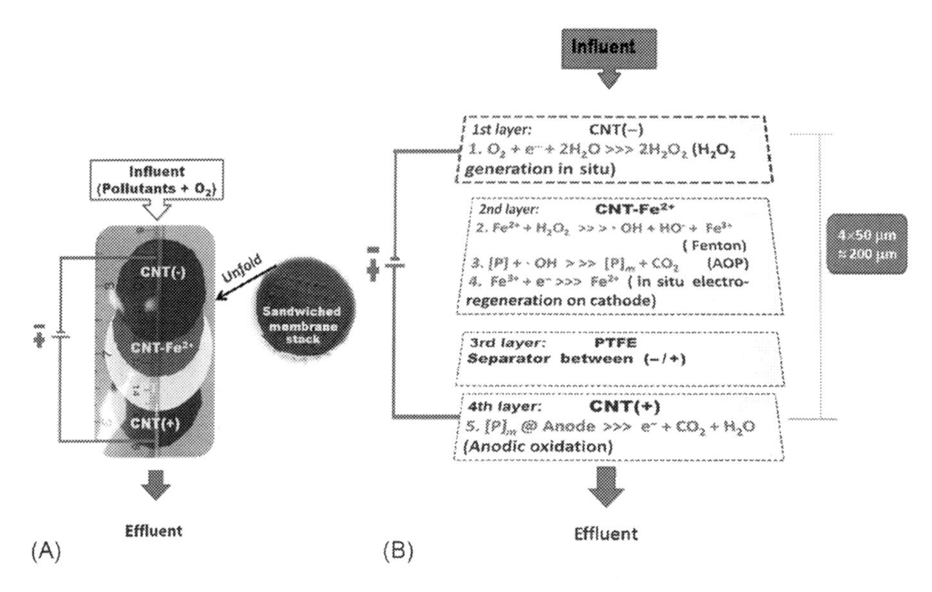

Fig. 10.6 Sandwiched electro-Fenton system based on CNT membrane stacks. (A) Images of the unfolded sandwich membrane stacks including four layers and (B) schematic representation of main roles of every layer in membrane stacks, and [P] and [P]*m* are pollutants and their oxidation intermediates, respectively [56].

10.3 Potential application of CNTs in membrane desalination

The structure of CNTs can be described just as the modified version of a fullerene with single dimension; having a micrometer scale graphene sheet being rolled into a cylinder—the diameter of which is in nanoscale range. The CNTs can be either single walled or multiwalled depending on the methods of synthesis. Synthesis methods and applications of CNTs have been widely reported by researchers [57–61]. The emerging need to further upgrade the desalination technique has pushed this nanostructured material to the versatile edge of uses because of their extraordinary water transport properties.

10.3.1 Mechanism for water and ion transport through CNTs

The CNTs have nano-confined inner pores which enable them to be used as model framework for producing membrane. The membrane system thus obtained can ensure water ion transportation and water ion interactions also [42,62]. The structures of CNTs can closely mimic the biological pores based on their hydrophobicity and size of the pores [63]. The fluid transport through CNT channel is almost similar to water transport through biological membranes [64,65]. There is still a lot to be learned regarding the rapid water transport properties of CNTs. However, some propositions have been made that the smooth nature of their walls with hydrophobic structure contributed to high water flux during the desalination process [65]. This enables frictionless water transport throughout the process [65]. Suggestions have also been made that the shielding effect due to liquid water molecule layer along the walls was the primary reason for the high water transport [66]. To fully utilize CNTs as a structural material, the comprehensive understanding of the flow of fluid in the cylindrical channel is highly desirable. To this end, computer modeling approach has been developed and this has provided engineering approach for using CNTs as transport medium especially in desalination applications [67]. A large number of theoretical approaches to CNTs transportation characteristics have been modeled and reported [64,68,69]. Of these, molecular dynamic modeling has revealed that the magnitude of flux through CNTs is greater than the conventional pores [64,70]. It has been discovered that the electrostatic charge distribution plays a significant role in water and ion intake by CNTs and it has been observed that changing the pattern of the alternating charge on the CNTs channel would favor water intake but prevent ion intake into the nano-channels [71,72]. It has been revealed that the hollow structure of CNTs will allow passage of water molecules rapidly but will hinder the passage of ions [72]. The enhancement of water flux by utilizing CNT-based membrane will certainly reduce the cost of desalination process. The surface curvature of CNTs has also been reported to play an important role in determining their friction coefficient [73]. The curvature regulates the energy with which water molecules interact with CNT wall [73,74]. Earlier membrane desalination was carried out by immobilizing MWCNTs onto the porous hydrophobic PVDF membranes (Fig. 10.7) [75]. The CNTs served

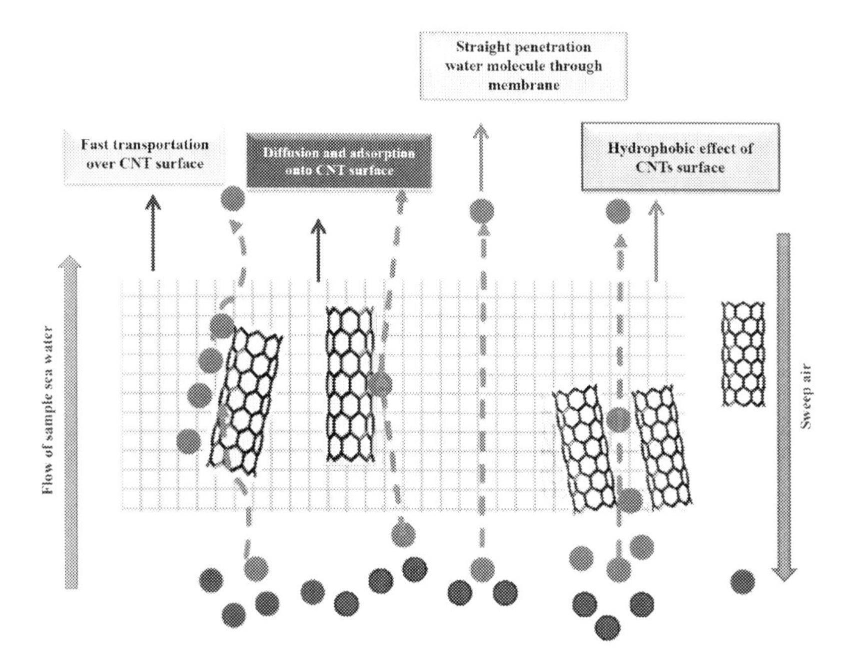

Fig. 10.7 Typical process for CNT-based membrane desalination [75].

as adsorbent to initiate solute transport and it stimulated vapor transportation white retaining liquid water molecule penetration through the pores of the membrane due to its hydrophobic characteristics [75].

10.3.2 Modification of CNTs

Even though CNTs have remarkable properties which make them appealing candidates for numerous form of membrane filtration, their lack of solubility and processability in most common solvents has imposed barriers on their large-scale industrialization for precise applications. The CNT bundles typically form large aggregates due to van der Waals interactions which make them insoluble in common organic solvents and aqueous solutions [76]. The CNTs can be dispersed in solvents through ultrasonication, but precipitation straightaway happens in most instances. This phenomenon interrupts overall desalination process. Surface modification of CNTs by covalent functionalization or noncovalent wrapping and adsorption of various functionalized molecules onto the surfaces of CNTs, both are well-known strategies for easing their dispersion into solution [77].

Various functional groups such as —COOH, —COH, —NH_2, and —OH can be attached onto the surface of CNTs using covalent bonds (Fig. 10.8) [78]. The sidewall functionalization of CNTs can change the hybridization of carbon from sp2 to sp3 with concurrent loss of π-conjugation in the graphene layer. The presence of these functional groups allow CNTs to go for additional chemical reactions, such as silanation, polymer grafting, esterification, thiolation, alkylation, arylation, and addition of other biomolecules onto their surface [77,79,80]. The attachment of these functional groups makes CNTs soluble in many organic solvents. These polar groups alter their normally hydrophobic nature to hydrophilic one. Chemically functionalized CNTs can produce

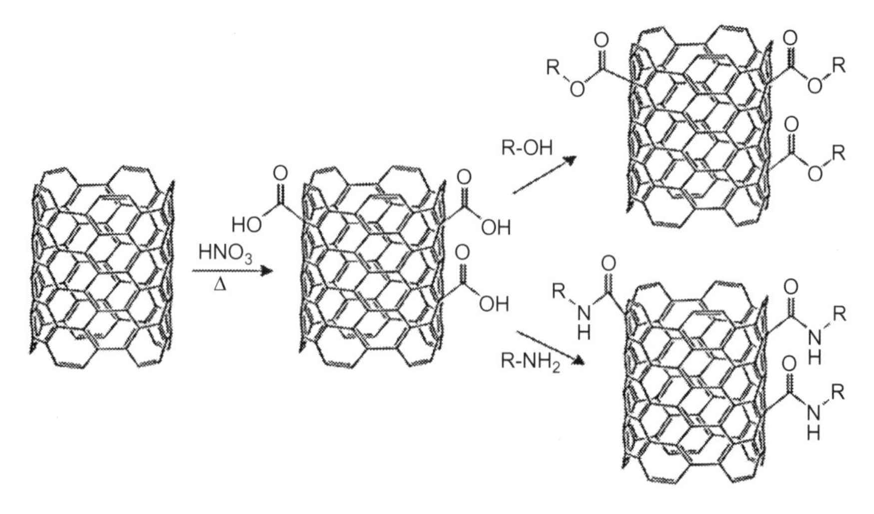

Fig. 10.8 Surface functionalization of CNTs by thermal acidic oxidation and esterification or amidation of the carboxyl groups [78].

robust interfacial bonds with many polymers, permitting the practice of CNT-based nanocomposites that demonstrate notable mechanical properties. A prime disadvantage of this technique is the extensive damage to the sp2 hybridized carbon structure that takes place because of the advent of different functional groups. Therefore, the widespread effort has been dedicated to developing techniques for solubilizing of CNTs that are convenient to apply and cause less damage to their structure.

Noncovalent surface alteration is an alternative scheme for amendment of the interfacial traits of CNTs. This approach is appropriate as it does not compromise the physical characteristics of CNTs, but does mend their solubility and processability. Noncovalent functionalization principally implicates enfolding the external layer of CNTs with polymer, bio-macromolecular or surfactant molecules. The aptitude to disperse CNTs into solution by using polymers such as poly(phenylene vinylene) and polystyrene, was described earlier to be the result of covering the tubes to form supra-molecular complexes [81,82]. The polymer wrapping process includes van der Waals and π-π interactions between the CNTs and polymer chains having aromatic rings.

Recent research findings have revealed that a variety of proteins, comprising bovine serum albumin and lysozyme, are correspondingly proficient of forming regular aqueous dispersions of CNTs [82,83]. The use of protein dispersants is of unique importance due to their privation of toxicity related to surfactant and different representative dispersant molecules collectively with their biocompatibility [84] but proteins incorporate special kinds of reactive functional groups including hydroxyls, carboxylic acids, amines, and thiols, which commendably provide sites for additional surface change of CNTs [84]. The dispersion of CNTs wrapped by using protein molecules encompasses an electrostatic interaction mechanism [83,84]. The technique is distinctly reliant on the charge distribution present alongside the protein and pH of the solution [83,84].

Carbohydrates consisting of chitosan and gelatin gum have also been shown to be surprisingly effective at wrapping themselves around CNTs to facilitate the formation of aqueous dispersions of the latter [85]. Usually, biopolymers are either protonated or deprotonated in aqueous medium. The presence of biopolymers on the surface of CNTs minimizes aggregation of CNTs due to electrostatic repulsion and steric hindrance mechanisms [86]. Chitosan was observed to be very beneficial for setting apart SWCNTs on the basis of differences in their size. It was reported earlier that only the smaller diameter nanotubes will be noncovalently functionalized through the biopolymer [87]. It has been suggested that chitosan chains wrap alongside the nanotube axis as shown schematically in Fig. 10.9.

Gelatin gum is another polysaccharide which can act as fantastic biomolecules for making both SWNTs and MWNTs disperse in aqueous solution. For instance, it has been observed that solutions containing gelatin gum at concentrations as low as 0.0001% (*w*/*v*) have improved the dispersion properties of MWNTs [89]. The DNA has also been proven to be able to disperse CNTs into aqueous solution. This was attributed to the capability of the DNA bases to bind over the nanotubes through π-π interactions. When the polar backbone of DNA molecule is exposed to the solvent, dispersion of CNTs becomes easier.

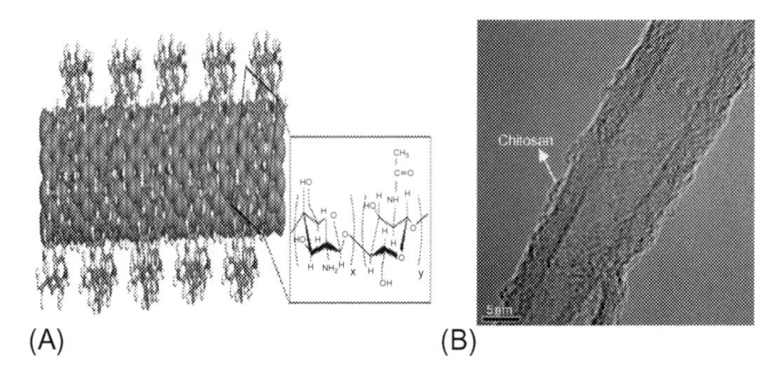

Fig. 10.9 (A) Schematic representation of chitosan molecule enfolding over the CNT surface. (B) TEM image of individual CNT tube wrapped by chitosan [88].

Surfactants are amphiphilic in nature and they have been proven to be highly effective dispersing agents for CNTs [90–92]. The surfactant sodium dodecylsulfate (SDS) has hydrophilic head groups, which can help in the dispersion of CNTs by electrostatic repulsion between micellar domains [93]. Another surfactant, polyoxyethylene octylphenylether (Triton X-100), a commonly used nonionic surfactant attaches itself around the individual nanotubes. Its hydrophilic ends form a large solvation shell around the assembly [94]. The type of interaction between surfactant molecules and CNTs depends on the structure and properties of the surfactant, including its alkyl chain length, bulkiness of head group and charge dominates the attachment of surfactant molecule over the surface of CNTs. That is why Triton X-100 and sodium dodecylbenzene sulfonate (SDBS) display stronger interactions with the surfaces of nanotubes than SDS. This phenomenon is attributed to the presence of the benzene rings in the former surfactants [94].

There is a general need for continuous modification of the characteristics of CNTs for multidisciplinary applications and these have remained a great challenge due to certain inconsistencies [95]. Different strategies have been employed to tackle these problems through chemical modification and functionalization of CNTs to improve their selective and controlled transport. Several reports have been published on various purification processes and effective functionalization of CNTs [96–99] but the findings indicated that there is still room for improvement in the structural modification for the optimization of the process. Technological solutions are also suggested for functionalization of the pore tips to improve the selective characteristics, thereby increasing the surface hydrophilicity. Other form of modifications which include oxidation and chemical modifications using hydrophilic functional groups can be done to enhance the binding energy between the water and the nanotubes which will invariably improve water flux [100,101].

10.4 Potential application of CNTs water purification

Adsorption is an easy and green technique for the elimination of organic and inorganic compounds from aqueous stream. Activated carbons (ACs) are considered as one of the best option for adsorptive removal of pollutants due to its chemical inertness and thermal stability. However, regeneration and recycling of ACs are difficult. Furthermore, the sorption kinetics is slow when ACs are used for removal process. Activated carbon fibers (ACFs) had been developed and is considered as the second-generation carbon-based sorbents [102]. The diffusion distance of pollutants to the adsorption sites of ACFs is relatively smaller as the micropores of ACFs are located over the external surface of the carbon [103]. Consequently, ACFs characteristically exhibit better sorption kinetics than the powdered or granular activated carbons (GACs). The CNTs can be considered as the contracted version of ACFs [4]. The CNTs are hollow tubes having versatile surface chemistry which can be modified based on end application. Hypothetically, CNTs are basically the advanced generation of carbonaceous adsorbents. The outstanding capacity of CNTs to work as adsorptive medium is ascribed to its "structural and functional properties." Furthermore, it has microbial cytotoxicity that has fractional impact on concentrating biological contaminants for removal process.

10.4.1 Adsorptive removal of organic contaminants by CNTs

During adsorption of organic compound, several chemical forces act concurrently. These include electrostatic interaction, π-π interaction, hydrophilic-hydrophobic effect, π-π electron-donor-acceptor (EDA) effect, and hydrogen bonding. The outer surface of CNTs is hydrophobic which make them bind the nonpolar organic compounds [104,105]. The π electrons on the surface of CNTs enable to go for π-π coupling reactions with benzene ring. Previous studies concluded that adsorption of polar and nonpolar compounds onto the surface of CNTs do not take place by hydrophobic effect [106]. Same study showed that the adsorption of $C_6H_5NO_2$ was stronger than C_6H_6, $C_6H_5CH_3$, and C_6H_5Cl, despite its less hydrophobic nature. In those cases, π-π EDA interaction between the nitrobenzene ring (π acceptors) and the graphene sheets (π donors) of CNTs played the vital role for adsorption. The curving and chirality of CNTs is anticipated to have an exquisite impact on the sorption of organic molecules. The nanoscale curvature effect was predominant for differences in sorption percentages of tetracene and phenanthrene onto the surface of CNTs [107]. Due to the variation in chiral angle of CNTs, the sorption capacities of benzene varied significantly [108].

The aggregation tendency of CNTs reduces the nano-curvature of their wall. Usually, the aggregation of CNTs follows the order: SWCNTs > double-walled CNTs (DWCNTs) > MWCNTs [102]. The SWCNTs typically exist as packs of wires while MWCNTs are arbitrarily entwined as isolated tubes as illustrated in Fig. 10.9 [109]. Due to aggregation, the external surface is decreased which might lead to the development of different sorption sites inside the CNTs channel (Fig. 10.10) [109]. Thus, adsorption of some synthetic organic compounds (SOCs) become difficult on aggregated CNTs. Earlier researchers have proposed some methods to calculate the reduction in pore volume and surface area for aggregation. Ultrasonication considerably enhances the dispersion of CNTs resulting in better sorption kinetics of SOCs onto the surface of CNTs [110].

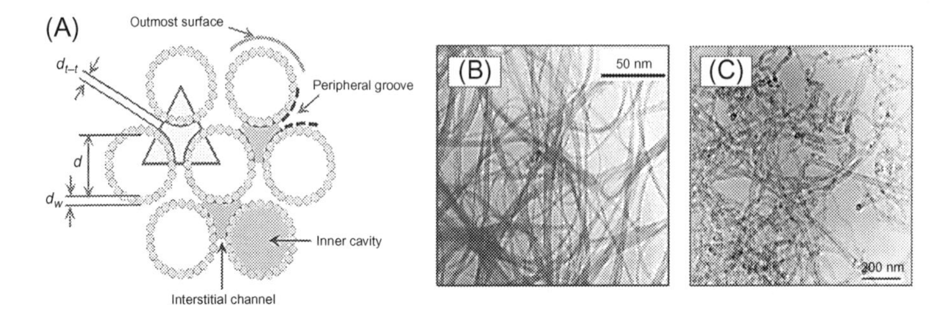

Fig. 10.10 (A) Schematic illustration of SWCNTs bundle. (B) TEM image of SWCNTs. (C) TEM images of MWCNTs.

Modification of CNTs surface by attaching some specific functional groups can make it hydrophilic resulting in greater removal efficiencies of low-molecular weight and polar organic compounds such as phenol [103] and 1,2-dichlorobenzene [111]. It was reported earlier that increasing the number of oxygen containing functional groups will reduce the adsorption of relatively nonpolar chemicals like naphthalene [112].

10.4.2 Catalytic photodegradation of organic compounds using CNTs

Photocatalysis has been an interesting issue for the degradation of organic contaminants for quite a few years [113]. Conventional photocatalysts incorporate some semiconductors oxide such as TiO_2, CdS, Fe_2O_3, ZnO, etc. Application of these semiconductors has some adverse effects. Due to vast band gap in TiO_2, it has to be intensively energized by UV light. Therefore, it cannot be viably used under daylight. CdS and ZnO hold the disadvantage of photo-corrosion. Overall, it can decline the photoactivity and stability. The electrons and holes generated during photocatalysis have tendency to recombine rapidly without initiating the photocatalytic activities.

Attributable to their outstanding physiochemical properties, CNTs can fill in as a perfect building block for composites. It can improve the photocatalytic properties of the resultant composites. The CNTs can substantially store electrons. It was observed that 32 carbon molecules in SWCNTs can supply one electron [114]. Once it is incorporated with TiO_2, it can incite electron exchange from the conduction band of TiO_2 to the surface of CNTs because of their lower Fermi level [114]. Hence, CNTs can collect photo-generated electrons and restrain the recombination of charges. These stored electrons, later on, can be exchanged with another electron acceptor, for example, subatomic oxygen. Subsequently, this can form different oxygen species (O_2, H_2O_2, and OH) which initiates photo-degradation and additionally minimize the organic pollutants in aqueous stream.

The CNTs are coated with thin and uniform TiO_2 nanoparticles which results in carbon diffusion into oxide phase. The light adsorption capacity under visible region can be enhanced by carbon doping. The presence of carbon can produce a medium band gap adjacent to the valance band of the TiO_2 particles (Fig. 10.11) [115,116]. The CNTs can be used as photosensitizers and introduce the photoexcited electrons to the conducting band of TiO_2 due to their semiconductor property [117]. The photocatalytic oxidation activity of phenol was enhanced by CNT/TiO_2 composites. The SWCNTs can improve the photocatalytic properties of TiO_2 more than MWCNTs as the surface of SWCNTs can ensure more contact with TiO_2 nanoparticle [118]. Organic chemicals such as dyes [119], aromatic compounds [120], and carbamazepine [121] can be competently degraded by CNT-TiO_2 composites.

Incorporation with metal can significantly improve the performance of CNTs as photocatalyst. It was revealed that the stacking of Ag to CNTs clearly upgraded the photocatalytic properties of CNTs [122]. The Ag/CNT composite displayed enhanced photocatalytic properties for RhB and the mechanism is illustrated in Fig. 10.12 [122]. At initial stage, RhB molecules are adsorbed onto the surface of CNTs. After that, due to the presence of visible light, RhB molecules were excited and the electrons thus engendered are transported along CNT surface. These electrons were confined by Ag particles. Those confined electrons help in the reduction of adsorbed oxygen to form superoxide anion radicals. This further initiates the degradation of RhB.

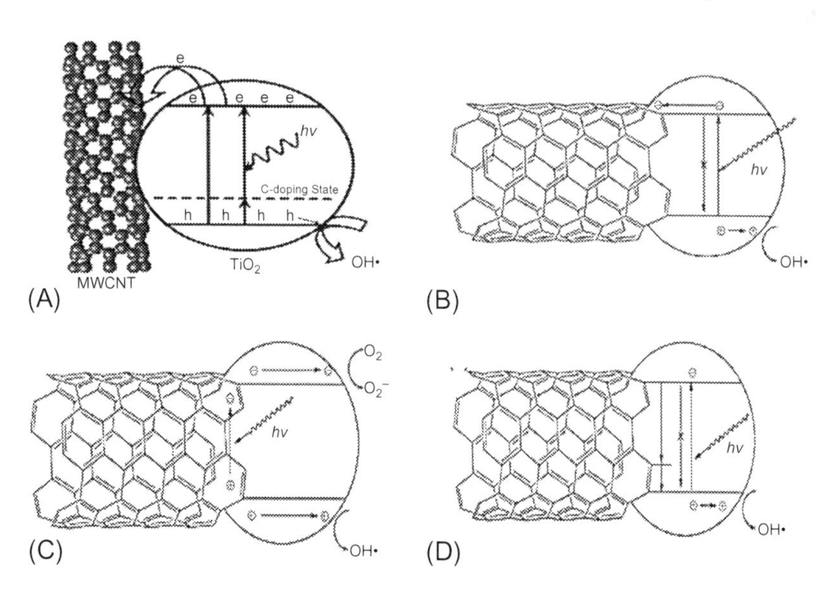

Fig. 10.11 (A) Mechanism for overall photocatalytic degradation using CNTs. (B) Photo-initiated electrons by TiO_2 will be stored by CNTs which in turn prevent charge recombination. (C) Generation of electron-hole pair in CNTs. (D) The role of CNTs as dopant by the formation of Ti—O—C bonds [102,115,116].

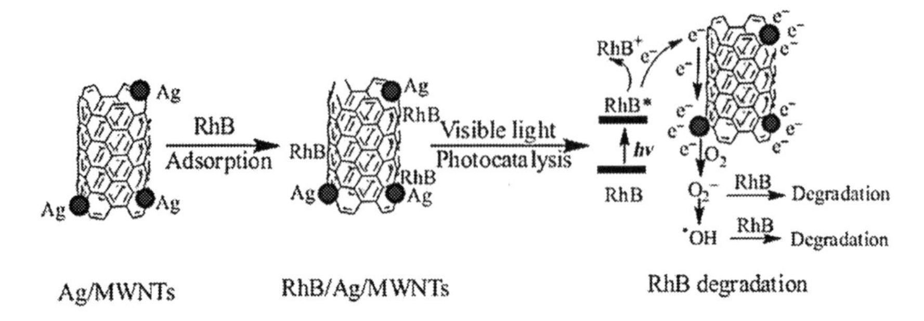

Fig. 10.12 Photocatalytic degradation of RhB using Ag/CNTs [122].

10.4.3 Microbial decontamination using CNTs

Elimination of bacteria and other microbes from drinking water is essential. Both SWCNTs and MWCNTs have shown strong antimicrobial activities. The SWCNT-based filters showed high bacterial retention and MWCNTs has removed virus even at low pressure [123,124]. Recently, hybrid filter incorporating both types of SWCNTs and MWCNTs were used for the inactivation of bacterial and viral retention using low pressure [125]. The transportation of viral particles was enhanced by applying external electric field [33]. Silver nanoparticles and nanowires were incorporated to achieve enhanced antibacterial activity [33,126]. Recently, membrane with aligned CNTs has been fabricated by spray pyrolysis process which has uniform nanoporous texture [127]. This makes it favorable for filtration process [127]. It has been used for the elimination of *Escherichia coli* bacteria from aqueous effluents successfully [127]. Filters fabricated with CNTs have several advantages than the conventional membrane. Ultrasonication and autoclaving are used for cleaning CNT-based membranes. After regeneration, the filter can be used for subsequent cycles showing full filtering efficiency [127].

Microbial fuel cell (MFC) using CNTs as electrodes uses self-sustained electricity using microorganisms for wastewater treatment. Until now the anode of MFC was produced from carbon cloth, carbon paper, and carbon foam. But these carbonaceous materials have less electro-catalytic properties for microbial reaction at electrode. The CNTs can increase the surface area of electrode which consequently can facilitate the charge transfer process. Due to cellular toxicities of CNTs, it is recommended to use it by coating it with conductive polymers. Some commonly used conductive polymers are polyaniline and polypyrrole [128]. Some macroscale porous substrate such as textile or sponge can be coated by CNTs to produce anode for MFCs [129,130]. The anode constructed in this way can exhibit decreased charge transfer resistance and improves the performance of MFCs as illustrated in Fig. 10.13 [129]. The electrons are efficiently accepted by oxygen at the cathode surface of MFCs. However, the oxygen reduction reactions under common operation conditions are hindered sometimes. The presence of some bacteria in the cathode will catalyze oxygen reduction reactions. The CNTs can interact with the redox proteins of the bacteria [131]. Accordingly, the cathode using CNT will enable the electron transferal easily and increase the reduction of oxygen simultaneously [131].

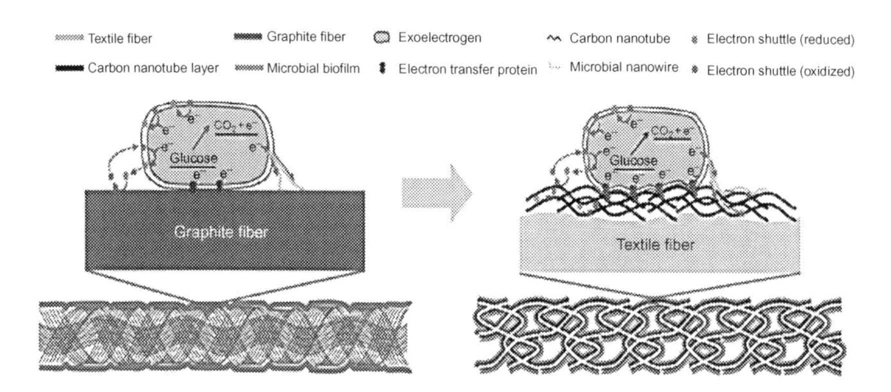

Fig. 10.13 Schematic illustration of the electrode material and electron-transfer mechanisms for the CNT-textile anode and carbon cloth anode [129].

10.4.4 Oil-water

Disposal of oil-water emulsion generated from oil refineries, sewage waste and crude oil production can affect aquatic environment and cause severe harm to human health. To address this problem, microfiltration (MF), UF, and nanofiltration (NF) have been extensively used with discrete benefits. These processes can ensure high-quality purified water with low energy consumption [132–134]. However, these techniques are dependent on size exclusion separation of oily droplets. The CNT-based composite membrane was earlier fabricated by inserting CNTs inside the porous channels of an Y_2O_3-ZrO_2 (YSZ) membrane [135]. The presence of CNTs inside the YSZ membranes exhibited 100% rejection rate of oil particles with water flux of 0.6 L/m^2/min having pressure drop of only 1 bar during 3 days of operation. The lipophilic soft layers produced on CNTs could ensure adsorption as well as size exclusion separation as compared to size exclusion separation for YSZ membranes. Incorporation of CNTs provide improved mechanical properties such as tensile strength, Young's modulus, and toughness to the membrane and make it durable for oil-water treatment [136]. Moreover, a superhydrophobic CNTs-polystyrene composite membrane was fabricated for emulsified oil-water separation which could effectively separate an inclusive array of surfactant stabilized water-in-oil emulsions [136]. The membrane showed high rejection efficiency ($>$99.94%) for oil droplets with water flux of 5000 L/m^2/h/bar [136] (Fig. 10.14).

10.4.5 Heavy metal

Careless disposal of waste effluents from many industries such as mining operation, metal plating facilities, chemical manufacturing, and battery industries has toxic cations [137]. Adsorptive removal of these divalent cations using CNTs-based filter medium have gained attraction recently [138,139]. A recent literature has demonstrated almost 100% removal of metallic cations using CNT-based composite membrane [97]. Correspondingly, another type of membrane was fabricated for Ni^{2+} ions removal from water where CNTs were incorporated inside the pore channels of

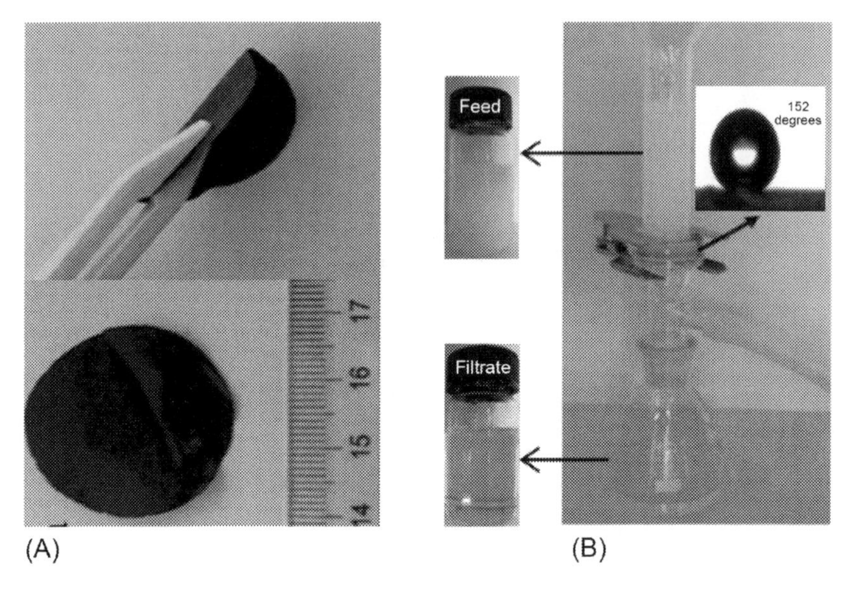

Fig. 10.14 Oil-water separation for superhydrophobic CNT-polystyrene composite membranes. (A) Composite membrane. (B) Oil emulsion separation [136].

ceramic membrane using chemical vapor deposition (CVD) technique [140]. The sorption process followed well with Langmuir and Freundlich isotherm models. However, ceramic-based CNT membrane cannot be used for continuous operation because the efficiency of the membrane was low due to the larger pore diameter of ceramic [140]. Previously, phase inversion method was used to fabricate CNTs-PSF composite membranes [141]. The membrane showed excellent performance due to its smaller pore size of 20–30 nm and was successfully used for continuous filtration process [141]. The composite membrane has even efficiently removed 94.2% of Cr(VI) and 78.2% of Cd^{2+} whereas the PSF membrane alone could eliminate only 10.2% of Cr(VI) and 9.9% of Cd^{2+} for the bare PSF membranes, respectively [141].

10.5 Factors affecting electrochemical purification of water using CNTs

10.5.1 Electrochemical properties of cathode

Different types of cathodic materials were used earlier for electrochemical systems, such as graphite felt [142], carbon felt [143], and MWNT [144]. The oxidation rate was significantly different for different carbonaceous materials. Simply 50% of ferrocyanide was oxidized at 0.3 V anode potential using Ti cathode, whereas >80% of the ferrocyanide was oxidized under the identical conditions using the CNT cathode [142].

10.5.2 Electrochemical properties of anode

The potential of anode plays a vital role for electrooxidation of organic pollutant using the CNT filters. The potential of anode should be higher than the redox potential to facilitate the oxidation process. Nevertheless, the phenomenon of water splitting should be taken under consideration by applying high anode potential. In that case, there is an uninterrupted competition between the oxidation of the organic contaminants and evolution of oxygen species as illustrated by following equation:

$$1/2H_2O \rightarrow 1/4O_2 + H^+ + E^- \quad E_0 = +1.03\,V\,vs\,Ag/AgCl \tag{10.3}$$

Overpotential of anode may cause degradation of CNTs [145] and confines the uses of CNTs for electrochemical treatment of waste stream. When the applied potential was 3 V using electrochemical CNT filtration, the MS-2 virus having concentration of 106 m/L was completely eliminated and it could not be detected in the effluent [144]. Previous literature suggested that higher anode potentials caused higher anoxic oxidation when total applied potential was below 3 V. This trend was further confirmed for electrooxidation of ferroxyanide where rate of anoxic oxidation increased under higher anode potentials [146,147]. Current densities were found to increase with anode potentials [148,149]. The CNT reactive sites are also affected by anode potential. The sp2-conjugated sidewall CNT sites govern the electrochemical filtration process when the anode potentials are low (≤ 0.2 V) [148,149]. Even the CNT tips were also electro-active when anode potentials are high (≥ 0.3 V) [148,149]. Recently, electrooxidative CNT filter has been developed which can effectively define the correlation among target species, adsorption, mass transport, and electron transfer with subsequent product desorption [50]. Recently, binder-free bismuth doped conductive three-dimensional CNT network was fabricated and it was uniformly coated with tin oxide (BTO) nanoparticles [149]. The BTO-coated CNT filters showed better stability of anode and efficiency for organic electrooxidation [149]. The BTO-coated CNT filters showed overpotential of 1.71 V which was 0.44 V higher than that for uncoated CNT filters [149]. Even electroxidation using BTO-coated CNT filter for other compounds such as ethanol, methanol, formaldehyde, and formate were also very successful and demonstrated four to five times less energy consumption than those of the uncoated CNT anode [149].

10.5.3 Flow rate

The reaction kinetics for electro oxidation using CNT filters are greatly affected by flow rate. The rate of ferroxyanide electrooxidation and flow rate was proportional provided the flow rate was up to 4 mL/min and anode potential was 0.3 V [146]. Current efficiency was observed to increase slightly with increasing flow rate from 0.5 to 4.0 mL/min and reached a plateau at the higher flow rates [146]. This was expected due to increased mass transfer to the surface of anode with increasing flow rates [146].

10.6 Advantages of using CNTs in membrane

The major disadvantage of polymeric membrane is its fouling tendencies which results from the interaction between the surface of the membrane and the foulants. Modification of the membrane surface by enhancing the hydrophilicity of the surface can resist fouling. The CNTs are itself hydrophobic, but its chemistry can be altered to hydrophilic one by oxidation using acid. Blending of CNTs with some polymers like polysulfone [150] and polyethersulfone [151] can make it more resistant to fouling. This is due to the presence of —COOH groups on the surface of CNTs. The fouling can be also resisted using functional groups like hydrophilic isophthaloyl chloride groups [152] and amphilic polymer groups with protein-resistant ability [152] onto the surface of CNTs. Even biofouling, that is, the growth of biofilm onto the surface of membrane can be prevented by integration of CNTs in membrane [153]. Due to antibacterial properties of CNTs, it can impart biocidal properties to the polymeric membrane [153]. Ag nanoparticles incorporated with CNTs can ensure more direct contact with bacteria which in turn shows more effective antibacterial activity and is illustrated in Fig. 10.15 [154].

Incorporation of CNTs provides better mechanical strength to the polymer membrane. Adding 2wt% of CNTs in polyacrylonitrile membranes [155] and chitosan-based membrane [156] can provide increased tensile strength of 97% and 90%, respectively. It can also enhance the permeability of water through the polymer membrane. High flux filtration can be achieved using surface-oxidized MWCNTs [157]. Increasing the amount of MWCNT could increase the water flux as well as maintain the rejection of net organic content. The rate of flux was improved by nano-channels of CNTs for passage of water inside the composite membranes [155]. The presence of ionic groups in seawater cause scaling of membrane during desalination process. This happens due to the precipitation of sparingly dissolvable salts. The cytotoxic nature of CNTs can forbid the microorganisms development [158].

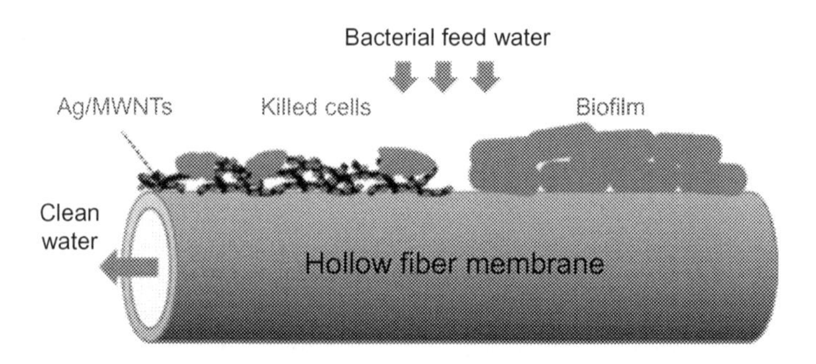

Fig. 10.15 Polyacrylonitrile fiber coated by CNTs/Ag showing antibacterial activity [154].

10.7 Current status and challenges of using CNTs

The distinctive and adjustable morphological, chemical, and electrical characteristics of hollow CNTs made them suitable for producing unique and high-performance materials for water desalination and treatment. To obtain synergistic effect, CNTs can be incorporated with other nanomaterials for water treatment. The limiting factor for using CNT is its high cost. However, contemporary advances have confirmed that it is promising to fabricate high-quality CNTs at low prices. Catalytic CVD method using fluidized bed reactor can yield 595 kg/h of CNT [159]. The large-scale manufacturing may reduce the price of CNTs up to a greater scale. It was observed that CNT-based filters can ensure high magnitude of water flux while maintaining the low pressure for the process [80]. The tendency to form biofilm can be reduced due to its cytotoxicity. The CNT-based filters are more easily regenerated than GAC-based filters. Even CNT filters have shown their ability to remove large range of organic, inorganic pollutants as well as bacteria and viruses. Consequently, they are used as disinfectant also. However, filters carrying CNTs should be used carefully before its large-scale utilizations in terms of the leakage of CNTs into the drinking water.

The CNT membranes comprising carbonaceous nanomaterials represent an emerging part of membrane separation process in the case of filtration and seawater desalination. Basically, CNT-based membranes are easier to regenerate using sonication and autoclaving. Due to their excellent electrical conductivity, the flux regeneration properties can be improved using principles of electrochemistry. The CNT pores can ensure high water flux as well as high permeability and water rejection. The cumulative effects of these advantageous properties are suitable for the preparation of membrane for desalination process. For electrochemically activated CNTs in filter, when the voltage applied is >2 V, and then hydrogen is produced at cathode continuously and oxygen at anode. Formation of bubbles blocks the pores of CNTs, increases back pressure, lessens the mass transfer efficiencies, and subsequently the performance of the filter drops down. Agglomeration tendencies of CNTs are other major drawbacks when the applied potential is higher than 2.5 V [29]. The electrochemical process is hindered if the overall filtration process is carried out for 3 h due to the accumulation of organic polymer; inorganic sodium per sulfate deposition on anodic plates [29] increases the resistance of electron transfer. The specific configuration of CNTs with dead-end system may limit the overall water flux. The filters using CNTs are fragile and can be fractured. After partial oxidation some toxic metabolites might be produced which are difficult to treat as their chemical properties are still not well understood.

Until recently more effort has been given for CNT-supported catalysts, whereby the catalytic activity of those types of catalyst has been monitored using some model pollutants, such as phenol and dyes. Nevertheless, greater emphasis should be given for the degradation of more recalcitrant organic pollutants using CNT-based composite materials. The catalytic activity to produce degraded chemicals should be carefully analyzed as some intermediate products formed during the process might be more hazardous. In that case additional methods should be carried out for further degradation of those intermediates.

Until now many obstacles still exist and make the overall process challenging. For synthesis of desalination membrane, the capacity for salt rejection should be >95% and it requires <1 pore in 100 having diameter > 1 nm [108]. Another obstacle is membrane fouling by algae and other contaminants' accumulation during water treatment as well as desalination.

The perusal of literature clearly reflects that CNT-based nanomaterials have numerous advantages over conventional materials for desalination and purification after necessary modification. To overcome the challenges, expansion of cost-effective technological methods may show the proper way to integrate CNTs into the composite matrix for membrane fabrication. Surface modification with definite functional groups and manipulation of pore size distribution of CNTs are perhaps the appropriate ways to entirely take the advantages of unique physical, chemical, and electrical properties of CNTs. Although the expedition for flux enhancement has controlled the investigation and progress in the field of CNT-based membranes, these constituents have other favorable characteristics that could further enhance the overall efficiency of the process. It is estimated that 20%–30% of operating cost for seawater desalination comes from fouling of membrane [2]. Due to antibacterial properties, it can prevent the growth of biofilm. Simple electrochemical approaches can be made to remove fouling from CNT-based membrane [24].

With the approaching era of nano, CNT-based membrane or filtration medium will be widely used in future but the process still suffer from lot of disadvantages. The electrochemical CNT filters are still expensive to fabricate especially CNT sheets having larger sizes with porous texture. The cost for synthesizing CNTs are still comparatively high than the other commonly used water treatment materials, such as AC.

10.8 Conclusion

Currently, polymeric or inorganic membrane separation procedures have achieved an edge at which it is difficult to enhance the separation performance just by means of further optimization of the fabrication conditions of the membrane itself. The blend of nanotechnology and membrane separation offers new ways to deal with the challenge. The presence of inorganic functional groups on the surface, sidewalls, or tips of CNTs improves the dispersion and performance of CNT-based composite membrane during desalination and purification. The CVD method is quite efficient to fabricate inorganic (ceramic)/CNT-based composite membranes. Moreover, the blending method is relatively flexible for polymeric/CNT-based membranes. The CNT-based composite membranes have better performance compared to the polymeric and inorganic membranes for desalination and purification process. Nevertheless, penetrating research and development on CNT-based membranes endures with substantial industrial involvement.

The CNT membranes can ensure high flux which makes it a promising candidate to compete with the existing commercial NF and RO polymeric systems, if new economical fabrication methods can be developed. Moreover, the applications of CNT-based composite membranes should be monitored carefully due to its potential toxicity effects. Thus, some studies emphasizing the adherence between CNTs and the matrix should be performed in the near future. The membrane separation process coupled with other techniques such as adsorption, catalysis, and electrochemistry should be studied elaborately for the comprehensive understanding of the mechanisms using CNT-based composite membranes. Therefore, more investigations are needed in terms of water chemistry, process parameters for membrane separation process, physiochemical properties of CNTs as well as the function of composite membranes under precise circumstances. There is no hesitation to say that the incorporation of this nanostructured tube in composite membranes has prominent role in desalination and purification of aqueous effluents. It can help in addressing critical water issues to a greater extent in the future.

References

[1] Shannon MA, Bohn PW, Elimelech M, Georgiadis JG, Marinas BJ, Mayes AM. Science and technology for water purification in the coming decades. Nature 2008;452:301–10.
[2] Elimelech M, Phillip WA. The future of seawater desalination: energy, technology, and the environment. Science 2011;333:712–7.
[3] Surwade SP, Smirnov SN, Vlassiouk IV, Unocic RR, Veith GM, Dai S, et al. Water desalination using nano-porous single-layer graphene. Nat Nanotechnol 2015;10:459–64.
[4] Qu XL, Brame J, Li QL, Alvarez PJJ. Nanotechnology for a safe and sustainable water supply: enabling integrated water treatment and reuse. Acc Chem Res 2013;46:834–43.
[5] Hu M, Mi BX. Enabling graphene oxide nano-sheets as water separation membranes. Environ Sci Technol 2013;47:3715–23.
[6] Eugster HP, Harvie CE, Weare JH. Mineral equilibria in a six-component seawater system, Na–K–Mg–Ca–SO4–Cl–H2O, at 25°C. Geochim Cosmochim Acta 1980;44:1335–47.
[7] Oren Y. Capacitive deionization (CDI) for desalination and water treatment—past, present and future (a review). Desalination 2008;228:10–29.
[8] Duke MC, Mee S, Da Costa JCD. Performance of porous inorganic membranes in non-osmotic desalination. Water Res 2007;41:3998–4004.
[9] Gin DL, Noble RD. Designing the next generation of chemical separation membranes. Science 2011;332:674–86.
[10] Liu GP, Jin WQ, Xu NP. Graphene-based membranes. Chem Soc Rev 2015;44:5016–30.
[11] Mahmoud KA, Mansoor B, Mansour A, Khraisheh M. Functional graphene nanosheets: the next generation membranes for water desalination. Desalination 2015;356:208–25.
[12] Lau WJ, Ismail AF. Polymeric nanofiltration membranes for textile dye wastewater treatment: preparation, performance evaluation, transport modelling, and fouling controls—a review. Desalination 2009;245:321–48.
[13] Geise GM, Lee HS, Miller DJ, Freeman BD, Mcgrath JE, Paul DR. Water purification by membranes: the role of polymer science. J Polym Sci Polym Phys 2010;48:1685–718.
[14] Marchetti P, Solomon MFJ, Szekely G, Livingston AG. Molecular separation with organic solvent nanofiltration: a critical review. Chem Rev 2014;114:10735–806.
[15] Van der Bruggen B, Vandecasteele C, Van Gestel T, Doyen W, Leysen R. A review of pressure-driven membrane processes in wastewater treatment and drinking water production. Environ Prog 2003;22:46–56.

[16] Joshi R, Alwarappan S, Yoshimura M, Sahajwalla V, Nishina Y. Graphene oxide: the new membrane material. Appl Mater Today 2015;1:1–12.
[17] Tsuru T. Nano/subnano-tuning of porous ceramic membranes for molecular separation. J Sol-Gel Sci Technol 2008;46:349–61.
[18] Brown AJ, Brunelli NA, Eum K, Rashidi F, Johnson JR, Koros WJ, et al. Interfacial microfluidic processing of metal-organic framework hollow fiber membranes. Science 2014;345:72–5.
[19] Baek Y, Seo DK, Choi JH, Lee B, Kim YH, Park SM, et al. Improvement of vertically aligned carbon nanotube membranes: desalination potential, flux enhancement and scale-up. Desalin Water Treat 2016;57:28133–40.
[20] Joshi RK, Carbone P, Wang FC, Kravets VG, Su Y, Grigorieva IV, et al. Precise and ultrafast molecular sieving through graphene oxide membranes. Science 2014;343:752–4.
[21] Ismail AF, Goh PS, Sanip SM, Aziz M. Transport and separation properties of carbon nanotube-mixed matrix membrane. Sep Purif Technol 2009;70:12–26.
[22] Holt JK, Park HG, Wang YM, Stadermann M, Artyukhin AB, Grigoropoulos CP, et al. Fast mass transport through sub-2-nanometer carbon nanotubes. Science 2006;312:1034–7.
[23] Konduri S, Tong HM, Chempath S, Nair S. Water in single-walled aluminosilicate nanotubes: diffusion and adsorption properties. J Phys Chem C 2008;112:15367–74.
[24] Zhu A, Christofides PD, Cohen Y. On RO membrane and energy costs and associated incentives for future enhancements of membrane permeability. J Membr Sci 2009;344 (1–2):1–5.
[25] Lee KP, Arnot TC, Mattia D. A review of reverse osmosis membrane materials for desalination—development to date and future potential. J Membr Sci 2011;370(1–2):1–22.
[26] Whitby M, Quirke N. Fluid flow in carbon nanotubes and nanopipes. Nat Nanotechnol 2007;2:87–94.
[27] Chan W-F, et al. Zwitterion functionalized carbon nanotube/polyamide nanocomposite membranes for water desalination. ACS Nano 2013;7(6):5308–19.
[28] Saito R, Dresselhaus G, Dresselhaus MS. Physical properties of carbon nanotubes. London: Imperial College Press; 1998.
[29] Liu YB, Liu H, Zhou Z, Wang TR, Ong CN, Vecitis CD. Degradation of the common aqueous antibiotic tetracycline using a carbon nanotube electrochemical filter. Environ Sci Technol 2015;49:7974–80.
[30] Brady-Estévez AS, Schnoor MH, Vecitis CD, Saleh NB, Elimelech M. Multiwalled carbon nanotube filter: improving viral removal at low pressure. Langmuir 2010;26:14975–82.
[31] De Lannoy CF, Jassby D, Gloe K, Gordon AD, Wiesner MR. Aquatic biofouling prevention by electrically charged nanocomposite polymer thin film membranes. Environ SciTechnol 2013;47:2760–8.
[32] Sun X, Wu J, Chen Z, Su X, Hinds BJ. Fouling characteristics and electrochemical recovery of carbon nanotube membranes. Adv Funct Mater 2013;23:1500–6.
[33] Rahaman MS, Vecitis CD, Elimelech M. Electrochemical carbon-nanotube filter performance toward virus removal and inactivation in the presence of natural organic matter. Environ Sci Technol 2012;46(3):1556–64.
[34] Vecitis CD, Gao G, Liu H. Electrochemical carbon nanotube filter for adsorption, desorption, and oxidation of aqueous dyes and anions. J Phys Chem C 2011;115:3621–9.
[35] Li H, Zhang D, Han X, Xing B. Adsorption of antibiotic ciprofloxacin on carbon nanotubes: pH dependence and thermodynamics. Chemosphere 2014;95:150–5.
[36] Deng S, Zhang Q, Nie Y, Wei H, Wang B, Huang J, et al. Sorption mechanisms of perfluorinated compounds on carbon nanotubes. Environ Pollut 2012;168:138–44.

[37] Gao GD, Vecitis CD. Electrochemical carbon nanotube filter oxidative performance as a function of surface chemistry. Environ Sci Technol 2011;45:9726–34.
[38] Ajmani GS, Goodwin D, Marsh K, Fairbrother DH, Schwab KJ, Jacangelo JG, et al. Modification of low pressure membranes with carbon nanotube layers for fouling control. Water Res 2012;46:5645–54.
[39] Boo C, Elimelech M, Hong S. Fouling control in a forward osmosis process integrating seawater desalination and wastewater reclamation. J Membr Sci 2013;444:148–56.
[40] Zhang Q, Vecitis CD. Conductive CNT-PVDF membrane for capacitive organic fouling reduction. J Membr Sci 2014;459:143–56.
[41] Vecitis CD, Schnoor MH, Rahaman MS, Schiffman JD, Elimelech M. Electrochemical multiwalled carbon nanotube filter for viral and bacterial removal and inactivation. Environ Sci Technol 2011;45:3672–9.
[42] Hinds BJ, Chopra N, Rantell T, Andrews R, Gavalas V, Bachas LG. Aligned multiwalled carbon nanotube membranes. Science 2004;303:62–5.
[43] National Research Council (US) Committee on Advancing Desalination Technology and National Academies Press (US). Desalination: a national perspective. vol. XIV. Washington, DC: National Academies Press; 2008. p. 298.
[44] Khawaji AD, Kutubkhanah IK, Wie JM. Advances in seawater desalination technologies. Desalination 2008;221:47–69.
[45] Hahn WJ. Measurements and control in freeze desalination plants. Desalination 1986;59:321–41.
[46] Sonune A, Ghate R. Developments in wastewater treatment methods. Desalination 2004;167:55–63.
[47] Kabay NM, Bryjak S, Schlosser M, Kitis S, Avlonitis Z, Matejka I, et al. Adsorption-membrane filtration (AMF) hybrid process for boron removal from seawater: an overview. Desalination 2008;223:38–48.
[48] Kabay NM, Sarper S, Mithat Y, Arar OZ, Marek B. Removal of boron from seawater by selective ion exchange resins. React Funct Polym 2007;67:1643–50.
[49] Probstein RF. Physicochemical hydrodynamics: an introduction. 2nd ed. vol. XV. Hoboken, NJ: Wiley-Interscience; 2003. p. 400.
[50] Balasubramanian K, Burghard M. Chemically functionalized carbon nanotubes. Small 2005;1:180–92.
[51] Szpyrkowicz L, Kaul S, Neti R. Tannery wastewater treatment by electro-oxidation coupled with a biological process. J Appl Electrochem 2005;35:381–90.
[52] Martinez-Huitle CA, Ferro S. Electrochemical oxidation of organic pollutants for the wastewater treatment: direct and indirect processes. Chem Soc Rev 2006;35:1324–40.
[53] Liu X, Wang M, Zhang S, Pan B. Application potential of carbon nanotubes in water treatment: a review. J Environ Sci 2013;25:1263–80.
[54] Pan B, Xing B. Adsorption mechanisms of organic chemicals on carbon nanotubes. Environ Sci Technol 2008;42:9005–13.
[55] Dichiara AB, Harlander SF, Rogers RE. Fixed bed adsorption of diquat dibromide from aqueous solution using carbon nanotubes. RSC Adv 2015;5:61508–12.
[56] Gao G, Zhang Q, Hao Z, Vecitis CD. Carbon nanotube membranestack for flow-through sequential regenerative electro-Fenton. Environ Sci Technol 2015;49:2375–83.
[57] Chavan R, Desai U, Mhatre P, Chinchole R. A review: carbon nanotubes. Int J Pharm Sci Rev Res 2012;13(1):124–34.
[58] Ying LS, Bin Mohd Salleh MA, Mohamed Yusoff HB, Abdul Rashid SB, Abd Razak JB. Continuous production of carbon nanotubes—a review. J Ind Eng Chem 2011;17 (3):367–76.

[59] Liu Q, Ren W, Liu B, Chen Z-G, Li F, Cong H, et al. Synthesis, purification and opening of short cup-stacked carbon nanotubes. J Nanosci Nanotechnol 2009;9(8):4554–60.
[60] Pillai SK, Ray SS, Moodley M. Purification of multi-walled carbon nanotubes. J Nanosci Nanotechnol 2008;8(12):6187–207.
[61] Ismail AF, Goh PS, Tee JC, Sanip SM, Aziz M. A review of purification techniques for carbon nanotubes. Nano 2008;3:127–43.
[62] Yu M, Funke HH, Falconer JL, Noble RD. Gated ion transport through dense carbon nanotube membranes. J Am Chem Soc 2010;132:8285–90.
[63] Lu D, Li Y, Ravaioli U, Schulten K. Empirical nanotube model for biological applications. J Phys Chem B 2005;109:11461–7.
[64] Hummer G, Rasaiah JC, Nowotyta JP. Water conduction through the hydrophobic channel of a carbon nanotube. Nature 2001;414:188–90.
[65] Noy A, Park HG, Fornasiero F, Holt JK, Grigoropoulos CP, Bakajin O. Nanofluidics in carbon nanotubes. Nano Today 2007;2:22–9.
[66] Kotsalis EM, Walther JH, Koumoutsakos P. Multiphase water flow inside carbon nanotubes. Int J Multiph Flow 2004;30:995–1010.
[67] Chan Y, Hill JM. A mechanical model for single-file transport of water through carbon-nanotube membranes. J Membr Sci 2011;372:57–65.
[68] Whitby M, Quirke N. Fluid flow in carbon nanotubes and nanopipes. Nat Nanotechnol 2007;12:87–94.
[69] Giovambattista N, Rossky PJ, Debenedetti PG. Effect of pressure on the phase behavior and structure of water confined between nanoscale hydrophobic and hydrophilic plates. J Phys Chem C 2007;111:1323–32.
[70] Kolesnikov AI, Zanotti JM, Loong CK, Thivagarajan P, Moravsky AP, Loutfv RO, et al. Anomalously soft dynamics of water in a nanotube: a revelation of nanoscale confinement. Phys Rev Lett 2004;93. 035503–1.
[71] Banerjee S, Murad S, Puri IK. Preferential ion and water intake using charged carbon nanotube. Chem Phys Lett 2007;434:292–6.
[72] Song C, Corry B. Intrinsic ion selectivity of narrow hydrophobic pores. J Phys Chem B 2009;113:7642–9.
[73] Falk K, Sedlmeier F, Joly L, Netz RR, Bocquet L. Molecular origin of fast water transport in carbon nanotube membranes: superlubricity versus curvature dependent friction. Nano Lett 2010;10:4067–73.
[74] Melillo M, Zhu F, Snyder MA, Mittal J. Water transport through nanotubes with varying interaction strength between tube wall and water. J Phys Chem Lett 2011;2:2978–83.
[75] Gethard K, Sae-Khow O, Mitra S. Water desalination using carbon nanotube enhanced membrane distillation. Appl Mater Interfaces 2011;3:110–4.
[76] Green M. Analysis and measurement of carbon nanotube dispersions: nanodispersion *versus* macrodispersion. J Polym Int 2010;59:1319–22.
[77] Tasis D, Tagmatarchis N, Bianco A, Prato M. Chemistry of carbon nanotubes. Chem Rev 2006;106:1105–36.
[78] Kim S, Kim W, Kim T, Choi YS, Lim HS, Yang SJ, et al. Surface modifications for the effective dispersion of carbon nanotubes in solvents and polymers. Carbon 2012;50:3–33.
[79] Ma PC, Kim JK, Tang BZ. Functionalization of carbon nanotubes using a silane coupling agent. Carbon 2006;44:3232–8.
[80] Ma PC, Siddiqui NA, Marom G, Kim JK. Dispersion and functionalization of carbon nanotubes for polymer-based nanocomposites: a review. Compos A: Appl Sci Manuf 2010;41:1345–67.

[81] Star A, Stoddart JF, Steuerman D, Diehl M, Boukai A, Wong EW, et al. Preparation and properties of polymer-wrapped single-walled carbon nanotubes. Angew Chem Int Ed 2001;40:1721–5.
[82] Hill DE, Lin Y, Rao AM, Allard LF, Sun YP. Functionalization of carbon nanotubes with polystyrene. Macromolecules 2002;35:9466–71.
[83] Nepal D, Geckeler KE. Proteins and carbon nanotubes: close encounter in water. Small 2007;3:1259–65.
[84] Nepal D, Geckeler KE. pH-sensitive dispersion and debundling of single-walled carbon nanotubes: lysozyme as a tool. Small 2006;2:406–12.
[85] Gorityala BK, Ma J, Wang X, Chen P, Liu XW. Carbohydrate functionalized carbon nanotubes and their applications. Chem Soc Rev 2010;39:2925–34.
[86] Rinaudo M. Chitin and chitosan: properties and applications. Prog Polym Sci 2006;31:603–32.
[87] Yang H, Wang SC, Mercier P, Akins DL. Diameter-selective dispersion of single-walled carbon nanotubes using a water-soluble, biocompatible polymer. Chem Commun 2006;1425–7.
[88] Peng F, Pan F, Sun H, Lu L, Jiang Z. Novel nanocomposite pervaporation membranes composed of poly(vinyl alcohol) and chitosan-wrapped carbon nanotube. J Membr Sci 2007;300:13–39.
[89] Panhuis MIH, Heurtematte A, Small WR, Paunov VN. Inkjet printed water sensitive transparent films from natural gum–carbon nanotube composites. Soft Matter 2007;3:840–3.
[90] Geng Y, Liu MY, Li J, Shi XM, Kim JK. Effects of surfactant treatment on mechanical and electrical properties of CNT/epoxy nanocomposites. Compos A: Appl Sci Manuf 2008;39:1876–83.
[91] Yu J, Grossiord N, Koning CE, Loos J. Controlling the dispersion of multi-wall carbon nanotubes in aqueous surfactant solution. Carbon 2007;45:618–23.
[92] Kim TH, Doe C, Kline SR, Choi SM. Water-redispersible isolated single-walled carbon nanotubes fabricated by in situ polymerization of micelles. Adv Mater 2007;19:929–33.
[93] O'Connell MJ, Bachilo SM, Huffman CB, Moore VC, Strano MS, Haroz EH, et al. Band gap fluorescence from individual single-walled carbon nanotubes. Science 2002;297:593–6.
[94] Moore VC, Haroz EH, Hauge RH, Smalley RE, Schmidt J, Talmon Y. Individually suspended single-walled carbon nanotubes in various surfactants. Nano Lett 2003;3:1379–82.
[95] Frizzell CJ, Panhuis M, Coutinho DH, Balkus KJ, Minett AI, Blau W, et al. Reinforcement of macroscopic carbon nanotube structures by polymer intercalation: the role of polymer molecular weight and chain conformation. Phys Rev B Condens Matter Mater Phys 2005;72. 245420-1–245420-8.
[96] Tanaka T. Formation of microporous membranes of poly (1, 4-butylene succinate) via nonsolvent and thermally induced phase separation. Desalin Water Treat 2010;17:176–82.
[97] Whitby RLD, Fukuda T, Maekawa T, James SL, Mikhalovsky SV. Geometric control and tuneable pore size distribution of buckypaper and buckydiscs. Carbon 2008;46:949–56.
[98] Gou J. Single-walled nanotube bucky paper and nanocomposite. Polym Int 2006;55:1283–8.
[99] Coleman JN, Blau WJ, Dalton AB, Muñoz E, Collins S, Kim BG, et al. Improving the mechanical properties of single-walled carbon nanotube sheets by intercalation of polymeric adhesives. Appl Phys Lett 2003;82:1682.

[100] Boge J, Sweetman LJ, Panhuis M, Ralph SF. The effect of preparation conditions and biopolymer dispersants on the properties of SWNT bucky-papers. J Mater Chem 2009;19:9131–40.
[101] Viswanathan G, Kane DB, Lipowicz PJ. High efficiency fine particulate filtration using carbon nanotube coatings. Adv Mater 2004;16:2045–9.
[102] Liu X, Wang M, Zhang S, Pan B. Application potential of carbon nanotubes in water treatment: a review. J Environ Sci 2013;25(7):1263–80.
[103] Lin DH, Xing BS. Adsorption of phenolic compounds by carbon nano-tubes: role of aromaticity and substitution of hydroxyl groups. Environ Sci Technol 2008;42(19):7254–9.
[104] Gotovac S, Song L, Kanoh H, Kaneko K. Assembly structure control of single wall carbon nanotubes with liquid phase naphthalene adsorption. Colloids Surf A Physicochem Eng Asp 2007;300(1–2):117–21.
[105] Yang K, Zhu LZ, Xing BS. Adsorption of polycyclic aromatic hydrocarbons by carbon nanomaterials. Environ Sci Technol 2006;40(6):1855–61.
[106] Chen W, Duan L, Zhu DQ. Adsorption of polar and nonpolar organic chemicals to carbon nanotubes. Environ Sci Technol 2007;41(24):8295–300.
[107] Gotovac S, Honda H, Hattori Y, Takahashi K, Kanoh H, Kaneko K. Effect of nanoscale curvature of single-walled carbon nanotubes on adsorption of polycyclic aromatic hydrocarbons. Nano Lett 2007;7(3):583–7.
[108] Tournus F, Charlier JC. Ab initio study of benzene adsorption on carbon nanotubes. Phys Rev B 2005;71(16).
[109] Zhang SJ, Shao T, Bekaroglu SSK, Karanfil T. The impacts of aggregation and surface chemistry of carbon nanotubes on the adsorption of synthetic organic compounds. Environ Sci Technol 2009;43(15):5719–25.
[110] Zhang SJ, Shao T, Kose HS, Karanfil T. Adsorption kinetics of aromatic compounds on carbon nanotubes and activated carbons. Environ Toxicol Chem 2012;31(1):79–85.
[111] Peng XJ, Li YH, Luan ZK, Di ZC, Wang HY, Tian BH, et al. Adsorption of 1, 2 -dichlorobenzene from water to carbon nanotubes. Chem Phys Lett 2003;376 (1–2):154–8.
[112] Cho HH, Smith BA, Wnuk JD, Fairbrother DH, Ball WP. Influence of surface oxides on the adsorption of naphthalene onto multiwalled carbon nanotubes. Environ Sci Technol 2008;42(8):2899–905.
[113] Hoffmann MR, Martin ST, Choi WY, Bahnemann DW. Environmental applications of semiconductor photocatalysis. Chem Rev 1995;95(1):69–96.
[114] Kongkanand A, Kamat PV. Electron storage in single wall carbon nano-tubes. Fermi level equilibration in semiconductor-SWCNT suspensions. ACS Nano 2007;1(1):13–21.
[115] Lu SY, Tang CW, Lin YH, Kuo HF, Lai YC, Tsai MY, et al. TiO_2-coated carbon nanotubes: a redshift enhanced photocataly-sis at visible light. Appl Phys Lett 2010;96(23).
[116] Cong Y, Li XK, Qin Y, Dong ZJ, Yuan GM, Cui ZW, et al. Carbon-doped TiO2 coating on multiwalled carbon nanotubes with higher visible light photocatalytic activity. Appl Catal B Environ 2011;107(1–2):128–34.
[117] Wang WD, Serp P, Kalck P, Faria JL. Visible light photo-degradation of phenol on MWNT-TiO2 composite catalysts prepared by a modified sol-gel method. J Mol Catal A Chem 2005;235(1–2):194–9.
[118] Yao Y, Li G, Ciston S, Lueptow RM, Gray KA. Photoreactive TiO_2/carbon nanotube composites: synthesis and reactivity. Environ Sci Technol 2008;42(13):4952–7.
[119] Yu Y, Yu JC, Chan CY, Che YK, Zhao JC, Ding L, et al. Enhancement of adsorption and photocatalytic activity of TiO2 by using carbon nanotubes for the treatment of azo dye. Appl Catal B Environ 2005;61:1–11.

[120] Silva CG, Faria JL. Photocatalytic oxidation of benzene derivatives in aqueous suspensions: synergic effect induced by the introduction of carbon nano-tubes in a TiO_2 matrix. Appl Catal B Environ 2010;101(1–2):81–9.
[121] Martinez C, Canle M, Fernandez MI, Santaballa JA, Faria J. Kinetics and mechanism of aqueous degradation of carbamazepine by heterogeneous photo-catalysis using nanocrystalline TiO_2, ZnO and multi-walled carbon nanotubes-anatase composites. Appl Catal B Environ 2011;102(3–4):563–71.
[122] Yan Y, Sun HP, Yao PP, Kang SZ, Mu J. Effect of multi-walled carbon nanotubes loaded with Ag nanoparticles on the photocatalytic degradation of rhodamine B under visible light irradiation. Appl Surf Sci 2011;257(8):3620–6.
[123] Brady-Estevez AS, Kang S, Elimelech M. A single-walled-carbon-nano-tube filter for removal of viral and bacterial pathogens. Small 2008;4(4):481–4.
[124] Brady-Estevez AS, Schnoor MH, Vecitis CD, Saleh NB, Ehmelech M. Multiwalled carbon nanotube filter: improving viral removal at low pressure. Langmuir 2008;26 (18):14975–82.
[125] Brady-Estevez AS, Schnoor MH, Kang S, Elimelech M. SWNT/MWNT hybrid filter attains high viral removal and bacterial inactivation. Langmuir 2010;26(24):19153–8.
[126] Akhavan O, Abdolahad M, Abdi Y, Mohajerzadeh S. Silver nanoparticles within vertically aligned multi-wall carbon nanotubes with open tips for antibacterial purposes. J Mater Chem 2011;21(2):387–93.
[127] Srivastava A, Srivastava ON, Talapatra S, Vajtai R, Ajayan PM. Carbon nanotube filters. Nat Mater 2004;3(9):610–4.
[128] Qiao Y, Li CM, Bao SJ, Bao QL. Carbon nanotube/polyaniline composite as anode material for microbial fuel cells. J Power Sources 2007;170(1):79–84.
[129] Zou YJ, Xiang CL, Yang LN, Sun LX, Xu F, Cao Z. A mediator less microbial fuel cell using polypyrrole coated carbon nanotubes composite as anode material. Int J Hydrog Energy 2008;33(18):4856–62.
[130] Xie X, Hu LB, Pasta M, Wells GF, Kong DS, Criddle CS, et al. Three-dimensional carbon nanotube-textile anode for high-performance microbial fuel cells. Nano Lett 2011;11 (1):291–6.
[131] Liu XW, Sun XF, Huang YX, Sheng GP, Wang SG, Yu HQ. Carbon nanotube/chitosan nano-composite as a biocompatible bio-cathode material to enhance the electricity generation of a microbial fuel cell. Energy Environ Sci 2011;4(4):1422–7.
[132] Padaki M, Murali RS, Abdullah MS, Misdan N, Moslehyani A, Kassim MA, et al. Membrane technology enhancement in oil–water separation. A review. Desalination 2015;357:197–207.
[133] Zhu L, Chen M, Dong Y, Tang CY, Huang A, Li LA. Low-cost mullite-titania composite ceramic hollow fiber microfiltration membrane for highly efficient separation of oil-in-water emulsion. Water Res 2016;90:277–85.
[134] Chen M, Zhu L, Dong Y, Li L, Liu J. Waste-to-resource strategy to fabricate highly porous whisker-structured mullite ceramic membrane for simulated oil-in-water emulsion wastewater treatment. ACS Sustain Chem Eng 2016;4:2098–106.
[135] Chen X, Hong L, Xu Y, Ong ZW. Ceramic pore channels with inducted carbon nanotubes for removing oil from water. ACS Appl Mater Interfaces 2012;4:1909–18.
[136] Gu JC, Xiao P, Chen J, Liu F, Huang Y, Li G, et al. Robust preparation of superhydrophobic polymer/carbon nanotube hybrid membranes for highly effective removal of oils and separation of water-in-oil emulsions. J Mater Chem 2014;2:15268–72.
[137] Fu FL, Wang Q. Removal of heavy metal ions from wastewaters: a review. J Environ Manag 2011;92:407–18.

[138] Rao GP, Lu C, Su F. Sorption of divalent metal ions from aqueous solution by carbon nanotubes: a review. Sep Purif Technol 2007;58:224–31.
[139] Mubarak NM, Sahu JN, Abdullah EC, Jayakumar NS. Removal of heavy metals from waste water using carbon nanotubes. Sep Purif Rev 2014;43:311–38.
[140] Tofighy MA, Mohammadi T. Nickel ions removal from water by two different morphologies of induced CNTs in mullite pore channels as adsorptive membrane. Ceram Int 2015;41:5464–72.
[141] Shah P, Murthy CN. Studies on the porosity control of MWCNT/polysulfone composite membrane and its effect on metal removal. J Membr Sci 2013;437:90–8.
[142] Abda M, Oren Y, Soffer A. The electrodeposition of trace metallic impurities: Dependence on the supporting electrolyte concentration—a comparison between bipolar and monopolar porous electrodes. Electrochim Acta 1987;32:1113–5.
[143] Delanghe B, Tellier S, Astruc M. The carbon-felt flow-through electrode in waste water treatment: the case of mercury (II) electrodeposition. Environ Technol 1990;11:999–1006.
[144] Vecitis CD, Schnoor MH, Rahaman MS, Schiffman JD, Elimelech M. Electrochemical multiwalled carbon nanotube filter for viral and bacterial removal and inactivation. EnvironSci Technol 2011;45:3672–9.
[145] Ohmori S, Saito T. Electrochemical durability of single-wall carbon nanotube electrode against anodic oxidation in water. Carbon 2012;50:4932–8.
[146] Schnoor MH, Vecitis CD. Quantitative examination of aqueous ferrocyanide oxidation in a carbon nanotube electrochemical filter: effects of flow rate, ionic strength, and cathode material. J Phys Chem C 2013;117:2855–67.
[147] Liu H, Vecitis CD. Reactive transport mechanism for organic oxidation during electrochemical filtration: mass-transfer, physical adsorption, and electron-transfer. J Phys Chem C 2012;116:374–83.
[148] Liu H, Vajpayee A, Vecitis CD. Bismuth-doped tin oxide-coated carbon nanotube network: improved anode stability and efficiency for flow-through organic electrooxidation. ACS Appl Mater Interfaces 2013;5:10054–66.
[149] Liu H, Liu J, Liu YB, Bertoldi K, Vecitis CD. Quantitative 2D electrooxidative carbon nanotube filter model: insight into reactive sites. Carbon 2014;80:651–64.
[150] Choi JH, Jegal J, Kim WN. Fabrication and characterization of multiwalled carbon nanotubes/polymer blend membranes. J Membr Sci 2006;284(1–2):406–15.
[151] Celik E, Park H, Choi H, Choi H. Carbon nanotube blended polyether-sulfone membranes for fouling control in water treatment. Water Res 2011;45(1):274–82.
[152] Qiu S, Wu LG, Pan XJ, Zhang L, Chen HL, Gao CJ. Preparation and properties of functionalized carbon nanotube/PSF blend ultrafiltration membranes. J Membr Sci 2009;342(1–2):165–72.
[153] Liu YL, Chang Y, Chang YH, Shih YJ. Preparation of amphiphilic polymer functionalized carbon nanotubes for low-protein-adsorption surfaces and protein-resistant membranes. ACS Appl Mater Interfaces 2010;2(12):3642–7.
[154] Tiraferri A, Vecitis CD, Elimelech M. Covalent binding of single-walled carbon nanotubes to polyamide membranes for antimicrobial surface properties. ACS Appl Mater Interfaces 2011;3(8):2869–77.
[155] Gunawan P, Guan C, Song XH, Zhang QY, Leong SSJ, Tang CY, et al. Hollow fiber membrane decorated with Ag/MWNTs: toward effective water disinfection and biofouling control. ACS Nano 2011;5(12):10033–40.
[156] Majeed S, Fierro D, Buhr K, Wind J, Du B, Boschetti-Fierro De A, et al. Multi-walled carbon nanotubes (MWCNTs) mixed polyacrylonitrile (PAN) ultrafiltration membranes. J Membr Sci 2012;403:101–9.

[157] Tang CY, Zhang Q, Wang K, Fu Q, Zhang CL. Water transport beheavier of chitosan porous membranes containing multi-walled carbon nanotubes (MWNTs). J Membr Sci 2009;337(1–2):240–7.
[158] Upadhyayula VKK, Gadhamshetty V. Appreciating the role of carbon nanotube composites in preventing biofouling and promoting biofilms on material surfaces in environmental engineering: a review. Biotechnol Adv 2010;28:802–16.
[159] Agboola AE, Pike RW, Hertwig TA, Lou HH. Conceptual design of carbon nanotube processes. Clean Techn Environ Policy 2007;9(4):289–311.

Further reading

[1] Wu P, Xu YZ, Huang ZX, Zhang JC. A review of preparation techniques of porous ceramic membranes. J Ceram Process Res 2015;16:102–6.
[2] Ji C, Tian Y, Li YD, Lin YS. Thin oriented AFI zeolite membranes for molecular sieving separation. Microporous Mesoporous Mater 2014;186:80–3.
[3] Liu Y, Dustin Lee JH, Xia Q, Ma Y, Yu Y, Lanry Yung LY, et al. A graphene-based electrochemical filter for water purification. J Mater Chem A 2014;2:16554–62.

Beneficial uses and valorization of reverse osmosis brines

Domingo Zarzo
Valoriza Agua, Madrid, Spain

11.1 Introduction

Desalination technologies have been expanded worldwide in the last decades as a new source of water. Along with the major advantages and benefits resulting from its use, there are still aspects to improve such as energy consumption or the reduction in environmental impacts, mainly focused on brine discharge and management.

Brine is simply the concentrate from desalination plants with different chemical characteristics and concentrations depending on water source, quality, and plant recovery. It is composed of high concentrations of inorganic salts and compounds from raw water and small amounts of some chemicals used in the industrial process (such as antiscalants, biocides, cleaning chemicals, etc.).

While brines from seawater desalination plants are mainly discharged to the sea with the corresponding diffusion and dilution, not representing a serious environmental threat if it is done properly, brines from inland brackish water have different characteristics that may involve nonviable technical or economical solutions.

One of the first aspects to evaluate the environmental impact or the potential uses of brines is the chemistry and characteristics of brine. Brine, as concentrated seawater, is similar in different parts of the world (with the logical local peculiarities), but brackish water brines may be completely different for each case and containing in some cases toxic or hazardous compounds such as heavy metals, pesticides, or emerging pollutants, which will be concentrated in brine generating additional problems for their disposal.

In this scenario, the search of technologies able to recover salts or commercially profitable compounds from brines is a priority research field, along with other applications such as energy recovery and energy production. These applications additionally produce the positive effect of the reduction of brine volume discharged to the environment.

In this chapter, these applications will be outlined, besides other management systems for desalination of brines. Although it will be focused on brines produced by reverse osmosis (RO) plants, most of the applications and uses mentioned could be applied for brines from other desalination processes.

Many different technologies will be listed and commented but the description of the technologies will be only detailed in the cases of emerging or nonconventional technologies, due to a complete comprehensive description would require an entire book.

Emerging Technologies for Sustainable Desalination Handbook. https://doi.org/10.1016/B978-0-12-815818-0.00011-4

11.2 Brine characteristics

The term "brine" is used in desalination to define the concentrated stream containing the salts and compounds extracted from the system. Although this term is very commonly used and will also be extensively used in this chapter (and its title), it would be preferable to use the term "concentrate", because for the general population "brine" is a dismissive term and, according to the popular belief, it describes something dirty, nontransparent, and pollutant. In this way, efforts are needed for changing this perception with measures such as visual demonstration in desalination plants, as it is shown in Fig. 11.1. Another alternative term, "reject," neither supply a positive image of the brine nor it would be avoided by the industry.

Brine characteristics basically depend on raw water quality and plant recovery. For seawater RO plants, recovery is typically in the range of 40%–45%, which means that brine salinity will be approximately double than that of seawater. In the case of brackish water, considering a typical average recovery of 75%, concentration of salts in brine will be multiplied by four.

Ionic composition of brines is very variable, depending on raw water type and chemicals used in desalination process. The process itself should not cause enrichment of any specific ion because the relationship between ions in raw water is generally preserved in brine characteristics.

In general terms, brine from seawater desalination plants contains high levels of sodium and chloride (such as seawater), with other major ions such as calcium, magnesium, and sulfates, as shown in Table 11.1, with small differences depending on the regional characteristics of seawater and plant recovery.

Brines contain natural water components and chemical additives used in the treatment system. In addition, brackish water concentrates (including those generated from

Fig. 11.1 Brine canal at the Muchamiel SWRO Plant (Alicante, Spain). Courtesy of Carlos Fernandez.

Table 11.1 **Chemical characteristics of seawater brines from different origins (average values)**

Plant		Aguilas	SSDP	Mantoverde	Las Palmas III	Marina Baja
Location		Spain (Mediterranean)	Australia	Chile	Spain (Canary Islands, Atlantic ocean)	Spain (Mediterranean)
Plant type		SWRO	SWRO	SWRO	SWRO	SWRO
Raw water type		Seawater	Seawater	Seawater	Seawater	Seawater
Brine characteristics						
Calcium	mg/L	790		845	873	
Magnesium	mg/L	2479		2550	2900	
Sodium	mg/L	21,921		21,070	23,200	
Potassium	mg/L	743		784	918	
Ammonium	mg/L					
Barium	mg/L					
Strontium	mg/L	8.9		15		
Iron	mg/L					
Bromides	mg/L					
Chlorides	mg/L	38,886		38,014	43,790	
Sulfates	mg/L	5316		5342	5964	
Nitrates	mg/L	1.8			4.0	5.11
Bicarbonates	mg/L	173		274	283	
Carbonates	mg/L	155		19.5		
Fluoride	mg/L	2		1.8	6.3	
Silica	mg/L	0.5			0.42	
Boron	mg/L	8.7		8.6	9.3	
Phosphate	mg/L					<0.01
TDS	mg/L	70,488	63,000	68,967		63,580
Conductivity	μS/cm		91,000		84,500	
pH		7.9	7.8	7.92	7.8	7.88

Data supplied by O&M managers.

wastewater) can include organic pollutants, heavy metals, nutrients, and other components that could increase the toxicity of the brine and in consequence a negative environmental impact.

In addition, desalination process can provide brines (depending on the plant process) chemical additives such as coagulants (typically aluminum and ferric chloride), antiscalants (polyphosphates, phosphonates, polymaleates, etc.), chlorine, and derived compounds and products from membrane chemical cleaning (detergents, biocides, etc.) and from the effluent-treatment plants (coagulants and flocculants).

Brackish water-treatment plants produce brines with very different characteristics depending on the origin of raw water, concentration of salts, and recovery. In Table 11.2, different examples are shown; most of the cases showed that high concentrations of nitrates could cause problems for discharges to the sea as well as high concentrations of boron could invalidate potential uses for agriculture.

Brine in brackish water plants can also be time varying, depending on raw water evolution. In Fig. 11.2, the evolution of sulfate concentration over time (months) in an RO brackish water plant feed by groundwater from a coastal aquifer with marine intrusion is shown. This variability will also be logically transferred to brine.

Other constituents of brine, which could be present in brine discharges (although not directly generated by brine itself), are suspended solids and organic matter coming from pretreatment processes or chemical cleaning (Fig. 11.3). Although some modern large-scale plants include effluent-treatment plants, some plants are still discharging these effluents blended with brine without any pretreatment. In some countries, such as Australia (and according to the environmental regulation), these effluents can be discharged with brine if this stream do not contain chemicals outside seawater natural components.

Latteman et al. [1] reported the possibility of the presence of trace amounts of metals (iron, chromium, nickel, and molybdenum) in RO brines due to metal corrosion of stainless steel, although this is not a major concern due to the exceptional nature of its occurrence and the concentration of these substances that should be negligible. In Table 11.3, the analysis made for different metals at the Southern Seawater Desalination Plant in Western Australia is shown. Only a few metals have increased significantly their concentration, although always below the discharge limits. These analyses are made quarterly for brine control. In similar analysis of metals in brine made at Marina Baja SWRO plant (Spain), only Zinc was detected over detection limits and was only 1 μg/L over this value (9 μg/L).

Similarly, temperature (which in general terms can be considered as a pollutant factor) in RO brines is very similar to raw water and, with the exception of a few very specific cases, it should not raise concerns (obviously this is not the case of thermal desalination plants).

In the specific case of plants treating wastewater by membranes (reclamation plants), the high quantity of nutrients (nitrogen and phosphorus) in brine can cause eutrophication in natural water bodies [2,3], and the potential presence of contaminants of emerging concerns (CECs) in wastewater is of increasing concern due to the high concentration factor in these brines [4,5]. For the elimination of these compounds from wastewater brines, different advanced treatment technologies have been

Table 11.2 **Chemical characteristics of brackish water brines from different origins (average values)**

Plant		La Toma	Abrera	Xeresa Golf	Cuevas de Almanzora	University of Alicante
Location		Spain	Spain	Spain	Spain	Spain
Plant type		BWRO	EDR	BWRO	BWRO	BWRO
Raw water type		Groundwater	River	Wastewater	Groundwater from a coastal aquifer	Groundwater
Brine characteristics						
Calcium	mg/L	1371	960	1336	1855	825
Magnesium	mg/L	1348	344	59	1556	659
Sodium	mg/L	3858	1150	1795	7359	2976
Potassium	mg/L	33	422	134	241	81
Ammonium	mg/L				0.04	
Barium	mg/L	0.4			0.6	
Strontium	mg/L	103			32	
Iron	mg/L				0.5	
Bromides	mg/L		10.9			
Chlorides	mg/L	8018	3443	2742	14,428	3388
Sulfates	mg/L	4811	1344	1551	8366	4715
Nitrates	mg/L	90	104	1.4	155	345
Bicarbonates	mg/L	1362	885	2648	863	1025
Carbonates	mg/L	4.7	909	6.3	5	
Fluoride	mg/L	2.9			1.5	
Silica	mg/L	27	4.9		18	
Boron	mg/L	3.8			2.9	5.46
Phosphate	mg/L	0.8			0.635	<0.1
TDS	mg/L	21,035	9579	10,275	34,885	
Conductivity	μS/cm	24,915	13,060		38,650	15,360
pH		7.9	6.9	8.4	7.4	7.7

Data supplied by O&M managers.

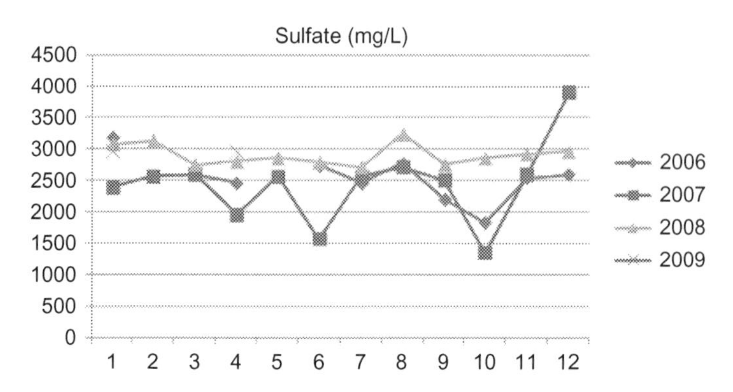

Fig. 11.2 Sulfate concentration evolution in Cuevas de Almanzora BWRO plant (Almeria, Spain) between 2006 and 2009 (monthly changes).

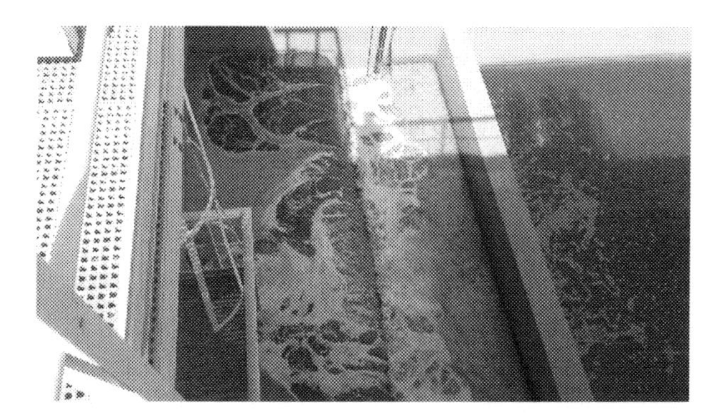

Fig. 11.3 Effluent stream after filter backwash at the Aguilas SWRO plant (Spain). Courtesy of Domingo Zarzo.

developed and tested such as ozonization, Fenton processes, photocatalysis and photooxidation, sonolysis, electrochemical oxidation, and adsorption [6]. In some coastal reclamation plants, the apparently simple discharge of brine to the sea can be complicated or even not feasible without previous treatment due to the regulation of nutrient discharges to the sea.

11.3 Brine management strategies

Although there is a growing interest for the valorization of brines from desalination plants, they have been traditionally discharged to the ocean or inland superficial water bodies along with other brine management strategies.

The lack of an adequate and feasible (technically and economically) method for the management or disposal of brines is often one of the main causes of the unfeasibility of many failed desalination projects.

Table 11.3 Analysis of different metals in seawater and brine (March 2017) at the SSDP plant in Australia

Element	Units	Seawater	Brine outfall 1	Brine outfall 2	Brine outfall 3
V	μg/L	1.5	2.7	2.9	2.9
Cr	μg/L	<0.4	<0.4	<0.4	<0.4
Mn	μg/L	1.7	1	1	1
Ni	μg/L	<0.3	0.8	0.8	0.7
Cu	μg/L	0.2	4.5	4.7	4.4
Zn	μg/L	<1	6	6	4
As	μg/L	1.9	3	3	3
Se	μg/L	<1	< 2	<2	<2
Mo	μg/L	11	19	19	19
Ag	μg/L	<0.2	<0.2	<0.2	<0.2
Cd	μg/L	<0.2	<0.2	<0.2	<0.2
Pb	μg/L	<0.2	<0.2	<0.2	<0.2
Hg	mg/L	<0.0001	<0.0001	<0.0001	<0.0001
Al	μg/L	<5	<10	<10	<10
Fe	μg/L	<1	2	<2	<2

Data supplied by the O&M manager.

Feasibility of brine disposal depends on many factors such as quality, distance to the discharge point, and economical and environmental aspects. Brine disposal costs can be a limiting factor for the implementation of desalination plants, and they can be in the rate of 5% and 33% of total desalination costs [7] (although this is from a relatively old study, the values are reasonably actual). In a recent study conducted in Australia [8], it was concluded that for those plants away from the coast, energy supply, brine disposal, and civil work costs are the major costs for producing water, while for seawater desalination plants, the operational costs for this component are minimal.

The most common strategies for brine management are discharges into the sea (for seawater plants and other plants near the coast), superficial discharge, combined discharge in water-treatment plants (wastewater-treatment plants, sewerage networks), deep well injection, land application, and evaporation ponds.

Over 90% of the large desalination plants in operation (mainly seawater plants) dispose of their concentrates through ocean outfalls specifically designed and built for that purpose [9], by one of the following methods; discharge through new ocean outfall, through existing wastewater-treatment plant outfall, or through existing power plant outfall.

While global data regarding brine management in inland regions are insufficient, some authors have reported data for countries such as the United States [10], where main alternatives were superficial discharge (45%), discharge to sewer networks (27%), deep well injection (13%), land application (8%), evaporation ponds (5%), and others (2%) or in Australia [11], the main strategies reported were the freshwater discharge (48%), sewer disposal (17%), ocean disposal (12%), deep well injection (12%), and land application (9%).

Mezher et al. [12] also reported the main disposal methods for a list of different countries, including evaporation ponds (Australia, Saudi Arabia, Kuwait), land application and evaporation (Qatar, Jordan, Oman), surface water (UAE, Spain, Japan, Algeria), and sewer system blending and land application (United Kingdom).

There are other possibilities for brine management and volume minimization, and if it is possible, its reuse and valorization. These processes are in general looking for the zero liquid discharge (ZLD) concept in different extensions or at least to increase plant recovery and reduction of brine volumes. Some of these technologies include evaporation-crystallization, salinity-gradient solar ponds, processes with two stages and interim chemical precipitation or biological reduction, RO with previous softening and high recovery (HERO process among others), nanofiltration (NF) in two stages, seeded slurry precipitation and recycle (SPARRO), membrane distillation (MD) and pervaporation (PV), capacitive deionization (CDI), dewvaporation (DW) (humidification-dehumidification), forward osmosis (FO), wind-aided intensified evaporation (WAIV), salt solidification and sequestration, vibratory shear-enhanced membrane process (VSEP), and hybrid systems. Many of these technologies will be commented later because they are also processes used for brine valorization.

In the scheme showed in Fig. 11.4, the main brine management strategies are listed and related, making this a resume of this complete chapter.

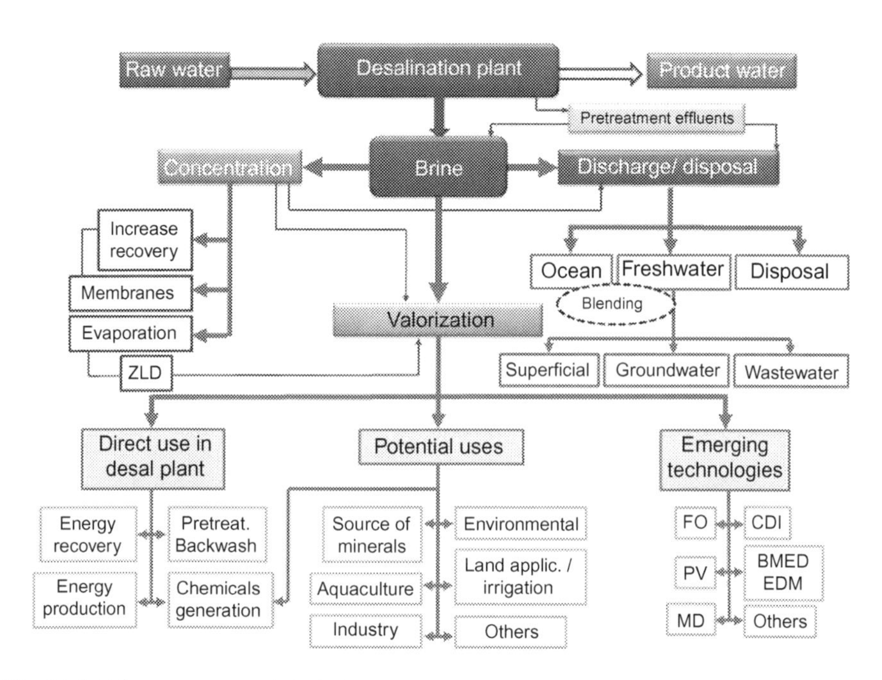

Fig. 11.4 Brine management strategies.

11.4 Beneficial uses and valorization of brines

With the advent of environmental protection and circular economy concepts, desalination industry cannot reconcile with the simple brine discharge to environmental matrix, but it have to be also committed to the valorization of brines in different ways, which additionally improves the environmentally friendly image of desalination.

11.4.1 Uses of brines inside the desalination plants

The first measure to apply before considering the discharge or disposal of brines is to attempt their use inside the desalination plants in different stages and parts of the process.

One of the main current uses of brines inside desalination plants is as a source of water for pretreatment backwash; this was primarily applied for conventional pretreatments (filtration) and more recently for membrane pretreatments [micro- (MF) and ultrafiltration (UF)]. Although this is not a disposal or management method, it is actually a method for saving raw water or filtered water and consequently is a beneficial use of brines, increasing the global plant recovery and saving water and energy. Additionally, brines have disinfectant power because they can cause osmotic shock to the microorganisms if the salinity is high enough.

Another potential use of brines that can be applied in desalination plants is the generation of chlorine by means of electrochlorination [13]. Salt and saline solutions are the source of chlorides for producing sodium hypochlorite in situ by means of the application of electrical current, and brines can be one of these source solutions, taking advantage of the production in the very desalination plant saving transport and chemical costs for chlorine and/or biocide dosage (Fig. 11.5). The use of brine for this

Fig. 11.5 Electrochlorination at the Sagunto (Spain) power plant. Courtesy of Domingo Zarzo.

purpose has some drawbacks, and it requires further research and testing. In a similar way, the insite production of acids (HCl) and basic compounds (NaOH) by means of electrodialysis with bipolar membranes applied to the brine [14] could reduce the purchase of chemicals in the plant reducing operational costs, although this is also an incipient technology.

Without any doubt, the main and better known use of brine inside desalination plants is the application of brine streams for energy recovery, taking advantage of its high flow and pressure, as it will be explained later.

11.4.2 Brine as a source of minerals

Salt is the oldest condiment used by humans for thousands of years and its importance is such as it influenced the human development in different times with social, economic, and gastronomic implications. As an example, in Roman times, salt was so valuable that it was used as a currency or payment method, reason for the origin of the word "salary."

Brines from desalination plants can be a potential source for the extraction of minerals due to the high concentration of some salts of commercial interest. This application can be an alternative for salt extraction among the existing ones, such as extraction from natural brines (sea and inland salt) (Fig. 11.6) or salts obtained from mining extraction. In general, these traditional techniques have required evaporation processes combined with solar action and/or air kinetics, which are also the basis of the technologies used for brines.

Among others, main minerals extracted evaporating saline water and seawater are [15] anhydrite ($CaSO_4$), bischofite ($MgCl_2\ 6H_2O$), calcite ($CaCO_3$), carnallite ($MgCl_2\ KCl\ 6H_2O$), dolomite ($CaMg(CO_3)_2$), epsomite ($MgSO_4\ 7H_2O$), gypsum ($CaSO_4\ 2H_2O$), halite (NaCl), Hexahydrite ($MgSO_4\ 6H_2O$), kieserite ($MgSO_4\ H_2O$), langbeinite ($K_2SO_4\ 2MgSO_4$), mirabilite ($Na_2SO_4\ 10H_2O$), sylvinite (KCl+NaCl),

Fig. 11.6 Saltworks at Santa Pola (Spain).
Courtesy of Domingo Zarzo.

sylvite (KCl), and thenardite (Na_2SO_4). In Fig. 11.7, different crystals obtained by evaporation–crystallization of brine from a brackish RO plant are shown.

Some authors have proposed the extraction of other interesting different compounds from brines such as the study by Le Dirach et al. [16], which identified eight main elements (phosphorus, cesium, indium, rubidium, germanium, magnesium, sodium chloride and potassium chloride), also studying how they could be extracted in a technical and economical feasible way. Other works by Jeppesen et al. [17] also analyzed the potential extraction of rubidium and phosphorus.

The extraction of different minerals and compounds from seawater is not new and many scientists have considered their extraction from seawater or even processed seaweeds [18]. Salt extraction from seawater is reported in China from 2000 years BC, and regarding precious metals, it is well-known that Svante Arrhenius was one of the first scientists making an estimate of the amount of gold in seawater (thinking about its extraction), with a further estimation by his colleague Fritz Haber in 1923 recognizing that there was not enough quantity for recovering the extraction costs.

More recently, the English writer Hugh Aldersey-Williams wrote in 2011 a book [19] on the story and cultural associations of the chemical elements, where he calculated not only the potential value of gold contained in seawater around EUR 400 billions but also recognizing that extraction costs would be too great to consider it right

Fig. 11.7 Electron microscopy of salts from evaporation of BWRO brine. (A) $CaCO_3$, (B) $CaSO_4$, (C) ClNa, and (D) different salts of sodium chloride, sulfate, magnesium, and potassium.
Courtesy of University of Alicante (Spain).

now. Other commercially interesting compounds identified in seawater (and concentrated in brines) to their potential extraction could be germanium, lithium, uranium, barium, molybdenum, and nickel.

There are many different salts with numerous specifications for each application; for softening, for swimming pools, for the chlor-alkali industry, as table salt, for detergent production, hide salt, butchers salt, flake and gourmet salt, for oil perforation, for dying, and so on, being the brines optimum for some applications and not recommendable for others depending on their chemical characteristics.

Sodium chloride, the most common salt, is the main raw material for chlorine (and derived compounds) production used for chlor-alkali industry, for processing food, as additive or preservative, for livestock farming, and for general application in industry (pharmaceutical, textile, as a melting, ceramic, oil and gas, water treatment, freezing control in roads, etc.).

Turek et al. described in a paper [20], a preliminary study exploring the possibility of the application of RO, NF-RO, and NF-RO-MED seawater desalination brines as a sodium chloride source for membrane electrolysis; they proposed a new concept in which electrodialysis with univalent permselective membranes would be used to enrich chlor-alkali lean brine with sodium chloride and simultaneously to desalinate brine discharge.

There are many technologies for salt recovery from brines that have been used in different extension or in development and they will be described later on. These processes fall in general within the field of ZLD systems that are looking for the processing of liquid streams into a solid waste and could be additionally used for the extraction of commercially applicable salts.

11.4.2.1 Evaporation-crystallization

The main, better known, and commercially extended ZLD system is based on evaporation-crystallization technologies combining both techniques to produce a solid residue from a liquid stream. This technology is used traditionally for different applications such as the reduction of volumes in different effluents; brines from membrane technologies, purges from cooling systems and refrigeration circuits, manufacturing process of pickles, waste gas cleaning in power stations, coal or steel industry, wastes from metallurgical processes, leachates, and other effluents difficult to treat by conventional processes.

ZLD process by evaporation-crystallization includes as the main stages a brine concentrator (evaporator) followed by a crystallizer or alternatively a spray dryer (which uses forced air and heat), as shown in Fig. 11.8.

Other different systems for ZLD are the thermo-ionic technologies, which combine any desalination system with a concentration unit for reducing concentrate volumes. There are two main commercial systems: AquaSel by GE (with a nonthermal brine concentrator based on electrodialysis reversal (EDR)) and the thermo-ionic desalination process by Saltworks technologies.

Although ZLD systems based on evaporation-crystallization are apparently the most technically feasible systems for the complete elimination of brines by means

Fig. 11.8 Evaporation-crystallization pilot plant for a R&D project. Courtesy of Domingo Zarzo.

of the formation of a solid waste (or a profitable solid residue), these systems have high investment and operational costs, reason why they are hardly feasible from the economical point of view even with the provision of a source of residual heat.

Anyway, it seems that there is still space to improve the technology and, for example, Chung et al. [21] concluded that the brine concentration step has more potential for improvement that the crystallization step, which is a more traditional process difficult to optimize.

11.4.2.2 Evaporation ponds

The use of evaporation ponds is, to date, the most extended method for brine disposal for inland desalination plants due to its reduced costs if there is enough available space with low land cost. This method is especially effective in regions with low rainfall and where climatic conditions are favorable to maintain a high and continuous evaporation rate. In Fig. 11.9, evaporation ponds are shown.

In general, evaporation ponds are limited for relatively small plants (below 20,000 m^3/day) [22] located in arid weather conditions.

11.4.2.3 Desalination plants for combined water and salt production

Although the combined production of water and salt from brines close to seawater desalination plants could seem a very evident and a reasonable possibility, the reality is that this is not very common due to some reasons; the proximity of the desalination plant to a salt exploitation, the availability of land for the process, required and adequate concentration of brine and as a deterrent, the presence of chemicals in brine (such as antiscalants).

Fig. 11.9 Evaporation ponds in Spain.
Courtesy of Domingo Zarzo.

Fig. 11.10 Saltworks at Santa Pola (Spain).

Anyway there are some reported cases such as the Eilat desalination plant in Israel (10,000 m^3/day), where brine is sent to evaporation ponds in order to produce table salt [23], and other works studying the feasibility of this application have been done in countries such as Spain and Greece [24], although the feasibility is limited due to brine characteristics, concentration and presence of chemicals from desalination process. In the case of Spain, during the planning stage of the large desalination program in the Mediterranean coast, this possibility was studied for plants close to salt exploitations such as the case of SWRO Torrevieja, where it was discarded, and in Santa Pola (Fig. 11.10), the owners of saltworks planned an associated desalination plant which finally was not built.

11.4.2.4 Salt solidification and sequestration (SAL-PROC process)

Systems based on salt solidification and sequestration such as the SAL-PROC process are commercial processes combining evaporation and concentration techniques used in small-scale plants [25]. The SAL-PROC process was commercialized by the company Geoprocessors and is a process to obtain chemicals in crystalline state with a combined multievaporation process including cooling stages and chemical treatment. Different process combinations were tested in different plants in Oman producing as main products gypsum, magnesium hydroxide, and halite [25].

11.4.2.5 Intensive evaporation processes

Intensive evaporation processes are evaporation processes improved by different ways and including processes such as WAIV and advanced solar dryer (ASD).

WAIV was originally developed by The Ben Gurion University in Israel and commercialized by the company Lesico Group [26] with some small reported references in Mexico, Australia, and Israel.

The WAIV system is based on the vertical evaporation over surfaces installed in modules. Water is pumped to the top and flows by gravity over the vertical strips. Wind is used to drive away the excess humidity from surfaces to intensify the evaporation process, and according to its developers, by a magnitude of 15–20 times more than in a conventional pond.

ASD is a technology using a hybrid solar/gas desalination system reducing the number of stages required in conventional evaporation systems [27]. In this system, an air current flows by a pressure gradient between the inlet and the outlet (with a chimney) of an evaporation channel, proportional to differences in air density.

11.4.2.6 Use of electrodialysis for salt recovery

EDR generally works at higher recoveries than RO not being influenced by many water components, which can seriously affect RO systems. EDR has also been applied to reduce the volume of brine concentrates from RO systems, reclaiming high water recoveries of 96%–98% [8,28]. In Fig. 11.11, the largest EDR plant in the world is shown; in this plant, very high recovery is achieved thanks to the recirculation of off-spec water and water from electrodes.

Electrodialysis has also been redesigned for specific brine applications by means of the use of different membranes or configurations (bipolar membranes and electrodialysis metathesis (EDM)).

Bipolar membrane electrodialysis (BMED) is a variation of the EDR technology able to produce acids and bases from the corresponding salts from saline streams such as brines. In this process, mono-polar cationic and anionic exchange membranes are installed together with bipolar membranes in alternating series under the influence of an electrical potential gradient.

EDM is another different concept in which there are two diluting streams and two concentrating streams. The combination of the four streams and membranes is called a quad, and the commercial stacks can include 100 or more quads. This system has the

Fig. 11.11 Drinking water-treatment plant Abrera (Barcelona, Spain), the largest EDR in the world (200,000 m^3/day).
Courtesy of Domingo Zarzo.

advantage of less membrane fouling potential due to the production of soluble salts and allowing the return of the product to the RO feed or mixed directly with RO product increasing the recovery and the efficiency of the ZLD system.

There are several studies with regard to different configurations using electrodialysis with bipolar membranes or EDM for the production of NaCl and derivates such as acids and alkaline compounds [6,13,29] with different pretreatments including precipitation or ionic exchange and crystallizers as the final step.

11.4.2.7 Use of ion exchangers for salt recovery

Ion exchange has been traditionally used for the removal of selected ions from water. Some authors have also reported the use of polymeric ion exchangers for recovering ions from brines [30]. The authors tested the extraction of metals by ion exchange chelation although they recognize that the feasibility of the process is limited due to the frequent exhaustion of the resins.

11.4.2.8 Solvent extraction

Zarzo et al. [31] reported the extraction of divalent salts from brine using liquid-liquid extraction by means of the use of organic solvents. The process is based on the capacity of some organic solvents (such as ethanol) to extract salts (mainly divalent) from brines reducing their solubility, which could be separated later by different processes. This process can increase the recovery of the remaining brine stream, and it can afford the extraction and purification of these divalent salts if a reliable process for separation is developed. Peterskova et al. [32] also studied the potential extraction of different metals from RO brines, getting good results with hexacyanoferrate-based extractant for the extraction of cesium, rubidium, and lithium. An alternative for metal extraction is their concentration followed by precipitation,

11.4.2.9 Other processes

There are other less common processes for mineral extraction from brines. One of these reported methods for salt extraction is the removal of inorganic compounds via supercritical water [33]. Supercritical water appears when water is over its critical pressure and temperature, which produces a behavior between gases and liquids. This technology includes different stages of precipitation and separation with energy recovery and temperature recovery.

Chemical precipitation of salts was also proposed by Jibril and Ibrahim [34] for the extraction of chemicals from brines by means of processes involving chemical reaction to convert NaCl into Na_2CO_3, $NaHCO_3$, and NH_4Cl or techniques such as the eutectic freeze crystallization as a novel technique for salt separation from water streams [35].

Although it was not exactly designed as a method for mineral production, the SPARRO process (slurry precipitation and recycling RO) forces the controlled precipitation of soluble salts with seeded crystals and then it could be combined with any of the above-mentioned technologies.

Other technologies for salt removal or production are combined or hybrid processes including RO, NF, and precipitation. Almarsi et al. [36] obtained $Ca_6Al_2(SO_4)_3(OH)_{12}$ and $CaSO_4$ from brine in a two-stage RO with an intermediate NF with precipitation, and Telzhensky et al. [37] also proposed NF to separate magnesium ions for post RO stabilization. Alternatively, concentrate from NF could be used for struvite precipitation from wastewater.

Finally, mineral extraction of brines by chemical and biological precipitation processes and pelletized lime softening have been also tested by different researchers.

11.4.3 Use of brines for energy recovery and energy production

Since energy consumption is one of the main issues for desalination technologies, often representing >50% of water production costs, the reduction of energy consumption in desalination has become one of the main research topics. In this way, the application of brine in different ways for energy reduction or production is a matter of growing interest.

11.4.3.1 Energy recovery in desalination plants

The use of residual energy from brine as a source of energy recovery is very well known and extended in all the modern desalination plants, being more frequent in seawater plants due to the higher flows and pressures. A seawater desalination plant without energy recovery could consume >8 kW h/m^3, whereas current seawater desalination plants have reduced energy consumptions below 3 kW h/m^3 with the new pressure exchanger devices in their different configurations (rotating, displacement,

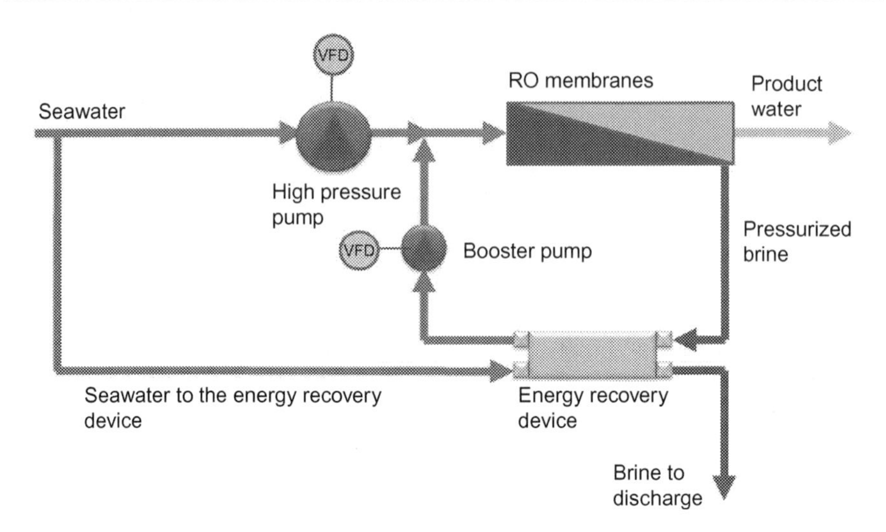

Fig. 11.12 Scheme of energy recovery from brine in a seawater RO system.

etc.). The historical evolution of the different energy recovery devices taking advantage of brine pressure from Francis and Pelton turbines to the most recent pressure exchanger devices (also called isobaric or hyperbaric) has supposed besides the membrane developments the most important factors in the reduction of desalination costs, improving the extension of technology worldwide. In Fig. 11.12, the typical energy recovery system configuration in a seawater RO train is schematized.

Less frequent is the use of energy recovery devices for brackish water installations due to the reduced pressure and lower flows. In this case, energy recovered can be supplied at the inlet of the high-pressure pump or as an interstage booster for balancing the flows between stages. As an example of this application, Garcia et al. [38] reported energy recoveries over 30% for a brackish water plant in Spain treating water with variable salinities between 9000 and 20,000 μS/cm using an interstage turbocharger.

11.4.3.2 Energy production with turbines

The production of energy by means of turbines fed by water streams from the pressure generated by height differences is well-known for a long time, and it is the basis of the hydroelectric power production. Desalination can take advantage of this kind of energy when there are important differences of height between the desalination plant and the coast and/or the product water tank or brine discharge. This is the case of Adelaide SWRO plant, where two Francis turbines were installed in the brine outfall for producing energy.

In recent times, there are also a lot of manufacturers of microturbines, which have been used for producing energy in irrigation systems, wastewater treatment, etc., although they are difficult to implement in desalination due to the large volumes involved (except in small plants) and the fact that main streams in desalination plants are pressurized and the pumping energy in general very optimized by means of variable frequency drives (VFDs), etc.

11.4.3.3 Energy production using the osmotic potential energy

Thermodynamically speaking, the necessary energy for any desalination process is equal to the produced energy dissolving the salts in water. This means that if we are able to reverse the process by means of technologies such as FO [and the different variants such as pressure retarded osmosis (PRO)], we should have a positive balance of energy. The increase of pressure can be exploited for generating energy by means of turbines or other devices. Although it is difficult to generalize and to quantify the produced energy, for example, Chung et al. [39] reported a recovery of 0.42 kW h/m^3 in a PRO system combined with desalination and Ordonez et al. [40] reported 0.78 kW h/m^3 in experiences blending seawater with wastewater.

Although FO can be considered as an emerging technology, the first idea about recovering energy from the blending of different salinity solutions arose when Pattle [41] suggested this possibility in 1954. A little later, Loeb [42] also studied the process in 1974 by means of what he called PRO. However, the development of adequate membranes for this process and the growing interest in its application are much more recent, with the emergence of some membrane manufacturers all over the world although still producing membranes on a small scale.

Other conventional technologies such as EDR can be used for energy production taking advantage of salinity gradients [43]. EDR can be used directly or in combination with other processes such as the one described in a research work from the Pennsylvania State University [44], where a new technology (microbial reverse-electrodialysis cell (MRC)) was developed to produce energy from salinity gradients combining EDR and microbial fuel cells (MFC). Although strictly speaking, the application was developed capturing the salinity gradient energy from ammonium bicarbonate salt solutions and not from desalination brines, this is a potential use to be explored.

Brines from seawater desalination plants can also be used as the draw solution for producing energy using PRO using as feed solutions different sources of low salinity water (river water, wastewater, etc.) [45]. PRO has also been tested for producing energy with brines from evaporation technologies [46] with the differences caused by the different technology applied (salinity, high temperature, etc.).

Other reported technology for energy production by salinity gradient which it could be used with brines is the hydrocratic generation, which consists of the use of a hydrocratic generator without membranes.

11.4.3.4 Technologies based on solar ponds

Saline solar ponds have been traditionally used to store thermal energy, which can be used directly as heat or to produce electrical energy.

The use of a salt gradient in solar ponds as a source of renewable energy is reported for years, and although its use has been limited to arid regions with high rates of solar radiation [47].

The feasibility of these technologies for energy storage/production requires solar exposition throughout the year, large volumes of brine and a freshwater source, flat land with low permeability and high thermal and structural stability, location away from aquifers that might be potentially affected, relatively light wind currents, and a continuous energy demand.

A salinity gradient solar pond (SGSP) consists of three layers (as shown in Fig. 11.13):

- Upper convective zone (UCZ): the most superficial layer (convective) formed by water with low salinity, ideally with 30 cm width.
- Lower convective zone (LCZ): a gradient or nonconvective layer, with widths typically between 0.5 and 1.5 m
- Nonconvective zone (NCZ): the area of heat storage or with lesser convection, which is the deepest and it is formed by a saline solution almost saturated (achieved by the direct pumping of concentrated brine), with widths between 1 and 2 m.

Energy is produced by heat transfer between fluids through heat exchangers. There are two main methods: the first one pumps off hot brine from the deepest layer to an external heat exchanger and the second one is based on the use of a pumped cooling fluid (usually freshwater) through a heat exchanger located in the deepest area. According to the reported experiences, it seems that the most cost-effective method is the first one, which was developed in Texas.

An interesting variant of this process is the thermal desalination with solar ponds. This method is based on the combination of different units of multistage evaporation (MED) and membrane desalination besides a solar pond and a brine concentrator and recovery system (BCRS).

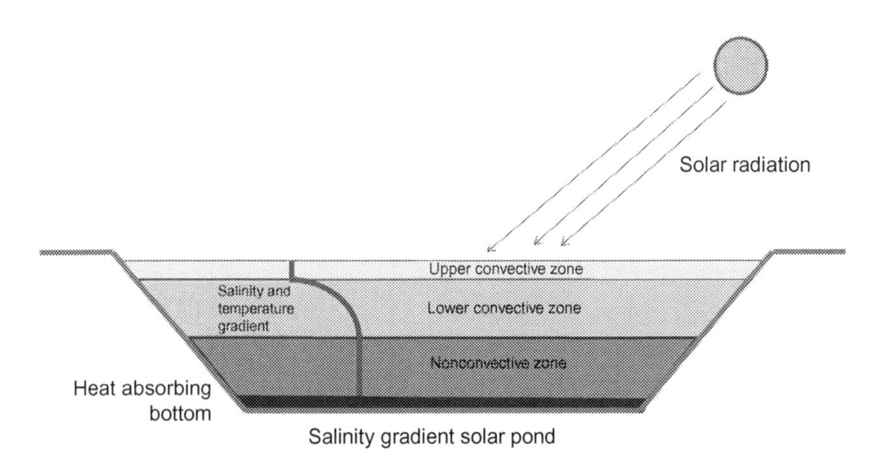

Fig. 11.13 Scheme of a salinity gradient solar pond.

The BCRS system operates by means of the thermal energy obtained from the solar pond, producing freshwater and a saline product in the form of a slurry, which can be used for the recharge of the solar pond or processed as a chemical for its sale.

Other emerging variants of solar desalination techniques are multieffect humidification, humidification-dehumidification (also known as dewvaporation), and greenhouse distillation.

11.4.4 Environmental applications

An interesting field for the brine application is the improvement of the environment. Since brine discharge is in general interpreted as a threat, to turn this into an environmentally friendly practice could suppose a turning point for the misconception of the general population.

11.4.4.1 Land application

Although this application can be interpreted as a method for brine disposal apparently not very environmentally friendly, the concentrate could actually be used for irrigation of parks, lawns, golf courses, or farmland. This potential use depends on the availability and land costs, irrigation needs, water quality, and tolerance to salinity of target vegetation, although this can be only used for small plants using techniques such as sprinkler systems or hydroponic irrigation.

11.4.4.2 Regeneration of degraded areas

Brine can be used for the improvement of water quality in polluted or degraded areas, such as estuaries, canals, wetlands, etc. The benefits of these practices include oxygenation, changes in temperature, salinity, etc., and there are reported cases in Spain, Oman, and United Arab Emirates [47,48]. In the case of Javea (Spain) desalination plant [48], brine discharge increased dissolved oxygen in a canal open to the sea that was producing bad odor and sediment accumulation before, being solved by this application. Another example is the case of Alicante SWRO plant (Spain) where brine discharge has had beneficial effects for the marine area degraded by previous impacts (sewages, harbor, anchoring, etc.) [49].

Recovery of degraded wetlands or creation of new wetlands can also be beneficial from many points of view; it can provide recreational activities (boating, birdwatching, etc.) and to assist in wildlife conservation. The combination between two apparently incompatible (brine discharge and environmental regeneration) activities with countervailing environmental consideration can be feasible taking into account the similarity between brines and some natural wetlands, where this application could be the perfect and most sustainable solution. As an example, this solution has been implemented (besides seawater pumping) in the regeneration of saline wetlands (called Agua amarga) close to the SWRO Plant Alicante II in Alicante (Spain) with the aim to compensate the decrease in the groundwater table and to ensure the wetland flooding.

An example of environmental regeneration by brine is the so called Red Sea-Dead Sea project, agreed between the governments of Israel, Jordan, and the Palestinian authorities with the support of the World Bank, where a large desalination plant in Aqaba port (Jordan) will produce water for the region and the brine produced will be pumped to the Dead Sea, in order to reverse its shrinkage, besides seawater transfers. The project will also include the installation of a hydroelectric power installation.

11.4.5 Use of brines in aquaculture and fish farming

Due to the salinity of brine, it could be an ideal medium for cultivation of some microalgae species, including species with high added value such as *Spirulina Platensis* and *Dunaliella Salina*, which could be developed in evaporation ponds as described above or in specifically designed reactors.

Other reported species able to grow in saline environment are *Artemia Salina*, *Lates calcarifer*, *Aconthopagrux Butcheri*, *Pagrus auratus*, *Chanos chanos*, *Mugil Cephulux*, and *Oreochromis mossambicus*.

However, there is a lack of literature references about cultivation of microalgae in brines (as shown in Fig. 11.14) from desalination or hypersaline water although there are many research studies about microalgae cultivation in other media (both in seawater and freshwater or even in wastewater) and its application.

The use of microalgae as biomass for the removal of certain salts (such as nutrients) opens a good possibility for the treatment of some kind of brines. Brines from RO plants treating wastewater have typically high concentrations of nutrients that are difficult to manage (even discharged to the sea) due to the regulation of these substances to avoid eutrophication. Zarzo et al. [50] found that species such as *Tetraselmis suecica* were able to reduce in >45% the concentration of nitrates in brines from different origins, although it would require the further separation of the algae species

Fig. 11.14 Microscopy of microalgae growing in brine from a BWRO plant. Courtesy of Jose Antonio Garcia, University of Alicante (Spain).

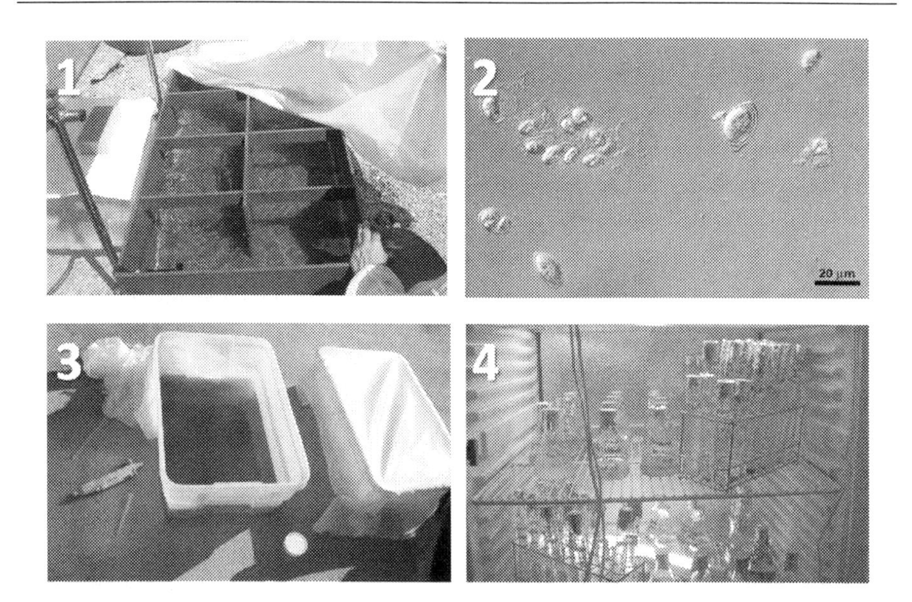

Fig. 11.15 Microalgae cultivation for brine treatment. (A) Pilot raceway tank, (B) microscopy of Oocystis cf. nephrocytioides species, (C) external small reactors, (D) laboratory cultivation.
Courtesy of the University of Alicante (Spain).

from water and it is necessary to valorize it to guarantee the economical feasibility. In Fig. 11.15, different scale experiments of this work are shown.

After separation, microalgae have potential valorization depending on the species, such as production of fertilizers, animal feed, antioxidants, and energy production.

Other marine species of fish and crustaceans have been reported growing in salt concentrations in the range of 25–40,000 ppm of Total Dissolved Solids (TDS), although there are not data about the higher concentrations that are typical in brines, requiring then some kind of dilution or even desalination. These species are marine shrimps, rainbow trout, hybrid striped bass, catfish, or brine shrimp.

Some experiences have been reported for inland saline aquaculture in Australia [51]. Small projects included fish farms with barramundi, trout, snapper, silver perch, and other species such as *Artemia*, *Dunaliella*, and *Penaeus monodon* [52], although brines from desalination plants were not studied.

11.4.6 Other potential uses

11.4.6.1 Use in agriculture

Mass media have recently reported that some vegetables have been irrigated with sea-water in the Atacama desert (Chile), including crops such as tomatoes and chards, with good results. The experiences have been promoted by the Catholic University of the North of Chile and they have concluded that nutrients present in seawater reduced the

Fig. 11.16 Hydroponic crops partially irrigated with desalinated water. Primaflor installations, Almeria (Spain).
Courtesy of Domingo Zarzo.

need of fertilizers. Seawater was ascending to the crops from different depths in the soil, ascending by capillarity.

Different sources reported that desalination for agriculture is one of the less frequent uses of desalinated water, representing <3% of total desalination production [53,54] with a small number of countries with much higher values (such as Spain, with 22%). However, these data are considering the product water application (as shown in Fig. 11.16), with very few examples about the direct use of brines for agriculture irrigation. Some of these works reported the use of brine for olives and date trees irrigation in some in areas in Palestine [55] and different vegetable species (lettuce, tomato, pepper, sunflower, and others) in a study done in Brazil [56], which was considered successful although reporting crop yield losses between 25.3% and 26.7% irrigating with a brines with around 6000 μS/cm compared to the results with freshwater.

Sanchez et al. [57] also reported uses of the reject brine from inland desalination for fish farming, Spirulina cultivation, and irrigation of halophyte forage shrub and crops using hydroponic cultivation in South America.

Problems derived from desalinated water quality for this application (toxicity of boron for different crops and imbalanced sodium adsorption ratio (SAR), which causes soil degradation and reduced permeability, among others, are much more serious with brines, in which high salinity is the main issue.

11.4.6.2 Hydrotherapy

A potential use for brines is the use in hydrotherapy. The hypersaline swimming pools promote the interchange of mineral salts with the body, which it is useful as disinfectant and healing, for the treatment of health problems such as rheumatism and arthrosis, skin conditions such as acne or psoriasis, and flotation is ideal for people with reduced mobility, being also an efficient relaxer.

11.4.6.3 Secondary recovery of oil through deep well injection of brine and/or CO_2

This application consists of the injection of brine in oil wells in order to improve the well yield. This technology is well-established and reliable, but It requires to be reasonably free from sulfates, carbonates, and bicarbonates and an in depth knowledge about hydrogeology in site.

11.4.6.4 Food industry

In food Industry, brines (not specifically from desalination) are traditionally used for salting pickles saving also energy for salt production, and it is also with homogeneous quality meeting the health administration requirements. It could be also used as cooling fluid because the salt reduces freezing point.

11.4.6.5 Growing of halophiles

Halophiles are microorganisms including bacteria, archaebacteria, and some eukaryotic organisms that live in hypersaline environments with different salinities from moderate to extreme halophiles such as *Salinibacter* species. These species have interesting applications such as the production of enzymes, biopolymers, pigments, and polyunsaturated fats.

11.4.6.6 CO_2 retention technologies

Some retention technologies of CO_2 such as mineral sequestration by means of carbonation with reaction of Mg and Ca compounds could be other potential uses for brines [58].

11.4.6.7 Deicing and dust suppression

Saline solutions and salts are well-known for its use in deicing of roads and other infrastructures. Brines could be used for this purpose, saving costs in salt production. However, due to their corrosive characteristics, besides the substitution by other more specific products and the transport costs, it could be only economically feasible when applied in short distances.

Brine could also be used for dust suppression in large mine sites as it has been reported for other kinds of brines.

11.4.6.8 Other uses

Different products obtained from brines could be potentially used for the production of cattle food, fire retardants, fireproof construction materials, products for the paper industry, plastics, paintings, dyes and waterproof products, road paving, and the production of flocculants for water treatment.

11.5 Emerging technologies

It is difficult to define the limits for the term "emerging" when we speak about technologies due to the fast development speed of technology and research. Anyway, it is conventionally accepted in the industry that the definition of emerging technologies to describe those that are not absolutely available and operating in the market at a large scale for years.

Inside this category, we can mention new developments such as FO, MD, PV, and capacitive deionization (CDI) with their different variations.

These technologies were born with the aim to develop new desalination technologies with less energy consumption, but the reality is that it does not seem that any of these technologies could oust RO as the main desalination technology in the near future.

However, the use of these technologies alone or combined between them or with RO or evaporation technologies (as hybrid systems) seems very promising for the reduction of brine volumes, energy production, or mineral production.

11.5.1 Forward osmosis

FO process and its different variants (pressure enhanced osmosis, PRO) are based on the use of a draw solution with high osmotic pressure for extracting product water from seawater or high salinity water passing through a membrane, taking advantage of the osmotic pressure generated to equilibrate chemical potential on both sides of the membrane. Further, water is separated from the draw solution by means of different processes (mainly thermal or membranes).

The process has numerous potential applications although only a few have been commercialized [59]; water-treatment plants for emergency situations, power generation, enhanced oil recovery, produced water treatment, fluid concentration, thermal desalination feedwater softening, water substitution, osmotic MBR (membrane bioreactor), and hybrid systems.

The main challenges of FO are the reduced flux (and in consequence increased membrane area), the lack of effective draw solutions, and effective separation techniques. For drinking water applications, the draw solution has to be nontoxic and compatible with the application. In Fig. 11.17, a laboratory scale pilot plant is shown.

The application of FO and PRO technologies for the treatment of desalination brines will require more membrane development, more studies about the uses with brine and integration of RO, PRO with pressure exchangers [39]. Until date, the most promising application is for energy production by means of blending of brines with water from different origins.

11.5.2 Pervaporation

PV (vacuum PV and PV with carrier gas technologies) is membrane separation processes with nonporous membranes applied to miscible liquids. Separation is produced by means of the application of vacuum in the membrane side where permeate is collected as vapor, which is finally condensed as the product.

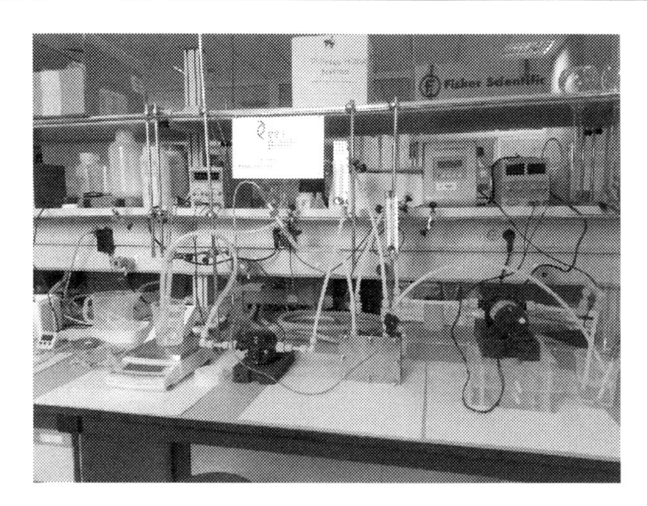

Fig. 11.17 FO pilot plant at the University of Alicante R&D project.
Courtesy of Domingo Zarzo.

One of its advantages is the lower latent heat compared to evaporation techniques, which is very useful for temperature-sensitive compounds.

The main drawbacks of pervaporation technology are the low-permeate flow rate, water flux, and reduced membrane stability. It is commercially used for dehydration of ethanol and isopropanol, and its use in desalination has been is restricted to pilot plants and small installations.

11.5.3 Membrane distillation

MD and PV are thermal processes involving a phase change with the water flux being depending on the vapor pressure difference across the membrane.

In membrane distillation technologies (direct contact MD, air gap MD, and vacuum MD), the preheated saline water and permeate are in both sides of a hydrophobic membrane that maintains liquid streams out of the membrane. Due to ΔT and vapor pressure driving force, water is vaporized from feed side, diffused through the membrane, and finally condensed into the cold permeate side, leaving salts in the feed side.

This technology has a promising perspective to its large-membrane contact area, high-salt rejection, small footprint, and mild-operating conditions, being also capable of integrating renewable energies from different sources.

The challenges are low-permeate flow rate and water flux, membrane fouling and pore wetting, long-term performance and uncertain economics, and energy costs.

A continuous membrane distillation crystallization (CMDC) was tested for obtaining NaCl products from desalination brines, which resulted in increased water and solid fluxes [60]. Different authors have tested this technology [61,62] as well as the scaling effects of $CaCO_3$ and $CaSO_4$ over the membranes [63,64] with the major advantage to concentrate the salt up to the saturated state, which allows its crystallization.

11.5.4 Capacitive deionization

CDI is an electrochemically induced alternative approach for removing ions from concentrated aqueous solutions by means of forcing charged ions into the electrical double layer at an electrode-solution interface when the electrode is connected to an external power supply.

Main strengths of this technology are the low operating costs, reduced pretreatment, high recovery, and the reduced fouling due to the reversal charge, while the most critical component is the carbon electrode materials due to electrosorptive capacity depends strongly on the physical properties such as surface area and conductivity of the electrode.

11.5.5 Other technologies

Other emerging desalination technologies of interest are the use of nanoporous graphene, nanotechnologies applied to membranes, technologies coupled to renewable energy (solar humidification-dehumidification, greenhouse distillation, geothermal desalination), biomimetic membranes (such as aquaporins), and the use of ultrasound or cavitation. There are not many reported experiences in their application for brine treatment.

As a remarkable example, research about nanoporous graphene has showed that the permeability of this material could be several orders of magnitude higher than conventional membranes, although there are only laboratory studies or modeling.

Table 11.4 Comparison of emerging technologies

Technology	Strengths	Challenges	Current situation
Forward osmosis	High rejection, low fouling, less energy consumer	Low flux (increased area). Draw solutions	Promising for energy production. Small applications, membranes in development
Pervaporation	Lower latent heat	Low permeate flow rate, water flux, membrane stability	Commercially used for dehydration of ethanol and isopropanol. For desalination, pilot and small plants
Membrane distillation	Large membrane contact area, small footprint, integration with renewables	Low permeate flow rate, water flux, membrane fouling	Laboratory and small pilot plants
Capacitive deionization	Low operating costs, low fouling	Electrode materials	Laboratory and small pilot plants
Others			Under research

We can also consider as emerging the different EDR variations (metathesis, bipolar membranes) for chemical production, although they have been mentioned before.

In Table 11.4, the main strengths and challenges of the main emerging technologies are summarized.

Morillo et al. [65] concluded in a paper, after comparing most of the available technologies and their pros and cons, that MD, FO, electro-separation processes, and metal recovery were the most promising ones for the management of desalination of brines.

11.6 Conclusions

It is clear that brine management and/or its disposal is one of the main issues of desalination from the environmental point of view, and it may even suppose the failure of a desalination project if it is not planned and solved from the very first beginning of the project.

In general, brines from seawater plants are disposed to the ocean without significant impact if it is done properly, with all the previous environmental studies and the further monitoring schemes during the plant operation. However, brines from brackish water plants, with a wide casuistry, due to the different water quality and even including is some cases the presence of toxic substances, nutrients, organic matter, or contaminants of emerging compounds, generally represent very complicated or not feasible solutions for brine management.

Therefore, there is a growing interest in the valorization of brines by means of salt production, energy recovery, and production or other uses such as aquaculture, agriculture, or environmental applications.

Traditional technologies used for this application such as ZLD with evaporation crystallization and solar ponds have demonstrated their technical feasibility, although the economic feasibility is limited and there are difficulties to apply them at large scale.

The emerging technologies can be developed further for brine treatment, special applications, or energy production, where they may be combined with conventional technologies as hybrid systems.

To date, almost all the brine treatment or management systems looking for mineral extraction or energy production are at the level of research, pilot plant, or small scale demonstrations. This means that it will be very difficult to apply any of them to the current large-scale seawater desalination plants in near future, although there are many research works and published papers about this subject, and new companies developing the necessary equipment (mainly membrane manufacturers) for their development.

To conclude, it is also necessary to mention the strong link of the described technologies with the use of renewable energies, which is one more aspect of the water and energy nexus.

Acknowledgments

The author would like to express his gratitude to the following people and institutions:

- Valoriza's O&M managers who supplied operational data, brine analysis and photos: Francisco Domenech, Carlos Fernandez, Ignacio Graciano, Raul Lemes, Pablo Lobo, Francisco Molina and Tom Ransome.
- Valoriza's technical team who supplied information and knowledge; Mercedes Calzada, Elena Campos, Carlos Garcia and Patricia Terrero.
- Jose Antonio Garcia, Jose Luis Sanchez Lizaso and Daniel Prats of the University of Alicante for supplying photos and valuable information.

References

[1] Latteman S, Hopner T. Environmental impact and impact assessment of seawater desalination. Desalination 2008;220(1–3):1–15.

[2] Walker T, Roux A, Owens E. Western corridor recycled water project—the largest recycled water scheme in the southern hemisphere. In: Khan SJ, Stuetz RM, Anderson JM. Water reuse and recycling. Sydney: UNSW Publishing; p. 498.

[3] Davis JR, Koop K. Eutrophication in Australian rivers, reservoirs and estuaries—a southern hemisphere perspective on the science and its implications. Hydrobiologia 2006;559:23.

[4] Khan SJ, Wintgens T, Sherman P, Zaricky J, Schafer AI. Removal of hormones and pharmaceuticals in the advanced water recycling demonstration plant in Queensland, Australia. Water Sci Technol 2004;50:15.

[5] Drewes JE, Bellona C, Oedekoven M, Xu P, Kim TU, Amy G. Rejection of wastewater-derived micropollutants in high-pressure membrane applications leading to indirect potable reuse. Environ Prog 2005;24:400.

[6] Perez-Gonzalez A, Urtiaga AM, Ibanez R, Ortiz I. State of the art and review on the treatment technologies of water reverse osmosis concentrates. Water Res 2012;46:267–83.

[7] Ahmed M, Shayya WH, Hoey D, Al-Handaly J. Brine disposal from reverse osmosis desalination plants in Oman and the United Arab Emirates. Desalination 2001;133:135.

[8] Burn A, Hoang M, Zarzo D, Olewniak F, Campos E, Bolto B, et al. Desalination techniques—a review of the opportunities for desalination in agriculture. Desalination 2015;364:2–16. Special Issue: Desalination for agriculture.

[9] WHO. Desalination for safe water supply. Guidance for the health and environmental aspects applicable to desalination. Geneva: Public health and the environment world health organization; 2007.

[10] Mickley MC. Membrane concentrate disposal: practices and regulation. Desalination and water purification research and development program report n 123 (2nd edition). Denver: US Department of the interior, bureau of reclamation; 2006.

[11] NSW public works. Brackish groundwater: a viable community water supply option? Waterlines report series, vol. 66. Camberra: National Water Commission; 2011.

[12] Mezher T, Fath H, Abbas Z, Khaled A. Techno-economic assessment and environmental impacts of desalination technologies. Desalination 2010;266:263–73.

[13] Badruzzaman M, Oppenheimer J, Adham S, Kumar M. Innovative beneficial reuse of reverse osmosis concentrate using bipolar membrane electrodialysis and electrochlorination processes. J Membr Sci 2009;326:392–9.

[14] Fernandez C, Dominguez A, Ibanez R, Irabien A. Electrodialysis with bipolar membranes for valorization of brines. Sep Purif Rev 2016;45(4):275–87.
[15] Aral H, Sleigh R, LI S. Salt recovery strategies for new value-added salt products. Aust J Dairy Technol 2006;61(2):143–6.
[16] Le Dirach J, Nisan S, Poletiko C. Extraction of strategic materials from the concentrated brine rejected by integrated nuclear desalination systems. Desalination 2005; 182(1–3):449–60.
[17] Jeppesen T, Shu L, Keir G, Jegatheesan V. Metal recovery from reverse osmosis concentrate. J Clean Prod 2009;17:703–7.
[18] Mero JL, editor. Elsevier oceanography series, vol. 1. Amsterdam, London, New York: Elsevier; 1965.
[19] Aldersey-Williams H. Periodic tales: a cultural history of the elements, from arsenic to zinc. New York: Harper Collins; 2011.
[20] Turek M, Krzusztof M, Chorazewska M, Dydo P. Use of the desalination brines in the saturation of membrane electrolysis feed. Desalin Water Treat 2013;51(13–15):2749.
[21] Chung HW, Nayar KG, Swaminathan J, Chehayeb KM, Lienhard JH. Thermodynamic analysis of brine management methods: zero-discharge desalination and salinity-gradient power production. Desalination 2017;404:291–303.
[22] Glater J, Cohen Y. Brine disposal from land based membrane desalination plants: a critical assessment. Metropolitan water district of Southern California: Los Angeles; 2003.
[23] Ravizkya A, Nadav N. Salt production by the evaporation of SWRO brine in Eilat: a success story. Desalination 2007;205:374–9.
[24] Laspidou C, Hadjibiros K, Gialis S. Minimizing the environmental impact of sea brine disposal coupling desalination plants with solar saltworks: a case study for Greece. Water 2010;2:75–84.
[25] Ahmed M, Arakel A, Hoey D, Thumarukudy MR, Goosen MFA, Al-Haddabi M, et al. Feasibility of salt production from inland RO desalination plant reject brine: a case study. Desalination 2003;158:109–17.
[26] Gilron J, Folkman Y, Savliev R, Waisman M, Kedem O. WAIV—wind aided intensified evaporation for reduction of desalination brine volume. Desalination 2003;158:205–14.
[27] Collares-Pereiraa M, Mendesa JF, Hortaa P, Korovessisb N. Final design of an advanced solar dryer for salt recovery from brine effluent of an MED desalination plant. Desalination 2007;211:222–31.
[28] Sal R, Segura C, Zarzo D. In: Towards a near zero liquid discharge in a solar thermal power industry. Proceedings of the International desalination association world congress on desalination and water reuse. Tianjin (China); 2013.
[29] Perez-Gonzalez A, Ibanez R, Gomez P, Urtiaga AM, Ortiz I, Irabien JA. Recovery of desalination brines: separation of calcium, magnesium and sulfate as a pretreatment step. Desalin Water Treat 2015;56(13):3617–25.
[30] El-Sayed MMH, Hani HA, Mohamed H. Polymeric ion exchangers for the recovery of ions from brine and seawater. Chem Eng Process Tech 2014;2(1):1020.
[31] Zarzo D, Campos E. Project for the development of innovative solutions for brines from desalination plants. Desalin Water Treat 2011;31:206–17.
[32] Peterskova M, Valderrama C, Gilbert O, Cortina JL. Extraction of valuable metal ions (Cs, Rb, Li, U) from reverse osmosis concentrate using selective sorbents. Desalination 2012;286:316–23.
[33] Leusbrock I. Removal of inorganic compounds via supercritical water: fundamentals and applications. Groningen: University of Groningen; 2011.

[34] Jibril BE, Ibrahim AA. Chemical conversions of salt concentrates from desalination plants. Desalination 2001;139:287–95.
[35] Cob SS, Genceli FE, Hofs B, Van Spronsen J, Witkanp GJ. Three strategies to treat reverse osmosis brine and cation exchange spent regenerant to increase system recovery. Desalination 2014;344:36–47.
[36] Almarsi D, Mahmoud KA, Abdel-Wahab A. Two-stage sulfate removal from reject brine in inland desalination with zero-liquid discharge. Desalination 2015;362:52–8.
[37] Telzhensky M, Birnhack K, Lehmann O, Windler E, Lahav O. Selective separation of seawater Mg2+ ions for use in downstream water treatment processes. Chem Eng J 2011;136:136–43.
[38] Garcia C, Molina F, Zarzo D. 7 year operation of a BWRO plant with raw water from a coastal aquifer for agricultural irrigation. Desalin Water Treat 2011;31:331–8.
[39] Chung T, Luo L, Wan FW, Cui Y, Amy G. What is next for forward osmosis (FO) and pressure retarded osmosis (PRO). Sep Purif Technol 2015;156(2:856–60.
[40] Ordonez A, Gutierrez B. El aprovechamiento energético de la salmuera mediante osmosis directa (in Spanish). AEDyR (Spanish desalination and reuse association), Madrid (Spain); 2012.
[41] Pattle RE. Production of electric power by mixing fresh and salt water in the hydroelectric pile. Nature 1954;174:660.
[42] Loeb S. Osmotic power plants. Science 1974;189:654–5.
[43] Tufa RA, Curcio E, Baak WV, Veerman J, Grasman S, Fontananova E, et al. Potential of brackish water and brine for energy generation by salinity gradient power-reverse electrodialysis (SGP-RE). RSC Adv 2014;4:42617–23.
[44] Cusick RD, Kim Y, Logan BE. Energy capture from thermolytic solutions in microbial reverse-electrodialysis cells. Science 2012;335:1474–7.
[45] Helfer F, Sahin O, Lemckert CJ, Anissimov YG. Salinity gradient energy: a new source of renewable energy in Australia. Water Utility J 2013;5:3–13.
[46] Akram W, Sharqawy MH, Leinhard JH. In: Energy utilization of brine from an MSF desalination plant by pressure retarded osmosis. Proceedings of the International desalination association world congress on desalination and water reuse. Tianjin (China); 2013.
[47] Ahmed M, Arakel A, Hoey D, Coleman M. Integrated power, water and salt generation: a discussion paper. Desalination 2001;134:37–45.
[48] Malfeito JJ, Diaz J, Farinas M, Fernandez Y, Gonzalez JM, Carratala A, et al. Brine discharge from the Javea desalination plant. Desalination 2005;185:87.
[49] Torquemada Y, Sanchez JL. Monitoring of brine discharges from seawater desalination plants in the Mediterranean. Int J Environ Health 2007;1(3):1–13.
[50] Zarzo D, Campos E, Prats D, Hernandez P, Garcia A. Microalgae production for nutrient removal in desalination brines. IDA J Desalin Water Reuse 2014;6(2):61–8.
[51] Allan GL, Banens B, Fielder S. Developing commercial inland saline aquaculture in Australia; parts 1 and 2. Deakin: Fisheries Research & Development corporation; 2001.
[52] Khan SJ, Murchland D, Rhodes M, Waite TD. Management of concentrated waste streams from high-pressure membrane water treatment systems. Crit Rev Environ Sci Technol 2009;39(5):367–415.
[53] FAO. In: Desalination for agricultural applications. Proceedings of the FAO expert consultation on water desalination for agriculture applications. Rome; FAO; 2004.
[54] Zarzo D, Campos E, Terrero P. Spanish experience in desalination for agriculture. Desalin Water Treat 2012;1–14.
[55] Al-Agha MR, Mortaja RS. Desalination in the Gaza strip: drinking water supply and environmental impact. Desalination 2005;173:157–71.

[56] De Souza FI, Da Silva N, De Sousa ON, Cruz J, Medeiros AC, Nascimento VC, et al. Agricultural potential of reject brine from water desalination. Afr J Agric Res 2015; 10(51):4713–7.
[57] Sanchez AS, Nogueira IBR, Kalid RA. Uses of the reject brine from inland desalination for fish farming, Spirulina cultivation, and irrigation of forage shrub and crops. Desalination 2015;364:96–107.
[58] De Vito C, Ferrini V, Mignardi S, Cagnetti M, Leccese F. Progress in carbon dioxide sequestration via carbonation of aqueous saline wastes. Periodico di Mineralogia 2012;81(3):333–44.
[59] Nicoll PG. In: Forward osmosis-a brief introduction. IDA world congress in desalination and reuse. Tianjin (China); 2013.
[60] Chen G, Lu Y, Krantz WB, Wang R, Fane AG. Optimization of operating conditions for a continuous membrane distillation crystallization process with zero salty water discharge. J Membr Sci 2014;450:1–11.
[61] Gryta M. Concentration of NaCl solution by membrane distillation integrated with crystallization. Sep Sci Technol 2002;37(15):3535–58.
[62] Ji X, Curcio E, Al Obaidani S, Di Profio G, Fontananova E, Drioli E. Membrane distillation-crystallization of seawater reverse osmosis brines. Sep Purif Technol 2010;7(1):703–7.
[63] Curcio E, Ji X, Di Profio G, Sulaiman AO, Fontananova E, Drioli E. Membrane distillation operated at high seawater concentration factors: role of the membrane on CaCO3 scaling in presence of humic acid. J Membr Sci 2010;346(2):263–372.
[64] He F, Gilron J, Lee H, Song L, Sirkar KK. Potential for scaling by sparingly soluble salts in crossflow DCMD. J Membr Sci 2008;311:68–80.
[65] Morillo J, Usero J, Rosado D, El Bakouri H, Riaza A, Bernaola FJ. Comparative study of brine management technologies for desalination plants. Desalination 2014;336:32–49.

Desalination of shale gas wastewater: Thermal and membrane applications for zero-liquid discharge

12

Viviani C. Onishi, Eric S. Fraga†, Juan A. Reyes-Labarta‡, José A. Caballero‡*
*University of Coimbra, Coimbra, Portugal, †University College London, London, United Kingdom, ‡University of Alicante, Alicante, Spain

Nomenclature

AGMD	air gap membrane distillation
DCMD	direct contact membrane distillation
EC	evaporative crystallization
ED	electrodialysis
EDR	electrodialysis reversal
EIA	energy information administration
FO	forward osmosis
GHG	greenhouse gas
MEE	multiple-effect evaporation
MED	multieffect distillation
MD	membrane distillation
MSF	multistage flash distillation
MVC	mechanical vapor compression
NF	nanofiltration
NORM	naturally occurring radioactive materials
RO	reverse osmosis
SEE	single-effect evaporation
TDS	total dissolved solids
TOC	total organic carbon
TSS	total suspended solids
TVC	thermal vapor compression
VMD	vacuum membrane distillation
ZLD	zero-liquid discharge

12.1 Introduction

Shale gas is currently the natural gas resource whose production exhibits the largest worldwide growth. Especially in the last decade, technological developments in horizontal drilling and hydraulic fracturing ("*fracking*") have boosted large-scale gas extraction from previously inaccessible unconventional shale reservoirs.

Emerging Technologies for Sustainable Desalination Handbook. https://doi.org/10.1016/B978-0-12-815818-0.00012-6

Recent projections from the US Energy Information Administration (EIA) [1,2] draw attention to the global increase in natural gas exploitation from 342 in 2015 to 554 billion cubic feet per day (Bcf day^{-1}) by 2040. The almost 62% rise in total natural gas production is mainly due to the intensification in shale gas exploration. Actually, shale gas production is expected to grow by more than 125 Bcf day^{-1} over the forecast period, reaching 30% of all natural gas produced in the world by 2040 [1,2].

Along with the depletion of conventional natural gas reserves, supply reliability and energy independence have emerged as driving forces for further development of shale gas exploration [3]. Notwithstanding, the latent advance of shale gas production around the globe, notably in the United Kingdom, Argentina, Brazil, Australia, Algeria, and Poland, to name a few [4], has also prompted serious concerns about environmental and social implications associated with greenhouse gas (GHG) emissions [5,6], induced seismic events [7], and quantity and quality of natural water resources and wastewater discharges [8–11]. Regarding water-related impacts alone, shale gas production from tight shale formations usually requires impressive freshwater volumes and generates large amounts of polluting hypersaline wastewater. Consequently, water management is nowadays one of the biggest challenges faced by the shale gas industry for maintaining process cost-effectiveness, while accounting for the environmental and human health protection [12,13].

Environmental, public health, and socioeconomic risks can be significantly reduced by adequate high-salinity wastewater treatment for allowing water reuse (i.e., water reinjection in new wells or existing ones), water recycling (i.e., water reuse in other activities not related to hydraulic fracking operations), and safe disposal. Decreasing total dissolved solids (TDS) is the key consideration to attain water quality required for internal and/or external reuse or discharge [13]. Within this framework, the application of effective desalination technologies is imperative to enhance overall shale gas process efficiency and sustainability [14,15]. The main strength of desalination resides in its ability to achieve salt concentrations that comply with strict regulations, promoting cleaner shale gas production [16,17]. In this chapter, the most promising thermal- and membrane-based desalination alternatives for shale gas wastewater management are summarized and examined in detail. Energy and economic analyses of potential zero-liquid discharge (ZLD) processes are presented as well, to evaluate the best desalination options for more sustainable shale gas development.

12.2 Water consumption, wastewater generation, and management options

12.2.1 Water consumption in shale gas operations

Contrarily to conventional natural gas production from geological formations such as porous sandstone and carbonate reservoirs, shale gas extraction is strongly impaired by the low shale rock permeability that compels the use of additional engineered solutions for attaining cost-effective production rates [9,18]. Economically viable gas exploitation from shale reservoirs is facilitated through the combined application of horizontal drilling and fracking processes [19]. These techniques together have allowed access to major shale deposits and improved permeability for releasing natural gas entrapped into tight rock formations [13].

In shale gas production, water-based fracturing fluid at very high pressure (about 480–680 bar) is injected in the shale well to unlock the existing fissures and create new artificial fracture networks, increasing the contact surface between reservoir and wellbore [20,21]. The chemical composition of the hydrofracturing fluid is conditioned by geological shale formations and water supply features, as well as the gas extraction operators [20,22]. Recent reports suggest that horizontal drilling and well-completion technologies demand about 7570–30,000 m^3 (~2–8 million US gallons) of water per well operation [23,24]. The hydraulic fracturing process requires ~90% of the total water amount, while the remaining (~10%) is used for horizontal drilling [25]. As a result of the exhaustive water consumption, progress in shale gas industry is greatly restricted by water availability, particularly in water-stressed regions [26,27]. In these areas, the effects of water shortages can be controlled by enhancing water usage efficiency in the shale gas process. The latter can be achieved via more rigorous regulations on water conservation and reuse and, finally, through the implementation of effective desalination plants.

12.2.2 Wastewater generation in shale gas operations

Shale gas wastewater encompasses both flowback water and produced water (also referred as formation water). Depending on the geologic setting and the well characteristics, US shale basin exploration indicates that around 10%–80% of the injection fluid returns to the surface as flowback water within the first 2 weeks following the hydraulic fracturing operation [23,28]. Afterwards, with the beginning of gas production, flowback water gradually decreases—usually, it remains in a range from ~210 to 420 US gallons h^{-1}, as has been observed in prominent shale plays from North America, including Marcellus, Fayetteville, Haynesville, and Barnett [29]—and high-salinity produced water is recovered over the well lifetime (~20–40 years) [30]. Recently, Kondash et al. [31] have estimated wastewater quantities ranging from 0.5 to 3.8 million US gallons per well over a period of 5–10 years of shale gas production. Among other pollutants, the high-salinity nature (average values typically higher than 100,000 ppm TDS) of shale gas wastewater is extremely hazardous to the environment and human health [32], and demands energy-intensive desalination processes. Table 12.1 displays the average water amounts required for horizontal drilling and fracking operations, and shale gas wastewater data from important US shale plays.

12.2.3 Wastewater management options

Different management options are available for dealing with the wastewater from shale gas operations. In the United States, it is estimated that almost 95% of all wastewater generated in shale gas industry is currently disposed in Class II saltwater disposal wells through deep underground injection [22,37]. Concerning the latter procedure, waste brine can be released into the environment with or without water

Table 12.1 Water amount required per well for drilling and hydrofracturing processes, and shale gas wastewater information from prominent US shale plays

Data source	US shale play	Water amount (m^3)	Wastewater recovery (%)	Average TDS (kppm)
Hayes [33]	Marcellus	11,356 – 15,142	25	157[a]
Acharya et al. [34]	Fayetteville	11,368		13
	Woodford	–		30
	Barnett	12,719	15 – 40[b]	80
	Marcellus	14,627		120
	Haynesville	14,309		110
Galusky and Hayes [35]	Barnett	11,356 – 18,927	25–40	~92
Hayes and Severin [36]	Marcellus	–	–	120[a]
	Barnett	–	–	50.55[c]
Slutz et al. [28]	–	12,700–19,000	10 – 40	–
Vidic et al. [9]	Marcellus	7570–26,500	9–53	–
Zammerilli et al. [24]	Marcellus	7570 – 22,712	30 – 70	70
Rosenblum et al. [22]	Niobrara	11,000	~3–30[d]	18.6–18.8[d]
Hammond and O'Grady [23]	–	10,000–30,000	40–80	–

[a]TDS average values for the shale gas flowback water in 14th day following hydraulic fracturing.
[b]Overall produced water recovery after 90 days.
[c]TDS average values for the shale gas flowback water in 10th to 12th day following hydraulic fracturing.
[d]Average values in 15th and 220th days following hydraulic fracturing.

treatment [38]. Although underground injection is the preferred practice for managing wastewater due to its economic benefits, it has lately been associated with potential induced seismic activity, and groundwater and soil contamination [7,37]. Moreover, capacity of Class II disposal wells is becoming progressively more limited and, consequently, it might not be able to accommodate all produced shale gas wastewater [39]. Besides water conservation policies and severe environmental regulations on discharges quality, disposal capacity constraints have also emphasized the importance of developing new alternatives for high-salinity wastewater desalination, mainly to allow its reuse or recycling [40]. Fig. 12.1 presents the main options for wastewater management in shale gas industry.

Reusing wastewater in hydraulic fracking operations, commonly classified as "*internal reuse*" [13], is an economically advantageous management strategy to address current concerns about the considerable freshwater consumption and wastewater pollution risks. However, direct water reuse is unsuitable due to the high concentration of contaminants that can compromise the well exploration [36]. For this reason, the onsite portable units for wastewater pretreatment—which comprises

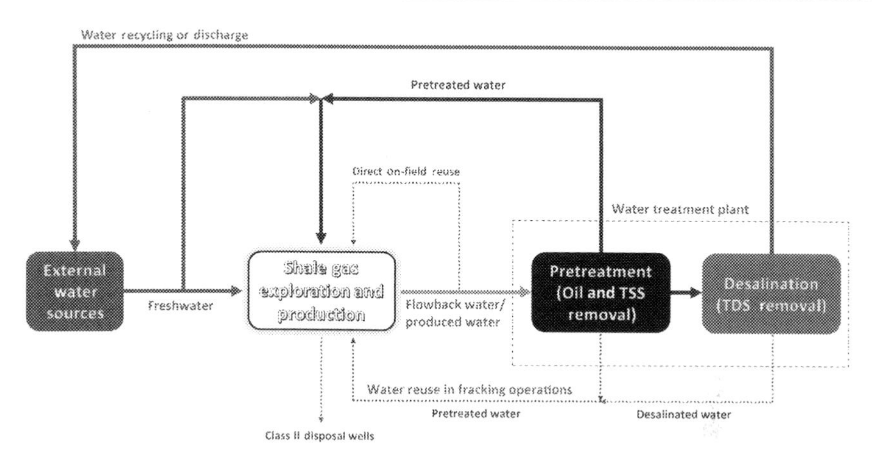

Fig. 12.1 Wastewater management alternatives for shale gas industry.

primary and secondary treatment options such as filtration, physical and chemical precipitation, flotation, sedimentation, and softening—are generally used to avoid operational problems [39].

The onsite treatment plants usually include established technologies to remove total suspended solids (TSS), oil and greases, and scaling materials [41]. Typically, the onsite treated wastewater is blended with freshwater to reduce the high TDS contents (which are responsible for negative viscosity effects on the hydraulic fluid), allowing its reuse in hydraulic fracturing operations [13]. Nevertheless, even if transportation costs are not considered in onsite plants, capacity and practical constraints alone restrict the application of this treatment alternative [39]. It is also worth noting that wastewater composition and water treatment technologies employed in the corresponding system are crucial to the process cost-effectiveness. Moreover, internal reuse practice is dependent on the demand for new well exploitation and ultimately, on the industry expansion.

With the maturity of shale gas industry, drilling and fracking operations will eventually decrease, transforming the activity in a potential wastewater producer. At this point, the application of effective desalination processes will become inevitable [9,42]. In this context, centralized (offsite) plants for wastewater pretreatment followed by effective desalination emerge as other options for water management. In fact, they are appealing alternatives to achieve high water quality, permitting its reuse for other beneficial purposes—for instance, water recycling for agricultural activities [43]—or even safe release to surface water bodies.

12.3 Challenges of shale gas wastewater desalination

Shale gas wastewater produced by hydraulic fracturing operations present physical and chemical properties that varies according to different factors, such as formation geology and geographic location, fracking fluid composition, and the water's time of contact with shale deposits [13,44,45]. Note that the fracturing fluid is a complex mixture, predominantly composed of water and proppant (sand suspension ~99.5%, *v*/*v*) and chemical enhancers that embrace surfactants, inorganic acids, biocides, friction reducers, scale and corrosion inhibitors, flow improvers, etc. [20,46,47]. Furthermore, chemical contents in shale gas wastewater may also vary throughout the time of well exploitation [13].

The selection of most suitable treatment alternatives is strongly influenced by the physicochemical composition of the wastewater [45]. Apart from the chemical additives utilized within hydrofracturing fluids, shale gas wastewater is also composed of formation-based constituents, which comprises salt and other minerals (i.e., scale-forming ions: Ba^{2+}, Ca^{2+}, and Mg^{2+}), organic matter, and naturally occurring radioactive materials (NORM) [48–51]. Table 12.2 presents the typical composition ranges for critical components in shale gas wastewater from Marcellus play.

Among all contaminants, removal of the high TDS contents from shale gas wastewater is especially challenging due to the intensive energy consumption needed to accomplish with the severe regulations on water quality (particularly on water recycling and safe disposal). In addition, besides the variations in wastewater compositions throughout the well lifetime, another complicating factor is associated with the considerable differences observed in wastewater from distinct shale basins, and even in different wellbores from the same well pad (see Table 12.1) [30]. Fig. 12.2 displays conceptual profiles for TDS concentration and wastewater flowrate after hydraulic fracturing operations.

Table 12.2 Typical concentration ranges for critical constituents found in shale gas wastewater from Marcellus play[a]

Constituent	Minimum ($mg L^{-1}$)	Maximum ($mg L^{-1}$)	Average ($mg L^{-1}$)
Total dissolved solids (TDS)	680	345,000	106,390
Total suspended solids (TSS)	4	7600	352
Total organic carbon (TOC)	1.2	1530	160
Chloride	64.2	196,000	57,447
Sulfate	0	763	71
Sodium	69.2	117,000	24,123
Calcium	37.8	41,000	7220
Barium	0.24	13,800	2224
Strontium	0.59	8460	1695
Iron, total	2.6	321	76
Alkalinity (as $CaCO_3$)	7.5	577	165
Bromide	0.2	1990	511
Magnesium	17.3	2550	632
Oil and grease	4.6	802	74

[a]Data compiled from Barbot et al. [20] for flowback water samples collected between day 1 and day 20 following hydraulic fracturing.

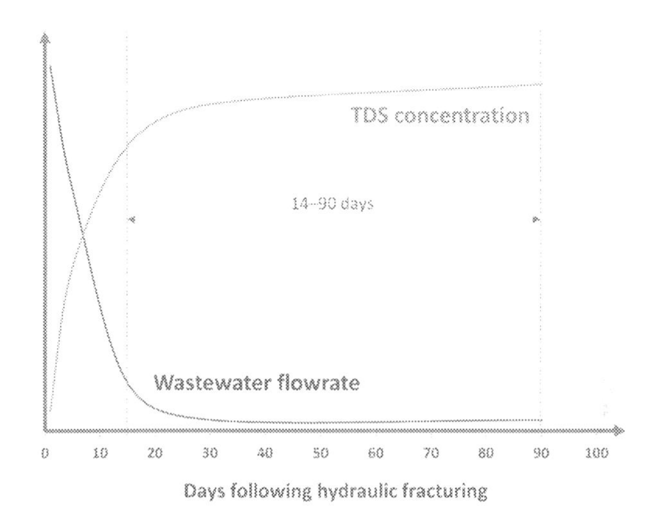

Fig. 12.2 Conceptual profiles for total dissolved solids (TDS) concentration and wastewater flowrate in function of time from hydraulic fracturing operations.

Hayes and Severin [36] have showed TDS contents in wastewater samples from Barnett shale play ranging from 5850 to 31,400 ppm (average value of 25,050 ppm) in day 1 following hydraulic fracturing; and, values between 16,400 and 97,800 ppm (average value of 50,550 ppm) for 10–12 days from the beginning of well exploration. As reported by Acharya et al. [34], TDS concentrations in shale gas wastewater can widely vary from average values of 13,000 ppm for Fayetteville shale play (maximum value of 20,000 ppm), to 120,000 ppm for Marcellus shale play (maximum value >280,000 ppm TDS). Also, chemical composition analyses performed by Thiel and Lienhard [30] have indicated TDS amounts in wastewater from Permian and Marcellus basins ranging from 120,000 to ~250,000 ppm. Results presented by Barbot et al. [20] reveal even higher maximum TDS concentrations of 345,000 ppm (data from Northeast Pennsylvania basins).

Several desalination processes can be applied to treat the hypersaline shale gas wastewater, for ensuring the strict composition constraints in accordance with specific wastewater-desired destinations (i.e., water reuse, water recycling, or disposal). Desalination technologies include thermal- and membrane-based desalination processes. Thermal technologies comprise multistage flash distillation (MSF); multieffect distillation (MED); and, single- or multiple-effect evaporation (SEE/MEE) systems, which can be coupled to mechanical or thermal vapor compression (MVC/TVC). The membrane-based group includes processes such as membrane distillation (MD), forward osmosis (FO), reverse osmosis (RO), and electrodialysis (ED). Fig. 12.3 displays the schematic representation of main thermal- and membrane-based processes for shale gas wastewater desalination.

High TDS contents in shale gas wastewater pose specific desalination challenges, mostly related to high-energy consumption and operational problems produced by scaling, fouling, and corrosion [52,53]. Actually, deposition of scale-forming ions on the equipment surface can compromise the system energy performance of both

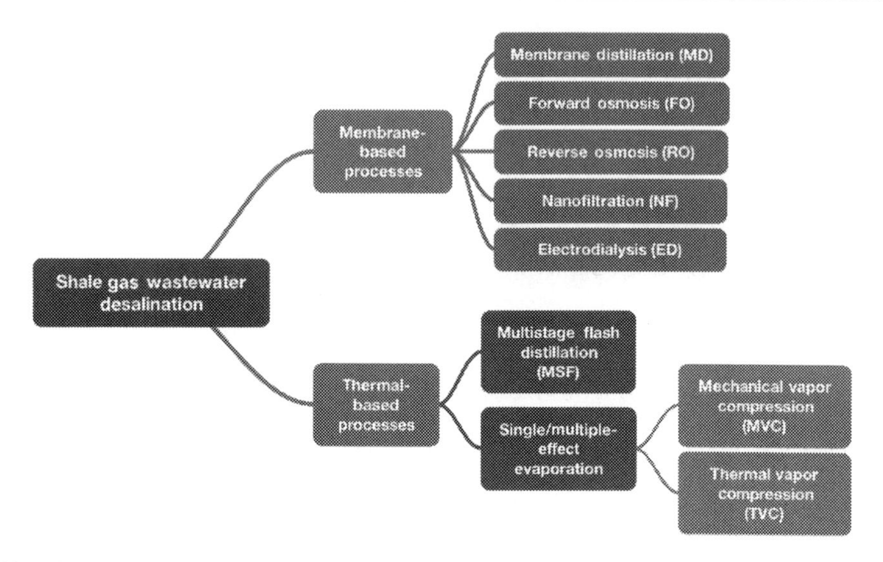

Fig. 12.3 Schematic representation of major thermal- and membrane-based processes for shale gas wastewater desalination.

thermal- and membrane-based technologies. Due to changes in process conditions (i.e., composition, pH, and temperature) during desalination, fouling, and scaling surface-growth phenomena can reduce heat transfer in thermal evaporation technologies and mass transfer in membrane-based systems [53]. In the last case, the presence of scaling compounds in the wastewater can severely decrease permeate flux across the membrane [54,55].

12.4 ZLD desalination for wastewater management

12.4.1 Drivers and benefits of ZLD systems

In recent years, ZLD desalination has attracted increased interest by the scientific community and industry as an auspicious strategy for wastewater management. This is mainly due to its ability to enhance water usage efficiency, while reducing brine discharges and water and disposal-related environmental impacts [56,57]. From general efficiency and environmental protection viewpoints, the ambitious goal of zero-emission desalination could be a game changer for the entire shale gas industry.

ZLD desalination systems are high-recovery processes that allow the production of high-quality treated water (i.e., freshwater) and concentrate brine, by decreasing liquid contents present in the brine waste [58]. In this work, although evaporation ponds or brine crystallizers are required to literally achieve zero-discharge operation, brine discharges salinity close to salt saturation conditions is considered as ZLD operation. Thus, ZLD alternatives are usually operated to recover ~75%–90% of the total amount of water from the wastewater. The remaining water contents can be eliminated by including brine crystallizers or evaporation ponds into the system. Consequently, almost the water totality in the wastewater can be reclaimed for internal reuse in shale gas operations. In this way, ZLD desalination enhances water sustainability and diminishes the environmental pollution and social risks related to wastewater and brine disposals, as well as depletion of freshwater resources [14,56].

Although widely recognized as an important approach for reducing water impacts and improving water supply sources, the implementation of ZLD desalination systems is still limited by their intensive energy consumption and high associated processing costs [56,59]. However, recent studies have demonstrated the economic viability of thermal-based ZLD desalination systems applied to shale gas wastewater treatment [3,14,15,17]. In Onishi et al. [15], for instance, electric-driven SEE/MEE-MVC technologies for ZLD desalination (by considering brine discharges at 300,000 ppm or 300 g kg^{-1}) have presented specific energy consumption in a range of 28.12–50.47 kWh_e m^{-3}, with operational expenses estimated between 2.73 and 4.90 US\$ m^{-3} for 77% conversion ratio (freshwater production ratio at 7.99 kg s^{-1}). Also, the authors have shown freshwater production costs ranging from 6.7 US\$ m^{-3} (MEE-MVC with thermal integration) to 10.9 US\$ m^{-3} (SEE-MVC with thermal integration). It should be noted that disposal costs in Class II saline water injection wells (i.e., conventional deep-well injection) are estimated to be in the range of ~8–25 US\$ m^{-3} (~0.03–0.08 US\$ $gallon^{-1}$)—water disposal cost for locally available injection sites in Barnett shale play [34]. These results emphasize the need for developing more realistic energy performance and cost analysis for ZLD desalination systems, to evaluate the best trade-off between their benefits, energy consumption and capital and operating costs.

Future progress in ZLD applications to shale gas wastewater will ultimately be achieved by stricter regulations on water quality and brine discharges, as well as by incrementing regulatory incentives to compensate eventual economic shortcomings [56]. These factors, allied to the rising in wastewater disposal costs, will ultimately drive shale gas industry towards the implementation of cleaner ZLD desalination systems. Tables 12.3 and 12.4 present the freshwater production cost and energy consumption of promising thermal- and membrane-based ZLD desalination technologies for shale gas wastewater.

12.4.2 Environmental impacts

Since both thermal and electric power used in desalination systems are usually produced from fossil fuel energy sources, the elevated energy consumption related to ZLD systems is also responsible for significant pollutant emissions to the atmosphere. These emissions are predominantly composed by GHG (carbon dioxide), acid rain gases (nitrogen oxides and sulfur dioxide) and fine particulate matter [65]. According to EIA [66], around 939 g of CO_2 per kWh_e are generated from burning coal. Under the latter consideration, MEE-MVC systems operating at ZLD conditions will produce ~26.4–47.4 kg of CO_2 per cubic meter of treated water—considering an energy consumption in a range of 28.12–50.47 kWh_e m^{-3} [15]. Carbon footprint and other air pollutant releases directly (e.g., thermal sources as steam) or indirectly (e.g., energy

Table 12.3 Freshwater production cost and specific energy consumption of thermal-based systems for shale gas wastewater desalination

Desalination system	ZLD operation	Freshwater production cost (US$ m^{-3})	Specific energy consumption	Reference
SEE-MVC (electric-driven system with single-stage compression)	Brine salinity at 300k ppm and 76.7% of conversion ratio	10.90	50.47 kWh m^{-3} (4.90 US$ m^{-3})	Onishi et al. [15]
SEE-MVC (electric-driven system with multistage compression)	Brine salinity at 300k ppm and 76.7% of conversion ratio	10.85	49.85 kWh m^{-3} (4.84 US$ m^{-3})	Onishi et al. [15]
SEE-MVC (rigorous heat transfer coefficients estimations)	Brine salinity at 300k ppm and 76.7% of conversion ratio	10.07	49.78 kWh m^{-3} (4.83 US$ m^{-3})	Onishi et al. [14]
SEE-MVC	Not ZLD, 26% of brine salinity	–	23–42 kWh m^{-3}	Thiel et al. [60]
MEE (steam-driven system)	Brine salinity at 300k ppm and 76.7% of conversion ratio	12.85	214.19 kWh m^{-3} (10.24 US$ m^{-3})	Onishi et al. [15]
MEE-MVC (electric-driven system with single-stage compression)	Brine salinity at 300k ppm and 76.7% of conversion ratio	6.70	28.63 kWh m^{-3} (2.78 US$ m^{-3})	Onishi et al. [15]
MEE-MVC (electric-driven system with multistage compression)	Brine salinity at 300k ppm and 76.7% of conversion ratio	6.83	28.84 kWh m^{-3} (2.80 US$ m^{-3})	Onishi et al. [15]
MEE-MVC (rigorous heat transfer coefficients estimations)	Brine salinity at 300k ppm and 76.7% of conversion ratio	6.55	28.33 kWh m^{-3} (2.75 US$ m^{-3})	Onishi et al. [14]

Table 12.3 Continued

Desalination system	ZLD operation	Freshwater production cost (US$ m^{-3})	Specific energy consumption	Reference
MEE-MVC (hybrid steam and electricity energy sources)	Brine salinity at 300k ppm and 73.3% of conversion ratio	5.25	23.25 kWh m^{-3} (2.26 US$ m^{-3})	Onishi et al. [3]
MEE-MVC	Not ZLD, 26% of brine salinity	–	20 kWh m^{-3}	Thiel et al. [60]

from electricity grids) associated with ZLD schemes can be mitigated by developing higher energy efficiency technologies, and incorporating renewable (e.g., solar, wind, and geothermal energy) and/or low-grade energy sources [17,56].

Additional polluting risks linked to ZLD systems are connected to brine waste production. Concentrate management strategies can include brine disposal in landfills and evaporation ponds. Apart from soil contamination possibility, the deposition of solid wastes in landfills can also compromise groundwater by leaching chemicals through the soil matrix. Likewise, brine storage in evaporation ponds can cause environmental and social impacts, due to leakage risks, odors generation, and wildlife depletion [67]. These negative effects on water and soil and their consequences can be prevented by the implementation of reliable monitoring systems, as well as the use of impermeable linings to isolate surface zones [56].

Major thermal- and membrane-based process for ZLD desalination of shale gas wastewater are presented in the following sections.

12.5 ZLD desalination technologies for shale gas wastewater

Desalination systems for the ZLD treatment of high-salinity shale gas wastewater can comprise thermal- and membrane-based technologies such as SEE/MEE (with MVC or TVC), MD, FO, and RO (see Fig. 12.3). As described before, these technologies are able to produce high-quality water by accomplishing with the severe regulations on salt contents required for recycling opportunities (e.g., irrigation, livestock watering, or industrial uses). In addition, their modular feature and simple scale-up are propitious for the implementation of onsite treatment plants at shale plays constrained by infrastructure limitations [13]. Thermal-based evaporation systems coupled to MVC are comparatively well-established processes, whereas MD, FO, RO, and ED are promising technologies for high-salinity shale gas wastewater applications. Table 12.5 shows the main advantages and limitations of thermal- and membrane-based ZLD desalination processes.

Table 12.4 **Freshwater production cost and specific energy consumption of membrane-based systems for shale gas wastewater desalination**

Desalination system	ZLD operation	Freshwater production cost	Specific energy consumption	Reference
Direct contact MD system (waste heat energy source)	Brine salinity at 300k ppm or 30% (*w/w*), water recovery ratio of 66.7%	–	527–565 kWh m^{-3} (depending on feed temperature)	Lokare et al. [61]
Direct contact MD system (waste heat and electricity heat energy sources)	Brine salinity at 300k ppm or 30% (*w/v*), water recovery ratio of 66.7%	0.74–5.70 US\$ m^{-3} and 61–66 US\$ m^{-3} (with transportation costs)[a]	–	Tavakkoli et al. [62]
Two-stage RO system	Not ZLD, 26% of brine salinity	–	4–16 kWh m^{-3}	Thiel et al. [60]
Hybrid EDR-RO with crystallizer system	Brine salinity at 239k ppm, water recovery ratio of ~77%	–	10–17 kWh_e m^{-3} (EDR-RO) and 40 kWh_e m^{-3} (crystallizer)	Loganathan et al. [57]
ED system	Not ZLD	–	49.7 kWh_e m^{-3} (wastewater with 70k ppm TDS) and 175.7 kWh_e m^{-3} (wastewater with 250k ppm TDS)	Ahmad and Williams [63]
Integrated coagulation and ED system	Not ZLD, 91% of salt removal	–	~7–14 kWh m^{-3} (depending on the ED voltage)	Hao et al. [64]

[a]Values estimated based on cubic meter of feed water (with salinity of 100k ppm).

Table 12.5 Comparison between thermal and membrane-based technologies for ZLD desalination of shale gas wastewater

Desalination technology	Advantages	Drawbacks	Reference
Multistage flash distillation (MSF)	– Well-established technology with application to shale gas wastewater with large range of TDS contents – High-quality water product (ultrapure water or freshwater) – Technical maturity – Possibility of using geothermal or solar energy sources	– Cost and energy-intensive process, not suitable for small scale operations [68] – Intensive use of scale inhibitors and cleaning agents	NA
Single/multiple-effect evaporation with mechanical vapor compression (SEE/MEE-MVC)	– Well-established technology with Application to shale gas wastewater with large range of TDS contents (10–>220k ppm) – Brine discharge salinity up to 300k ppm TDS – Use of less intensive pretreatment processes, when compared to membrane-based technologies – High energy efficiency – High-quality water product (ultrapure water or freshwater) – Technical maturity – Modular feature – Heat exchangers and flashing tanks can be used to further enhance energy recovery, reducing energy consumption	– Energy-intensive process – Usually operated by high-grade electric energy (for this reason, these systems present high operating expenses and indirect GHGs emissions) – High capital costs, due to the expensive materials (stainless steel or titanium) required to prevent rusting	Onishi et al. [3,14–17]

Continued

Table 12.5 **Continued**

Desalination technology	Advantages	Drawbacks	Reference
	– Possibility of using geothermal or other renewable energy sources, which allows to reduce carbon footprint		
Membrane distillation (MD)	– Application to shale gas wastewater with high TDS contents – Brine discharge salinity higher than 200k ppm TDS – Modular feature and operation at low temperature and pressure – Low fouling propensity – Possibility of using low-grade thermal energy, including geothermal or waste heat, which allows to reduce operating costs and carbon footprint	– Energy-intensive process with energy consumption higher than RO and ED/EDR (DCMD requires 40–45 $kWh_t\ m^{-3}$ for seawater desalination [56]) – Heat integration (by using heat exchangers and brine recycling) is critical to enhance energy efficiency to competitive levels with thermal systems [69] – Membrane wetting potential – Intensive pretreatment and use of cleaning agents and scale inhibitors [70,71] – Limited to commercial applications	Carrero-Parreño et al. [72] Boo et al. [73] Singh and Sirkar [74] Kim et al. [75] Chung et al. [76] Lokare et al. [61]
Forward osmosis (FO)	– Application to shale gas wastewater with TDS contents up to 180k ppm [77] – Brine discharge salinities higher than 220k ppm TDS – Modular feature – Can be used for preconcentrating and pretreating wastewater prior RO process – High rejection of many contaminants	– Intensive pretreatment processes (softening, pH adjustment, ultrafiltration, ion exchange, etc.) to prevent operating problems related to fouling and scaling (however, these processes are less intensive and more economical than those required prior RO) – Regular membrane cleaning	Salcedo-Díaz et al. [78] McGinnis et al. [77] Chen et al. [79] Hickenbottom et al. [80] Yun et al. [81]

	– Propensity to membrane fouling and scaling lower than RO process (with reversible membrane fouling) – Low electricity consumption – Possibility of using low-grade thermal energy, including geothermal or waste heat, which allows to reduce operating costs and carbon footprint		
Reverse osmosis (RO)	– Application to shale gas wastewater with TDS contents up to 40–45 k ppm [41,78] – High energy efficiency – Technical maturity – Modular feature and great adaptability to wastewater treatment plants with other technologies, including water pretreatment processes [41] – Can be used for preconcentrating wastewater prior energy-intensive thermal processes [56] – Low energy consumption of ~2 kWh$_e$ m^{-3}, for seawater desalination [82]	– High propensity to membrane fouling and scaling, which requires intensive pretreatment processes (softening, pH adjustment, coagulant/flocculant addition, ultrafiltration, ion exchange, etc.) to prevent operating problems [83] – Intensive use of antiscalants [84] – Inability to operate at high hydraulic pressure – Stand-alone RO systems are not able to operate at ZLD conditions: brine discharge salinity up to 70k ppm TDS (crystallizer/evaporator should be included in the system) [56]	Salcedo-Díaz et al. [78] Miller et al. [55]
Nanofiltration (NF)	– Effective as softening for subsequent wastewater treatment processes – High water recovery – Energy consumption lower than RO – Mature technology	– Not effective as stand-alone process for shale gas wastewater treatment – Intensive pretreatment and scale inhibitors	Michel et al. [85]

Continued

Table 12.5 Continued

Desalination technology	Advantages	Drawbacks	Reference
Electrodialysis (ED) and electrodialysis reversal (EDR)	– Application to high-salinity wastewater – Ability to achieve high brine salinities (TDS > 100k ppm) – Salt removal rate ~91% (product water meets the requirements on water reclamation) – Relatively simple operation and maintenance – Low propensity to fouling (especially with coagulation pretreatment) – Long-term operation – Modular feature	– High energy consumption and related operating costs when coupled to crystallizers/evaporators to achieve ZLD conditions – Regular membrane cleaning to maintain operational production ratios – Inability to remove noncharged contaminants	Loganathan et al. [57] McGovern et al. [86] Peraki et al. [87]

12.5.1 Thermal-based ZLD processes

12.5.1.1 ZLD evaporation systems

Despite the significant research efforts on the development of thermal-based MSF and MED processes for seawater desalination [88–90], their application in ZLD systems for shale gas wastewater has not been reported in the literature to date. In general, thermal evaporation systems with MVC can be more advantageous than membrane technologies for shale gas wastewater treatment [13]. Due to lower susceptibility to rusting and fouling problems, MEE-MVC demands lesser energy-intensive pretreatment processes than those required prior to membrane desalination. Furthermore, thermal systems are generally more robust and require lower cleaning frequency and intensity than membrane ones [60]. On the other hand, while low-grade thermal energy can be used in membrane systems [61,62], typical thermal evaporation schemes with MVC are driven by high-grade electrical energy. Besides the related high operating costs and GHG emissions, this is also a barrier for their operation in remote areas without easy access to power grids. To surpass these limitations, geothermal or other renewable energy sources can be incorporated into the thermal systems.

ZLD thermal evaporation systems for the desalination of hypersaline shale gas wastewater have been addressed by Onishi et al. [3,14–17]. In Onishi et al. [15], the authors have developed a mathematical optimization model for SEE/MEE systems design, considering single and multistage MVC and heat integration. Fig. 12.4 displays the MEE-MVC system proposed by Onishi et al. [15] for the ZLD desalination of shale gas wastewater. Their modeling approach is aimed at enhancing process energy efficiency, while reducing polluting brine discharges. The authors have performed a thorough comparison between the optimal systems configurations found (SEE/MEE with single or multistage mechanical compression) under a wide range of inlet wastewater salinities (10,000–220,000 ppm TDS), to evaluate their ability

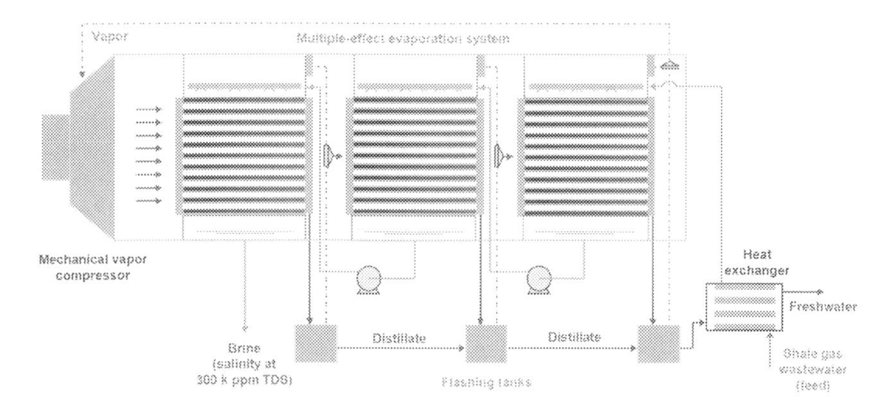

Fig. 12.4 Multiple-effect evaporation system with mechanical vapor compression (MEE-MVC) for the zero-liquid discharge (ZLD) desalination of shale gas wastewater as proposed by Onishi et al. [15].

to achieve high water recovery ratios and ZLD operation. Energy and economic analyses have revealed the MEE process with single-stage MVC as the most cost-effective system for treatment of shale gas wastewater. Further information on ZLD desalination process of shale gas wastewater via SEE/MEE-MVR systems can be found in references [14,15].

Based on the latter result, Onishi et al. [14] have proposed a new rigorous optimization approach for MEE-MVC systems design, by considering more precise estimation of the global heat transfer coefficient to minimize process costs. Furthermore, their method considers the modeling of major equipment features, including optimal number and length of tubes, and evaporator diameter. Their results indicate that the MEE-MVC system can be around 37% less expensive than SEE-MVC for recovering 76.7% of freshwater (brine discharge salinity at 300,000 ppm TDS). Afterwards, Onishi et al. [3] have focused on the high uncertainty related to well data (wastewater flowrates and salinities) from shale plays to support decision-makers in the implementation of more robust MEE-MVC systems. Distributions of energy consumption and operating expenses throughout different feeding scenarios are shown in Fig. 12.5.

Lastly, Onishi et al. [17] have developed a mathematical modeling approach for the optimization of solar energy-driven MEE-MVC systems. The authors have considered an integrated process composed by a solar-assisted Rankine cycle and a MEE-MVC desalination plant. The multiobjective optimization model allows to minimize environmental impacts, and investment and operating costs. Their trade-off Pareto-optimal solutions (especially intermediate solutions containing hybrid solar and electricity energy sources) reveal that renewable energy cogeneration in desalination ZLD plants can promote significant environmental and cost savings for shale gas industry. Fig. 12.6 presents the zero-discharge MEE-MVC system driven by solar energy proposed by Onishi et al. [17] for the desalination of high-salinity shale gas wastewater.

12.5.1.2 Crystallizers

Solid waste produced by thermal evaporation systems can be further concentrated in brine crystallizers. In this case, all remaining water can be recovered from waste brine. Analogously to SEE/MEE-MVC concentrators, electrically driven mechanical compressors are used in large-scale crystallizers (i.e., for treating brine flows higher than 6 gal min^{-1}) to superheat vapor and supply heat required for driving the evaporation process. For lower brine flows ranging 2–6 gal min^{-1}, steam-driven crystallizers are generally more economical [91]. While horizontal-tube falling film evaporators are preferred in SEE/MEE-MVC schemes, thermal evaporative crystallizers are generally operated through forced-circulation. Crystallization of concentrate brines is an energy-intensive process, which usually demands a range of 52–66 kWh$_e$ m^{-3} of treated water [56,67]. This is mainly due to the higher salt concentration and viscosity that characterize brine wastes. However, crystallizer technology can be especially appropriate for shale gas exploration areas in which deep-well injection is not allowed or costly, the solar irradiance is low, or cost of evaporation ponds construction is excessively high [92].

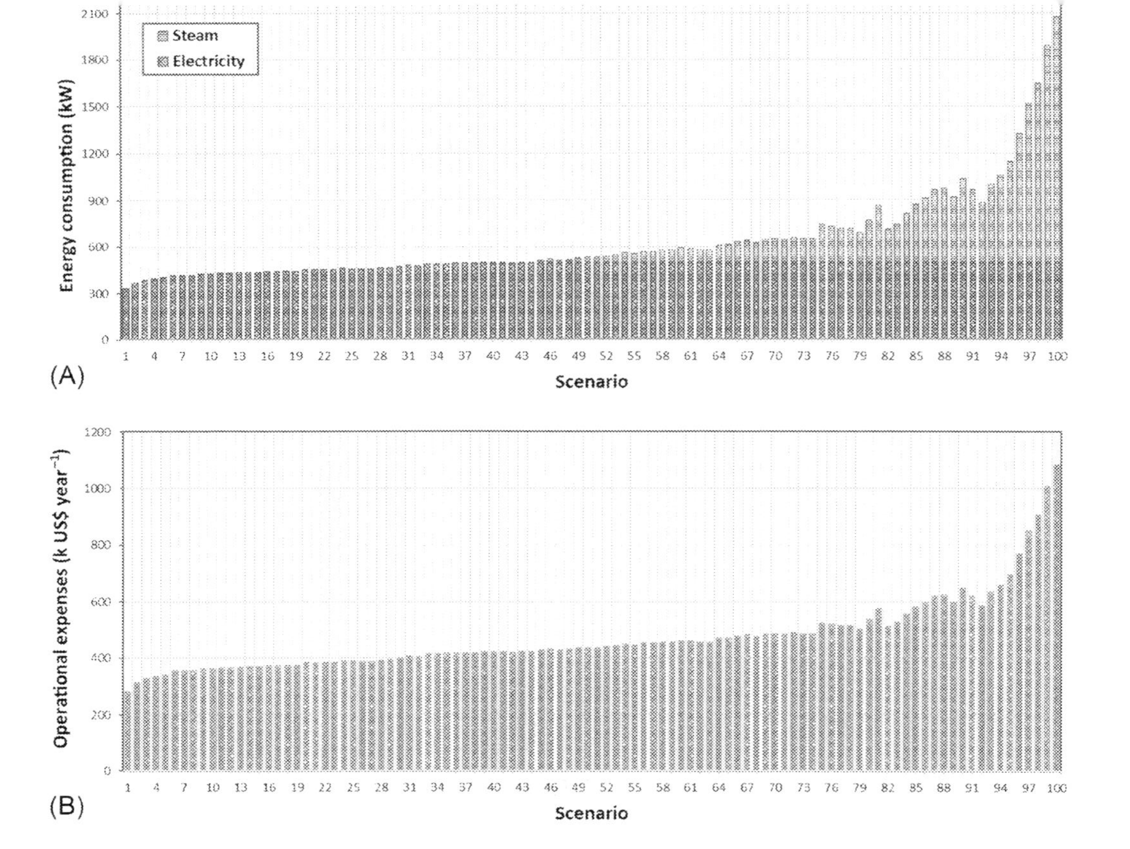

Fig. 12.5 Distributions throughout different feeding scenarios of zero-discharge MEE-MVC system for: (A) energy consumption and (B) operational expenses.

Data retrieved from Onishi VC, Carrero-Parreño A, Reyes-Labarta JA, Fraga ES, Caballero JA. Desalination of shale gas produced water: a rigorous design approach for zero-liquid discharge evaporation systems. J Clean Prod 2017;140:1399–414.

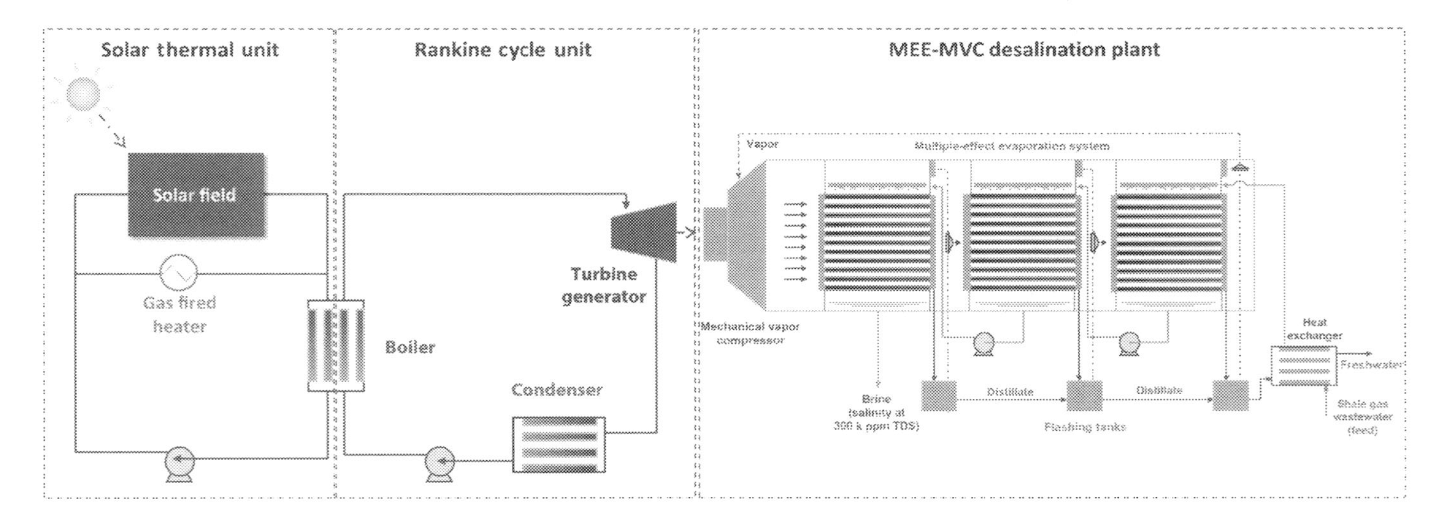

Fig. 12.6 Zero-discharge MEE-MVC system driven by solar energy for the desalination of high-salinity shale gas wastewater.

12.5.1.3 Evaporation ponds

Evaporation ponds are competitive disposal alternatives to thermal brine crystallizers. This technology uses natural solar irradiance to drive the evaporation process and eliminate water contents from the brine waste. Although the operational expenses are low, evaporation ponds implementation is constrained by its high capital investment and environmental concerns related to brine waste leakage risks [56]. Additionally, since the process allows to recover only solid wastes, the remaining water from waste brines cannot be reclaimed for recycling or reuse in shale gas operations. As a consequence, water usage efficiency in shale gas industry cannot be further improved (more than that achievable by thermal systems) by applying evaporation ponds into the desalination system. Also, evaporation ponds coupled to ZLD desalination systems should be designed to ensure the deposition of all precipitated solids over the zero-discharge plant lifetime, or even the construction of new ponds [91]. Fig. 12.7 depicts the schematic representation of a thermal-based ZLD evaporation plant coupled to the pretreatment system and crystallization or evaporation ponds.

12.5.2 Membrane-based ZLD processes

Membrane-based technologies have recently arisen as promising alternatives for ZLD desalination of high-salinity wastewater from shale gas production. Membrane systems usually present great potential for shale gas wastewater applications due to their high efficiency, operational and control simplicity, elevated permeability, and selectivity for some critical components, simple scale-up and possibility of using low-grade waste energy [93,94]. Table 12.6 presents process characteristics and applications of major membrane-based systems for ZLD desalination of shale gas wastewater.

Analogously to MVC concentrators, membrane-based technologies are able to achieve ZLD conditions with brine discharges salinity higher than 300,000 ppm or 30% weight-to-volume fraction (*w*/*v*) [61,72,78]. Note that, although these systems can theoretically concentrate the feed stream until the salt saturation conditions (~350,000 ppm or 35%, *w*/*v*), near-ZLD operation is preferable to prevent operational difficulties related to salt crystallizing in the system [62]—in this case, brine crystallizer units or evaporation ponds can be considered to recover the remaining water and valuable by-products [56]. Also, recent studies indicate that the energy requirements and associated capital and operating costs of membrane technologies can be competitive when compared to conventional thermal-based ZLD desalination systems and disposal alternatives [72,78]. However, the elevated pretreatment costs are still an obstacle for the broad application of membrane-based schemes in shale gas industry [60].

12.6 Outlook and future directions

Shale gas industry is responsible for elevated freshwater consumption and generation of large amounts of hazardous wastewater, which is comprised by flowback and produced waters. Developing more effective desalination systems for the treatment of

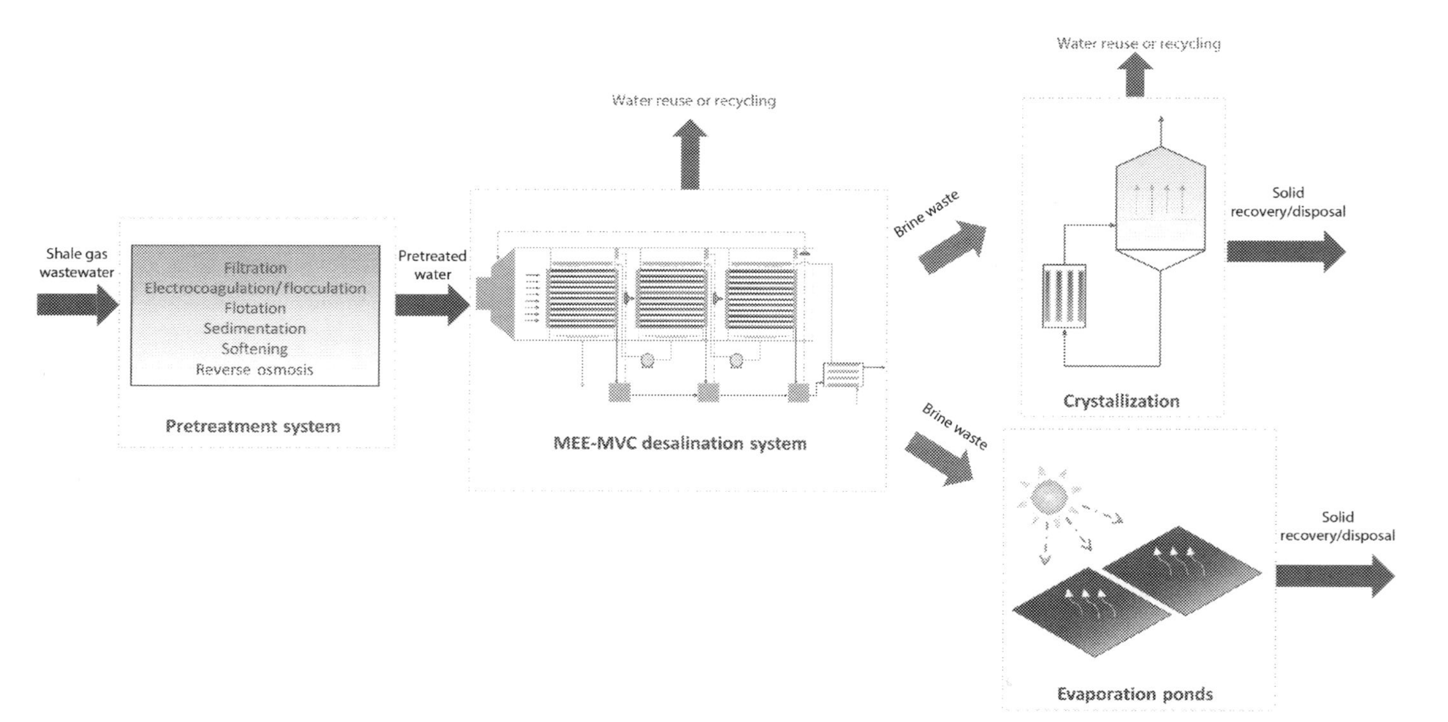

Fig. 12.7 Schematic representation of a thermal-based ZLD evaporation plant coupled to the pretreatment system and crystallization or evaporation ponds.

Table 12.6 Process characteristics and applications of membrane-based technologies for ZLD desalination of shale gas wastewater

Desalination technology	Driving force and process characteristics	High-salinity application
Membrane distillation (MD)	MD is a thermal-driven membrane desalination process, in which vapor pressure difference across the membrane acts as driving force. The vapor pressure gradient is caused by the temperature difference between the hot wastewater stream (feed stream) and the cold permeate stream (distillate) [73]. In recent years, MD has gained increased attention by the literature due to its potential to efficiently deal with high-salinity wastewater from shale gas production. High purity water can be expected by applying MD treatment to the shale gas wastewater. This is due its high removal rate of salts, metals, and nonvolatile components. Also, MD systems present several advantages over standard thermal and pressure-based membrane processes, including their ability to achieve higher brine concentrations (ZLD operation) and potential use of low-grade waste heat or renewable energy sources (e.g., wind, solar, geothermal, wave, etc.) [93]. Typically, MD processes can be operated at temperatures ranging 40–80°C (at atmospheric pressure) and driven by a low temperature difference of 20°C between the feed and distillate streams. For these reasons, waste grade heat can provide the thermal energy required by the MD desalination process [95]	Singh and Sirkar [74] have performed an experimental study on the desalination of brine and produced water through direct contact membrane distillation (DCMD) at high temperature and above-ambient pressure, using hollow fibers membranes. Their results emphasize that DCMD is a cost-competitive desalination process for high-salinity wastewaters applications, especially when compared to conventional RO. This is because of the DCMD process does not require feed cooling at the operating conditions considered by the authors. Chung et al. [76] have proposed a multistage vacuum membrane distillation (VMD) for ZLD[a] desalination of high-salinity wastewater applications. The latter authors have used a finite differences-based method for numerical process simulations, by allowing brine discharge salinity near to salt saturation conditions. Their results indicate that multistage VMD systems can be as cost-efficient as MSF schemes for a large range of feed water salinities. Tavakkoli et al. [62] have studied the techno-economic suitability of MD at ZLD operation (brine discharge salinity at 30%, *w/v*) for desalinating produced water from Marcellus shale play. Their results reveal that the freshwater production cost is

Continued

Table 12.6 Continued

Desalination technology	Driving force and process characteristics	High-salinity application
		significantly affected by the initial TDS contents on wastewater, as well as by the thermal energy prices. Lastly, Carrero-Parreño et al. [72] have successfully reach ZLD operation (brine discharge salinities) by applying the DCMD system for the shale gas wastewater desalination
Forward osmosis (FO)	FO is an osmotically driven membrane-based technology, in which a chemical potential difference between the concentrated draw solution and a wide range of solutions (e.g., shale gas wastewater) acts as driving force for salt separation [80]. FO is a promising membrane process for the desalination of high-salinity shale gas wastewater. In fact, this technology presents several advantages over other membrane alternatives, such as its ability to operate at higher salt concentrations (mainly when draw solutes regeneration is considered) [77], and easier fouling reversibility when compared to RO treatment [96]. FO systems can also be operated at low pressure, which can prevent fouling and reduce pretreatment requirements and maintenance. In this process, concentrate brine can be sent to a crystallizer (or evaporation ponds) to achieve ZLD operation, while treated water is separated from draw solutes to regenerate the draw solution [56]. For shale gas wastewater desalination, RO and MD can be	Hickenbottom et al. [80] have studied the suitability of FO for the treatment of fracturing wastewater from shale gas operations. Bench-scale experiments performed by the authors reveal that the FO system can achieve a water recovery efficiency of ~80%, with high rejection of organic and inorganic contaminants. Yun et al. [81] have investigated the application of pressure-assisted FO and air gap membrane distillation (AGMD) for the desalination of shale gas wastewater. Their experimental results indicate that the water flux across the membrane can be increased to 10%–15% for wastewaters with low and medium TDS contents, by considering an external pressure of 10 bar. However, the effect of the external pressure is considerably reduced for high-salinity wastewaters. Also, the authors have shown that AGMD can be an effective process to re-concentrate draw solutes. McGinnis et al. [77] have tested a pilot-scale FO system for the desalination of high-salinity

Table 12.6 Continued

Desalination technology	Driving force and process characteristics	High-salinity application
	coupled to the FO system to re-concentrate the draw solution and produce high quality water. Despite recent advances, further improvement in the development of membrane materials and draw solutions, as well as operating conditions optimization, will be critical to enhance process cost-effectiveness, and make FO a competitive alternative for high-salinity applications [42]	shale gas wastewater from Marcellus shale play. The authors have considered a NH_3/CO_2 draw solution to treat wastewaters with ~73 k ppm TDS (and hardness of 17 k ppm $CaCO_3$). The process proposed by the authors include pretreatment (softening, media filtration, activated carbon and cartridge filtration), post-FO thermal desalination, RO and brine stripper. Their results indicate water recovery of ~64% (brine discharge salinity of ~180 k ppm), with an energy consumption 42% lower than conventional MVC process
Reverse osmosis (RO)	RO is a pressure-driven desalination process characterized by the separation of dissolved salts from a (pressurized) saline water solution through a semipermeable membrane. In this way, the flow across the membrane occurs due to a pressure differential established between the high-pressure feed water and the low-pressure permeate. In the RO process, water molecules are transferred from a high salt concentration region to the permeate side owed to an osmosis pressure. For this reason, feed water should be pressurized above osmotic condition, whilst the permeate should be at near-atmospheric pressure [83]. RO is an energy-intensive process, in which the major energy requirement is related to the feed water mechanical pressurization.	Jang et al. [37] have experimentally evaluated the applicability of three different techniques for the desalination of high-salinity shale gas wastewater: MD, RO, and evaporative crystallization (EC). Their results indicate relatively higher efficiencies for MD and EC (>99.9%) than the RO technology (97.1%–99.7%). Despite the elevated removal rates presented by the RO process, the latter has been significantly affected by the TDS levels on the wastewater, requiring four times more dilution before operation than MD and EC. In a recent study, Salcedo-Díaz et al. [78] have proposed a ZLD desalination system composed RO and FO technologies for shale gas wastewater application. The authors have developed a mathematical model for the

Continued

Table 12.6 **Continued**

Desalination technology	Driving force and process characteristics	High-salinity application
	The efficiency of RO separation process can severely be impaired by membrane fouling and scaling. These problems can be prevented by effective wastewater pretreatments and the consideration of different membrane processes in the system [93]. Salt concentrations in shale gas wastewater are critical for RO desalination [37]. RO systems are cost-effective for wastewaters with TDS contents lower than 30k ppm [42]. In addition, RO can be included into ZLD desalination systems to enhance process cost-effectiveness. Almost 80% of wastewater volume can be reduced by using RO technology [47]. Usually, RO processes are operated at low temperatures <45°C (at 20–60 atm)	optimal design of onsite RO-FO systems, to minimize freshwater consumption and specific fracturing water cost. Their results show that is technically possible to reduce to zero the amount of freshwater used for re-injection in shale gas operations. However, due to the high freshwater production cost presented by the desalination system—in which, the cost of the cubic meter of treated water is about 100 times higher than the same amount of freshwater—an intermediate solution can be more affordable for shale gas industry
Electrodialysis (ED)/ electrodialysis reversal (EDR)	ED and EDR are electrochemical charge-driven membrane-based processes for the desalination of high-salinity shale gas wastewater. These technologies are characterized by dissolved ions separation across ion-selective membranes, in which the electrical potential gradient works as driving force [87,93]. In EDR process, membranes polarity is changed to fouling and scaling control [93]. ED and EDR systems can be used for removing salts from RO treated waters [97]. The performance of ED and EDR processes is significantly affected by several factors, including applied voltage, wastewater flowrate and ions	McGovern et al. [86] have proposed a 10-stage ED system for the treatment of high-salinity shale gas wastewater. The authors have experimentally evaluated the optimal equipment size and energy requirements to desalinate wastewater with salinities up to 192k ppm TDS. Their results emphasize the process effectiveness and the need for further investigating fouling and operating conditions (stack voltage) to minimize desalination costs. Hao et al. [64] have developed an integrated process of coagulation and ED for the treatment of fracturing wastewater. Coagulation is

Table 12.6 **Continued**

Desalination technology	Driving force and process characteristics	High-salinity application
	concentration, membrane density, diffusion, etc. The main disadvantages are related to high energy consumption and water production costs, and fouling propensity [63]. In addition, these processes require regular membrane cleaning (alkalis or dilute acidic solutions) to keep operating conditions. The latter drawbacks must be addressed to improve competitiveness of ED/EDR for the industrial scale application to high-salinity shale gas wastewaters [93]	used for removing organic contaminants from the wastewater, while its desalination is performed by the ED system. Their results show ion removal rates up to 91%, reaching water reclamation regulations. Peraki et al. [87] have investigated the ED efficiency as a pretreatment alternative for desalination of high-salinity shale gas wastewater from Marcellus shale play. Their results indicate a reduction of ~27% in the wastewater TDS contents after 7 h of application of a low direct current electric field

[a]Although evaporation ponds or brine crystallizers are required to literally achieve zero-liquid discharge operation, brine discharges salinities close to salt saturation conditions (e.g., 300,000 ppm TDS) are considered as ZLD operation in this work.

high-salinity wastewater to allow its reuse and/or recycling is critical to alleviate environmental and public health impacts, and enhance the overall sustainability of shale gas process. Among all pollutants in shale gas wastewater, removal of high TDS contents (usually >100,000 ppm) is particularly challenging due to the intensive energy consumption needed to comply with strict regulations on water quality (especially on water recycling and safe disposal).

ZLD desalination systems have recently emerged as an interesting alternative for shale gas wastewater management. The main advantages of ZLD processes relies in their ability to enhance water usage efficiency in shale gas production, while reducing brine discharges and water-related environmental impacts. Since ZLD desalination systems are typically able to achieve water recovery ratios up to 90% (note that the remaining water contents from brine wastes can be eliminated via incorporating either brine crystallizers or evaporation ponds into the desalination system), almost the totality of water from wastewater can be reclaimed for internal reuse or recycling opportunities.

Several desalination technologies can be used in ZLD systems for high-salinity wastewater applications, including thermal- and membrane-based processes. While thermal evaporation systems with MVC are relatively well-established processes, membrane-based schemes containing MD, FO, RO, and ED/EDR technologies are promising desalination systems for high-salinity shale gas wastewater. In general, membrane desalination systems present high efficiency, operational and control simplicity, easy scale-up and possibility of using low-grade waste energy.

Although widely accepted as an important wastewater management option to reduce water-related impacts, the implementation of ZLD systems in shale gas industry is still constrained by high-energy demands and associated processing costs. Nevertheless, a critical review of literature has revealed the cost-competitiveness of ZLD thermal evaporation systems for shale gas wastewater desalination. Advances in membrane materials, fouling control and optimization of operating conditions should increase the application of membrane-based ZLD systems in the shale gas desalination market. More generally, the wide employment of ZLD systems depends on further development of effective and sustainable desalination technologies, regulatory incentives to compensate economic limitations, and stricter regulations on brine discharges and water quality.

Acknowledgments

This project has received funding from the European Union's Horizon 2020 Research and Innovation Programme under grant agreement No. 640979.

References

[1] EIA. Annual energy outlook 2016 with projections to 2040. Washington, DC: U.S. Energy Information Administration; 2016.
[2] EIA. International energy outlook 2016. Washington, DC: U.S. Energy Information Administration; 2016.
[3] Onishi VC, Ruiz-Femenia R, Salcedo-Díaz R, Carrero-Parreño A, Reyes-Labarta JA, Fraga ES, et al. Process optimization for zero-liquid discharge desalination of shale gas flowback water under uncertainty. J Clean Prod 2017;164:1219–38.
[4] Cooper J, Stamford L, Azapagic A. Shale gas: a review of the economic, environmental, and social sustainability. Energy Technol 2016;4(7):772–92.
[5] Staddon PL, Depledge MH. Fracking cannot be reconciled with climate change citigation policies. Environ Sci Technol 2015;49(14):8269–70.
[6] Stephenson T, Valle JE, Riera-Palou X. Modeling the relative GHG emissions of conventional and shale gas production. Environ Sci Technol 2011;45(24):10757–64.
[7] NRC. Induced seismicity potential in energy technologies. Washington, DC: National Academies Press; 2013. 300 p.
[8] Thomas M, Partridge T, Harthorn BH, Pidgeon N. Deliberating the perceived risks, benefits, and societal implications of shale gas and oil extraction by hydraulic fracturing in the US and UK. Nat Energy 2017;2(5):17054.
[9] Vidic RD, Brantley SL, Vandenbossche JM, Yoxtheimer D, Abad JD. Impact of shale gas development on regional water quality. Science 2013;340(6134):1235009.
[10] Prpich G, Coulon F, Anthony EJ. Review of the scientific evidence to support environmental risk assessment of shale gas development in the UK. Sci Total Environ 2016;563–564:731–40.

[11] Warner NR, Kresse TM, Hays PD, Down A, Karr JD, Jackson RB, et al. Geochemical and isotopic variations in shallow groundwater in areas of the Fayetteville shale development, north-central Arkansas. Appl Geochem 2013;35:207–20.
[12] Kausley SB, Malhotra CP, Pandit AB. Treatment and reuse of shale gas wastewater: electrocoagulation system for enhanced removal of organic contamination and scale causing divalent cations. J Water Process Eng 2017;16:149–62.
[13] Shaffer DL, Arias Chavez LH, Ben-Sasson M, Romero-Vargas Castrillón S, Yip NY, Elimelech M. Desalination and reuse of high-salinity shale gas produced water: drivers, technologies, and future directions. Environ Sci Technol 2013;47 (17):9569–83.
[14] Onishi VC, Carrero-Parreño A, Reyes-Labarta JA, Fraga ES, Caballero JA. Desalination of shale gas produced water: a rigorous design approach for zero-liquid discharge evaporation systems. J Clean Prod 2017;140:1399–414.
[15] Onishi VC, Carrero-Parreño A, Reyes-Labarta JA, Ruiz-Femenia R, Salcedo-Díaz R, Fraga ES, et al. Shale gas flowback water desalination: single vs multiple-effect evaporation with vapor recompression cycle and thermal integration. Desalination 2017;404 (C):230–48.
[16] Onishi VC, Ruiz-Femenia R, Salcedo-Díaz R, Carrero-Parreño A, Reyes-Labarta JA, Caballero JA. Optimal shale gas flowback water desalination under correlated data uncertainty. In: Computer aided chemical engineering, vol. 40. Amsterdam, Netherlands: Elsevier; 2017. p. 943–8 Available from: https://www.sciencedirect.com/science/article/pii/B9780444639653501598?via%3Dihub.
[17] Onishi VC, Ruiz-femenia R, Salcedo-díaz R, Carrero-Parreño A, Reyes-Labarta JA, Caballero JA. Multi-objective optimization of renewable energy-driven desalination systems. In: Computer aided chemical engineering, vol. 40. Amsterdam, Netherlands: Elsevier; 2017. p. 499–504, Available from: https://www.sciencedirect.com/science/article/pii/B9780444639653500854?via%3Dihub.
[18] Clark CE, Horner RM, Harto CB. Life cycle water consumption for shale gas and conventional natural gas. Environ Sci Technol 2013;47(20):11829–36.
[19] Bilgen S, Sarıkaya İ. New horizon in energy: shale gas. J Nat Gas Sci Eng 2016;35 (A):637–45.
[20] Barbot E, Vidic NS, Gregory KB, Vidic RD. Spatial and temporal correlation of water quality parameters of produced waters from devonian-age shale following hydraulic fracturing. Environ Sci Technol 2013;47(6):2562–9.
[21] Ghanbari E, Dehghanpour H. The fate of fracturing water: a field and simulation study. Fuel 2016;163:282–94.
[22] Rosenblum J, Nelson AW, Ruyle B, Schultz MK, Ryan JN, Linden KG. Temporal characterization of flowback and produced water quality from a hydraulically fractured oil and gas well. Sci Total Environ 2017;596–597:369–77.
[23] Hammond GP, O'Grady Á. Indicative energy technology assessment of UK shale gas extraction. Appl Energy 2017;185:1907–18.
[24] Zammerilli A, Murray RC, Davis T, Littlefield J. Environmental impacts of unconventional natural gas development and production. National Energy Technology Laboratory (NETL), Report No. DOE/NETL-2014/1651; 2014, Available from: https://www.netl.doe.gov/File%20Library/Research/Oil-Gas/publications/NG_Literature_Review3_Post.pdf.
[25] Yang L, Grossmann IE, Manno J. Optimization models for shale gas water management. AICHE J 2014;60(10):3490–501.
[26] Wan Z, Huang T, Craig B. Barriers to the development of China's shale gas industry. J Clean Prod 2014 Dec;84:818–23.

[27] Nicot J-P, Scanlon BR. Water use for shale-gas production in Texas, U.S. Environ Sci Technol 2012;46(6):3580–6.
[28] Slutz, J.; Anderson, J.; Broderick, R.; Horner P. Key shale gas water management strategies: an economic assessment tool. SPE/APPEA international conference on health safety, environment in oil and gas exploration and production, Perth, Australia, 2012.
[29] Takahashi S, Kovscek AR. Spontaneous countercurrent imbibition and forced displacement characteristics of low-permeability, siliceous shale rocks. J Pet Sci Eng 2010;71 (1–2):47–55.
[30] Thiel GP, Lienhard JH. Treating produced water from hydraulic fracturing: composition effects on scale formation and desalination system selection. Desalination 2014;346:54–69.
[31] Kondash AJ, Albright E, Vengosh A. Quantity of flowback and produced waters from unconventional oil and gas exploration. Sci Total Environ 2017;574:314–21.
[32] Vengosh A, Jackson RB, Warner N, Darrah TH, Kondash A. A critical review of the risks to water resources from unconventional shale gas development and hydraulic fracturing in the United States. Environ Sci Technol 2014;48(15):8334–48.
[33] Hayes T. Sampling and analysis of water streams associated with the development of Marcellus shale gas. Rep by Gas Technol Institute, Des Plaines, IL, Marcellus Shale Coalit; 2009. p. 10.
[34] Acharya HR, Henderson C, Matis H, Kommepalli H, Moore B, Wang H. Cost effective recovery of low-TDS frac flowback water for re-use. GE Global Research; 2011. p. 1–100.
[35] Galusky P, Hayes TD. Feasibility assessment of early flowback water recovery for reuse in subsequent well completions RPSEA. Report No 08122-05.07; 2011.
[36] Hayes T, Severin BF. In: Barnett and Appalachian shale water management and reuse technologies RPSEA Report No 08122-05. Project report by Gas Technology Institute for Research Partnership to Secure Energy for America (RPSEA); 2012. p. 1–125.
[37] Jang E, Jeong S, Chung E. Application of three different water treatment technologies to shale gas produced water. Geosys Eng 2017;20(2):104–10.
[38] Rahm BG, Bates JT, Bertoia LR, Galford AE, Yoxtheimer DA, Riha SJ. Wastewater management and Marcellus shale gas development: trends, drivers, and planning implications. J Environ Manag 2013;120:105–13.
[39] Gao J, You F. Design and optimization of shale gas energy systems: overview, research challenges, and future directions. Comput Chem Eng 2017;1–20.
[40] Zendehboudi S, Bahadori A. Shale oil and gas. In: Shale oil and gas handbook. Amsterdam, Netherlands: Elsevier; 2017. p. 357–404, Available from: https://www.elsevier.com/books/shale-oil-and-gas-handbook/zendehboudi/978-0-12-802100-2.
[41] Carrero-Parreño A, Onishi VC, Salcedo-Díaz R, Ruiz-Femenia R, Fraga ES, Caballero JA, et al. Optimal pretreatment system of flowback water from shale gas production. Ind Eng Chem Res 2017;56(15):4386–98.
[42] Estrada JM, Bhamidimarri R. A review of the issues and treatment options for wastewater from shale gas extraction by hydraulic fracturing. Fuel 2016;182:292–303.
[43] Camarillo MK, Domen JK, Stringfellow WT. Physical-chemical evaluation of hydraulic fracturing chemicals in the context of produced water treatment. J Environ Manag 2016;183:164–74.
[44] GWPC (Ground Water Protection Council). ALL consulting. Modern shale gas development in the United States: a primer. Washington, DC: United States Department of Energy, Office of Fossil Energy; 2009. p. 96.
[45] Lester Y, Ferrer I, Thurman EM, Sitterley KA, Korak JA, Aiken G, et al. Characterization of hydraulic fracturing flowback water in Colorado: implications for water treatment. Sci Total Environ 2015;512–513:637–44.

[46] Stringfellow WT, Domen JK, Camarillo MK, Sandelin WL, Borglin S. Physical, chemical, and biological characteristics of compounds used in hydraulic fracturing. J Hazard Mater 2014;275:37–54.
[47] Gregory KB, Vidic RD, Dzombak DA. Water management challenges associated with the production of shale gas by hydraulic fracturing. Elements 2011;7(3):181–6.
[48] Vengosh A, Warner N, Jackson R, Darrah T. The effects of shale gas exploration and hydraulic fracturing on the quality of water resources in the United States. Proc Earth Planet Sci 2013;7:863–6.
[49] Dale AT, Khanna V, Vidic RD, Bilec MM. Process based life-cycle assessment of natural gas from the Marcellus shale. Environ Sci Technol 2013;47(10):5459–66.
[50] Zhang T, Gregory K, Hammack RW, Vidic RD. Co-precipitation of radium with barium and strontium sulfate and its impact on the fate of radium during treatment of produced water from unconventional gas extraction. Environ Sci Technol 2014;48(8):4596–603.
[51] Rahm BG, Riha SJ. Toward strategic management of shale gas development: regional, collective impacts on water resources. Environ Sci Pol 2012;17:12–23.
[52] Xiong B, Zydney AL, Kumar M. Fouling of microfiltration membranes by flowback and produced waters from the Marcellus shale gas play. Water Res 2016;99:162–70.
[53] Kaplan R, Mamrosh D, Salih HH, Dastgheib SA. Assessment of desalination technologies for treatment of a highly saline brine from a potential CO_2 storage site. Desalination 2017;404:87–101.
[54] Lee S, Kim YC. Calcium carbonate scaling by reverse draw solute diffusion in a forward osmosis membrane for shale gas wastewater treatment. J Membr Sci 2017;522:257–66.
[55] Miller DJ, Huang X, Li H, Kasemset S, Lee A, Agnihotri D, et al. Fouling-resistant membranes for the treatment of flowback water from hydraulic shale fracturing: a pilot study. J Membr Sci 2013;437:265–75.
[56] Tong T, Elimelech M. The global rise of zero liquid discharge for wastewater management: drivers, technologies, and future directions. Environ Sci Technol 2016;50(13):6846–55.
[57] Loganathan K, Chelme-Ayala P, Gamal El-Din M. Treatment of basal water using a hybrid electrodialysis reversal–reverse osmosis system combined with a low-temperature crystallizer for near-zero liquid discharge. Desalination 2015;363:92–8.
[58] López DE, Trembly JP. Desalination of hypersaline brines with joule-heating and chemical pre-treatment: conceptual design and economics. Desalination 2017;415:49–57.
[59] Xu P, Cath TY, Robertson AP, Reinhard M, Leckie JO, Drewes JE. Critical review of desalination concentrate management, treatment and beneficial use. Environ Eng Sci 2013;30(8):502–14.
[60] Thiel GP, Tow EW, Banchik LD, Chung HW, Lienhard JH. Energy consumption in desalinating produced water from shale oil and gas extraction. Desalination 2015;366:94–112.
[61] Lokare OR, Tavakkoli S, Rodriguez G, Khanna V, Vidic RD. Integrating membrane distillation with waste heat from natural gas compressor stations for produced water treatment in Pennsylvania. Desalination 2017;413:144–53.
[62] Tavakkoli S, Lokare OR, Vidic RD, Khanna VA. Techno-economic assessment of membrane distillation for treatment of Marcellus shale produced water. Desalination 2017;416:24–34.
[63] Ahmad M, Williams P. Assessment of desalination technologies for high saline brine applications—discussion paper. Desalin Water Treat 2011;30(1–3):22–36.
[64] Hao H, Huang X, Gao C, Gao X. Application of an integrated system of coagulation and electrodialysis for treatment of wastewater produced by fracturing. Desalin Water Treat 2015;55(8):2034–43.

[65] Ghaffour N, Lattemann S, Missimer T, Ng KC, Sinha S, Amy G. Renewable energy-driven innovative energy-efficient desalination technologies. Appl Energy 2014;136:1155–65.
[66] EIA (U.S. Energy Information Administration). How much carbon dioxide is produced per kilowatthour when generating electricity with fossil fuels?; 2016.
[67] Burbano A, Brandhuber P. Demonstration of membrane zero liquid discharge for drinking water systems: a literature review. Alexandria, VA: Water Environment Research Foundation; 2012. 72 p.
[68] Gude VG. Energy storage for desalination processes powered by renewable energy and waste heat sources. Appl Energy 2015;137:877–98.
[69] Lu Y, Chen J. Integration design of heat exchanger networks into membrane distillation systems to save energy. Ind Eng Chem Res 2012;51(19):6798–810.
[70] Cho H, Choi Y, Lee S, Sohn J, Koo J. Membrane distillation of high salinity wastewater from shale gas extraction: effect of antiscalants. Desalin Water Treat 2016;57(55):26718–29.
[71] Cho H, Jang Y, Koo J, Choi Y, Lee S, Sohn J. Effect of pretreatment on fouling propensity of shale gas wastewater in membrane distillation process. Desalin Water Treat 2016;57(51):24566–73.
[72] Carrero-Parreño A, Onishi VC, Ruiz-Femenia R, Salcedo-Díaz R, Caballero JA, Reyes-Labarta JA. Multistage membrane distillation for the treatment of shale gas flowback water: multi-objective optimization under uncertainty. In: Computer aided chemical engineering. vol. 40. Amsterdam, Netherlands: Elsevier; 2017. p. 571–6.
[73] Boo C, Lee J, Elimelech M. Omniphobic polyvinylidene fluoride (PVDF) membrane for desalination of shale gas produced water by membrane distillation. Environ Sci Technol 2016;50(22):12275–82.
[74] Singh D, Sirkar KK. Desalination of brine and produced water by direct contact membrane distillation at high temperatures and pressures. J Membr Sci 2012;389:380–8.
[75] Kim J, Kwon H, Lee S, Lee S, Hong S. Membrane distillation (MD) integrated with crystallization (MDC) for shale gas produced water (SGPW) treatment. Desalination 2017;403:172–8.
[76] Chung HW, Swaminathan J, Warsinger DM, Lienhard VJH. Multistage vacuum membrane distillation (MSVMD) systems for high salinity applications. J Membr Sci 2016;497:128–41.
[77] McGinnis RL, Hancock NT, Nowosielski-Slepowron MS, McGurgan GD. Pilot demonstration of the NH_3/CO_2 forward osmosis desalination process on high salinity brines. Desalination 2013;312:67–74.
[78] Salcedo-Díaz R, Ruiz-femenia R, Carrero-parreño A, Onishi VC, Reyes-Labarta JA, Caballero JA. Combining forward and reverse osmosis for shale gas wastewater treatment to minimize cost and freshwater consumption. In: Computer aided chemical engineering. vol. 40. Amsterdam, Netherlands: Elsevier; 2017. p. 2725–30.
[79] Chen G, Wang Z, Nghiem LD, Li X-M, Xie M, Zhao B, et al. Treatment of shale gas drilling flowback fluids (SGDFs) by forward osmosis: membrane fouling and mitigation. Desalination 2015;366:113–20.
[80] Hickenbottom KL, Hancock NT, Hutchings NR, Appleton EW, Beaudry EG, Xu P, et al. Forward osmosis treatment of drilling mud and fracturing wastewater from oil and gas operations. Desalination 2013;312:60–6.
[81] Yun T, Koo J-W, Sohn J, Lee S. Pressure assisted forward osmosis for shale gas wastewater treatment. Desalin Water Treat 2015;54(4–5):829–37.
[82] Elimelech M, Phillip WAA. The future of seawater desalination: energy, technology, and the environment. Science 2011;333(6043):712–7.

[83] Greenlee LF, Lawler DF, Freeman BD, Marrot B, Moulin P. Reverse osmosis desalination: water sources, technology, and today's challenges. Water Res 2009;43(9):2317–48.
[84] Salvador Cob S, Yeme C, Hofs B, Cornelissen ER, Vries D, Genceli Güner FE, et al. Towards zero liquid discharge in the presence of silica: stable 98% recovery in nanofiltration and reverse osmosis. Sep Purif Technol 2015;140:23–31.
[85] Michel MM, Reczek L, Granops M, Rudnicki P, Piech A. Pretreatment and desalination of flowback water from the hydraulic fracturing. Desalin Water Treat 2016;57 (22):10222–31.
[86] McGovern RK, Weiner AM, Sun L, Chambers CG, Zubair SM, Lienhard VJH. On the cost of electrodialysis for the desalination of high salinity feeds. Appl Energy 2014;136:649–61.
[87] Peraki M, Ghazanfari E, Pinder GF, Harrington TL. Electrodialysis: an application for the environmental protection in shale-gas extraction. Sep Purif Technol 2016;161:96–103.
[88] Lappalainen J, Korvola T, Alopaeus V. Modelling and dynamic simulation of a large MSF plant using local phase equilibrium and simultaneous mass, momentum, and energy solver. Comput Chem Eng 2017 Feb;97:242–58.
[89] Piacentino A. Application of advanced thermodynamics, thermoeconomics and exergy costing to a multiple effect distillation plant: in-depth analysis of cost formation process. Desalination 2015;371:88–103.
[90] Gabriel KJ, Linke P, El-Halwagi MM. Optimization of multi-effect distillation process using a linear enthalpy model. Desalination 2015;365:261–76.
[91] Mickley M. Survey of high-recovery and zero liquid discharge technologies for water utilities. WateReuse Foundation; 2008. 156 p.
[92] US Department of the Interior Bureau of Reclamation. Evaluation and selection of available processes for a zero-liquid discharge system for the Perris, California, Ground Water Basin. Desalin Water Purif Res Dev Progr Rep No 149. 2008.
[93] Drioli E, Ali A, Lee YM, Al-Sharif SF, Al-Beirutty M, Macedonio F. Membrane operations for produced water treatment. Desalin Water Treat 2016;57(31):14317–35.
[94] Tufa RA, Curcio E, Brauns E, Van Baak W, Fontananova E, Di Profio G. Membrane distillation and reverse electrodialysis for near-zero liquid discharge and low energy seawater desalination. J Membr Sci 2015;496:325–33.
[95] Camacho L, Dumée L, Zhang J, Li J, Duke M, Gomez J, et al. Advances in membrane distillation for water desalination and purification applications. Water 2013;5(1):94–196.
[96] Lee S, Boo C, Elimelech M, Hong S. Comparison of fouling behavior in forward osmosis (FO) and reverse osmosis (RO). J Membr Sci 2010;365(1–2):34–9.
[97] Fakhru'l-Razi A, Pendashteh A, Abdullah LC, Biak DRA, Madaeni SS, Abidin ZZ. Review of technologies for oil and gas produced water treatment. J Hazard Mater 2009;170(2–3):530–51.

Fertilizer drawn forward osmosis for irrigation

*Peter Nasr**, *Hani Sewilam*†
*American University in Cairo, Cairo, Egypt, †RWTH Aachen University, Aachen, Germany

13.1 Introduction

13.1.1 Global irrigation water consumption

One of the most crucial challenges for the 21st century is coping with the large increase in demand for food supplies to meet the enormous expected population growth. The United Nations expects the world population would reach 9 billion by the year 2050 [1]. Freshwater availability in this case would greatly affect food production. The agricultural sector continues to consume the largest amount of water accounting for at least 80% of all water worldwide [2], compared to domestic use of about 10% and industry 21% [3]. For example, irrigation water represents 80% of total water used in the MENA region where agricultural water use rose from about 165 BCM in 1995 to more than 218 BCM in 2012/2013, a 32% increase in a period of about 15 years [4].

Water preservation, infrastructure repair, conservation of catchment areas, and development of distribution systems could help ease water stress to a certain level. Yet, such measures would merely help develop the existing water sources and not create new ones [5].

Thus, one of the possible water management solutions is desalination, which creates new water resources for various water uses. Although desalinated water is the main source of potable water in the Gulf countries and in many islands around the world, it could also be used in certain countries to irrigate high-value crops [6].

13.1.2 Membrane-based desalination

Nowadays, membrane-based desalination is increasingly considered as a reliable option for providing water supplies to augment available water sources [7,8]. Yet, membrane desalination as currently practiced fails to sustainably supplement freshwater to meet future enormous demand [9]. Although the performance of membrane technology in terms of energy consumption has significantly improved in the last few decades, reverse osmosis (RO) cannot generate water in a sustainable fashion as long as the energy needed is produced from fossil fuels. More emissions would cause more water scarcity, demanding even more energy consumption, causing an unstoppable downward spiral. Researchers have been trying hard to avoid this problem by using novel energy sources for desalinating by thermal techniques.

Emerging Technologies for Sustainable Desalination Handbook. https://doi.org/10.1016/B978-0-12-815818-0.00013-8

These technologies may make use of the reject thermal energy from other processes (such as industrial and geothermal) and generate freshwater from saline. Yet, the heat temperature required by such technologies for feasible operation is too high; so a significant amount of energy in the form of fossil fuels is inevitable [9]. A significant amount of electrical energy is needed as well in addition to the high-quality heat requirement [1]. Moreover, seawater RO desalination (SWRO) operates at high hydraulic pressure ranging from 50 to 80 bar requiring special high-pressure pumps, high-quality, and high-pressure rating materials such as pressure vessels for housing membrane modules, pipes, and fittings all of which increases the capital cost of the SWRO desalination plants. Fouling still proves to be a major challenge for membrane processes such as SWRO [10–12]. Although various approaches are adopted to reduce the severity of membrane fouling such as pretreatments, physical, and chemical cleaning, most methodologies involve supplementary processes, thus increasing the capital and operational costs of the process as a whole [13]. The most energy-efficient RO desalination plant equipped with the energy recovery system has been reported to consume 3.2 kWh/m^3 excluding the energy consumed by the pretreatment processes [14].

Since energy and climate change issues are strongly interrelated, tackling global water-scarcity problems calls for robust and innovative methods of purifying water at minimum energy and cost [13]. Energy-efficient and low-cost desalination technologies could have a major impact on humanity by efficiently using the abundant saline water sources available on earth.

13.2 Forward osmosis

13.2.1 Basic principles

Osmosis has been used in nature for rather a long time by plants, trees, sharks, and human cells. Osmosis is defined as "the natural diffusion of solvents or water through a semipermeable membrane while preventing the passage of solutes" [7]. If a solution and a solvent are segregated by a semipermeable membrane, the solution starts to be diluted via attracting the solvent through the membrane. In case an external force is applied on the solution preventing the passage of solvent through the membrane and sustaining an equilibrium, this force is termed "osmotic pressure" [15]. The osmotic pressure (π) is given by the Van't Hoff's relation:

$$\pi = nMRT \tag{13.1}$$

where

n = the Van't Hoff factor (the number of particles of compounds dissolved in the solution, for example n = 2 for NaCl),
M = molar concentration of the solution,
R = the universal gas constant (0.0821 L atm/mol/K).
T = absolute temperature (in K) of the solution.

It is worth noting that the Van't Hoff relation is only relevant to dilute and ideal solutions in which ions are independent. However, at higher ionic concentrations the solution becomes nonideal as the electrostatic interactions between the ions increase, decreasing the activity coefficient of ions, and the osmotic pressure of solution [15]. The process in its pure form takes place at atmospheric pressure, yet variations such as pressure-enhanced osmosis and pressure-retarded osmosis are also possible. These are simply illustrated in Fig. 13.1.

In the FO process, the driving force is naturally created due to the differential concentration between a saline solution and a concentrated draw solution across a semipermeable membrane [7]. FO makes use of the osmotic differential ($\Delta\pi$) across the membrane (Fig. 13.2), and not the hydraulic pressure differential (as in the case of RO), to transfer pure water across the selective membrane [18]. Being a semipermeable membrane, the FO membrane permits the permeation of just water molecules,

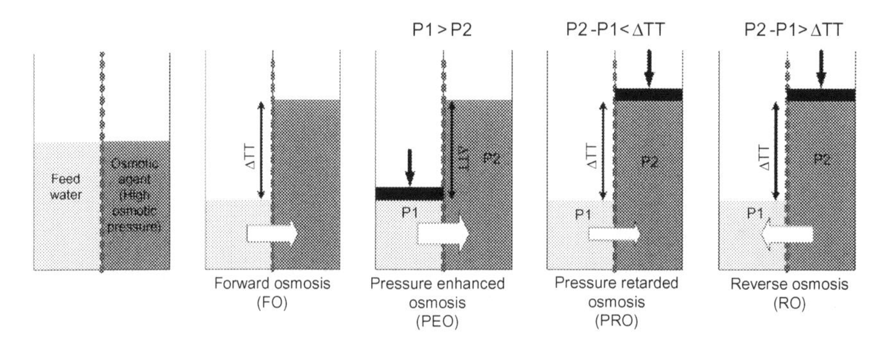

Fig. 13.1 Osmotic processes [16].

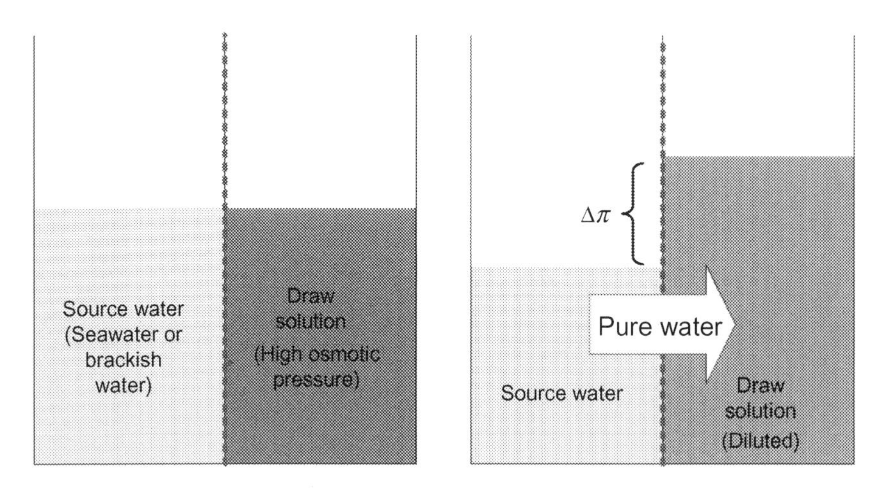

Fig. 13.2 Osmotic pressure differential ($\Delta\pi$) in the FO process [7,17].

and rejects most solute ions [7]. Freshwater diffuses from feedwater towards the draw solution, resulting in concentration of the feed solution (FS) (producing a highly saline solution or brine) and dilution of the draw solution, as presented in Fig. 13.2 [19].

According to Cath et al. [7], the relation describing water transport in FO is

$$J_w = A(\sigma\Delta\pi - \Delta P) \tag{13.2}$$

where

J_w = the water flux (negative values indicates reverse osmotic flow)
A = water permeability constant of the membrane
σ = reflection coefficient
$\Delta\pi$ = the differential osmotic pressures through the membrane (between the draw and feed solution) (Fig. 13.2)
ΔP = applied pressure (for FO: ΔP is zero, for RO: $\Delta P > \Delta\pi$)

Since for the FO process ΔP is zero, and σ is assumed unity, Eq. (13.2) can be rewritten as follows:

$$J_w = A\,\Delta\pi = A[\pi_{DS} - \pi_{FS}] \tag{13.3}$$

where

π_{DS} = bulk osmotic pressure of the DS
π_{FS} = bulk osmotic pressure of the FS.

13.2.2 Draw solution

The key factor of any successful FO process is the choice of an appropriate draw solution. There are different words used in publications to identify this solution, such as "draw solution," "osmotic agent," "osmotic media," "driving solution," "osmotic engine," "sample solution" or "brine" [7]. For clarity purposes, the term "draw solution" or "DS" will be used entirely in this work. A draw solution could be any aqueous solution exhibiting high osmotic pressure. It should provide sufficient driving force to cause a forward permeation of water across the membrane and therefore it is an essential part of the FO process. The osmotic pressure is a function of concentration, number of species in the solution, the molecular weight (MW) of the solute, and temperature. Osmotic pressure is independent of the types of species created in the solution (colligative property). A solute with small MW and highly soluble is expected to generate higher osmotic pressure and thus can result in better water flux [18]. Many types of DS have been studied in the past and they can be generally classified as inorganic-based DS, organic-based DS, and other compounds such as magnetic nanoparticles, RO brine, ionic polymer hydrogels, and dendrimes [20].

Over the past few years, many draw solutions were considered. A review of different draw solutions and their recovery techniques is shown in Table 13.1.

Table 13.1 Summary of the draw solutions tested in FO investigations and their recovery techniques

Year	Draw solute/solution	Recovery method	References
1964	Ammonia and carbon dioxide	Heating	[21]
1965	Volatile solutes (e.g., SO_2)	Heating or air stripping	[22]
1965	Mixture of H_2O and another gas (SO_2) or liquid (aliphatic alcohols)	Distillation	[23]
1972	Al_2SO_4	Precipitation by doping $Ca(OH)_2$	[24]
1975	Glucose	None	[25]
1976	Glucose-Fructose	None	[26]
1989	Fructose	None	[27]
1992	Glucose	Low-pressure RO	[28]
1997	$MgCl_2$	None	[29]
2002	KNO_3 and SO_2	SO_2 was recycled through standard means	[30]
2005–07	NH_3 and CO_2 (NH_4HCO_3) or NH_4OH-NH_4HCO_3	Moderate heating ($\sim 60°C$)	[18,31]
2007	Magnetic nanoparticles	Captured by a canister separator	[32]
2007	Dendrimers	Adjusting pH or UF	[32]
2007	Albumin	Denatured and solidified by heating	[32]
2008	Salt, ethanol	Pervaporation-based separations	[33]
2010	2-Methylimidazole-based solutes	Membrane distillation (MD)	[34]
2010	Magnetic nanoparticles	Recycled by external magnetic field	[35,36]
2011	Stimuli-responsive polymer hydrogels	Deswelling of the polymer hydrogels	[37,38]
2011	Hydrophilic nanoparticles	UF	[39]
2011	Fertilizers	None	[40]
2011	Fatty acid-polyethylene glycol	Thermal method	[41]
2012	Sucrose	NF	[42]
2012	Polyelectrolytes	UF	[43]
2012	Thermo-sensitive solute (Derivatives of Acyl-TAEA)	Not studied	[44]
2012	Urea, ethylene glycol, and glucose	Not studied	[45]
2012	Organic salts	RO	[46]
2012	Hexavalent phosphazene salts	Not studied	[47]
2014	Hydro acid complexes	Recycled	[48]

Data from Ge Q, Fu F, Chung T-S. Ferric and cobaltous hydroacid complexes for forward osmosis (FO) processes. Water Res 2014;58:230–8; Ge Q, Ling M, Chung T-S. Draw solutions for forward osmosis processes: Developments, challenges, and prospects for the future. J Membr Sci 2013;442:225–7; Zhao S, Zou L, Tang CY, Mulcahy D. Recent developments in forward osmosis: Opportunities and challenges. J Membr Sci 2012;396:1–21.

13.2.3 Draw solution selection criteria

An effective DS solute must have the following distinctive properties [8,18,49]:

1. High osmotic driving force, which leads to high water flux and recovery rates (zero-liquid discharge or "ZLD").
2. Soluble in water.
3. Small MW to produce a high osmotic pressure.
4. Nontoxic, since limited amounts might exist in produced water after separation. Sometimes, the solute is for eating or drinking, such as sucrose or fructose.
5. Chemically well-matched with the membrane, since the DS can react and deteriorate the membrane.
6. Easily and economically separated from the FS and recycled (if needed).

13.3 Fertilizer drawn forward osmosis

13.3.1 Basic concept

The FO process is more promising if the diluted DS can be directly used without further processing as draw solutes' presence in the water adds value to the final product water [26,50,51]. The FO process has been applied for nutritious drinks such as the hydration bags commercialized by Hydration Technology Innovations (HTI) company where a highly sugary solution is used as a DS that draws water from impaired water sources [7]. In this application, the diluted sugar can be directly used as a nutritious drink. Thus, in this approach there was no need for a posttreatment process to separate the draw solutes from the diluted DS. Similarly, in the FDFO process, a soluble fertilizer solution is used as DS and the diluted fertilizer draw solution after water permeation containing the valuable plant nutrients can be directly used for fertigation [51]. Fertigation is defined as "the application of fertilizer nutrients (dissolved form or suspended form) to the crops with irrigation water instead of broadcast application" [52].

Fig. 13.3 illustrates the basic concept of FDFO for fertigation, where two different solutions are used on each side of the semipermeable membrane: saline feedwater on one side and highly concentrated fertilizer DS on the other side [51]. The solutions are kept in contact with the membrane surfaces through a crossflow system minimizing the influence of concentration polarization (CP) effects [53]. Due to the osmotic pressure difference across the membrane, water flows from the FS (lower concentration) towards the DS (highly concentrated fertilizer), thus desalting the saline feedwater. The fertilizer DS finally becomes diluted, where the dilution extent is a function of the osmotic pressure of the feedwater. If it meets the water-quality standards for irrigation, the final fertilizer solution can be used directly for fertigation, and if not, then further dilution is necessary before applying it for fertigation [54].

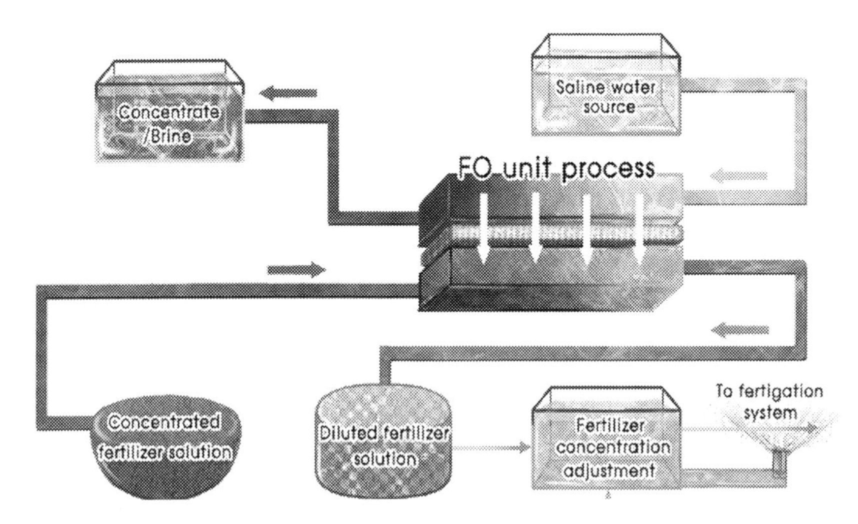

Fig. 13.3 Basic concept of FDFO for direct fertigation [40].

13.3.2 Advantages

13.3.2.1 Energy efficiency

FO is mainly operated by concentration difference between the DS and the FS as no external force is needed to push the water through the membrane. Yet, energy is only needed to maintain the cross-flow of the FS and DS making sure they are in contact with the membrane surface and providing sufficient shear force to minimize the CP. Fig. 13.4 shows the relative energy requirement for different desalination technologies.

Since the recovery of draw solutes from the diluted draw solution is not necessary, the estimates in Fig. 13.4 point out that the energy required for FDFO is significantly lower than other desalination technologies. From Fig. 13.4, it can be concluded that FDFO needs less than half the energy needed for ammonium bicarbonate FO application with DS feed recovery. Regarding this amount of energy, when compared to other current desalination technologies, up to 85% of energy can be saved and used for other applications [56].

As FO desalination is not energy intensive, it could be easily driven by renewable energy, such as wind and solar energy, rendering it a green desalination technology (with no carbon footprint). Renewable energy, especially solar energy, is abundant in many countries such as the MENA region, and therefore can be easily utilized for such purposes [57].

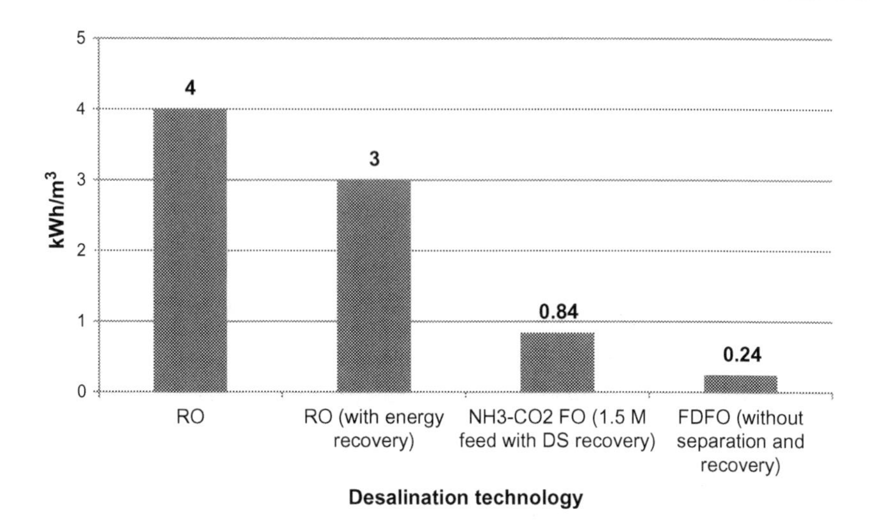

Fig. 13.4 Comparison of average energy requirements for different desalination technologies [55].

13.3.2.2 Fertilized irrigation (fertigation)

Agricultural productivity is most sensitive to fertilizer and water availability [58]. As agriculture is by far the largest consumer of freshwater, any savings in agricultural water through improved techniques will provide huge quantities of water for the community and the environment [59].

Fertigation is an agricultural irrigation process where water-soluble fertilizers or soil-improvement products are added to the irrigation water. According to [52], fertigation has some pros in contrast with the use of water and fertilizers independently. The benefits of using fertigation compared to traditional fertilization methods include:

- optimum dosage of nutrients leading to increased nutrient absorption of plants, leading to improved root systems
- reduction in water consumption due to improved root systems
- reduction in the amount of nutrients needed due to controlled dosage
- reduced leaching of nutrients to water supplies
- substantial savings in labor and energy costs
- accommodating and flexible technology as it can be easily integrated in any already-existing fertigation scheme
- suitable for application in mixtures with other micronutrients such as pesticides

13.3.3 Limitations

13.3.3.1 Forward osmosis membranes

The most prominent limitation to the commercialization of the FO is the lack of a suitable high-flux membrane. The ideal FO membrane should have high water permeability and salt rejection, should be thin without a porous support layer minimizing the

CP effects, and should also have good mechanical strength [10]. However, providing a thin membrane without support layers is a challenge since it does not provide adequate mechanical strength to carry the water flow inside the membrane module [8]. Several advancements have been reported on membrane manufacturing recently. The thin film composite (TFC) FO membranes are reported to have much higher water flux and salt rejection than the existing CTA FO membrane [60]. Because of its exceptional properties, such as high salt rejection, high chemical resistance, and high mechanical strength, TFC membranes have been long used for RO desalination [61]. However, the thick and dense support layer used for TFC-RO is not suitable for the FO process as it causes severe CP. The innovative claim for this TFC has been the modification of the support layer which is thinner and porous rendering it more proper for the FO process. In particular, the hollow fiber TFC FO membrane is a significant breakthrough since flat sheet membranes are more complicated for the design of spiral-wound modules accommodating two different and independent flows in the module separately [62]. With the commercialization of TFC-FO membranes, the future prospects of FO process and its applications are certainly high.

13.3.3.2 Choice of suitable fertilizer and the performance of fertilizer draw solution

Phuntsho et al. [40] concluded that the majority of soluble fertilizers are candidate draw solution for FO desalination. However, pH compatibility of the fertilizer solution with the membrane used is of great importance. The wider the pH range of the membrane the better. Phuntsho et al. [40] anticipated that a unit kilogram of fertilizer has the ability to absorb 11–29 L of freshwater from seawater and 90–215 L of freshwater from brackish feed. As feed salinity drops, fertilizers have the ability to extract additional water.

The permeation of pure water through the membrane will take place until osmotic equilibrium is achieved [40]. Full recovery is not realistic as at higher DS concentration scaling of the FS starts to manifest itself, decreasing water flux. Knowing that water from natural sources such as sea or groundwater usually includes many dissolved elements such as calcium and magnesium, precipitation is expected earlier. In addition, more energy is needed to keep the fluid flowing due to the viscosity of the FS at high concentrations.

Reverse permeation of draw solutes also takes place during the FO process, affecting process performance as discussed previously [20]. The severity of reverse permeation depends on the formed species properties, pH, and membrane properties [40]. For that reason, it is vital to put in mind such aspects when choosing a candidate fertilizer DS.

13.3.3.3 Lower-than-expected water flux

Lower-than-expected water flux is a result of the CP phenomena. External concentration polarization (ECP) reduces the water flux considerably. The ECP effect is alleviated by insuring shear as well as turbulence on the membrane surface as a substitute to dead end filtration [49]. Internal CP is inherent to the FO process and is discovered to be significant as it takes place inside the membrane support layer [10]. In fact, it has been discovered that the key aspect in charge of reducing the water flux in the FO is ICP, particularly the dilutive form [63].

Also, dilutive CP is another reason for the lower-than-expected water flux in FO. This phenomenon decreases the osmotic potential of the DS close to the plane of the membrane. That being said, the differential osmotic pressure is reduced, which lowers the pure water flux [63]. On the other hand, with the continuous improvement in membrane design, it is feasible to avoid the polarization consequences to some degree.

Moreover, since the DS is diluted as it moves along the membrane module, the net differential pressure in the membrane is expected to be reduced. This in turn will decrease the flux; thus, the osmotic equilibrium between DS and FS might not reached by a single FO stage. Consequently, there may be a need for multiple FO stages, which will increase the total membrane area, raising the capital cost required.

13.3.3.4 Fouling and biofouling

Due to the nonexistence of high pressure, membrane fouling in the FO process is described as reversible fouling [11]. Such fouling is minimized by engineered design optimization of operating conditions [12]. Yet, there is rare information discussing FO fouling prosperity in the literature.

Biofouling is an additional important problem that requires concern in FO. Since the membrane is continuously in contact with water, microorganisms and biofilm eventually grow. Biofouling is deemed unavoidable as it is uninfluenced hydrodynamically [64]. Since nutrients are known to be precursors to biofouling, the latter is inevitable in FDFO implementation [65]. Biofouling is mainly due to the microbial activity; yet, modest literature is available about the topic [65].

13.3.3.5 Feed salt rejection and reverse permeation of draw solute

As the ideal FO membrane does not exist yet, the solute rejection is therefore expected to be slightly $<100\%$ [61]. Solute permeation can happen in one of two directions: (1) forward movement of feed salt, which is considered as rejection and (2) reverse permeation of draw solutes [7]. Reverse solute movement is mostly significant as fertilizer draw solution contains nitrogen and phosphorus. These elements could be damaging to the process of brine management. Such elements could possibly cause eutrophication of receiving water bodies in case they are discharged to the environment haphazardly [52]. The presence of sodium chloride in produced water would also cause sodium toxicity to plant life, as previously discussed [66].

The degree of salt rejection and reverse permeation of draw solute mainly rely on: (1) membrane characteristics and (2) the DS properties [61]. Unfortunately, the current commercially available CTA FO membrane exhibits low salt rejection [10]. Reverse solute flux differs significantly for each fertilizer, depending on the solute properties. It should be noted that DS containing ions of large hydrated diameter exhibited less reverse permeation than ions with smaller hydrated size [49,67].

13.3.3.6 Meeting irrigation water-quality standards

Any DS can extract freshwater from saline FS, provided that the fertilizer DS is soluble in water and has osmotic pressure more than the salty FS [51]. There is an ultimate limit to which the osmotic process can continue occurring [40]. In other words, each DS can extract water only up to the "osmotic equilibrium," which is defined as "the concentration where the DS osmotic potential equals that of the feedwater" [51]. Beyond this point, the DS cannot be further diluted. At this equilibrium point, depending on the feed salinity, the fertilizer concentration may be too high for direct fertigation. The fertilizer final nutrient concentration may possibly surpass the maximum limit and thus may cause problems to vegetation.

Depending on the osmotic pressure of the feedwater, the limit to which the DS could achieve its final concentration is established. The salinity of the feedwater is directly proportional to the final fertilizer DS concentration. The *optimum* nutrient content for fertigation relies on numerous aspects such as: crop type, season, soil nutrient conditions, etc. [52]. Using seawater as FS, it is expected that a large volume of water will be needed to reduce the nutrient content of the product water before fertigation. Thus, FDFO desalination is more appropriate for brackish water.

In case the nutrient concentration does not meet the fertigation standard, the DS must be further diluted to make the desalted water fit for fertigation. Dilution is achievable if the site has access to a source of potable water for irrigation. However, if this is not the case then this is a challenge. Since maintaining the required nutrient concentration is necessary for fertigation, an additional process could be augmented with the FO unit. According to Phuntsho et al. [40], to achieve lower nutrient concentration in the final FDFO product water, possible options are: (1) pretreatment of feedwater, (2) posttreatment of feedwater, (3) use of blended fertilizer, and (4) hybrid FO system. These four options are discussed below.

Pretreatment of feedwater

As shown in Fig. 13.5 the FDFO desalination process may be incorporated with a nanofiltration (NF) pretreatment process to decrease the TDS of the feedwater. NF is advantageous as it can reject up to 80% of monovalent and up to 99% of divalent ions [49]. Since brackish groundwater usually contains divalent ions such as Ca^{2+}, Mg^{2+}, SO_4^{2-}, etc., NF can be used to lessen the total dissolved solids and the osmotic pressure of the FS. In addition, any decrease in the divalent ions would reduce the scaling likelihood of the FS, improving the recovery rate [7]. So, feasible nutrient concentration is achievable and direct fertigation is possible.

Not only will NF achieve high water flux, but also will operate at low hydraulic pressure. NF is not energy intensive and thus has low operation and maintenance costs [67].

Posttreatment of feedwater

NF can be adopted as a posttreatment instead of a pretreatment option, as discussed previously. NF can be utilized to concentrate and reuse the DS. Permeate with considerably low-nutrient content can be deployed straightaway for fertigation and the concentrate with high-nutrient concentration is recycled as draw solution to desalinate more FS.

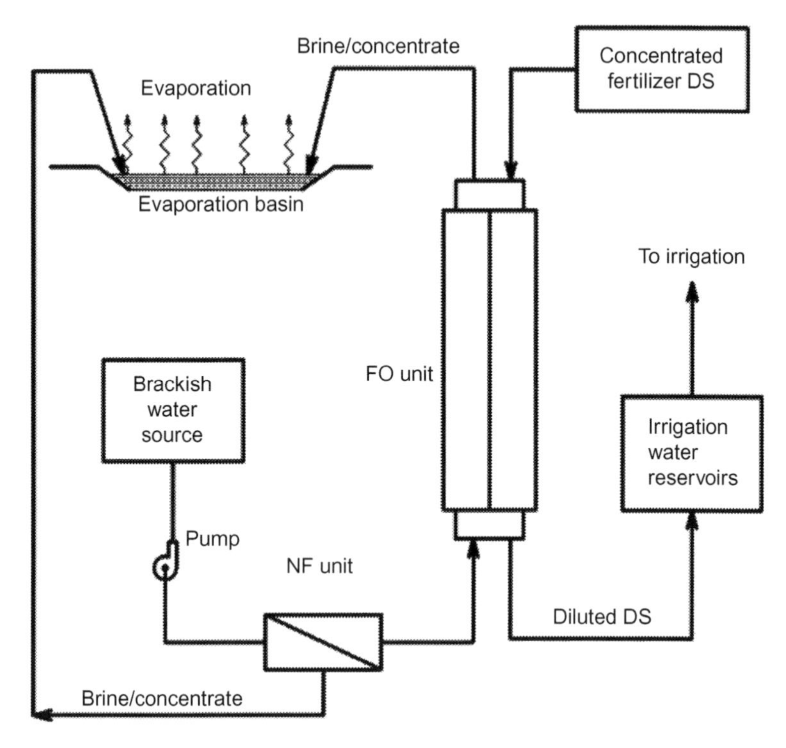

Fig. 13.5 FDFO desalination process integrated with NF pretreatment process [51].

It has been reported that two-staged NF posttreatment is capable of recovering divalent draw solutes meeting World Health Organization drinking water-quality standards [67]. One additional benefit of the NF posttreatment is the fact that NF is more efficient as the process effluent does not contain any foulants but contains just diluted fertilizer as any undesired foulant in the FS is eradicated in the previous FDFO step [67].

Blended fertilizers

Another potential alternative is to use a blend of thermolyte fertilizers as DS in the FDFO process (Fig. 13.6). Lower-nutrient content in the final DS is achievable by utilizing a DS with several ionic species. This can be done by mixing two or three fertilizers with other elements such as pesticides and insecticides. Doing that would significantly raise the osmotic potential of the draw solution as well as lower the final nutrient content.

Using blended fertilizer will overcome another problem related to the variable dilution factors required when fertilizers containing more than one nutrient are used as DS. For example, a fertilizer containing N and P may require a dilution factor of 2.5 for N concentration and 10 for P concentration. Such an issue exists with fertilizers like monoammonium phosphate (MAP), KNO_3, and KH_2PO_4.

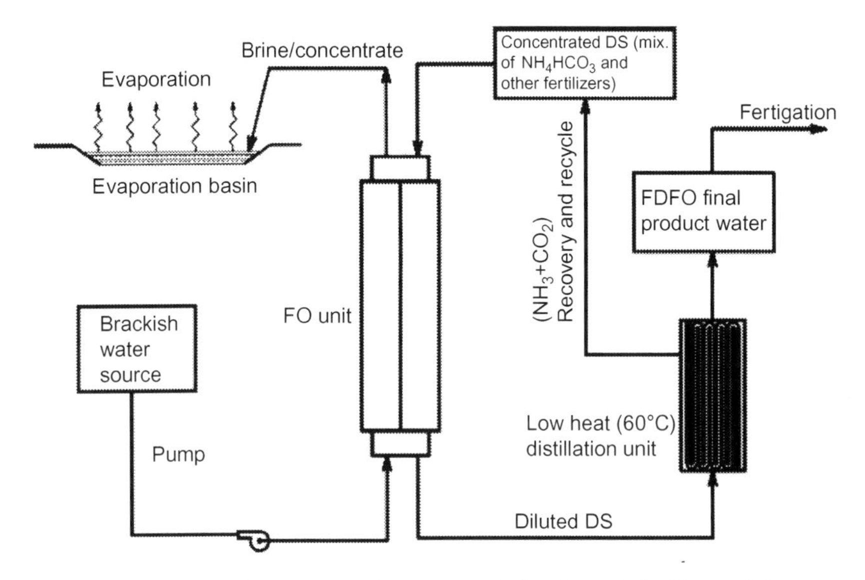

Fig. 13.6 FDFO desalination process using DS containing blended fertilizers [51].

Hybrid forward osmosis systems

Another option is to utilize wastewater effluent to dilute the fertilizer solution. The basic idea is to employ a multiple two-staged FO process for concurrent WW treatment and desalination of brackish water (Fig. 13.7) [7]. The brackish water passes by

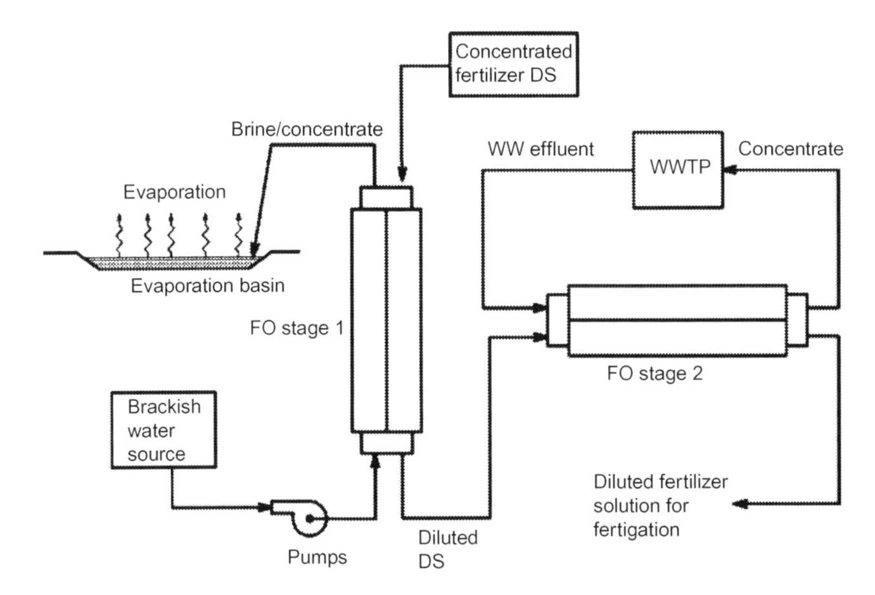

Fig. 13.7 Hybrid FDFO desalination process using a two-stage FO process with additional dilution water from a secondary WWTP effluent [51].

the first FO stage to be desalinated using a fertilizer as the DS. Then, the diluted fertilizer DS passes through FO stage 2 in which water is extracted from the WW effluent. FO stage 2 not only treats wastewater effluent to the required irrigation standard, but also provides additional dilution to the fertilizer solution decreasing its nutrient concentration deeming it fit for direct fertigation [51].

On the other hand, a second option would be designed differently. Brackish water could be employed as the DS in the first FO stage to absorb pure water from the WW effluent. The product of the first FO stage (diluted brackish water) can then be the FS of the second FO stage, with concentrated fertilizer as the DS. For either option, final nutrient concentration in product water is minimized.

13.4 Potential fertilizer draw solution

It is essential to understand which fertilizers are more suitable for the FDFO desalination process and how to screen and assess fertilizer candidates for use as DS in the FDFO process. The choice of fertilizer as DS would be a function of a number of factors, such as fertilizer availability, economics, and performance. Firstly, to have a sustainable FDFO process, the selected fertilizer should be available in the local market. Preferably, the fertilizer would be locally produced to avoid delays related to importing. Being a central aspect of the system, fertilizer scarcity would radically affect process efficiency. Secondly, with regard to fertilizer economics, fertilizers are a world-market commodity subject to global market forces, volatility, and risks. Yet, for FDFO to be cost-effective, the chosen fertilizer should not be expensive or costly [68]. Finally, the fertilizer should have physicochemical properties suitable for use as DS in the FO process, such as solubility, pH compatibility with selected FO membrane, MW, osmotic pressure, water extraction capability and final nutrient content in product water [20]. In addition, the DS should not chemically react with the FS to generate unwanted species impeding the osmotic process or the final intended deployment of the produced water (irrigation in the case of FDFO).

13.4.1 Types of fertilizers

Four groups of fertilizers are known to be essential for plant growth. The first group entails C, H, O, N, and S which are major constituents of organic substances. The second group containing P and B is needed for energy transfer reactions and carbohydrate movement. The third group contains M, Mg, Ca, and Cl, which are needed for sustaining ionic balance. Finally, the fourth group, containing Cu, Fe, Mn, Mo, and Zn, are required to enable electron transfer and function as enzyme catalysts [52]. Essential elements (C, H, O, and N) are derived directly or indirectly from the air making up more than 90% of plant material. The other six essential elements (Ca, Mg, P, K, Fe, and S) are derived from the soil. Crop type, cropping seasons and other factors affect plant requirements, although all these elements are essential for a healthy plant [69]. The elements that need special consideration are N, P, K, Ca, and S. Out of these, NPK are the main nutrients of great importance for mineral or synthetic fertilizers [54]. Depending on the types of major elements needed by plants, fertilizers are classified as nitrogen, phosphorous, or potassium fertilizers (NPK). The number of major elements present in each fertilizer determines their classification as single, compound, or mixed fertilizers [52].

13.4.1.1 Nitrogen-based fertilizers

Nitrogen forms a major component of proteins and chlorophyll in plants. N is essential for the healthy growth of the plant [58]. Not only is N responsible for increases in crop yield, but also it is taken up in large quantities among the major NPK nutrients. A frequent regulated amount of N is more desirable than large amounts with less frequency, maintaining healthy plant growth and reducing nutrient leaching. Excessive N results in excessive leaf growth with low fruit yield [69].

Almost 79% (by volume) of the Earth's atmosphere contains N in the form of nitrogen gas. Yet, only a limited number of plant types can make use of this N directly from the air. Thus, for most plants, N must be made available to the soil in a dissolved form for proper cropping [70]. Urea is the most widely used N fertilizer in the world. Inorganic N in urea is produced by fixing N from the atmosphere using natural gas [71]. Table 13.2 shows some of the most commonly used fertilizers as a main source of N for agricultural production.

13.4.1.2 Phosphorus-based fertilizers

Phosphorus is a fundamental component of every living cell. It has an important role in numerous physiological and biochemical processes as it cannot be replaced by other elements. P has more than one role at it is needed for stimulating cell division, promoting plant growth and root development, accelerating ripening, and improving the grain quality [72]. P, like N, is a nutrient that plants require in large quantities. P has low mobility in the soil; so its application is needed a few weeks before planting. Efficient use of P is vital as P is a nonrenewable resource and its irresponsible handling could lead to eutrophication of water bodies [54]. Table 13.2 shows some of the fertilizers used as a source of P for agricultural production.

13.4.1.3 Potassium-based fertilizers

Potassium (K) is the third major nutrient required for plant growth. K provides a number of important functions for the plants, such as activating enzyme actions facilitating the transport of nutrients; maintaining the structural integrity of plant cells; mediating the fixation of N in leguminous plant species; and protecting plants from certain plant pests and diseases [70]. In addition, K helps maintain an electrical balance within plant cells. Almost 95% of the K source in the world comes from potassium chloride (KCl) [70]. A variety of mineral fertilizers containing potassium are listed in Table 13.2.

There are many chemical fertilizers used for agriculture in many parts of the world. Yet, only those fertilizers that are water soluble and can exhibit osmotic pressure

Table 13.2 List of chemical fertilizers used worldwide [54]

Name of fertilizers	Chemical formula	Nutrients
Ammonia	NH_3	N
Ammonium bicarbonate	NH_4HCO_3	N
Ammonium carbamate	$NH_4CO_2NH_2$	N
Ammonium chloride	NH_4Cl	N
Ammonium hydrate	NH_4OH	N
Ammonium nitrate	NH_4NO_3	N
Ammonium nitrate ammonium sulfate	$(NH_4)_3NO_3SO_4$	N—S
Ammonium nitrate sulfate/bisulfate	$NH_4HNO_3SO_4$	N—S
Ammonium phosphate	$(NH_4)_3PO_4$	N—P
Ammonium sulfate	$(NH_4)_2SO_4$	N—S
Calcium nitrate	$Ca(NO_3)_2$	N—Ca
Diammonium hydrogen phosphate	$(NH_4)_2HPO_4$	N—P
Monocalcium phosphate monohydrade	$CaH_2(PO_4)_2 \cdot H_2O$	P—Ca
Monoammonium phosphate	$NH_4H_2PO_4$	N—P
Phosphoric acid	H_3PO_4	P
Potassium chloride	KCl	K
Potassium dihydrogen phosphate	KH_2PO_4	P—K
Potassium hydrogen phosphate	K_2HPO_4	P—K
Potassium nitrate	KNO_3	N—K
Potassium sulfate	K_2SO_4	K—S
Potassium thiosulfate	$K_2S_2O_3$	K—S
Single superphosphate	$Ca(H_2PO_4)_2$	P—Ca
Sodium nitrate	$NaNO_3$	N
Sodium tripolyphosphate	$Na_5P_3O_{10}$	P
Triammonium nitrate ammonium sulfate	$(NH_4)_5NO_3SO_4$	N—S
Tripotassium phosphate	K_3PO_4	P—K
Urea	$CO(NH_2)_2$	N

higher than that of the FS are appropriate for use as DS. The general properties of some of those selected fertilizer solutions are provided in Table 13.3.

Since nitrogen-based fertilizers come to be the most consumed type of fertilizer worldwide, this study will focus only on them. The selected nitrogen-based fertilizers include urea, ammonium nitrate, ammonium sulfate, and calcium nitrate. These fertilizers, which are commonly used in many parts of the world, could be possibly used as DSs for the FDFO desalination process.

13.4.2 Fertilizer performance

A performance screening of the four selected nitrogen-based fertilizers for the DS is assessed to determine basic properties (Table 13.4). OLI Stream Analyzer software was used to determine DS solubility, pH, speciation, and osmotic pressure.

Table 13.3 **Properties of selected fertilizer draw solutions [55]**

Fertilizers	MW	π @ 2.0 M (atm)	[a]Max π (atm)	Species formed in 2.0 M solution at 25°C and 1.0 atm pressure
$Ca(NO_3)_2$	164.1	108.5	601.0	NO_3^-: 3.47 M, Ca^{2+}: 1.47 M, $CaNO_3^-$: 0.53 M, total = 5.47 M
$(NH_4)_2HPO_4$	132.1	95.0	293.0	NH_4^-: 3.94 M, HPO_4(ion): 1.79 M, P_2O_7(ion): 0.07 M, H_2PO_4(ion): 0.02 M, HP_2O_7(ion): 0.02 M
$(NH_4)_2SO_4$	132.1	92.1	274.8	NH_4^+: 3.07 M, SO_4^{2-}: 1.07 M, $NH_4SO_4^-$: 0.93 M
NH_4Cl	53.5	87.7	356.0	NH_4^-: 2 M Cl^-: 2 M
$NH_4H_2PO_4$	115.0	86.3	181.3	NH_4^-: 2.0 M, $H_2PO_4^-$: 1.76 M, $H_2P_2O_7$(ion): 0.10 M, H_3PO_4(aq): 0.02 M), HP_2O_7(ion): 0.004 M
$NaNO_3$	85.0	81.1	417.9	Na^+: 1.92 M, NO_3^-: 1.92 M, $NaNO_3$(aq 0.08 M @ 2 M)
KCl	74.6	80.1	226.5	K^+: 1.99 M, Cl^-: 1.99 M, KCl (aq): 0.01 M @ 2 M
NH_4NO_3	80.04	64.9	3346.1	NH_4^+: 0.85 M, NO_3^-: 0.85 M, NH_4NO_3: aq 1.15 M
KNO_3	101.1	59.9	101.4	K^+: 2.0 M, NO_3^-: 2.0 M
[b]KH_2PO_4	136.09	58.0	58.0	1.8 M K^+, 1.735 M H_2PO_4(ion), 0.026 $H_2P_2O_7$(ion), H_3PO_4 (aq 0.0064 M), HPO_4(ion)
Urea	60.06	46.1	338.4	Urea

[a]Osmotic pressure at maximum solubility.
[b]Data refers to maximum solubility at 1.8 M concentration.

Table 13.4 **List of most popular nitrogenous fertilizers in Egypt**

Name of fertilizer	Chemical formula	Molecular weight (g/mol)	pH at 2 M	π at 2 M (atm)	Max. solubility ($C_{D,E}$)
Urea	$CO(NH_2)_2$	60.05	7.00	46.1	19.65 M
Ammonium nitrate	NH_4NO_3	80.04	4.87	64.9	Highly soluble
Ammonium sulfate	$(NH_4)_2SO_4$	132.1	5.46	92.1	5.7 M
Calcium nitrate	$Ca(NO_3)_2$	164.1	4.68	108.5	7.9 M

Solubility and osmotic pressure data obtained from OLI Stream Analyzer Software 9.1 [73]

13.4.2.1 *Osmotic pressure*

Osmotic pressure relies on the number of species formed rather than the species' nature [74]. Fig. 13.8 shows the osmotic pressure of the four fertilizers DS at variable concentrations. Calcium nitrate produces the largest osmotic pressure of 600 atm at its maximum solubility. This is because $Ca(NO_3)_2$ when dissolved generates the largest number of species in comparison to other fertilizers (Table 13.3).

Fixing the molar concentration of the DS (say at 2.0 M as per Table 13.4), the next highest osmotic pressure observed is that of ammonium sulfate (92.1 atm). The least osmotic pressure observed is for urea (46.1 atm at 2.0 M). Yet, as urea is readily soluble in water, it exhibits osmotic pressure in excess of 200 atm at concentrations more than 10 M (Figs. 13.8–13.11). Fig. 13.12 illustrate the type and concentration of each species present and the expected osmotic pressure at different concentrations of the fertilizers. Analysis was carried out using OLI Stream Analyzer 9.1 software.

It is worth noting that SWRO pressure ranges between 60 and 100 atm and that the typical osmotic pressure of seawater is around 28 atm [75–77]. Comparing these values to the osmotic pressures of the four fertilizers under study, it can be easily concluded that the four fertilizers generate osmotic pressure much more than that of seawater and SWRO (Fig. 13.8).

For ammonium sulfate, three dominant aqueous species exist, which are ammonium ion, sulfate ion, and ammonium sulfate ion. Ammonia and bisulfate ion are not considered as the dominant species (Fig. 13.9). The osmotic pressure of ammonium sulfate seems to increase as concentration increases up to 5.5 M concentration due to its maximum solubility [78].

Urea has only one dominant aqueous species (Fig. 13.10). The osmotic pressure lineally increases as urea concentration increases. Osmotic pressure reaches up to 150 atm at a 7 M concentration.

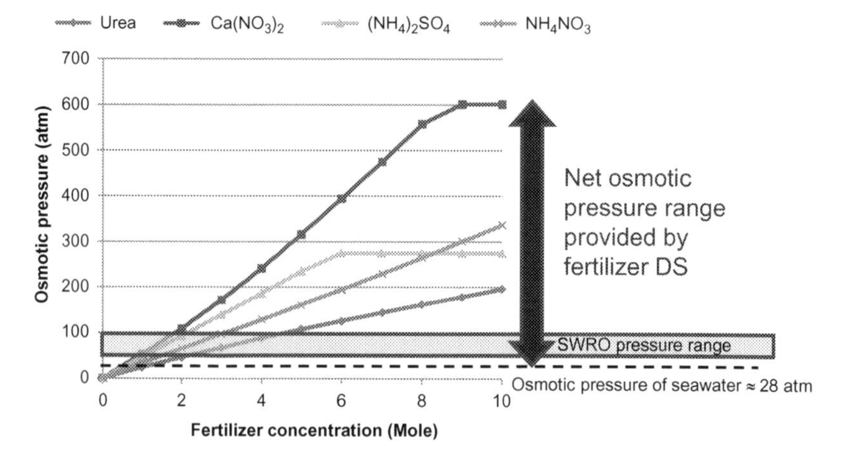

Fig. 13.8 Osmotic pressure of different nitrogenous fertilizers DS at 25°C analyzed using OLI Stream Analyzer 9.1 [68].

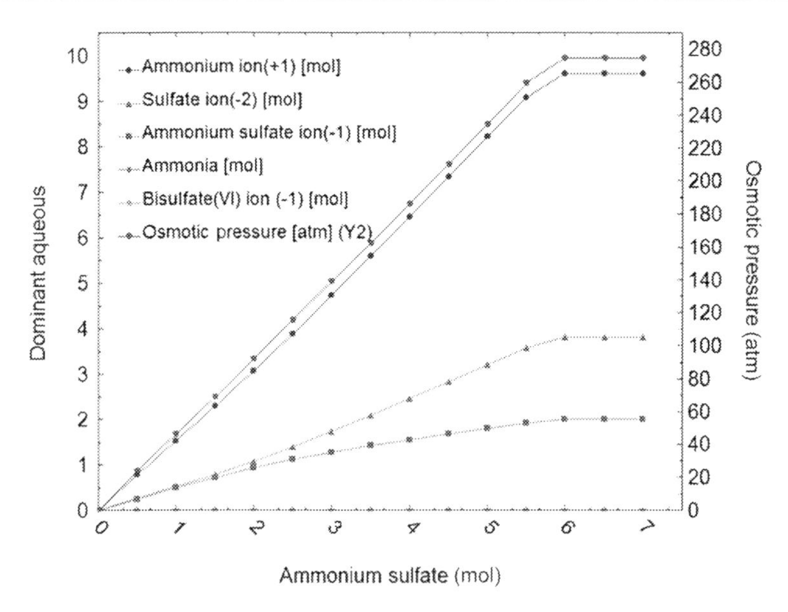

Fig. 13.9 Species generated and osmotic pressure of ammonium sulfate. Analysis carried out using OLI Stream Analyzer 9.1 at 25°C temperature and 1 atm pressure [73].

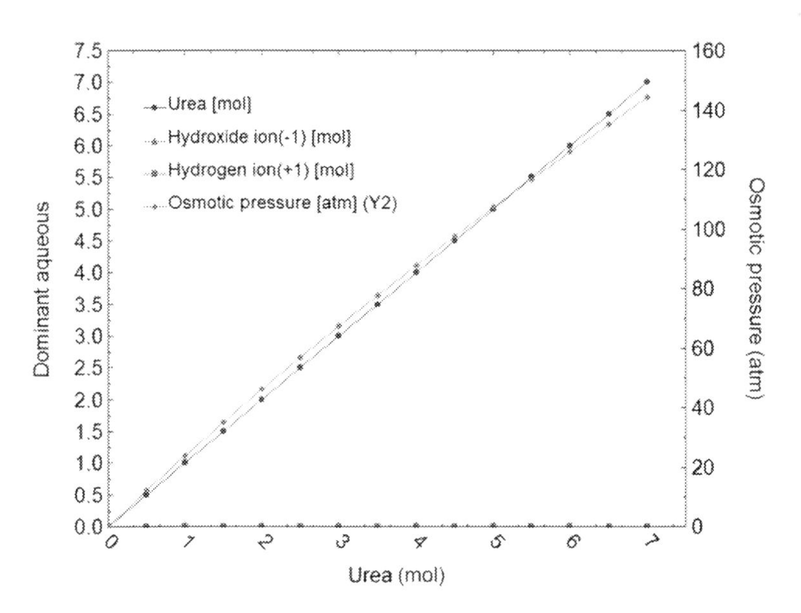

Fig. 13.10 Species formed and osmotic pressure of urea. Analysis carried out using OLI Stream Analyzer 9.1 at 25°C temperature and 1 atm pressure [73].

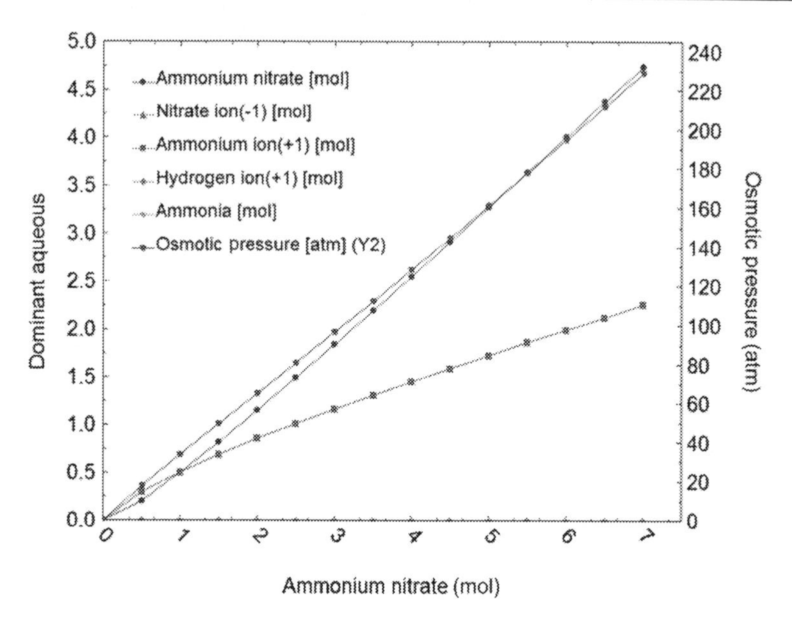

Fig. 13.11 Species formed and osmotic pressure of ammonium nitrate. Analysis carried out using OLI Stream Analyzer 9.1 at 25°C temperature and 1 atm pressure [73].

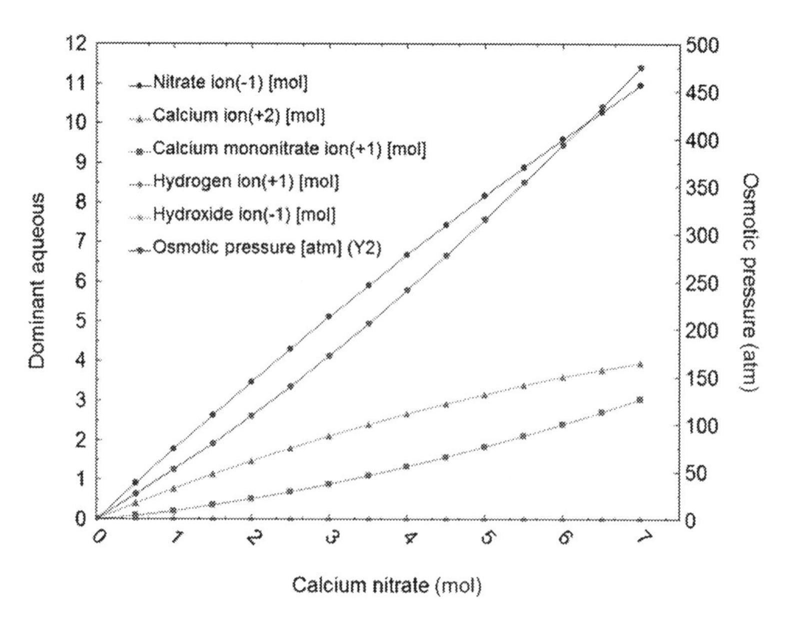

Fig. 13.12 Species formed and osmotic pressure of calcium nitrate. Analysis carried out using OLI Stream Analyzer 9.1 at 25°C temperature and 1 atm pressure [73].

For ammonium nitrate, two dominant aqueous species exist, which are ammonium nitrate and ammonium ion. Ammonia and nitrate ion are not considered as the dominant species (Fig. 13.11). Osmotic pressure of ammonium nitrate increases proportionally as concentration increases reaching 230 atm at 7 M concentration.

Calcium nitrate has three dominant aqueous species, which are nitrate ion, calcium ion, and calcium mononitrate ion (Fig. 13.12). The osmotic pressure of calcium nitrate increases proportionally as the concentration increases reaching 475 atm at 7 M concentration.

Any draw solute should generate higher osmotic pressure than that of the FS. For example, sweater has an osmotic pressure of 28 atm. So, if sweater is the FS, the DS must exhibit an osmotic pressure a lot more than 28 atm. Thus, since all investigated fertilizers produce osmotic pressure that is way higher than seawater or brackish water, they are suitable for use as an osmotic DS.

13.4.2.2 Water extraction capacity

The water extraction capacity of the draw solute has a key role in any FO process. DS can extract water from the FS until the osmotic pressure of the DS is in equilibrium with the osmotic pressure of the FS [15]. When different draw solutes are used, a number of species are formed in solution and the osmotic pressure of the DS is dependent on their osmotic coefficient. According to Phuntsho et al. [15], the volume of water (V) a kilogram of draw solute can extract from a FS can be estimated using the following equation:

$$V = \frac{1000}{M_w}\left[\frac{1}{C_{D,E}} - \frac{1}{C_{D,\text{Max}}}\right] \tag{13.4}$$

where

M_w is the molecular weight of the draw solute used (g/mol)—Table 13.4
$C_{D,E}$ is the molar concentration of the DS that generates equal bulk osmotic pressure (osmotic equilibrium condition) with the osmotic pressure of an FS (mol).
$C_{D,\text{Max}}$ is the maximum solubility of the draw solute (mol)—Table 13.4.

For comparison, six different TDS FSs are considered (1, 2, 5, 10, 20, and 35 g/L NaCl). The osmotic pressures of these FSs were predicted using OLI Stream Analyzer 9.1 and were found to be 0.8, 1.59, 3.91, 7.76, 15.52, and 28 atm, respectively.

For example, consider a case where urea DS is used with a 5 g/L NaCl FS. To calculate the volume of water extracted, $C_{D,E}$ is first estimated. The 5 g/L NaCl FS has an osmotic pressure equal to 3.91 atm and the equivalent concentration of urea at this osmotic pressure ($C_{D,E}$) is equal to 0.1607 M (Fig 13.10). OLI Stream Analyzer software 9.1 was utilized in these calculations. Substituting M_w, $C_{D,E}$, and $C_{D,\text{Max}}$ in Eq. 13.4 the volume of water extracted is 103 L/kg. The same could be applied to the other three fertilizers at different FS concentrations, producing Fig. 13.13.

As per Fig. 13.13, the water extraction capacity of the DS declines severely upon steady increase in feed TDS. It can also be inferred that the 4 fertilizers exhibit more or

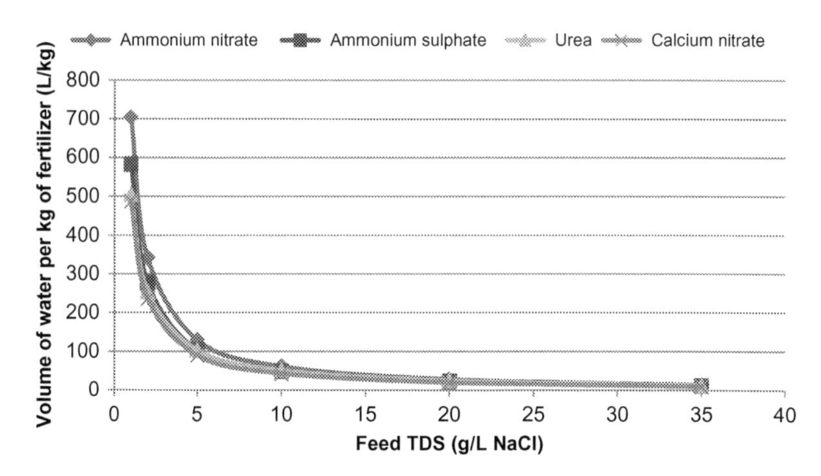

Fig. 13.13 Variation of water extraction capacities of the draw solutes by the FO process at different feed TDS using different draw solutes [68].

less comparable water extraction capacities. Yet, NH_4NO_3 exhibits slightly more water extraction, especially at low TDS feeds. For example, at a feed TDS equal to 1 g/L NaCl, while NH_4NO_3 extracts 700 L/kg of pure water, $Ca(NO_3)_2$ extracts only 488 L/kg. As the FS concentration increases from 1 to 35 g/L NaCl, the difference in the extraction capacities of the four fertilizers significantly decreases.

13.4.2.3 Expected final nutrient concentration in product water

Permeation will continue to take place until the osmotic pressure of the diluted DS is in equilibrium with the FS, regardless of the initial DS concentration used. Thus, the final molar concentrations of each fertilizer DS can be calculated according to the final osmotic pressure of the FS. Again, six different feedwaters (1, 2, 5, 10, 20, and 35 g/L NaCl) are considered to assess the expected nutrient content in the final product water after desalination.

The nutrient content is calculated in terms of nitrogen content and is presented in Fig. 13.14. For instance, urea's final concentration at 5 g/L NaCl as FS (osmotic pressure equal to 3.91 atm) is expected to be 0.1607 M. This concentration of urea contains (0.1607 mol/L $\times$ 28 g/mol) g/L of N_2, or 4.5 g/L of N_2. The same could be applied to the other three fertilizers at different FS concentrations, producing Fig. 13.14.

As per Fig. 13.14, the final nutrient concentrations in FDFO depend on the type of fertilizer used and the TDS of the FS. Feed TDS and final nutrient concentration of product water are directly proportional. The lowest N concentration was observed for $Ca(NO_3)_2$, with 349 mg/L with feed TDS of 1 g/L; however, this increases to 0.72, 1.87, 3.89, 8.2, and 14.8 g/L of N with 2, 5, 10, 20, and 35 g/L feed TDS, respectively. Urea will produce the highest N content in the final product water for all feed concentrations. These results indicate that when high N containing fertilizers such as urea are used as DS, the N content in the product water will be considerably higher

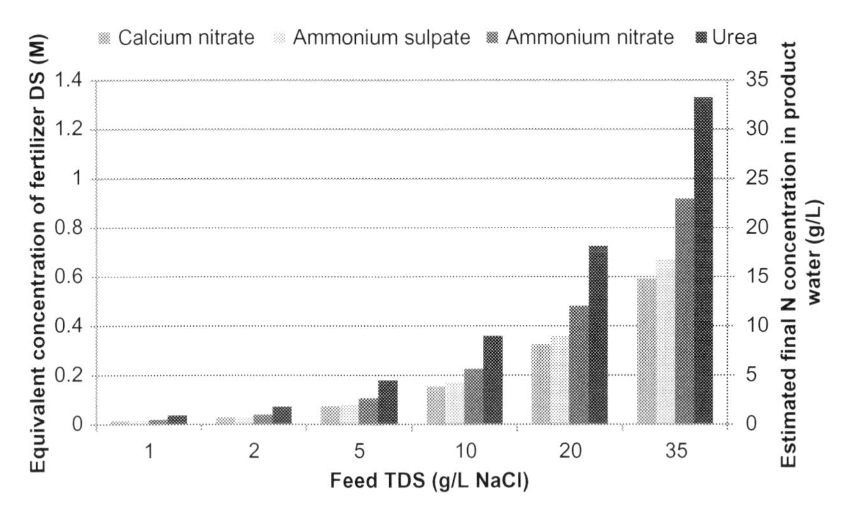

Fig. 13.14 Equivalent concentration of fertilizer DS and estimated final N concentration in product water for different feed TDS concentrations [68].

than in the other fertilizers containing low nitrogen [54]. Another reason for high N concentration with urea is that it exhibits one of the least osmotic pressures among all the fertilizers at equimolar concentration, in spite of its high solubility (Fig.13.8 and Table 13.4).

13.4.2.4 Dilution requirement

In case the final product water from the FDFO desalination process is to be used directly for fertigation, the nutrient concentration must meet the water-quality standards for irrigation. Therefore, it is important that the final FDFO produced water meets the nutrient concentration; otherwise, further dilution is required before fertigation. Excessive fertilizer nutrient can be detrimental to plants as it initiates soil salinity and causes plant toxicity [52]. In addition, leaching of fertilizer nutrients when excessive fertilizer is used in the water can cause undesired pollution of groundwater bodies [79].

Fig. 13.15 illustrates the highest recommended N concentrations for different types of plant crops. Plant requirement from nutrients varies depending on numerous factors, such as types of crop, cropping season, soil nutrient condition, etc. [69]. Generally, the required N nutrient concentrations ranges between 50 and 200 mg/L for N, function of the crop, and growing time of the year [80]. Comparing the information in Fig. 13.15 to that of Fig. 13.14, it can be easily deduced that it impossible to achieve the required water-quality standards by the FDFO desalination process only, especially if feed salinity is more than 1 g/L. The N concentrations are significantly higher, especially for feed with higher TDS, indicating that a high dilution factor is needed to achieve recommended concentrations. This means that the additional dilution required is of several orders of magnitude before it can be used for direct fertigation.

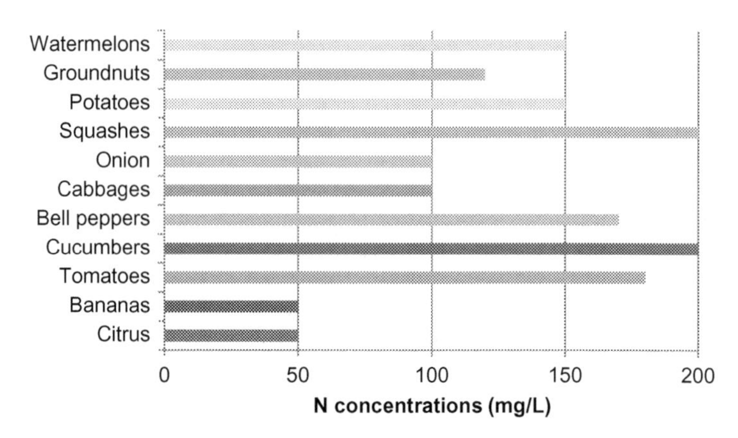

Fig. 13.15 Highest recommended N concentrations for different types of plant crops [80].

For example, if the target crop is potatoes, it is necessary for the N nutrient concentration to be 150 mg/L (Fig. 13.15). None of the four fertilizers achieve an acceptable N concentration for the potatoes without dilution before the fertilizer solution can be used for fertigation, even with the lowest FS concentration of 1 g/L NaCl. Using the selected four fertilizers as the DS will require a dilution factor of at least 4 to make the N concentration acceptable for the potatoes at 150 mg/L using feed with TDS of 2 g/L. The dilution factor for $Ca(NO_3)_2$, $(NH_4)_2SO_4$, NH_4NO_3, and urea are 4.8, 5.0, 6.8, and 12.2, respectively, when used with an FS TDS of 2 g/L. As the FS TDS increases, the dilution factor will increase.

13.5 Conclusion

In conclusion, if a low-cost desalination technology is viable, massive-scale desalination for irrigation may possibly be a reality. FDFO is a propitious technology that could possibly alleviate the water-scarcity problem. Not only is FDFO a sustainable desalination technology, but also it has numerous advantages over conventional desalination technologies, such as RO. The effect of such a technology on the agricultural sector is expected to be outstanding. Abundant saline water, in the form of groundwater or seawater, could be efficiently used to produce valuable nutrient-rich irrigation water, being the major freshwater consumer worldwide. Eventually, the proposed scheme could lead to a technology platform that would supply irrigation water, minimize soil salinity, control fertilizer application, and close the irrigation—brackish water—drainage vicious cycle.

References

[1] ESCWA. Role of desalination in addressing water scarcity. New York: United Nations Economic and Social Commission for Western Asia; 2009. (Water Development). Report No.: 3.

[2] Jury WA, Vaux H. The role of science in solving the world's emerging water problems. Proc Natl Acad Sci U S A 2005;102(44):15715–20.
[3] Beltrán JM, Koo-Oshima S. Water desalination for agricultural applications. Rome, Italy: FAO; 2004.
[4] Amer K, Adeel Z, Böer B, Saleh W, editors. The water, energy, and food security nexus in the Arab region. Cham: Springer International Publishing; 2017. [cited 2016 Dec 10]. (Water Security in a New World). Available from: https://doi.org/10.1007/978-3-319-48408-2.
[5] Elimelech M, Phillip WA. The future of seawater desalination: energy, technology, and the environment. Science 2011;333(6043):712.
[6] Sewilam H, Nasr P. Desalinated water for food production in the Arab region. In: The water, energy, and food security nexus in the Arab region. Cham: Springer International Publishing; 2017. p. 59–81 [Water Security in a Changing World].
[7] Cath T, Childress A, Elimelech M. Forward osmosis: principles, applications, and recent developments. J Membr Sci 2006;281(1–2):70–87.
[8] Zhao S, Zou L, Tang CY, Mulcahy D. Recent developments in forward osmosis: opportunities and challenges. J Membr Sci 2012;396:1–21.
[9] Danasamy G. Sustainability of seawater desalination technology: assessing forward osmosis as a potential alternative technology. London: Centre for Environmental Policy, Imperial College; 2009.
[10] Lay WCL, Chong TH, Tang CY, Fane AG, Zhang J, Liu Y. Fouling propensity of forward osmosis: investigation of the slower flux decline phenomenon. Water Sci Technol 2010;61(4):927–36.
[11] Lee S, Boo C, Elimelech M, Hong S. Comparison of fouling behavior in forward osmosis (FO) and reverse osmosis (RO). J Membr Sci 2010;365(1–2):34–9.
[12] Zhang J, Loong WLC, Chou S, Tang C, Wang R, Fane AG. Membrane biofouling and scaling in forward osmosis membrane bioreactor. J Membr Sci 2012;403–404:8–14.
[13] Semiat R. Energy issues in desalination processes. Environ Sci Technol 2008;42(22): 8193–201.
[14] Shon HK, Phuntsho S, Zhang TC, Surampalli RY. Forward osmosis: fundamentals and applications. Reston, VA: American Society of Civil Engineers; 2015.
[15] Phuntsho S, Hong S, Elimelech M, Shon HK. Osmotic equilibrium in the forward osmosis process: modelling, experiments and implications for process performance. J Membr Sci 2014;453:240–52.
[16] Nicoll PG. Forward osmosis: a brief introduction. IDA World Congress: China; 2013.
[17] Thompson NA, Nicoll PG. Forward osmosis desalination: a commercial reality. Perth: IDA World Congress; 2011.
[18] McCutcheon JR, McGinnis RL, Elimelech M. A novel ammonia—carbon dioxide forward (direct) osmosis desalination process. Desalination 2005;174(1):1–11.
[19] Elimelech M. Yale constructs forward osmosis desalination pilot plant. Membr Technol 2007;2007(1):7–8.
[20] Achilli A, Cath TY, Childress AE. Selection of inorganic-based draw solutions for forward osmosis applications. J Membr Sci 2010;364(1–2):233–41.
[21] Neff RA. Solvent extractor. US Patent 3130156; 1964. Available from: US Patent 3130156.
[22] Batchelder GW. Process of demineralization of water. US Patent 3171799; 1965.
[23] Glew DN. Process for liquid recovery and solution concentration. US Patent 3216930; 1965.
[24] Frank BS. Desalination of seawater. US Patent 3670897; 1972.

[25] Kravath RE, Davis JA. Desalination of sea water by direct osmosis. Desalination 1975;16(2):151–5.
[26] Kessler JO, Moody CD. Drinking water from sea water by forward osmosis. Desalination 1976;18(3):297–306.
[27] Stache K. Apparatus for transforming seawater, brackish water, polluted water or the like into a nutritious drink by means of osmosis. US Patent 4879030; 1989.
[28] Yaeli J. Method and apparatus for processing liquid solutions of suspensions particularly useful in the desalination of saline water. US Patent 5098575; 1992.
[29] Loeb S, Titelman L, Korngold E, Freiman J. Effect of porous support fabric on osmosis through a Loeb-Sourirajan type asymmetric membrane. J Membr Sci 1997;129(2):243–9.
[30] McGinnis RL. Osmotic desalination process. US Patent 6391205B1; 2002.
[31] McCutcheon JR, McGinnis RL, Elimelech M. Desalination by ammonia–carbon dioxide forward osmosis: influence of draw and feed solution concentrations on process performance. J Membr Sci 2006;278(1–2):114–23.
[32] Adham S, Oppenheimer J, Liu L, Kumar M. Dewatering reverse osmosis concentrate from water reuse applications using forward osmosis. Alexandria, USA: WateReuse Foundation; 2007.
[33] McCormick P, Pellegrino J, Mantovani F, Sarti G. Water, salt, and ethanol diffusion through membranes for water recovery by forward (direct) osmosis processes. J Membr Sci 2008;325(1):467–78.
[34] Yen SK, FMH N, Su M, Wang KY, Chung T-S. Study of draw solutes using 2-methylimidazole-based compounds in forward osmosis. J Membr Sci 2010;364(1–2): 242–52.
[35] Ge Q, Su J, Chung T-S, Amy G. Hydrophilic superparamagnetic nanoparticles: synthesis, characterization, and performance in forward osmosis processes. Ind Eng Chem Res 2011;50(1):382–8.
[36] Ling MM, Wang KY, Chung T-S. Highly water-soluble magnetic nanoparticles as novel draw solutes in forward osmosis for water reuse. Ind Eng Chem Res 2010; 49(12):5869–76.
[37] Li D, Zhang X, Yao J, Zeng Y, Simon GP, Wang H. Composite polymer hydrogels as draw agents in forward osmosis and solar dewatering. Soft Matter 2011;7(21):10048–56.
[38] Li D, Zhang X, Yao J, Simon GP, Wang H. Stimuli-responsive polymer hydrogels as a new class of draw agent for forward osmosis desalination. Chem Commun 2011;47(6):1710–2.
[39] Ling MM, Chung T-S. Desalination process using super hydrophilic nanoparticles via forward osmosis integrated with ultrafiltration regeneration. Desalination 2011; 278(1–3):194–202.
[40] Phuntsho S, Shon HK, Hong S, Lee S, Vigneswaran SA. Novel low energy fertilizer driven forward osmosis desalination for direct fertigation: evaluating the performance of fertilizer draw solutions. J Membr Sci 2011;375(1–2):172–81.
[41] Linda Y, Iyer S. Systems and methods for forward osmosis fluid purification using cloud point extraction. US Patent 8021553; 2011.
[42] Su J, Chung T-S, Helmer BJ, de WJS. Enhanced double-skinned {FO} membranes with inner dense layer for wastewater treatment and macromolecule recycle using sucrose as draw solute. J Membr Sci 2012;396:92–100.
[43] Ge Q, Su J, Amy GL, Chung T-S. Exploration of polyelectrolytes as draw solutes in forward osmosis processes. Water Res 2012;46(4):1318–26.
[44] Noh M, Mok Y, Lee S, Kim H, Lee SH, Jin G, et al. Novel lower critical solution temperature phase transition materials effectively control osmosis by mild temperature changes. Chem Commun 2012;48(32):3845–7.

[45] Yong JS, Phillip WA, Elimelech M. Coupled reverse draw solute permeation and water flux in forward osmosis with neutral draw solutes. J Membr Sci 2012;392–393: 9–17.
[46] Bowden KS, Achilli A, Childress AE. Organic ionic salt draw solutions for osmotic membrane bioreactors. Bioresour Technol 2012;122:207–16.
[47] Stone ML, Wilson AD, Harrup MK, Stewart FF. An initial study of hexavalent phosphazene salts as draw solutes in forward osmosis. Desalination 2013;312:130–6.
[48] Ge Q, Fu F, Chung T-S. Ferric and cobaltous hydroacid complexes for forward osmosis (FO) processes. Water Res 2014;58:230–8.
[49] Zhao S, Zou L, Mulcahy D. Brackish water desalination by a hybrid forward osmosis–nanofiltration system using divalent draw solute. Desalination 2012;284:175–81.
[50] Moody CD, Kessler JO. Forward osmosis extractors. Desalination 1976;18(3):283–95.
[51] Phuntsho S, Shon HK, Hong S, Lee S, Vigneswaran S, Kandasamy J. Fertiliser drawn forward osmosis desalination: the concept, performance and limitations for fertigation. Rev Environ Sci Biotechnol 2012;11:147–68.
[52] Kafkafi U, Tarchitzky J. Fertigation: a tool for efficient fertilizer and water management. 1st ed. Paris: International Fertilizer Industry Association and International Potash Institute; 2011.
[53] McCutcheon JR, Elimelech M. Influence of concentrative and dilutive internal concentration polarization on flux behavior in forward osmosis. J Membr Sci 2006;284(1–2): 237–47.
[54] Phuntsho S, Shon HK, Majeed T, El Saliby I, Vigneswaran S, Kandasamy J, et al. Blended fertilizers as draw solutions for fertilizer-drawn forward osmosis desalination. Environ Sci Technol 2012;46(8):4567–75.
[55] Phuntsho S. A novel fertiliser drawn forward osmosis desalination for fertigation [Doctoral of Philosophy thesis]. New South Wales: University of Technology, Sydney (UTS); 2012. [cited 2013 Dec 25]. Available from: http://epress.lib.uts.edu.au/research/handle/10453/21808.
[56] McGinnis RL, Elimelech M. Energy requirements of ammonia–carbon dioxide forward osmosis desalination. Desalination 2007;207(1–3):370–82.
[57] Nasr P, Sewilam H. Forward osmosis: an alternative sustainable technology and potential applications in water industry. Clean Techn Environ Policy 2015;17(7):2079–90.
[58] FAO. Fertilizer use by crop in Egypt. Rome: Food and Agriculture Organization of the United Nations; 2005.
[59] Nasr P, Sewilam H. The potential of groundwater desalination using forward osmosis for irrigation in Egypt. Clean Techn Environ Policy 2015;17(7):1883–95.
[60] Yip NY, Tiraferri A, Phillip WA, Schiffman JD, Elimelech M. High performance thin-film composite forward osmosis membrane. Environ Sci Technol 2010;44(10):3812–8.
[61] Phillip WA, Yong JS, Elimelech M. Reverse draw solute permeation in forward osmosis: modeling and experiments. Environ Sci Technol 2010;44(13):5170–6.
[62] Wang R, Shi L, Tang CY, Chou S, Qiu C, Fane AG. Characterization of novel forward osmosis hollow fiber membranes. J Membr Sci 2010;355(1–2):158–67.
[63] Gray GT, McCutcheon JR, Elimelech M. Internal concentration polarization in forward osmosis: role of membrane orientation. Desalination 2006;197(1–3):1–8.
[64] Yoon H, Baek Y, Yu J, Yoon J. Biofouling occurrence process and its control in the forward osmosis. Desalination 2013;325:30–6.
[65] Ivnitsky H, Minz D, Kautsky L, Preis A, Ostfeld A, Semiat R, et al. Biofouling formation and modeling in nanofiltration membranes applied to wastewater treatment. J Membr Sci 2010;360(1–2):165–73.

[66] Phuntsho S, Hong S, Elimelech M, Shon HK. Forward osmosis desalination of brackish groundwater: meeting water quality requirements for fertigation by integrating nanofiltration. J Membr Sci 2013;436:1–15.
[67] Tan CH, Ng HYA. Novel hybrid forward osmosis nanofiltration process for seawater desalination: draw solution selection and system configuration.Pdf. Desalination Water Treat 2010;13(1–3):356–61.
[68] Nasr P, Sewilam H. Selection of potential fertilizer draw solution for fertilizer drawn forward osmosis application in Egypt. Desalination Water Treat 2017;65:22–30.
[69] Kafkafi U, Kant S. Fertigation. In: Daniel H, editor. Encyclopedia of soils in the environment. Oxford: Elsevier; 2005.
[70] FAO. Report on world agriculture: towards 2015/2030. Manag Environ Qual 2004; 15(2):213.
[71] El-Gabaly S. Fertilizers industry in Egypt; 2015.
[72] Armstrong RD, Dunsford K, McLaughlin MJ, McBeath T, Mason S, Dunbabin VM. Phosphorus and nitrogen fertiliser use efficiency of wheat seedlings grown in soils from contrasting tillage systems. Plant Soil 2015;396(1–2):297–309.
[73] OLI Systems, Inc. OLI studio: stream analyzer. Cedar Knolls, NJ, USA; 2015.
[74] Hancock NT, Cath TY. Solute coupled diffusion in osmotically driven membrane processes. Environ Sci Technol 2009;43(17):6769–75.
[75] Altaee A, Zaragoza G, van Tonningen HR. Comparison between forward osmosis-reverse osmosis and reverse osmosis processes for seawater desalination. Desalination 2014; 336:50–7.
[76] Lenntech. Reverse osmosis desalination: brine disposal, Holland: Lenntech Water Treatment Solutions; 2014. [cited 2014 Apr 11]. Available from: http://www.lenntech.com/processes/desalination/brine/general/brine-disposal.htm.
[77] Shaffer DL, Yip NY, Gilron J, Elimelech M. Seawater desalination for agriculture by integrated forward and reverse osmosis: improved product water quality for potentially less energy. J Membr Sci 2012;415–416:1–8.
[78] Nasr P, Sewilam H. Investigating the performance of ammonium sulphate draw solution in fertilizer drawn forward osmosis process. Clean Techn Environ Policy 2016; 18(3):717–27.
[79] Freeze RA, Cherry JA. Groundwater. New Jersey: Prentice-Hall; 1979. p. 636.
[80] Phocaides A. Handbook on pressurized irrigation techniques. Rome: Food and Agriculture Organization of the United Nations; 2007.

Further reading

[1] Ge Q, Ling M, Chung T-S. Draw solutions for forward osmosis processes: developments, challenges, and prospects for the future. J Membr Sci 2013;442:225–37.

Seawater desalination for crop irrigation—Current status and perspectives

Victoriano Martínez-Alvarez, Manuel J. González-Ortega, Bernardo Martin-Gorriz, Mariano Soto-García, Jose F. Maestre-Valero
Technical University of Cartagena, Cartagena, Spain

14.1 Significance of the topic

Agriculture faces the challenge to produce more food to feed a growing population. The world population is expected to grow by 2.3 billion people by 2050, which will require extending the land equipped for irrigation by some 32 million hectares to guarantee raising overall food production by some 70%, since the per capita food consumption is also increasing. Under these circumstances, water withdrawals for irrigation to cope with the increasing food demands should grow by 11% for 2050 [1]. At the same time, water is becoming scarcer due to industrial development and a sustained increase in living standards. This unfavorable scenario is expected to be exacerbated by the negative effect of climate change on water resources under arid and semiarid climates. Consequently, the pressure on water resources is becoming more severe, leading to imbalances between renewable resources and total demands, thus jeopardizing irrigated agriculture and its resilience as a nonpreferential water use.

Agriculture drives global water demands, with irrigation accounting for 70% of water withdrawals, while industrial and domestic sectors account for the remaining 20% and 10%, respectively [2]. The importance of agricultural water use is even higher in arid and semiarid regions with a highly technical agriculture, reaching figures of over 85% in regions such as south-eastern (SE) Spain and Israel. Moreover, irrigated agriculture represents 20% of the total cultivated land, yet contributes 40% of the total food produced worldwide. This intensifying water scarcity scenario represents a threat to the long-term role of irrigated agriculture in global food security, which cannot meet current or future demands for irrigation by relying solely on conventional water sources. New solutions are required to maintain or enhance sustainable agricultural production, including new or alternative water resources, water conservation strategies, or more efficient and productive irrigation systems. In this context, an interesting technological approach is to increase the use of non-conventional water resources to guarantee long-term food security and socioeconomic stability. Whereas reclaimed water and desalinated brackish groundwater are usually limited by the domestic wastewater production and the exhaustion of aquifers, desalinated seawater (DSW) could be an abundant and steady water source able to sustain

Emerging Technologies for Sustainable Desalination Handbook. https://doi.org/10.1016/B978-0-12-815818-0.00014-X

agricultural production and effectively eliminate the climatological and hydrological constraints associated with other water sources. Moreover, the replacement of over-exploited water resources with DSW provides new opportunities to address environmental and socioeconomic problems and to carry out more sustainable water policies.

While the world production of DSW is continuously growing for domestic uses, the use of DSW has rarely been considered for crop irrigation. However, this situation is currently changing as a result of the aforementioned increasing pressure on renewable water resources and the lack of attainable solutions in arid and semiarid coastal regions. Seawater desalination has emerged in the past decade as a feasible option for irrigated agriculture in Spain and Israel, and its implementation is being considered in other states such as Egypt, Saudi Arabia, Australia, or California [3]. In the cases of Spain and Israel, DSW is mainly produced in large coastal desalination plants, and allocated inland through conveyance systems. In other countries such as Italy and Egypt, DSW for irrigation comes mainly from small on-farm desalination plants supplied from coastal aquifers with seawater intrusion, that is, transitional waters. Therefore, seawater desalination is increasingly being considered as an alternative water supply for crop irrigation and this trend is expected to intensify in the near future.

Reverse osmosis is widely considered to be the most adaptable technology for seawater desalination plants (SWDPs) devoted to agricultural supply because of its relatively low-energy consumption compared to other technologies [4]. This is endorsed by the fact that the current agricultural experiences in Spain and Israel rely on said technology. Nevertheless, reverse osmosis results in a very particular water composition that requires special attention to avoid agronomic problems, especially in comparison with other traditional water sources. These agronomic concerns, which are presented and analyzed in this chapter, could become a barrier to the expansion of DSW agricultural use, although most authors agree that DSW is becoming a technically and economically feasible solution for high-return agriculture, especially in the current scenario where the costs of surface water and groundwater are increasing.

Because of the potential role of DSW in meeting the increasing agricultural water demands in the context of growing water scarcity, it is of interest to review, analyze, and discuss the key issues revealed by current experiences of crop irrigation with DSW. Therefore, this chapter reviews the current irrigation experiences with DSW worldwide, focusing on (1) the main agronomic concerns, such as low nutrient concentration, crop toxicity risk due to high boron and chloride concentration, or the sodicity risk affecting soil physical properties; (2) the energy requirements for DSW production and allocation, the associated greenhouse gas emissions and the derived cost, as the current limiting factors for its agricultural application; (3) the on-farm recommended management strategies for promoting its sustainability; and (4) the future research and development perspectives. The chapter avoids the technological questions of desalination facilities and processes, which are widely discussed in other chapters of this book.

The chapter's contents will be useful for water managers and planners, field technicians, and farmers in a growing number of water-scarce coastal regions with highly technical agriculture, where a significant fraction of crop water requirements will rely on seawater desalination in the future.

14.2 Current irrigation experiences with DSW

At the outset, DSW was used to supply domestic and industrial demands. However, as desalination technology has improved and the DSW cost has decreased, its application is being extended to other sectors, especially to agriculture. Brackish-water desalination for crop irrigation is reported worldwide and has dramatically increased in recent years since its cost is usually below 50% of DSW costs [5]. On the contrary, the agricultural use of DSW is mainly limited to the past decade and to coastal regions with an arid climate and high-return agriculture, where water conveyance from SWDPs to the farmers is strategically and economically affordable. Another situation where DSW has been applied to crop irrigation is on small islands lacking conventional water sources.

A review of international databases has only confirmed the agricultural application of DSW in Spain and Israel, although its adoption is being analyzed and/or considered in other states such as Egypt, Saudi Arabia, Australia, or California [3]. The irrigation experiences with DSW are detailed below.

14.2.1 Spain

Two different regions where DSW is being applied to crop irrigation can be distinguished: south-eastern Spain and the Canary Islands.

South-eastern Spain's agriculture has systematically suffered from a lack of water that has worsened over time. In the mid-1990s, >200 brackish-water desalination plants were installed on-farm by farmers to resist a 4-year drought period [6]. In 2001, the Spanish government projected the Carboneras SWDP in Almeria province, with 44 Mm^3 $year^{-1}$ capacity and which was partially planned for agricultural supply. At that time, it was considered the biggest SWDP in the world. It has been in service since 2005, supplying irrigation water to 7000 greenhouse hectares in the Campo de Nijar Irrigation District and, more recently, to farmers of the Almanzora valley. A new distribution system that extends almost 60 km was built in order to allocate DSW for agriculture. In 2004, the construction of Valdelentisco SWDP in Murcia province began, with 48 Mm^3 $year^{-1}$ capacity and which was predominantly planned for agricultural use. It has been supplying water to farmers since 2008, with its distribution system still being developed.

The persistence of water scarcity problems on the SE Spanish Mediterranean coast, combined with the need to guarantee the existing high-return agriculture and to mitigate overexploitation of groundwater, led the Spanish government to sanction the so-called AGUA Program in 2004. The Program was implemented from 2005 to 2011 and was mainly aimed at building or upgrading 29 large SWDPs along the Spanish Mediterranean coast to reinforce water supply for agricultural, urban, and tourist use. It meant a reorientation of Spanish water policy toward seawater desalination as an alternative to interbasin water transfers. This program sought to produce 693 Mm^3 $year^{-1}$, of which 200 Mm^3 $year^{-1}$ would enhance the irrigation of about 10^5 ha. The current situation is that most of these SWDP are operational, but below their projected capacity, although some are still under construction or in the start-up period, and in some exceptional cases are unused or even unbuilt. The present drought period in SE Spain has allowed a quick rise of DSW demand for irrigation, which had remained fairly idle due to the absence of droughts since 2008, resulting in a clear unfulfillment of agricultural DSW supply prospects. The current situation is analyzed in a study case (Section 14.2.4).

Dealing with agricultural DSW use in the Canary Islands, it should be noted that it was the place that acted as the starting point of seawater desalination in Spain and Europe, first for domestic use (1965) and later for crop irrigation (1987) [7]. The question is that due to the extremely arid climate on some of these islands (e.g., Lanzarote) producing DSW is the only water source for irrigation, directly supplied from an SWDP or reclaimed after domestic use [8]. Therefore, crop irrigation with DSW has been progressively introduced in the Canary Islands over the past decades, first for traditional agriculture systems and nowadays for new high-return tropical crops. The first SWDPs for irrigation purposes were built in 1987 and 1988 for farmers' cooperative societies (Las Salinas and Llanos de Juan Grande, with 2.5 and 2.2 Mm^3 $year^{-1}$ capacity, respectively) [7]. The first large SWDP for domestic and agricultural use (Las Palmas, 30 Mm^3 $year^{-1}$ capacity) was built in 1990; since then about 10 SWDPs have been built with the sole purpose of providing water for irrigation, with an overall capacity of around 25 Mm^3 $year^{-1}$.

14.2.2 Israel

Israel has dealt with severe water scarcity throughout its history, which has led its agriculture to continuously innovate for optimizing water consumption and production. For these reasons, irrigation with underground brackish water has been prevalent. Additionally, since 2006 DSW has become a real option for irrigation, being the only country that has sanctioned specific recommendations for domestic and agricultural use [9].

Seawater desalination was first developed in Israel in 1965 to address the chronic lack of water of the city of Eilat and the surrounding agricultural communities. Since then the Israeli Water Authority has been made aware that the projected future water demand cannot be met by natural water resources, motivating increasing interest in seawater desalination. In such a way, The Israeli Desalination Master Plan was conceived in 1997 with the overarching goal of ensuring that water will be sustainable, available, and reliable in the required quantities, locations, and qualities [10]. The plan intended to meet all Israeli domestic water needs with DSW by enhancing existing facilities and building new large-scale ones [11]. As a result, Israel today has five large-scale operational SWDPs using the reverse osmosis technology, the location, inauguration date, and production capacity of which are displayed in Fig. 14.1.

The first SWDP in Ashkelon was built in 2005. The other facilities (Palmachim, Hadera, and Soreq) have been gradually launched since 2010, with the last facility in Ashdod beginning to operate in December 2015. In total, the five desalination plants along Israel's Mediterranean coast were producing over 600 MCM Mm^3 $year^{-1}$ by 2016, which accounts for 100% of domestic water demand. By 2020, expansion of

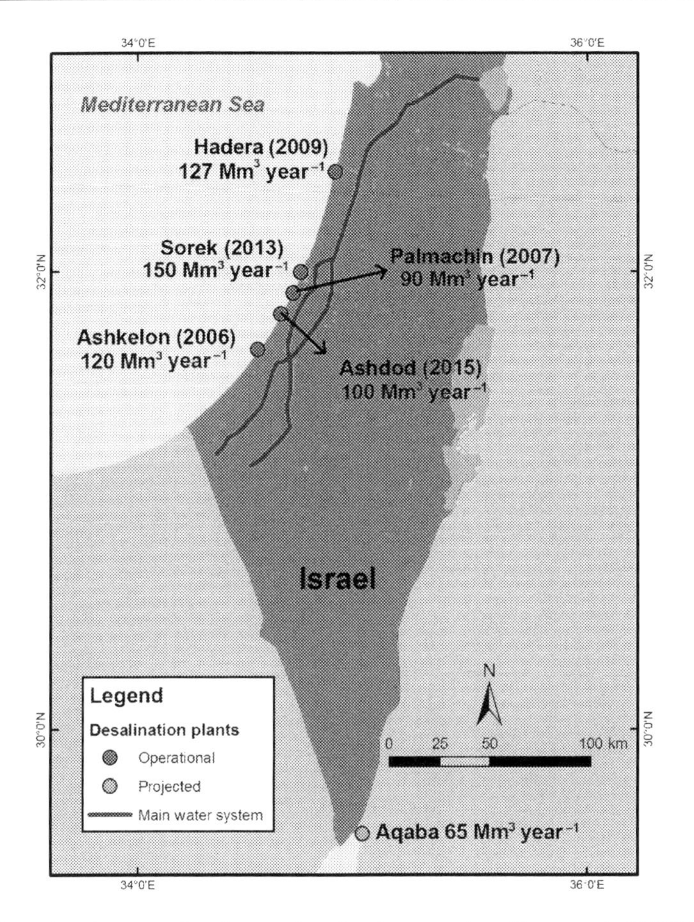

Fig. 14.1 Location, inauguration date, and production capacity of large-scale DSWPs in Israel.

existing plants will increase the total production capacity to 750 MCM annually, exceeding Israel's domestic water consumption. In addition, a new SWDP has been planned in the vicinity of Aqaba (Jordan) that will produce 65 Mm^3 $year^{-1}$, 35 of which will be supplied to Israel for crop irrigation.

The Israeli Desalination Master Plan and the construction of the National Water Carrier, which has been built in the past few years and is composed of a system of 100-in. pipelines that receive the large amount of water coming from SWDPs and distribute it to all regions, have enabled Israel to put an end to its water crisis and dependence on the weather.

The agricultural use of DSW is mainly focused in southern Israel, where the low population density has allowed for a substantial percentage of Ashkelon and Palmachin SWDP supply to be used by farmers [10]. This shift in irrigation water from natural water resources to DSW has had both positive and negative effects on the healthy growth of crops, as will be seen later. In this respect, the scientific monitoring of the initial farming experiences with DSW coming from Israeli facilities designed to provide water for human consumption has revealed important concerns to be considered when planning agricultural DSW use [12,13].

14.2.3 Other countries

Seawater desalination technology is being considered for irrigated agriculture in other regions such as the Middle East and North Africa (MENA) or the states of Florida and California in the United States. In Saudi Arabia, agriculture is the largest water-consuming sector, constituting >90% of the total water use [14]. As part of Saudi Arabia's strategy to cope with limited water resources, seawater desalination is proposed as one measure to redress the water shortage. Some studies are analyzing the potential of using DSW for growing crops in this country, which has a desalination capacity of nearly 2000 Mm^3 $year^{-1}$ [15]. For that matter, newly developed tools have been applied for optimizing cropping patterns under limited land and water availability for DSW use in agriculture by analyzing a case study with a 21 Mm^3 $year^{-1}$ capacity SWDP and about 84,753 ha of possible irrigated fields along the Arabian Gulf.

In Florida, the Commissioner of Agriculture stated that conservation and reuse alone will not be enough to guarantee future agricultural water supplies, and thus suggested DSW should be included in water planning as a reliable source that is resistant to droughts and shortages [16]. In California, where roughly 80% of the water supply goes to agriculture and where one of the most severe drought periods in its modern history persists, the Carlsbad SWDP was built and opened in December 2015 in San Diego, which is the largest in America with a 69 Mm^3 $year^{-1}$ capacity. It is the first intervention leading to the diversification of water resources in California, but 15 more SWDPs have been proposed [17]. In both American cases, the plan is that following the domestic use of DSW, the resulting wastewater will be reclaimed and reused for crop irrigation. Finally, it should be noted that national assessments of the agricultural applicability of desalination are currently underway in Chile, China, and Australia [5].

14.2.4 Case study: Agricultural DSW supply in the Segura River Basin (SE Spain)

The Segura River Basin (SRB) is the most water-scarce region in Spain, and subsequently in Europe. It suffers a structural water deficit that has been accounted in 429 Mm^3 $year^{-1}$ [18]. Water resources in the SRB are surface water (29.9%), groundwater (38.5%), reutilization (6.9%), desalinization (7.9%), and water transferred from central Spain (16.8%), whereas the water demand is characterized by a high agricultural use, representing >80% of the total. The extant pressure on water resources, together with the suitable climate conditions in the SRB, have led to a high-return agricultural development model which is a worldwide reference, usually named "the orchard of Europe." Thus, agriculture is one of the basic pillars of growth of the regional economy and exports to EU countries exceed 70% of the total production of horticultural crops (≈1,000,000 Tm = 3570 M€), which minimizes external dependency and maintains a very important socioeconomic sector. Therefore, agriculture resilience in the SRB is strategic from a regional economic and social point of view.

The application of DSW to overcome the water shortage for agricultural production in the area began in 1995, with the start-up of Mazarrón Irrigation District's SWDP, a private facility to guarantee the irrigation requirements of 3500 ha with tomato greenhouses and citrus trees. However, the real encouragement for DSW agricultural use came in 2004, when the Spanish government sanctioned the construction of several large-sized SWDPs in the SRB by means of the AGUA Program. The current situation (Fig. 14.2) presents 11 large-sized coastal SWDPs in operation; three of them were developed by different farmers' associations and are exclusively devoted to crop irrigation (Mazarrón Irrigation District, Marina de Cope Irrigation District, and Águilas Irrigation District); another four were planned for combined domestic and agricultural use (Torrevieja, Escombreras, Valdelentisco, and Águilas) although in practice nearly

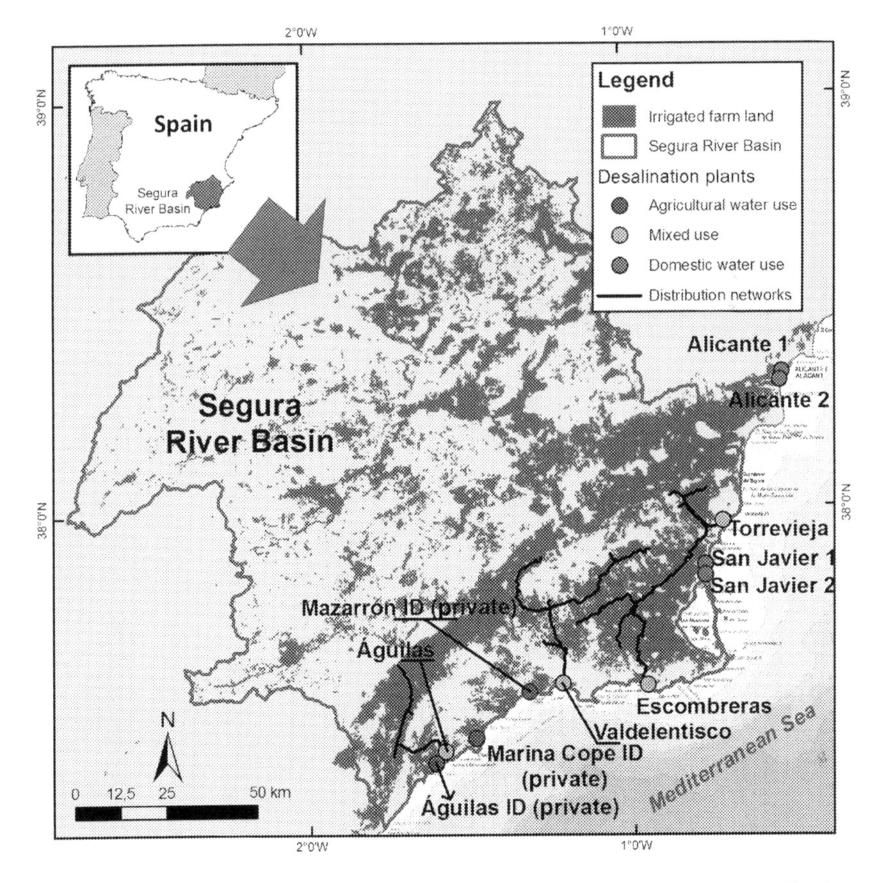

Fig. 14.2 Location of large-scale DSWPs in the Segura River Basin and new distribution schemes for DSW allocation to irrigated areas.

90% of their production is devoted to agricultural use; while the remaining four were established to guarantee the domestic supply for the also important tourist activity in the area (Alicante 1&2 and San Javier 1&2).

Table 14.1 includes the main technical characteristics of the SWDPs supplying agriculture in the SRB. All of them use reverse osmosis technology, although with a different number of stages and passes. Those that were designed before the AGUA Program was sanctioned have a required level in boron concentration of $<1\,\mathrm{mg\,L^{-1}}$ and one pass with one or two stages is enough to achieve this. The most modern SWDPs (Águilas and Torrevieja) have to comply with a boron concentration of $<0.5\,\mathrm{mg\,L^{-1}}$, as is stipulated in the AGUA Program contractual conditions; so they apply a reverse osmosis technology with two passes, and two stages in the second pass. A fraction of about 70% of the permeate produced in the first pass is sent to the second pass, where a 90% conversion is reached and the boron content falls below $0.2\,\mathrm{mg\,L^{-1}}$. Subsequently, the permeate of the second pass is blended with the remaining permeate of the first pass and the obtained boron concentration is $<0.5\,\mathrm{mg\,L^{-1}}$.

Table 14.1 also evidences how the current production capacity is lower than the final scheduled capacity for most SWDPs. The main reason is that other necessary facilities, such as the electricity lines to provide the required electrical power or the distribution pipes toward inland irrigation districts, are still under construction; so the SWDPs are being progressively upgraded as electricity supply and agricultural water demand increase. Fig. 14.2 shows the current development of distribution pipelines that, contrary to all prevailing water distribution schemes, go from the DSWPs at the coast to inland irrigation districts. Each SWDP has its own distribution system with hydrants for farmers that at the moment add up to about 250km and which will continue to develop in the coming years in order to obtain a completely interconnected DSW supply system. It should be noted that there is not a free water market, but that some farmers and irrigation districts propose temporary agreements with SWDPs for purchase transactions, which should be mandatorily assessed by the Regional Water Agency before approval, and should also target already existing irrigated lands suffering water shortage.

The beginnings of DSW for crop irrigation in the SRB have not been easy. On the one hand, the entry into operation of the SWDPs coincided with a humid period in which agricultural demand was well covered by conventional water resources. On the other hand, the irrigators associations were claiming that the DSW price was very expensive for most farms' economies, seeking the enactment of a "social price." As a result, most new SWDP remained fairly idle until 2013, when the water transferred from central Spain was considerably decreased due to a new drought period. This water shortage, together with the approval of a subsidized price for the DSW produced in Valdelentisco and Torrevieja SWDPs by the Spanish government as an adaptive drought measurement since autumn 2015, has allowed a quick rise in DSW demand for irrigation. Fig. 14.3 shows the evolution trend of DSW use for crop irrigation in the SRB over the past decade, reaching a total amount of $139\,\mathrm{Mm^3\,year^{-1}}$ in 2016, which is expected to be surpassed in 2017 with nearly $175\,\mathrm{Mm^3\,year^{-1}}$. This figure highlights the great potential of DSW as an alternative to traditional limited water resources in coastal high-return agricultural areas.

Table 14.1 **Technical characteristics of the SWDPs supplying agriculture in the SRB**

Desalination plant	Operational date	Current production capacity ($Mm^3\ year^{-1}$)	Scheduled enhanced capacity ($Mm^3\ year^{-1}$)	Percentage of production for crop irrigation	Type of intake	Number of stages	Energy recovery system	Contractual required level in boron concentration ($mg\ L^{-1}$)	Posttreatment configuration	Outfall pipe
Torrevieja	2015	50	120	>90%	Open, using one pipe (2000 m long and 2.4 m diameter) and a concrete box at Torrevieja harbor	Two passes. Second pass with two stages	Pressure exchanger	<0.5	Slaked lime slurry + CO_2 + hypochlorite	One pipe 2000 m long and 2 m diameter with diffusers at the breakwater of Torrevieja harbor
Escombreras	2009	21	24	>90%	Open, using one pipe (1600 m long and 1.4 m diameter) and a concrete box at Cartagena harbor	One pass with one stage	Pressure exchanger	<1	Calcite bed + CO_2 + hypochlorite	One pipe 1600 m long and 1.4 m diameter pouring at Cartagena harbor bottom
Águilas	2013	45	70	80%	Open, with a 5.5-diameter underwater cylindrical reinforced concrete tower 840 m offshore	Two passes. Second pass with two stages	Pressure exchanger	<0.5	Slaked lime slurry + CO_2 + hypochlorite	One pipe 2800 m long and 2 m diameter stretch on land. Underwater pipeline with diffusers
Valdelentisco	2008	50	70	100%	Open intake tower 5 m height	One pass with two stages	Francis turbine	<1	Slaked lime slurry + CO_2	One pipe 1250 m long and 1.25 m diameter with underwater diffusers
Mazarrón ID	1995	10	16	100%	–	–	–	–	–	–
Aguilas ID	2002	4	8	100%	–	–	–	–	–	–
La Marina-Cope ID	2006	2	8	100%	–	–	–	–	–	–

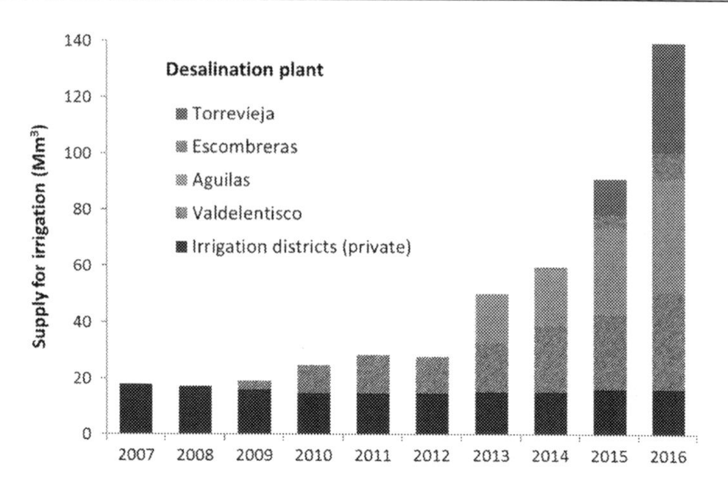

Fig. 14.3 Evolution of agricultural DSW use in the SRB since 2007.

14.3 Scientific and technical challenges of agricultural DSW use

14.3.1 Agronomic concerns

Agronomic water quality is a critical issue when planning to irrigate with DSW and should be specifically addressed. Permeate from RO processes generally has an acidic pH, a total dissolved solids concentration below $250\,mg\,L^{-1}$, and a very low hardness and buffering capacity [19]. Permeate in such conditions must be re-mineralized and ionically balanced prior to distribution for whatever application, since it is aggressive even toward distribution system components [20]. Therefore, the DSW chemical composition is quite variable depending on the RO technology and the type and intensity of the posttreatments.

It is usual for the posttreatments to be designed to conform to the national potable water regulations, which do not match the suitable characteristics for crop irrigation. This reliance on potable water regulations is ascribed to the fact that SWDPs are totally or partially destined for domestic use. For that reason, remineralization posttreatments are not very intense and the chemical composition of the resulting DSW is quite different from that of conventional water sources. In general, the composition of DSW is manifestly predominated by sodium (Na^+) and chloride (Cl^-) ions, with a somewhat low concentration of other ions such as magnesium (Mg^{2+}), calcium (Ca^{2+}), and sulfate (SO_4^{2-}). From an agronomic point of view, the high presence of boron (B^{3+}), a phytotoxic element, is also noteworthy.

Unlike other nonconventional water sources, such as the reclamation of urban wastewater for irrigation, there are no regulations for agricultural DSW use in most developed countries. To date, only the Israeli Water Authority has adopted specific water quality criteria for combined agricultural and domestic DSW use, since the production in some of their DSWPs may be applied in both uses. Lahav and Birnhack [9] detail the reasons for these specific criteria, which cover 10 water quality parameters: concentration ranges for a number of ions (Cl^-, Na^+, B^{3+}, Ca^{2+}, Mg^{2+}, and SO_4^{2-}), pH, electrical conductivity (EC), alkalinity, and calcium carbonate precipitation potential (CCPP).

To address the agronomic concern of DSW supplies for crop irrigation, certain quality standards for irrigation should be established. In this work, the chemical composition of DSW supplied in Spain and Israel for agricultural is assessed by comparison with (1) the Israeli Water Authority recommendations for combined agricultural and municipal DSW use, and (2) the standards reported by Ayers and Westcot [21], which are based on soil and crop protection criteria, and constitute the most widespread reference in the literature. A total of four SWDPs have been considered for that assessment, two from Israel (Ashkelon and Hedera; Fig. 14.1), and another two from SE Spain (Escombreras and Águilas; Fig. 14.2).

14.3.1.1 Effect of irrigation water salinity on crop production

Salinity is a property referring to the amount of soluble salt in the water. EC is the most common measure of water salinity. During the last century, many experiments have been carried out to determine the salt tolerance of crops. Maas and Hoffman [22] concluded that crop yield as a function of the soil salinity (salinity of saturated soil-paste water extract) could be well-described by a linear response function characterized by a salinity threshold value below which the yield is unaffected and above which yield decreases linearly with salinity. This threshold-slope model is crop and variety specific and has been widely used in crop irrigation management. Fig. 14.4 represents the model and provides threshold-slope values for some representative Mediterranean irrigated crops.

The EC of DSW salinity in single-stage RO plants is between 0.4 and 0.6 $mS\,cm^{-1}$, and in the double-stage between 0.1 and 0.4 $mS\,cm^{-1}$. In the case of selected DSWPs, the reported EC values are as follows: Ashkelon = 0.2–0.3 $mS\,cm^{-1}$; Hedera = 0.4–0.5 $mS\,cm^{-1}$; Escombreras = 0.5–0.6 $mS\,cm^{-1}$; and Águilas = 0.4–0.5 $mS\,cm^{-1}$. Therefore, EC values are always below the standards proposed by Ayers and Westcot ($EC < 0.7\,mS\,cm^{-1}$), although in some cases they exceed the Israeli Water Authority recommendations ($EC < 0.3\,mS\,cm^{-1}$).

The incorporation of DSW for irrigation can produce quite different agronomic effects, depending on the quality of the replaced water. When DSW replaces irrigation waters with an EC above the specific salinity threshold for a crop, an increase in the productivity and quality of crop yields can be expected due to reduced salinity stress. Moreover, meaningful decreases in the irrigation requirements can be achieved, since the supplementary amount of irrigation water for salt leaching can be significantly reduced or even dismissed. This explains why it is usual to blend DSW with other low-quality waters as a management strategy at the farm scale, establishing the adequate blending rate to obtain final irrigation water EC near the salinity threshold for the crop. Some examples of these beneficial effects have been reported; so when DSW replaced superficial (1.2–2.2 $mS\,cm^{-1}$) and brackish (5–7 $mS\,cm^{-1}$) irrigation waters in a Spanish citrus orchard the production increased by around 10% and by >50%, respectively, with 20% lower irrigation requirements [7]. In Israel, with a greenhouse

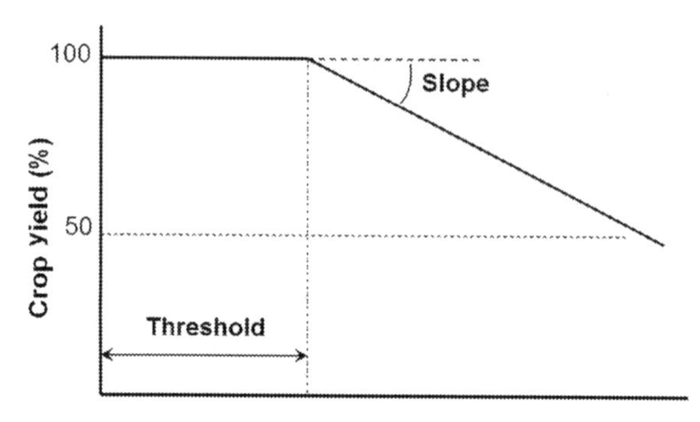

Crop	Threshold		Slope[b]	
	EC_s[a]	EC_w[a]	EC_s[a]	EC_w[a]
Lettuce	1.3	0.9	13.2	20.8
Melon	2.2	1.5	7.1	10.9
Pepper	1.5	1.0	13.9	20.8
Tomato	2.5	1.7	10.0	14.7
Orange trees	1.7	1.1	16.7	21.7

[a] EC_s refers to saturated soil-paste water extract and EC_w to irrigation water.
[b] Decrease in crop yield (%) per unit of increased electrical conductivity (mS cm^{-1}).

Fig. 14.4 Crop yield as a function of the soil salinity or of the irrigation water salinity.

pepper crop, replacing 3.20 mS cm^{-1} irrigation water with 0.40 mS cm^{-1} desalinated water increased maximum yields by almost 50% while allowing a reduction of applied irrigation water to half of that with the saline water [12].

Moreover, when no low-quality waters are replaced, the low salinity of DSW does not imply agronomic benefits, whereas it entails some agronomic concerns, as is analyzed in the following. A clear example is the current situation in SE Spain, where DSW is replacing high-quality water transferred from central Spain with EC about 1 mS cm^{-1}, and where some yield improvement or water savings are expected.

14.3.1.2 Lack of essential nutrients such as Mg^{2+}, Ca^{2+}, and SO_4^{2-}

Magnesium (Mg^{2+}), calcium (Ca^{2+}), and sulfate (SO_4^{2-}) are essential nutrients that play important roles in plant growth and disease resistance. Magnesium is of crucial importance to the photosynthesis and protein synthesis operations of plants; thus, crops suffering from Mg^{2+} deficiency symptoms could malperform. Calcium deficiency may harm the growth of terminal buds and of root apical tips. Sulfate deficiency reduces plant dry weight, photosynthetic rate, chlorophyll content, and total number of fruits.

Crop fertilization is generally programed as a function of its requirements in nitrogen, phosphorus, and potassium, whereas other nutrients such as Mg^{2+}, Ca^{2+}, and SO_4^{2-} play a secondary role because natural waters, together with soil mineral content, usually provide sufficient concentration to minimize additional fertilization with them. This circumstance is observed in Table 14.2, which compares representative nutritional requirements in Ca^{2+}, Mg^{2+}, and SO_4^{2-} for horticultural crops under SE

Table 14.2 Comparison among the typical nutritional requirements in Ca^{2+}, Mg^{2+}, and SO_4^{2-} for horticultural crops in SE Spain, their typical concentration in SE Spain and Israel conventional irrigation waters, and their concentrations in the selected SWDPs

Ion	Horticultural crops requirements[a]	Israeli recommendations for domestic and agricultural DSW use [13]	Concentration in conventional irrigation waters		Ashkelon [13]	Hedera [23]	Escombreras[c]	Águilas[c]
			SE Spain[b]	Israel [17]				
Ca^{2+} ($mg L^{-1}$)	80–120	32–48	90–110	45–60	40–46	32–34	18–22	15–20
Mg^{2+} ($mg L^{-1}$)	24–36	12–18	35–45	20–25	0	0	1.5–2.5	1–2
SO_4^{2-} ($mg L^{-1}$)	100–150	>94	200–350	60–80	0	60–80	2–6	–

[a] Representative values for intensive horticultural crops in SE Spain.
[b] Representative values for 48 irrigation districts in SE Spain.
[c] Own data.

Spain irrigated agriculture conditions with the typical concentration of these minerals in the conventional irrigation waters. In the SE Spain case, irrigation water presents very high water hardness, exceeding typical crop requirements in Ca^{2+}, Mg^{2+}, and SO_4^{2-}. Although typical Israeli irrigation waters are not so hard, they also provide most of these mineral requirements.

On the contrary, the concentrations of Ca^{2+}, Mg^{2+}, and SO_4^{2-} in DSW are considerably lower than the nutritional requirements, as Table 14.2 displays. The reverse osmosis process not only separates the undesirable salts from the seawater, but also removes those minerals that are essential crop nutrients. Table 14.2 also shows that Israeli SWDPs present a concentration of Ca^{2+} in agreement with the Israeli Water Authority recommendations, but lower for Mg^{2+} and SO_4^{2-}. The Spanish SWDPs have a lower Ca^{2+} concentration than the Israeli ones, whereas the values for Mg^{2+} and SO_4^{2-} are similar. The differences in the concentrations of these minerals among the four SWDPs in Table 14.2 clearly highlight the effect of the type and intensity of posttreatment processes in the concentration of these minerals in DSW and, consequently, the importance to reconsider the design of posttreatment processes when the DSWP supplies agriculture.

Therefore, the replacement of traditional irrigation waters with DSW could expose the crops to growth-limitation factors, affecting yield quality and quantity, as has already been reported for several crops in Israel [12,13]. To guarantee proper crop performance if the water supply switches to DSW, it is obvious that a significant amount of Ca^{2+}, Mg^{2+}, and SO_4^{2-} should be provided by fertilization at the farm level, especially in low mineralized soils or soilless farming, leading to an increase in the production costs for farmers [12].

There are three possible management options to accomplish the reintroduction of the required nutrients into DSW for agricultural use: they can be added at the SWDP as part of the posttreatment processing; they can be reintroduced by blending the DSW with other highly mineralized natural waters; or they can be provided by farmers as fertilizers. Careful assessment of the management option to be followed is essential, since, depending on the approach adopted, the remineralization costs are moved from the SWDP owners to the farmers, or even avoided if appropriate blending is possible.

14.3.1.3 *Crop toxicity risk due to high Cl^-, Na^+, and B^{3+} concentrations*

Irrigation with high chloride (Cl^-) and sodium (Na^+) concentrations exposes soils and crops to their accumulation. Increased concentration and exposure time is known to negatively affect soil characteristics, plant functioning, and to lead to the uptake and accumulation of plant-toxic ions in crops. Long-term leaf and soil analyses in Israel, where irrigation waters with high Cl^- and Na^+ concentrations have been applied in the past decades, indicate a steady increase in soil sodicity and crop salt uptake and accumulation [24].

Most orchards are sensitive to Cl^- and Na^+, whereas annual crops are less sensitive. Leaf burn, scorch, and dead tissue along the edges and tips of older leaves are common symptoms. Table 14.3 displays the toxicity risk from Cl^- and Na^+ concentration in

Table 14.3 Israeli Water Authority recommendations for chloride (Cl^-) and sodium (Na^+) concentration in DSW, toxicity risk following Ayers and Westcot's standards, and their concentrations in the selected SWDPs

Ion	Israeli recommendations for domestic and agricultural DSW use [13]	Low-risk level [21]	Moderate-risk level [21]	High-risk level [21]	Ashkelon [13]	Hedera [23]	Escombreras[a]	Águilas[a]
Cl^- ($mg\,L^{-1}$)	<20	<140	140–350	>350	15–20	<20	130–150	125–135
Na^+ ($mg\,L^{-1}$)	<20	<70	70–210	>210	9–10	<20	85–95	60–75

[a]Own data.

water for irrigation following the standards reported by Ayers and Westcot and the Israeli Water Authority recommendations.

The lower limits in Table 14.3 must be considered under sprinkler irrigation, since Cl^- and Na^+ can also be directly absorbed into the leaves causing foliar damage, especially during periods of high temperatures and low humidity. The wide variation range in the moderate risk category is because the toxicity to Cl^- and Na^+ is crop- and variety-specific. In general, sensitivity to Cl^- and Na^+ is related with crop salinity tolerance [22]; so salinity tolerance tables could be used as a first risk assessment.

DSW usually presents high chloride and sodium concentrations since about 55% and 31% of the salt content in seawater is Cl^- and Na^+. After the reverse osmosis process, these ions still predominate in its composition. Therefore, DSW is prone to produce specific phytotoxicity in sensitive crops. Table 14.3 shows the Cl^- and Na^+ concentrations in Spanish SWDPs, which in the case of Escombreras represents a moderate risk that could affect sensitive crops such as orchards, which are very abundant in its influence area. However, the values in Israeli SWDPs are much lower and imply a very low phytotoxicity risk, in agreement with the very restrictive threshold values established in the Israeli recommendations. In general, modern SWDPs apply a reverse osmosis technology with two passes to comply with boron concentration restrictions, which also gives rise to very low Cl^- and Na^+ concentrations, as happens in the Israeli case.

Boron (B^{3+}) can become highly toxic for plants if the amount is greater than that required for their growth and development [25]. The physiological adverse effects of B^{3+} on plants involve the reduction of root cell division, retarded shoot and root growth, the inhibition of photosynthesis, the deposition of lignin and suberin, a decrease in leaf chlorophyll, etc. The most habitual symptoms of B^{3+} toxicity are burned edges on the older leaves, yellowing of the leaf tips, and accelerated decay, which may result in plant expiration.

Several classifications have been proposed regarding B^{3+} tolerance by crops, usually depending on the concentration of boron in groundwater. For practical purposes, these classifications are usually directly applied to irrigation waters because the B^{3+} concentration in groundwater tends to equalize with its concentration in the irrigation water. Following the classification reported by Maas [26], the maximum permissible concentration of B^{3+} for a variety of crops is:

- Extremely sensitive crops ($B^{3+} < 0.5\,mg\,L^{-1}$): blackberry and lemon.
- Very sensitive crops ($0.5 < B^{3+} < 0.75$): avocado, grapefruit, orange, apricot, peach, cherry, plum, persimmon, Kadota fig, grape, walnut, pecan, onion, apple, and plum.
- Sensitive crops ($0.75 < B^{3+} < 1$): garlic, sweet potato, wheat, sunflower, mung bean, lupine, strawberry, Jerusalem artichoke, kidney bean, snap bean, and peanut.
- Moderately sensitive crops ($1 < B^{3+} < 2$): broccoli, red pepper, pea, carrot, radish, potato, cucumber, lettuce, pumpkin, spinach, tobacco, olive, and roses.
- Moderately tolerant crops ($2 < B^{3+} < 4$): cabbage, turnip, Kentucky bluegrass, barley, cowpea, oats, corn, artichoke, mustard, sweet clover, squash, muskmelon, and cauliflower.
- Tolerant crops ($4 < B^{3+} < 6$): alfalfa, purple vetch, parsley, red beet, sugar beet, tomato, cranberry, cotton, gladiolus, sesame, tulip, peppermint, and rye.
- Very tolerant crops ($6 < B^{3+} < 10$): sorghum, cotton, and celery.
- Extremely tolerant crops ($B^{3+} > 10\,mg\,L^{-1}$): asparagus.

DSW usually presents a high B^{3+} content due to (1) the high boron concentration in seawater (4.5–6 mg L^{-1}) in relation to natural waters (0–1.5 mg L^{-1}); and (2) the high boron membrane passage in the reverse osmosis processes [25]. As a result, DSW B^{3+} content is higher than in natural waters, so that irrigation with DSW can increase soil boron content substantially, triggering toxicity problems and leading to yield reductions in sensitive crops. This effect has already been found in Israel with B^{3+} concentrations of 0.6, 1.2, and 2.0 mg L^{-1} [13].

Therefore, specific technologies for B^{3+} reduction must be considered in SWDPs supplying agricultural use, such as a second reverse osmosis pass or the use of B^{3+} selective resins for ion exchange processes, although this entails increased investment and operational costs per cubic meter.

The current guideline value for B^{3+} content in potable water issued by the World Health Organization is 0.5 mg L^{-1}; so the regulations in relation to B^{3+} content in DSW follow this standard in most countries. It is an acceptable threshold value for DSW agricultural use, since only extremely sensitive crops would be affected. Whereas the Israeli recommendation (0.3 mg L^{-1}) would protect even the most sensitive crops, the requirement for DSW in Spain was 1 mg L^{-1} (the potable water standard in Spain) until the AGUA Program established a new threshold of 0.5 mg L^{-1} for the new SWDPs, which was clearly a requirement from the agriculture sector.

Boron content in Israeli SWDPs (Ashkelon = 0.2–0.3 mg L^{-1} and Hedera = 0.2–0.4 mg L^{-1}) is considerably lower than in Spanish ones, in agreement with their more stringent regulations. Values in Spain (Escombreras = 0.8–1.0 mg L^{-1} and Águilas = 0.6–0.8 mg L^{-1}) could affect orchards, which are very sensitive to B^{3+} phytotoxicity and are abundant in SE Spain.

It should be noted that higher water temperature produces a higher boron membrane passage; so B^{3+} phytotoxicity should be assessed with annual mean and variation range values of boron concentration, rather than punctual measurements.

14.3.1.4 Soil sodicity risk

Sodicity refers specifically to the amount of Na^+ present in irrigation water. Irrigating with water that has excessive amounts of Na^+ can deteriorate the soil structure due to soil sodicity, making plant growth difficult, and thus highly affecting crop yield. Apart from the aforementioned direct toxicity effects, the Na^+ concentration can produce an adverse impact on the soil's physical properties in the form of clay dispersion, leading to: deterioration of aggregate stability; decreased soil hydraulic conductivity; increased susceptibility to surface sealing and soil erosion problems; soil compaction; and decreased soil aeration [27].

The risk of soil sodicity is reduced if multivalent cations (primarily Ca^{2+} and Mg^{2+}) are also present in the irrigation water; so a reasonable assessment of the soil sodicity hazard should be conducted using the sodium adsorption ratio (SAR). The SAR is a quantitative indicator calculated from the concentrations of Na^+, Ca^{2+}, and Mg^{2+} ions in irrigation water. Following the Ayers and Westcot criteria [21], to analyze the mid-long-term soil sodicity hazard both the SAR and the EC values must be considered, resulting in three regions with different sodicity risks, as is shown in Fig. 14.5.

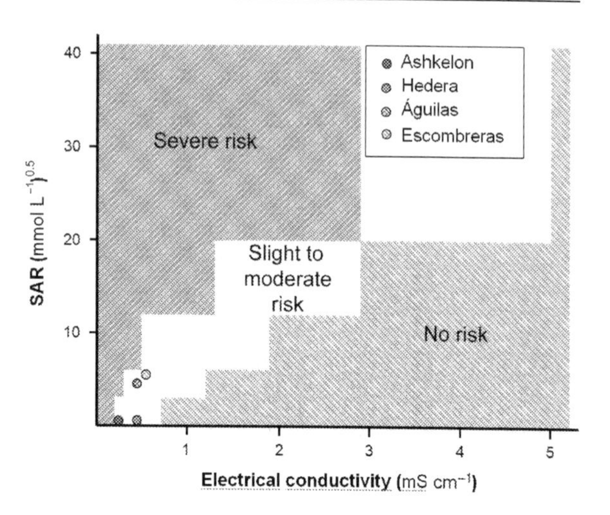

Fig. 14.5 Mid-long-term soil sodicity potential risk, evaluated using the sodium adsorption ratio (SAR) and the electric conductivity (EC) of irrigation water [28]. Dots represent the values for the selected SWDPs.

The reverse osmosis permeate usually presents a very low Ca^{2+} and Mg^{2+} concentration along with high Na^+, leading to SAR values of around 9 to 10 (SAR in seawater is above 10). The posttreatment processes always add important amounts of Ca^{2+}, and rarely some Mg^{2+}, decreasing the SAR value depending on their intensity. In the selected Spanish SWDPs the SAR is about 5, with EC around $0.5\,dS\,m^{-1}$, leading to a moderate soil sodicity hazard (Fig. 14.5) that could cause a drop in the soil infiltration rate in the mid-term and also hinder infiltration, drainage, and diffused atmospheric oxygen flux through soils. In the Israeli SWDPs, the very low Na^+ concentration threshold leads to SAR values close to zero, which combined with very low EC values also lead to a moderate soil sodicity risk (Fig. 14.5) and similar hazards to those in Spain.

Therefore, agricultural DSW use requires continuous soil monitoring to detect any deterioration of the soil structure. If the problem appears, it can be addressed with appropriate management practices, such as increasing the irrigation dose to ensure a higher leaching fraction or employing Ca^{2+} amendments to mitigate the effects of the Na^+ in soils and plants. It should be noted that the sodicity hazard is more important in soils with a high clay content, while in sandy soils it is less critical. Similarly, in areas where soil is washed regularly by abundant seasonal rainfall, the impact of high SAR values is lower.

14.3.1.5 Low buffering capacity risk

Alkalinity is a standardized measure of the buffering capacity of water, that is, the quantitative capacity of bases in the water to neutralize acids. Waters with low alkalinity are very sensitive to changes in pH; as alkalinity increases the ability of water to resist pH changes grows, enlarging its buffering capacity. The buffering bases are primarily the bicarbonate (HCO_3^-) and carbonate (CO_3^{2-}), and alkalinity is generally expressed as the equivalent calcium carbonate concentration ($CaCO_3$). Alkalinity is also directly related to water hardness, which is determined by the concentration of multivalent cations (primarily Ca^{2+} and Mg^{2+}) in the water. CCPP is a practical parameter that predicts the quantity of calcium carbonate that may precipitate in a given water.

Reverse osmosis permeate is characterized by very low pH, alkalinity, and water hardness because membranes are very efficient in removing bicarbonates and carbonates. These characteristics make reverse osmosis permeate aggressive even toward distribution system components [29]. To avoid the possible problems associated with chemical instability in DSW along the distribution schemes, alkalinity and water hardness should be attained through the posttreatment processes at the SWDP [30]. If DSW is supplied for crop irrigation, high alkalinity values are recommended for a number of reasons: to minimize corrosion in distribution systems, where metallic elements such as operational valves and devices are very frequent; to minimize the discharge of metallic ions into the water to prevent pipe rusting resulting in higher head losses in metallic pipes; and to stabilize pH where acidic or basic fertilizers are added at the farm scale [20].

Fertigation and soilless crops are common practices in high-technical agriculture, where DSW flow is usually subject to fertilizer addition, and sudden changes in DSW pH could have a deep impact on crop nutrient availability and ultimately on agricultural productivity. Then, enough DSW buffering capacity is especially relevant for agriculture when soils or growth media with low inherent buffering capacity are managed. Moreover, in irrigation schemes that have managed natural hard waters for long periods, as occurs in SE Spain, the formation of $CaCO_3$ scales is habitual, and the shift to low alkalinity waters could detach these scales, affecting the functioning of valves, filters, and flowmeters, as well as provoking pipes to become damaged.

Table 14.4 compiles the alkalinity, pH, and CCPP levels recommended by the Israeli Water Authority for combined agricultural and municipal use, together with observed values in the selected DSWPs.

DSW produced in the selected DSWPs presents pH values adequate for agriculture use, but alkalinity levels that are usually below the Israeli recommendations (Table 14.4). This data, together with the lack of essential nutrients such as Mg^{2+}, Ca^{2+}, and SO_4^{2-}, indicates that posttreatment processes should be generally intensified for agricultural use to minimize agronomic concerns and the appearance of specific problems in existing distribution systems, since DSW supplied with low alkalinity values is quite common [20,29].

14.3.2 Water-energy nexus of DSW and effects of agriculture carbon footprint

The energy demand from desalination is a key issue when planning supplying agriculture with DSW. It is critical to consider how much energy is required to produce the planned supply and to analyze how the electricity is going to be generated and transmitted to the SWDPs, which is usually a national issue [10].

Reverse osmosis is currently the most energy-efficient operational technology [31]; yet, it is still characterized by high-energy requirements, exacerbating the

Table 14.4 **Israeli Water Authority recommendations for alkalinity, pH, and calcium carbonate precipitation potential (CCPP) in DSW, and their values in the selected SWDPs**

Parameter	Israeli recommendations for domestic and agricultural DSW use [13]	Ashkelon [13]	Hedera [23]	Escombreras[a]	Águilas[a]
pH	<8.5	8.0–8.2	8.0–8.1	8.0–8.5	8.3–8.7
Alkalinity as $mg L^{-1}$ $CaCO_3$	>80	48–52	85	50–55	52–60
CCPP as $mg L^{-1}$ $CaCO_3$	3–10	0.7–1	–	–	–

[a]Own data.

water-energy nexus in comparison with conventional water sources. The specific energy consumption for reverse osmosis processes in new-construction SWDPs can be as low as 1.8–2.2 kWh m^{-3}, but its value is considerably larger in operating SWDPs because of the need for extensive pretreatment and posttreatment stages [32]. DSW for agriculture can be even more energy intensive than for potable use, due to the required posttreatment for boron and chloride removal (second reverse osmosis pass) to make their levels adequate for sensitive crops, which is estimated to require 0.50–0.77 kWh m^{-3} [33]. Table 14.5 shows the specific energy consumption of large-sized SWDPs supplying irrigated agriculture in Spain and Israel, which ranges from 2.9 to 4.3 kWh m^{-3}. This variation is related to several technical factors, such as the salinity of the feedwater, the targeted product's water quality, the wastewater disposal system, the capacity of the plant, the energy recovery systems, or the type of membrane technology used [35].

Energy consumption values in Table 14.5 are quite moderate since the included SWDPs are among those with the highest capacity in the world and their operational start-up dates are quite recent, which implies more efficient reverse osmosis technology and energy recovery systems. In addition to the figures in Table 14.5, the subsequent water delivery from SWDPs to the irrigation districts implies new energy

Table 14.5 Specific energy consumption, operational date, and total cost per cubic meter of SWDPs supplying irrigated agriculture in Spain and Israel

SWDP	Operational date	Specific energy consumption (kWh m^{-3})	Total cost[a] (€ m^{-3})	Reference
Escombreras (Spain)	2009	3.50	0.55	–
Valdelentisco (Spain)	2009	3.80	0.69	[34]
Torrevieja (Spain)	2013	4.30	0.68	[34]
Aguilas (Spain)	2013	4.09	0.66	[34]
Ashkelon (Israel)	2006	3.85	0.66	Adapted from [10]
Palmachin (Israel)	2007	2.91	0.61	Adapted from [10]
Hedera (Israel)	2009	3.50	0.64	Adapted from [10]
Sorek (Israel)	2013	3.50	0.50	Adapted from [10]
Ashdod (Israel)	2015	3.50	0.60	Adapted from [10]

[a]The total costs include capital and O&M costs only, they do not comprise delivery costs or environmental externalities.

requirements for pumping, which were estimated by Lapuente [34] as being between 0.59 and 1.00 kWh m^{-3} in the four SWDPs for agriculture of SE Spain. Furthermore, DSW should be allocated and managed at irrigation district and farm scales, which requires an additional averaged energy consumption of 0.16 and 0.17 kWh m^{-3}, respectively, for SE Spain [36].

The current average specific energy consumption of the different water sources for agriculture in SE Spain is 0.06 kWh m^{-3} for surface water; 0.48 kWh m^{-3} for groundwater; 0.72 kWh m^{-3} for reclaimed water; 1.21 kWh m^{-3} for desalinated brackish water; and 0.95 kWh m^{-3} for water transferred from central Spain [37]. Therefore, the high-energy requirement of DSW in comparison with other water sources remains the main limiting factor for its incorporation into crop irrigation. It is evident that the replacement of traditional water supplies with DSW would multiply the energy consumption linked to irrigated crop production, which entails important environmental concerns associated with the increased emission of greenhouse gases (GHG).

The carbon footprint of DSW production is substantial. Direct air emissions from SWDPs include only oxygen and nitrogen discharges associated with posttreatment processes, which do not contribute to GHG emissions. However, energy consumption of about 4 kWh m^{-3} in SWDPs implies important amounts of GHG emissions, varying depending on the electricity mix of each country. In such a way, an average value of 1.9 kg CO_2 equivalent by m^3 (kg CO_{2eq} m^{-3}) was reported in Spain [28], of 3.43 kg CO_{2eq} m^{-3} in Israel [10], and of 3.8 kg CO_{2eq} m^{-3} in Australia [38].

From an agricultural perspective, the analysis of GHG emissions linked to the intensive use of energy in SWDPs is of major relevance, since the increase in energy consumption to supply DSW could shift the CO_2 balance of irrigated agriculture from sink to source, thus exacerbating climate change. The current mean GHG emissions for traditional water sources in SE Spain's agriculture are 0.35 kg CO_{2eq} m^{-3}, whereas they reach 1.39 kg CO_{2eq} m^{-3} for DSW [3], which implies multiplying the GHG emissions linked to irrigation activity fourfold if traditional agriculture water sources are replaced by DSW.

Fig. 14.6 displays the effect of the increase in GHG emissions linked to irrigation activity on the CO_2 balance of the main irrigated crop groups in SE Spain, under

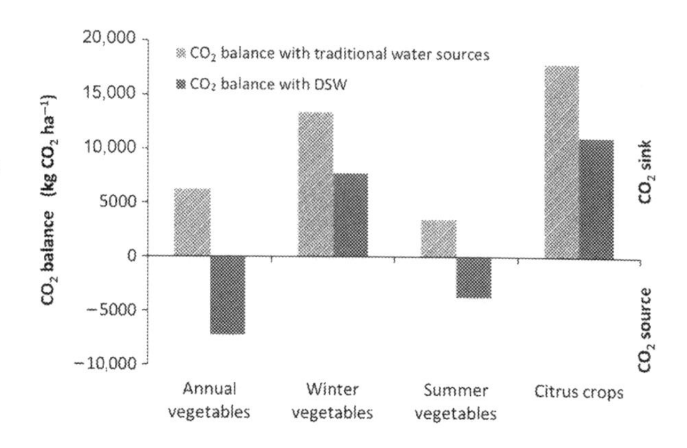

Fig. 14.6 CO_2 balance for irrigated crop groups in SE Spain under (1) the current water supply and (2) a hypothetical scenario where DSW replaces current water supply sources. Elaborated from own data.

current farming practices, by comparing a hypothetical scenario where traditional water sources are completely replaced with DSW. The results show how all crop groups would lessen their CO_{2eq} capture proportionally to their irrigation requirements, with annual vegetables and summer vegetables that are currently a CO_2 sink, becoming a CO_{2eq} source. Therefore, agriculture resilience in water-stressed regions by replacing traditional water sources with DSW could negatively contribute to climate change mitigation since it would enhance the processes that are generating water scarcity.

However, it should be noted that the maladaptation of agricultural DWS use to global climate change is directly related to the current high carbon footprint of the electricity mix. If renewable energy sources progressively replace fossil fuels in the electricity supply system [39], with the consequent relaxing in the water-energy nexus, the effect of agricultural DSW use on the crops' CO_2 balance would be considerably mitigated. In this sense, Shahabi et al. [40] indicated that renewable-energy powered SWDPs can achieve a GHG emissions reduction of 90%.

The use of renewable energy can mitigate some of the negative impacts of the agricultural DSW supply, although the current production cost from renewable-energy-coupled desalination systems is considerable higher than for conventional ones [35].

14.3.3 DSW cost and effects on crop economic viability

The total costs of DSW is site-specific, depending on total capacity, labor costs, energy sources, land availability, water salinity, and technological innovation [10]. The reported total costs of DSW from different SWDPs supplying irrigated agriculture in Spain and Israel are shown in Table 14.5. The total costs of DSW can be divided into two parts: capital costs, occurring as the initial SWDP investment, and operational and maintenance costs, occurring continuously during the operational period. The typical share of these costs is about 40% and 60%, respectively [34].

However, DSW costs are dynamic, since there is a direct relationship between energy consumption and the operational costs that is adjusted by the energy price [41]. Therefore, operational costs vary depending on the same variables affecting energy consumption, generating a volatility that will surely affect the total costs of DSW in the future. It is also important to note that DSW total costs increase more than proportionally when DSWPs do not operate at full working capacity [39].

As shown in Table 14.5, the total costs of DSW are moderately variable, which can be attributed to the use of different accounting criteria for cost analysis; to the aforesaid variables affecting energy cost and consumption; and to other local factors such as interest rates or subsidies. At present, the total costs in SWDPs supplying agriculture range from 0.50 to 0.66 € m^{-3} in Israel, and from 0.55 to 0.69 € m^{-3} in SE Spain. The final DSW cost will be even higher for farmers due to delivery costs from the SWDP to farms, especially if they are located far inland and/or at higher altitudes. In SE Spain, these delivery costs comprise [36]: the water transfer costs from SWDPs to the irrigation districts, estimated to be 0.04–0.06 € m^{-3}; the collective management and allocation costs at the irrigation districts, estimated to be 0.03–0.06 € m^{-3}; and the additional on-farm application costs if pressurized irrigation methods are managed.

As previously mentioned, the replacement of traditional water supplies with DSW multiplies the energy consumption linked to crop irrigation, which also entails important economic concerns associated with the profitability of the farming activity. DSW is far more expensive than conventional water resources. The mean price paid by the SE Spain farmers in the past decade varied between 0.12 and 0.24 € m^{-3} [42], whereas the DSW production price is much higher (0.55–0.69 € m^{-3}), even without considering the delivery costs. Such a difference implies that with DSW the profitability of all crops would be seriously reduced, or they would even become economically unsustainable. In that sense, the net margin per cubic meter of irrigation water (€ m^{-3}) for typical crops in SE Spain is included in Table 14.6. Data show that greenhouse products, pepper and tomato, provide greater net margins with 1.65 and 2.87 € m^{-3}, respectively; watermelon, melon, peach, and apricot crops (1.14, 0.70, 0.60, and 0.51 € m^{-3}, respectively) have intermediate net margins; and crops such as lettuce, broccoli, or citrus trees have lower net margins (0.12, 0.30, and 0.22 € m^{-3}, respectively). Taking into account that replacing traditional water supplies with DSW means an increase in water costs of nearly 0.50 € m^{-3}, only high-return crops such as greenhouse crops, water melon and melon, could afford current DSW costs. For most open-field crops, incorporating DSW lessens the mean net marginal value of water close to zero, which implies that if the DSW price assumes the total production costs, it is not a general solution for water scarcity in SE Spain's agriculture.

In spite of the current economic disadvantage for incorporating DSW to crop irrigation, the capital and operating costs of SWDPs are declining, largely due to technological advances, economies of scale of large-size plants, and an increased level of expertise among plant managers. Thus, it is expected that desalination will become a competitive water source for agriculture in the near future. However, a counterforce has emerged in the form of increases in the cost of raw materials and energy. As a result of these opposing forces, it is difficult to predict the actual cost of DSW in the future [10].

Table 14.6 Net margin (€ m^{-3}) for the main irrigated crops in SE Spain

Crop	Net margin (€ m^{-3})
Broccoli	0.12
Lettuce	0.30
Melon	0.70
Water melon	1.14
Greenhouse pepper	1.65
Greenhouse tomato	2.87
Apricot	0.51
Peach	0.60
Citrus trees	0.23

Data from Martínez-Alvarez V, Martin-Gorriz B, Soto-García M. Seawater desalination for crop irrigation—a review of current experiences. Desalination 2016;381:58–70.

14.4 Critical evaluation of DSW for crop irrigation

Food security concerns, climate change impacts, and the environmental and socioeconomic problems related to water resources overexploitation are encouraging the agricultural use of DSW in water-scarce regions. This innovative and technological approach may be a solution for coastal irrigated farmlands facing persistent water scarcity and for the development of high-return agriculture, where the most effective farming technologies have already been implemented. Under these conditions, where there is a lack of alternative water resources and the cost of water is small compared to the infrastructure investment and crop cash flow, DSW can represent an unlimited water source that additionally provides drought risk-buffering value.

The cost of DSW is the largest limitation to its general advance since it is still high for most open-field crops (e.g., Mediterranean vegetables and orchards), as well as for inland irrigated areas where delivery costs would add to the already high desalination costs, especially when compared to the price of other conventional water supplies. Another relevant issue to be considered is the agronomical concerns related to the chemical composition of DSW, which should be properly addressed to avoid unexpected adverse effects on agricultural productivity. In such a way, reverse osmosis processes and remineralization posttreatments should be adapted to better meet agricultural water-quality requirements by: (1) increasing Ca^{2+} and Mg^{2+} concentration to balance them with DSW Na^{+} content, minimizing deficiencies in essential nutrients required for crops and preventing soil sodicity; (2) reducing B^{3+} and Cl^{-} concentrations below the toxicity threshold for more sensitive crops in the area; and (3) providing enough buffering capacity to guarantee water pH and existing calcium carbonate scales stability along the distribution and fertigation systems.

The standardization of DSW quality to supply agriculture is needed in order to ensure the long-term sustainability of its agricultural use, as well as the optimization of joint management with other available water resources in each irrigation area. The observance of these quality standards would avoid the potential agronomic hazards imposed by DSW, which is essential for long-term agricultural sustainability; so the quality standards should be stipulated in all DSW supply contracts for irrigation.

The adaptation costs of remineralization to provide DSW with enough buffering capacity and Ca^{2+} are low and can be easily achieved by adding carbon dioxide in the traditional posttreatments of hydrated lime or limestone dissolution [43]. Novel posttreatments would also allow the addition of the required Mg^{2+} concentration at an acceptable cost [44,45]. Moreover, the use of multipass reverse osmosis membranes is an effective method for the removal of B^{3+}, Na^{+}, and Cl^{-} down to hazard thresholds for agriculture [25]. Therefore, newer remineralization and posttreatment technologies are able to cope with all agronomic and management concerns related to DSW supplies for crop irrigation, which could be technically dealt with at the SWDP. The incremental cost of tailoring agricultural requirements at the posttreatment stage has been stated as ranging from 5% to 10% of the total DSW costs in the newest SWDPs [31,41].

Irrigation with DSW might take place under two major scenarios: when it constitutes the only or the principal water supply, and when it is to be blended with other conventional water sources. The first implies that if the DSW agricultural requirements are not considered at the SWDP, farmers will continuously receive low-mineral content water and thus agronomical problems might appear. The second scenario concerns an irregular water-quality supply depending on the seasonal availability of other natural resources and the resulting blending rates. Under both scenarios, technical means may mitigate the adverse effects on agricultural production, but the farmers should take on the increased investment and operational costs for suitable fertilizing [3].

At the farm scale, blending with other water sources with high hardness and alkalinity is recommended rather than the adaptation of irrigation heads and fertigation programs. Several authors have analyzed these alternative strategies, concluding that blending is the most economical and viable management option [1,3]. For the development of blending strategies, the particular characteristics of each DSW supply should be detailed, modeling the blending with other water sources to optimize the mixing rates, and subsequently monitoring the irrigation water quality to confirm the required quality and chemical homogeneity over time.

Finally, valuing the economic effects of incorporating DSW to specific irrigation areas should be assessed in terms of collective and on-farm available water management options, considering the associated cost for fertilizing in each alternative, as well as the potential yield increases and irrigation water savings. This economic assessment must also consider the characteristics of the crop type in the area and the irrigation water that is replaced, resulting in a complex and site-dependant issue.

14.4.1 Case study: On-farm blending of DSW with surface water

This section develops a demonstration blending case study as the most recommended on-farm management alternative when DSW is supplied. The main goal is to analyze the effect of the mixing rate on the final irrigation water quality. It represents a current real scenario in SE Spain where farmers manage two different water inflows: (1) low mineralized DSW and (2) high hardness and alkalinity surface water. The farm combines orange orchards and lettuce plots, which are very sensitive and moderately sensitive crops to boron; so a maximum boron concentration of $0.5\,mg\,L^{-1}$ is established in the blended water to avoid yield reduction in the orange trees. The nutritional requirements in Ca^{2+}, Mg^{2+}, and SO_4^{2-} for both crops and under current agronomic practices are met with concentration values in the irrigation water of 90, 26, and $120\,mg\,L^{-1}$, respectively.

Table 14.7 compiles the concentration of different ions and the main water-quality parameters for both water sources (DSW and surface water) and the resulting values after different blending rates with incremental steps of 10%. The mixtures were modeled with the PHREEQC program [46], developed by the United States Geological Survey, although similar results can be obtained by applying direct saltwater balances.

According to boron tolerance in orange orchards, the maximum blending rate is 50% of DSW, since higher rates would reduce crop yield. For a 50% blending rate, the concentrations in Ca^{2+}, Mg^{2+}, and SO_4^{2-} are 64, 23, and $147\,mg\,L^{-1}$, respectively.

Table 14.7 Concentration of different ions and water quality parameters for different blending rates of DSW and surface water

Parameter	Units	Blending rate of DSW								
		0%	20%	40%	50%	60%	70%	80%	90%	100%
pH		8.4	8.4	8.4	**8.4**	8.4	8.4	8.4	8.4	8.3
EC	$dS\,cm^{-1}$	0.93	0.87	0.80	**0.77**	0.73	0.70	0.65	0.61	0.56
HCO_3^-	$mg\,L^{-1}$	195	162	139	**127**	116	105	93	82	71
Ca^{2+}	$mg\,L^{-1}$	98	85	71	**64**	57	50	43	36	29
Mg^{2+}	$mg\,L^{-1}$	42	35	27	**23**	19	16	12	8	4
Na^+	$mg\,L^{-1}$	49	57	64	**68**	72	75	79	82	86
Cl^-	$mg\,L^{-1}$	71	86	101	**109**	117	124	132	139	147
SO_4^{2-}	$mg\,L^{-1}$	279	229	173	**147**	119	90	62	34	7
B^{3+}	$mg\,L^{-1}$	0.1	0.26	0.42	**0.50**	0.58	0.66	0.74	0.82	0.90
Hardness as $CaCO_3$	$mg\,L^{-1}$	421	355	289	**256**	223	190	157	123	90
Alkalinity as $CaCO_3$	$mg\,L^{-1}$	168	147	125	**114**	103	93	82	71	60
SAR	$(mmol\,L^{-1})^{0.5}$	1.5	1.6	2.3	**2.6**	2.9	3.4	3.9	4.6	5.6
Cost	€ m^{-3}	0.15	0.25	0.35	**0.40**	0.45	0.50	0.55	0.60	0.65

These figures indicate that, although the mineralization of the irrigation water considerably improves with respect to DSW, some complementary Ca^{2+} and Mg^{2+} are required in the fertigation program. SAR and EC values indicate that there is no soil sodicity risk, whereas the high resulting alkalinity guarantees the absence of water-stability problems in the water distribution and fertigation systems. Cl^- and Na^+ also point to the absence of crop phytotoxicity risk due to these elements. The final price of irrigation water would be 0.40 € m^{-3}, 61.5% of the price if only DSW is managed.

Data in Table 14.7 show that other blending rates where DSW represents <50% are even better from an agronomic and economic point of view, but it should be recalled that the main reason for incorporating DSW is the scarcity of other water resources for irrigation. Therefore, a conclusion of this case study is that the on-farm water management should warrant 50% of irrigation requirements continuously coming from surface water to avoid agronomical hazards in crop performance.

14.5 Opportunities and future directions for research and development

The starting experiences using DSW for crop irrigation bring to light that a great deal of experience and further research is still needed to determine when, where, and how it could be a reliable agricultural water source. Nowadays, the agronomic impacts on crops and soils use are largely unknown and thus it is necessary to establish the knowledge foundations for its sustainable development. In that sense, experimental field trials with different irrigation treatments involving blended and sole DSW use are already demanded by implied stakeholders to analyze on-farm adaptive requirements, to define recommended blending strategies, and to develop innovative management technologies.

Among the research opportunities that this new agricultural irrigation source affords, the following can be highlighted:

- To monitor the quality of DSW currently produced at SWDPs supplying irrigated agriculture for assessing its agronomic suitability based on the quality standards for crop and soil protection.
- To survey the current and the expected irrigation areas using DSW, considering aspects such as crop types, irrigation requirements, on-farm water management, energy consumption at different management scales, associated GHG emissions, carbon footprint of crops, the farming costs and the profitability, etc. These knowledge foundations are essential to promote best practices for each specific agricultural DSW use case.
- To develop crop agro-physiological indices and soil quality sensitivity indicators for monitoring the crop performance and the soil fertility evolution in those areas where agricultural DSW use has already been introduced.
- To optimize the use of DSW for different crops in order to increase productivity and production quality, as well as to increase water use efficiency and productivity.
- To analyze the agronomic, socioeconomic, and environmental performance of the application of DSW to crop irrigation in closed soilless systems, as a new type of agriculture completely independent of natural water and soil resources, as an alternative future cropping system.

In addition, in managing concerns and risks related to agricultural DSW use the development of new specific technological measures is absolutely essential, with the following standing out:

- The development of new equipment able to continuously monitor the supplied DSW composition is essential for properly managing water blending, controlling phytotoxic elements, and providing data to optimize fertigation.
- The implementation of quick response on-demand fertilizing systems in the irrigation heads is necessary to deal with continuous water mineral content fluctuations, and allow increasing the economic and environmental sustainability of irrigated agriculture.

Finally, DSW for irrigation is concentrated in water-stressed areas where all available water resources have already been mobilized and where serious environmental problems dealing with natural water resources should be addressed by politicians in the short term. In such a situation, the addition of DSW to the pool of irrigation water sources will pose both challenges and opportunities for water policy making that have barely been analyzed in the literature. The assessment of the socioeconomic impact of agricultural DSW use is also a relevant research topic for policy makers.

14.6 Conclusions

Nowadays, agriculture is the major user of water resources in the world and its demand will increase in the near future to guarantee food security. In some Mediterranean regions extensive water withdrawals have come to exceed renewable water resources, leading to competition and potential bilateral conflicts among users that jeopardize irrigated agriculture. Seawater desalination for irrigated agriculture is beginning to be practised in some Mediterranean coastal regions (Israel and SE Spain) as an innovative and technological solution for irrigated agriculture resilience, and its spread across many other regions is expected as food security concerns and the effect of climate change intensify.

The first worldwide experiences reveal a high-energy requirement for agricultural DSW supply, which is leading to significant crop production costs and GHG emissions increases. Replacing traditional water supplies with DSW implies that the profitability of all agricultural productions is reduced, and in some cases even suppressed, exceeding the current economic capacity of many farms. At the moment, only high-return crops can afford total seawater desalination and allocation costs; so blending with other available water sources is the recommended on-farm management strategy to mitigate the high costs effects and make its use affordable in a wider range of crops.

Another relevant issue to be considered involves certain agronomical concerns related to the chemical composition of DSW, which could become a limitation to its widespread use in agriculture. Reverse osmosis processes and remineralization posttreatments must be adapted to better tailor agricultural water-quality requirements in order to avoid unexpected adverse effects on agricultural productivity. In the same sense, the standardization of DSW quality for agricultural supply has become necessary to ensure the long-term sustainability of its use and to optimize its management together with other available water resources.

Adapting DSW to crop irrigation requires adapting agricultural water management systems, especially where unblended water is planned to be used. Its wider application will depend on further developments in desalination technologies, reductions in desalination costs, increases in other agricultural water sources' prices, and improvements in agricultural productivity. A great deal of experience and further research are still required to promote reliable, sustainable, and profitable agricultural DSW use.

Acknowledgment

This chapter represents an adaptation and extension of the article [3] http://www.sciencedirect.com/science/article/pii/S001191641530134X to book-length form. The authors fully acknowledge Elsevier for the recognized permission for that considered personal use.

References

[1] FAO. Global agriculture towards 2050. In: Paper presented to high-level expert forum on "How to feed the world in 2050". Rome: FAO; 2009.

[2] WWAP (United Nations World Water Assessment Programme). The United Nations World Water Development Report 2014: water and energy. Paris: WWAP; 2014.

[3] Martínez-Alvarez V, Martin-Gorriz B, Soto-García M. Seawater desalination for crop irrigation—a review of current experiences. Desalination 2016;381:58–70.

[4] Greenlee LF, Lawler DF, Freeman BD, Marrot B, Moulin P. Reverse osmosis desalination: water sources, technology, and today's challenges. Water Res 2009;43:2317–48.

[5] Barron O, Ali R, Hodgson G, Smith D, Qureshi E, McFarlane D, et al. Feasibility assessment of desalination application in Australian traditional agriculture. Desalination 2015;364:33–45.

[6] Garcia C, Molina F, Zarzo D. 7 year operation of a BWRO plant with raw water from a coastal aquifer for agricultural irrigation. Desalin Water Treat 2011;31:331–8.

[7] Zarzo D, Campos E, Terrero P. Spanish experience in desalination for agriculture. Desalin Water Treat 2013;51:53–66.

[8] Díaz FJ, Tejedor M, Jiménez C, Grattan SR, Dorta M, Hernández JM. The imprint of desalinated seawater on recycled wastewater: consequences for irrigation in Lanzarote Island, Spain. Agric Water Manag 2013;116:62–72.

[9] Lahav O, Birnhack L. Quality criteria for desalinated water following post treatment. Desalination 2007;207:286–303.

[10] Spiritos E, Lipchin C. Water policy in Israel: context, issues and options. Dordrecht, Netherlands: Springer Publishing; 2013. [Chapter Desalination in Israel], p. 101–23.

[11] Tenne A. The master plan for desalination in Israel, 2020. Yuma, AZ: Water Authority, State of Israel Desalination Division; 2011.

[12] Ben-Gal A, Yermiyahu U, Cohen S. Fertilization and blending alternatives for irrigation with desalinated water. J Environ Qual 2009;38:529–36.

[13] Yermiyahu U, Tal A, Ben-Gal A, Bar-Tal A, Tarchitzky J, Lahav O. Rethinking desalinated water quality and agriculture. Science 2007;318:920–1.

[14] Al-Rashed MF, Sherif MM. Water resources in the GCC countries: an overview. Water Resour Manag 2000;14:59–75.

[15] Multsch S, Grabowski D, Lüdering J, Alquwaizany AS, Lehnert K, Frede HG, et al. A practical planning software program for desalination in agriculture—SPARE:WATER^{opt}. Desalination 2017;404:121–31.

[16] Putnam A. Keeping Florida's water flowing, Editorial in Tampa Bay Times; 2013, http://www.tampabay.com/opinion/columns/column-keeping-floridas-water-flowing/2114075. Accessed 10 January 2017.

[17] Odenheimer A, Nash J. Israel desalination shows California not to fear drought. Bloomberg; 2014. http://www.bloomberg.com/news/2014-02-13/israel-desalination-shows-california-not-to-fear-drought.html. Accessed 15 January 2017.

[18] Confederación Hidrográfica del Segura (CHS). Estudio General sobre la Demarcación Hidrográfica del Segura. Murcia: Confederación Hidrográfica del Segura; 2015.

[19] Birnhack L, Voutchkov N, Lahav O. Fundamental chemistry and engineering aspects of post-treatment processes for desalinated water—a review. Desalination 2011;273:6–22.

[20] Lahav O, Salomons E, Ostfelda A. Chemical stability of inline blends of desalinated, surface and ground waters: the need for higher alkalinity values in desalinated water. Desalination 2009;239:334–45.

[21] Ayers RS, Westcot DW. Water quality for agriculture. In: FAO irrigation and drainage paper 29. Rome: Food and Agriculture Organization of the United Nations; 1985.

[22] Maas EV, Hoffman GJ. Crop salt tolerance—current assessment. J Irrig Drain Div ASCE 1977;103:115–34.

[23] Lahav O, Kochva M, Tarchitzky J. Potential drawbacks associated with agricultural irrigation with treated wastewaters from desalinated water origin and possible remedies. Water Sci Technol 2010;61:2451–60.

[24] Raveh E, Ben-Gal A. Irrigation with water containing salts: evidence from a macro-datanational case study in Israel. Agric Water Manage 2016;170:176–9.

[25] Hilal N, Kim GJ, Somerfield C. Boron removal from saline water: a comprehensive review. Desalination 2011;273:23–35.

[26] Maas EV. Crop salt tolerance. In: Tanji KK, editor. Salinity assessment and management. New York: Amer. Society of Civil Engineers; 1990.

[27] Mandal UK, Bhardwaj AK, Warrington DN, Goldstein D, Tal AB, Levy GJ. Changes in soil hydraulic conductivity, runoff, and soil loss due to irrigation with different types of saline-sodic water. Geoderma 2008;144:509–16.

[28] Muñoz I, Fernández-Alba AR. Reducing the environmental impacts of reverse osmosis desalination by using brackish groundwater resources. Water Res 2008;42:801–11.

[29] Duranceau SJ, Pfeiffer-Wilder RJ, Douglas SA, Peña-Holt N, Watson IC. Post-treatment stabilization of desalinated water. Denver: Water Research Foundation; 2011.

[30] Birnhack L, Fridman N, Lahav O. Potential applications of quarry dolomite for post treatment of desalinated water. Desalin Water Treat 2009;1:58–67.

[31] Burn S, Hoang M, Zarzo D, Olewniak F, Campos E, Bolto B, et al. Desalination techniques—a review of the opportunities for desalination in agriculture. Desalination 2015;364:2–16.

[32] Elimelech M, Phillip WA. The future of seawater desalination: energy, technology, and the environment. Science 2011;333:712–7.

[33] Shaffer DL, Yip NY, Gilron J, Elimelech M. Seawater desalination for agriculture by integrated forward and reverse osmosis: improved product water quality for potentially less energy. J Membr Sci 2012;415–416:1–8.

[34] Lapuente E. Full cost in desalination. A case study of the Segura River Basin. Desalination 2012;300:40–5.

[35] Al-Karaghouli A, Kazmerski LL. Energy consumption and water production cost of conventional and renewable-energy-powered desalination processes. Renew Sustain Energy Rev 2013;24:343–56.

[36] Soto-García M, Martin-Gorriz B, García-Bastida PA, Alcón F, Martínez-Alvarez V. Energy consumption for crop irrigation in a semiarid climate (south-eastern Spain). Energy 2013;55:1084–93.
[37] Martin-Gorriz B, Soto-García M, Martínez-Alvarez V. Energy and greenhouse-gas emissions in irrigated agriculture of SE (southeast) Spain. Effects of alternative water supply scenarios. Energy 2014;77:478–88.
[38] Biswas WK. Life cycle assessment of seawater desalinization in Western Australia. World Acad Sci Eng Technol 2009;56:369–75.
[39] Ghaffour N, Bundschuh J, Mahmoudi H, Goosen MFA. Renewable energy-driven desalination technologies: a comprehensive review on challenges and potential applications of integrated systems. Desalination 2015;356:94–114.
[40] Shahabi MP, McHugh A, Anda M, Ho G. Environmental life cycle assessment of seawater reverse osmosis desalination plant powered by renewable energy. Renew Energy 2014;67:53–8.
[41] Ziolkowska JR. Is desalination affordable?—regional cost and price analysis. Water Resour Manage 2015;29:1385–97.
[42] Alcon F, Tapsuwan S, Brouwer R, de Miguel MD. Adoption of irrigation water policies to guarantee water supply: a choice experiment. Environ Sci Policy 2014;44:226–36.
[43] Fritzmann C, Löwenberg J, Wintgens T, Melin T. State-of-the-art of reverse osmosis desalination. Desalination 2007;216:1–76.
[44] Penn R, Birnhack L, Adin A, Lahav O. New desalinated drinking water regulations are met by an innovative post-treatment process for improved public health. Water Sci Technol 2009;9:225–31.
[45] Birnhack L, Nir O, Lahav O. Establishment of the underlying rationale and description of a cheap nanofiltration-based method for supplementing desalinated water with magnesium ions. Water 2014;6:1172–86.
[46] Parkhurst DL, Appelo CAP. Description of input and examples for PHREEQC version 3—a computer program for speciation, batch-reaction, one-dimensional transport, and inverse geochemical calculations: U.S. Geological Survey Techniques and Methods, book 6, chap. A43. 497 p.

Further reading

[1] March H, Saurí D, Rico-Amorós AM. The end of scarcity? Water desalination as the new cornucopia for Mediterranean Spain. J Hydrol 2014;519:2642–51.

Seawater for sustainable agriculture

Khaled Moustafa
Conservatoire National des Arts et Métiers, Paris, France

15.1 Introduction

Drylands make about 41% of the Earth's land surface.[1] In Africa, two-thirds of the land surface is deemed to be drylands. Drylands are also home to up to two billion people, half of them are threatened by desertification in about 100 countries.[1] So far, no accurate assessment of biodiversity in drylands was reported, making the biodiversity status in these regions unknown precisely. However, about 27,000 species seem to be lost each year due to desertification effects.[2] In the future, climate change and anthropogenic activities might exacerbate the desertification damages and reduce the biodiversity at an accelerated pace. Land desertification is also linked to farming practice, depletion of natural resources, food security, dust storms, migration, and poverty. By 2030, for example, water scarcity alone may displace up to 700 million people in the world.[2]

In this context, innovative and sustainable farming systems need to be developed to maintain viable ecosystems and to reduce current and future repercussions of climate change, drought, and water scarcity on agriculture and biodiversity. Vertical and urban farming systems are potential solutions toward this goal with many advantages [1]. Other solutions may include approaches that could be based on the sun and seawater as inexhaustible sources of energy and water to establish and maintain sustainable farming systems particularly in arid and dry regions worldwide. The potential approaches toward this objective will be discussed hereafter. The first will be centered on scaling up the solar still principle to establish large greenhouse-like structures, which we can call as "solar desert-houses" to evaporate seawater and produce little but permanent amount of freshwater that will be highly appreciated in arid environments [2]. The second approach will be focused on the establishment of floating farms on the sea surface, either as fixed stations or as movable farming ships. Both options (land or sea surfaces) fit well within the high interest of dry regions and countries with intense insolation and limited arable lands. Farming with natural insolation and seawater would allow extending the traditional land farming surfaces to new areas in the sea in an attempt to relieve food insecurity issues and to mitigate the effects of water shortage in the concerned regions. Such approaches could be seen as complementary

[1] http://www.un.org/en/events/desertification_decade/whynow.shtml [Accessed 27 June 2017].

[2] http://www.unccd.int/Lists/SiteDocumentLibrary/WDCD/DLDD%20Facts.pdf [Accessed 27 June 2017].

Emerging Technologies for Sustainable Desalination Handbook. https://doi.org/10.1016/B978-0-12-815818-0.00015-1

methods to the existing ones to enhance the food production systems in countries that suffer from freshwater scarcity and food penury. The future of humanity may be fashioned by the sea conditions so that humans need to take advantages of it while preserving it from similar degradation land conditions by taking maximum precaution for sustainable sea exploitation before starting to exploit it intensively.

15.2 Solar desert-houses (or "portable seawater ponds")

Greenhouses are well-established farming systems to produce different types of vegetables and flower crops worldwide. The optimal growth conditions of plant development and production are known and controllable inside the greenhouses over the whole plant life cycle [3]. In hot or cold regions, greenhouses could be built to produce crops that originally grow in environmental conditions other than their native conditions because farmers can easily control the temperature, water, and light requirements that plants need for their growth. However, greenhouses are energy-consuming systems that are also expensive to set up particularly in hostile environments when natural conditions are the limiting factors (i.e., light, temperature, and water) so that substantial costs are required to create, adjust, and supply the necessary growth conditions in an optimal balance with the surrounding climate conditions. Such environmental constraints may thus reduce the profitability of greenhouses under adverse environments. To palliate this problem, greenhouses need to be established at a high density and their products should be sold at higher prices to be beneficial from a long-term perspective, which unavoidably will raise other problems for the consumer and purchasing power that is beyond the scope of this chapter.

However, in hot and dry regions surrounded by seawater, such as the Arabian Peninsula, North Africa, and large areas in Asia and Africa, the availability of intense insolation over the year, particularly in summer, is a strong incentive to develop simple and durable solutions based on the abundance of the sun and seawater. Among the potential solutions that could be established at relatively low costs with simple technology tools for greening the desert and reviving its dry niches could be the approach of solar still that need to be scaled up to large levels that we can call "solar desert-houses" or "seawater ponds" as an artificial and movable still in the form of large greenhouses (Fig. 15.1) that hold seawater as a pond [2]. Solar seawater distillation is, in fact, a promising alternative for water provision with cheap energy and simple technology that keep the environment clean [4]. Future water desalination around the world should be increasingly powered by environmentally friendly systems such as solar, wind, and other clean natural resources at economic costs [5].

The concept of solar still (i.e., evaporation of water under sun radiation) is a proven principle known since a long time ago [6–8]. It was used in emergency situations to provide freshwater to save endangered lives. Ancient Greek sailors would have used it to evaporate salty water to condense the vapor into drinkable water when they are offshore [9]. In the 19th century, the invention of steam-powered engines and the need for safe water, free of corrosive salts, for boilers prompted the first desalination patent in England [9]. Prior to and during the Second World War, desalination systems based on

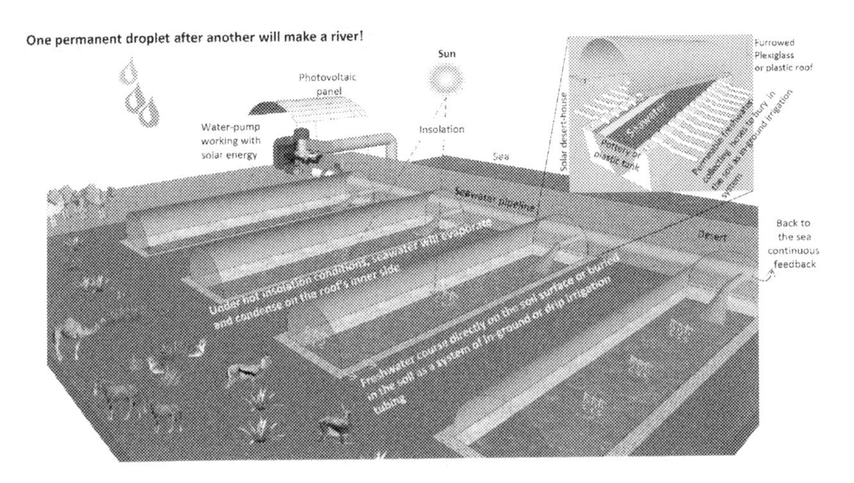

Fig. 15.1 An illustration of simple, reusable, and cost-effective approach of "solar desert-houses" for greening drylands and restoring biodiversity in the desert. Large solar desert-houses are filled up with seawater that will evaporate under hot desert conditions and condense into freshwater droplets on the solar house's roof. The freshwater droplets could then be collected and conducted in lateral tubes into the soil as a drip irrigation or in ground irrigation to meet the needs of some plant species growing in the desert (e.g., xerophyte and halophyte species). Such desert-houses could help creating natural oases in the desert and providing valuable food sources for people and desert plants and animals such as gazelles, camels, horses, birds, etc. Some crop species could eventually be grown. The seawater-conducting tubes could be buried in the soil in a way similar to the sewage or urban water network to avoid any interference with potential infrastructure on land. The accumulated salts in the solar houses could be recovered by discharging the container using specific trucks. The desert-houses should be manufactured from materials adapted to extreme temperatures and high salt concentrations to minimize maintenance operations. A full automation of the system could be envisioned from seawater pumping (using solar pump) to salt discharging using trucks. The excess of seawater brought to the solar houses could be returned back into the sea in a cyclic and sustainable pattern [2]. Reproduced with permission from Moustafa K. Greening drylands with seawater easily and naturally. Trends Biotechnol 2017;35:189–191.

saltwater evaporation were commonly used in boats on long trans-Atlantic expeditions [6,10,11]. However, the utilization of seawater solar-based systems is so far restricted to narrow and/or small-scale urgent applications, such as the production of low amounts of freshwater for drinking purposes in emergency situations when freshwater is rare or unavailable. Solar distillation of water could even be used, for example, to obtain water from soil and plant materials with a simple technique consisting of a hemispherical hole in the soil containing fresh-cut plant materials covered with a plastic film and held in a conical shape by a rock placed in the center. Then, under solar insolation, water droplets collect on the underside of the plastic film, run to the point of the cone, and drop into the container [11]. The utility of solar still at a small scale (in this case, to produce freshwater for drinking purposes) means that it could be scaled up to a higher level or a larger volume to produce larger amounts of freshwater for further applications such as irrigation and agricultural applications for plant growth and for raising animals.

Solar stills could thus be scaled up to an industrial or a large level as, for example, big "solar desert-houses" to purify large volumes of seawater in dry regions naturally under a sun shining most of the time and seawater abundant all around. In such regions, indeed, where temperature degrees are extremely high in summer, any permanent freshwater source, even if droplets only, would be highly precious to maintain the life of plant and animal species under such suffocating conditions. The usefulness of solar desert-houses to maintain minimal life conditions in adverse climate conditions would be undeniably valuable. Previously, a tube-type solar still was tested in vitro and in outdoor conditions as a conventional still and a water distributor for potential desert plantation [12,13]. A simple single basin solar still was also designed and tested in outdoor conditions to produce clean freshwater for drinking with an average daily output of about 1.7 L per day for a basin area of 0.54 m^2 [14]. It was also found that designing a single basin solar still with fins enhance productivity by up to 45% [15]. As such, the availability of solar radiation over the year and the inexhaustible seawater source make it possible to scale up solar stills to transform large areas of barren lands and wastelands from life-hostile environments into life-friendly environments. The cyclic nature of rain formation systems could be inspiring and replicated in drylands in a sustainable and cost-effective manner based on the widely available raw materials (sun and sea) at not much cost. For this, bulky solar stills could distill seawater naturally and sustainably to provide small but constant amounts of freshwater to fulfill the water requirements of at least some species living inherently in the desert and dry regions more generally. To build solar desert-houses, large and movable water reservoirs or tanks could be manufactured with adapted covers (i.e., transparent plastic or Plexiglas) in rectangular, square, circular, or dome shapes to be either fixed above the ground (as water cisterns) or buried in the soil near the coast or far from it (Fig. 15.1) [2]. Seawater could then be brought into these assemblies through an adapted and well-designed pipeline network that should support harsh conditions of high temperatures and saltwater corrosiveness. Left under hot and natural desert conditions, the saltwater in the solar houses will evaporate and transform into an elixir of life by the simple and lasting replication of a rain system sustainably and cost effectively. As the sun shines most of the time in hot drylands with an average temperature degree of up to ~40–50°C in summer (sometimes more depending on regions and latitudes), the seawater confined in the desert-house will evaporate slowly but permanently to condense onto the inner side of the roof as freshwater droplets. Under the effects of gravity the freshwater will drop down to be collected, stored, and conducted outside the desert-house via small grooves at the bottom lateral edges of the solar still (Fig. 15.1). Perforated hoses or semipermeable tubes will, then, deliver the produced freshwater into the soil similar to an in-ground irrigation system or a drip irrigation approach. As a result, a perpetual and miniaturized rain system is created naturally and easily with valuable benefits for agricultural purposes in dry and sunny lands for many species, including plants, domestic or wild animals, domestic or wild birds, insects and fungi, and nomadic people [2]. When temperature degrees drop down in the winter and autumn to lower levels than in the summer, the efficiency of seawater evaporation may decrease accordingly but is still useful to produce small but valuable and constant freshwater condensations to maintain a minimal level of life conditions in the desert. Furthermore, the development of a low temperature differential, phase-change desalination process that can be driven solely by solar and/or waste energy sources can also be a potential source for a sustainable desalination process, particularly in low-income rural and remote communities [16,17].

15.2.1 Advantages and inconveniences of solar desert-houses

Among the most obvious advantages of solar desert-houses approach (or artificial seawater ponds) are its simplicity, accessibility, affordability, and environmental friendliness. Solar desert-houses could also be built anywhere onshore in dry arable lands, including remote continental arid zones where sun is shining and seawater could be brought through special ducts or tubes in a way similar to urban water canalization [2].

Another major advantage of solar desert-houses is that solar energy could be used to feed the system with seawater automatically and permanently through solar pumps (working with solar photovoltaic panels) so that the desert-houses system is environmentally clean and completely independent of any fossil polluting energy. Many models have already been suggested for the implementation and optimizations of photovoltaic water pumping systems [18–20], which could be used in desert-house settings. Another nonnegligible advantage is that the desert-house could be transferred into new locations, removed, or ceased at any moment, should it be proven inefficient after fluctuating or prevailing conditions or after new urbanization plans or implementation of new infrastructure or strategies. Moreover, as the seawater in the desert-houses is maintained in controllable containers, the salts accumulating in the containers over saltwater evaporation could be easily recovered so that no soil salinization risk would exist for the surrounding lands or for any potential underground water near the location of a desert-house system. This is actually an important feature if the desert-house approach is intended to be built in lands with the underground water level where the desert-house will prevent the salts or saltwater from leaching outside the tank to contaminate the groundwater, if any.

In fact, there would be a double benefit from using a desert-house system (or artificial seawater ponds): from one hand, the desert-houses will help produce freshwater droplets for drylands permanently and sustainably for drinking and agriculture purposes, with all the subsequent benefits and advantages of water availability (i.e., increasing biodiversity, fixing the soil, increasing the organic matter and content, hence increasing the soil fertility, etc.), and on the other hand, solar desert-houses will help produce valuable mineral salts as an important byproduct for human consumption or industrial applications. To facilitate the recovery of the accumulating salts, the seawater containers need to be manufactured in such a way to be solid enough and adapted for such purposes, for example, in a way comparable to the modern underground waste collection systems with trucks that facilitate the collection, separation, and recycling of urban wastes using specialized machines. The obtained mineral salts could thus be collected using specific trucks and used or sold as a raw material for fertilization purposes in irrigated traditional agriculture systems after qualitative and quantitative chemical analyses are done to determine the identities and percentage of each mineral element present in the salty mixture. Doing so, optimal mixtures or adjustments could be made to reduce any potential toxicity or overdose effects of the utilized minerals.

Freshwater collecting tubes connected to the desert-house system and buried in the soil offer another important advantage to reduce the evaporation risks of the produced freshwater itself by delivering the harvested freshwater droplets directly to the soil, near the root areas, similar to a drip irrigation or an in-ground irrigation system, so that the freshwater is injected in the soil and plants can benefit it more efficiently than if the freshwater droplets are poured on the soil surface where it could be evaporated by the heat before it gets absorbed by the plants. A similar low-energy process for coupling water purification and drip irrigation has previously proposed to desalinize water and irrigate widely spaced row crops simultaneously [21]. In fact, under hot desert and dry conditions, it might be a bad idea to discharge freshwater droplets directly on the surface of the soil because the high temperatures in the desert will evaporate it before plants or animals can absorb it or benefit from it. So, it would be more judicious to burry or at least to half-burry the freshwater-conducting tubes in the soil to protect the freshwater droplets from evaporation risk. However, if the freshwater-collecting tubes will be designed to be buried in the soil, there might a need for regular checking to make sure that the tube perforations are not blocked by sand or soil particles; otherwise, the used tubes should be semipermeable so that the freshwater flowing inside the tubes could leach out into the surrounding soil by contact or capillarity effects.

While the efficiency of the outlined desert-house approach in producing freshwater droplets depends on many conditions such as the design and ambient temperatures, wind speed, seasonal fluctuations, cloudiness, size, and maybe shape of the desert-house itself, volume, depth, and thickness of the saltwater layer in the container, materials being used to capture the sunlight, or letting it passing through, etc., the proposed system of solar desert-houses could help exploiting the currently unexploited large areas more rationally when it is built at large scales and high density on lands that only need little amounts of freshwater to offer their full agricultural potentials and their farming abilities [2]. A strong rationale behind this ability is that many plant species already grow in the desert or drylands, or at least they can potentially grow and thrive in extremely dry conditions with only some traces of water or moisture in the atmosphere. Most desert species indeed are well adapted to take advantage of the minute dew droplets that accumulate during the night to develop their shoot or root systems. Their physiology and morphology in turn are also adapted to stock the water as long as possible.

Although little, the amount of freshwater that could be produced via the proposed system of solar desert-houses could be sufficient to meet the needs of many wild and domesticated plants and animal species including some crops species to produce food and nutrients to feed humans and animal species in their native dry niches.

Adapting the solar desert-house approach as a controllable day-night freshwater delivery system also permits to use freshwater drinkers for animals and humans living in the desert. That is, the system could be designed in a way that the freshwater condensed during the day could be collected and stored in specific and adapted drinkers placed preferably in the shaded side of the solar house or under parasols to reduce the evaporation risk of the collected freshwater. Animals, nomads, and passing people or travelers could then take advantage from the collected freshwater when they find themselves near a desert-house park. A vertical fountain with a tap system protected from insolation (e.g., by sun umbrella intended for this goal) could be envisioned for human drinking purposes as well.

Despite the several potential advantages discussed above, some minor drawbacks may hamper the optimal function of a solar desert-house system. However, the shortcomings are easily manageable and fully surmountable. As mentioned earlier, a minor hurdle that needs to be addressed is the risk of evaporation of the produced freshwater itself before it gets absorbed by plants or animals in a targeted location, particularly if the freshwater droplets are poured on the surface of the soil under burning heat in the summer. To alleviate this limitation and to increase the efficiency of the system, the freshwater-collecting tubes connected to the solar desert-house should be designed to be half-buried or completely buried in the soil so that the freshwater droplets are injected directly inside the soil, around the plant root areas, as soon as they are produced. Growing plants could thus absorb it before it evaporates or falls deeper into the soil. Another important enhancement to reduce the evaporation risk of the freshwater is to design the solar desert-house to retain or to hold the freshwater produced during the day in lateral reservoirs and to deliver it only during the night, when it is cooler so that the evaporation risk is significantly minimized [2]. Such an option could be achieved, for example, by using light sensors or mechanical adapters that could control the delivery of the freshwater during the dark only (i.e., by using light or optical sensors). Alternatively, a mechanical control (or control by weight) could also be used so that when the volume of the accumulated freshwater becomes heavy enough or when it reaches a certain threshold or weight, the collected freshwater is then spilled through regulating valves into reservoirs intended for this purpose. Solar still performance could also be enhanced by using cooling films on glass condensation covers, though poor combinations of film cooling parameters can lead to significant reductions in still efficiency [22].

Another possible hurdle is the potential effects of sandstorms on solar desert-houses. As dry and hot regions are sometimes prone to intense dust storms, and although they are temporal, sand and dust particles could eventually accumulate on the outside layer of the roof of solar desert-houses and thus reduce the penetration of sunlight into them. As a result, the efficacy of seawater evaporation would be reduced and the volume of freshwater would diminish. To reduce such risks, a programmable dust brush or sand broom that could be automatically triggered or activated—by solar energy when it comes back after a dust storm—in a way similar to the windscreen wipers could be attached to the solar desert-houses as an accessory design to remove any accumulated dusts or sand particles that might hinder the infiltration of light into the solar house. Such a potential design would also reduce human intervention to control the operability of the system while making it fully independent and functional under all conditions including sandy and dusty storms. Solar energy brooms could also be added onto the edge of any solar panels used in the outlined system or more generally any solar panel system built in stormy conditions where the dust particles could reduce the solar energy conversion rates.

Even though the yield of freshwater produced by solar desert-houses might be low at the first glance per units of time and space, but as the process is sustainable and scalable to large levels of container sizes and numbers per unit of surface, the final yield of freshwater produced at the long-term could be sufficient to fulfill the basic needs of many desert plant species, for instance, some xerophytes and halophytes whose needs for water are innately low. Consequently, many native desert plant and animal species can thrive and populate the surrounding areas around the established seawater desert-houses as a kind of artificial oasis in the heart of the desert (Fig. 15.1) [2]. Another important advantage is that, given the simplicity of the delineated approach and as there is no sophisticated infrastructure or equipment required, the maintenance procedures are relatively simple and cheap. They could be limited to switching the seawater delivery system on and off and to recovering the accumulated salts from the desert-house containers when they reach a critical threshold that would reduce the evaporation efficiency. When the salt level reaches such a threshold, which could be determined empirically, the provision of seawater could be interrupted temporally to allow special trucks or machines recover the accumulating salts and to permit the system work optimally again, and so on [2].

In fact the automation technologies and the availability of natural insolation in large areas of drylands in the world make it possible to fully automate the solar desert-house systems that could be built in targeted locations and left under the natural conditions to fulfill the goals of solar stills. A system of solar desert-house could thus be built and left under the natural conditions with only sporadic or minimum vetting to ensure the operability. For this, a programmable water pump functioning with solar energy (photovoltaic panels) could be used to connect a saltwater source to a desert-house system to pump seawater automatically and permanently from the sea to fill in the solar desert-houses in times of need. After that, all that would be needed is a simple, cheap, and regular checking of the installations to make sure that everything is functioning as expected. For further automation and self-control of the system, specific sensors could be used to monitor the salt and water levels in the containers in real time so that the solar-based water pump could pump the seawater when its level in the container is low and stop when its level is high. Subsequently, the cycle of seawater provision and evaporation could operate naturally and indefinitely as long as sun and saltwater are available [2].

In sum, a solar desert-house approach based on sustainable seawater provision-evaporation-condensation cycles would allow the instauration and restoration of a full and integrated functional ecosystem in drylands. With large solar desert-house systems, poor and sunny barren regions could be enriched with native and/or new species that were previously unimaginable to grow in such environments without freshwater supply.

15.2.2 Methodological and technical features of solar desert-houses

Depending on the size, the shape, and the sophistication level and automation degree required for the intended solar desert-house system, the structure of a seawater house system (or movable seawater ponds), as illustrated in Fig. 15.1, could be composed of different parts that should be easily replaceable or changeable in case of damage or attrition over time. In its simplest format and shape, a solar desert-house could be minimally composed of three indispensable parts as a general scheme: (1) seawater container, (2) curved or arched cover or roof, and (3) a set of perforated or semipermeable tubes fixable at the lateral side of the container (Fig. 15.1). All these parts should be manufactured from materials adapted to long exposure to harsh desert conditions and seawater corrosiveness. The bottom container part is destined to be filled up with seawater continuously through a special pipeline network connected to a solar water pump that draws seawater with solar energy power to minimize the reliance on fossil fuels and to keep the system fully clean and environmentally friendly. The arched roof of the desert-house should be light-transparent (i.e., made of suitable plastic or Plexiglas materials) to allow light pass through and reach the seawater surface inside the desert-house to heat it and evaporate it. The vapor will then condense onto the inner side of the roof and trickle down under the gravity effects (or the weight of the aggregated freshwater droplets). To increase the efficiency of vapor condensation and flowing, the desert-house's roof could be designed as a two-adjacent level roof covered with a hydrophobic layer to augment freshwater dribbling down to the collecting tubes as quickly as they are formed. The gradual evaporation and condensation cycles on the roof allows freshwater droplets to get cooler so that they could be used without burning harms for drinking animals or irrigated plants let alone that the freshwater produced during the hot days could be distributed during the night when it is cooler, as stated earlier.

The freshwater-collecting tubes, in turn, should be ideally manufactured from semipermeable materials and adjustable to different levels, optimally connected to the solar house in sloping or acute parallel angles to deliver freshwater under the sole effect of gravity and weight. Regarding seawater containers, they should be solid enough to resist seawater friction and salt corrosiveness. The ideal design might be to place the solar houses adjacently above the ground similar to water cisterns or they should be half-buried or buried in the soil as illustrated in Fig. 15.1. In all cases, the seawater containers need to be manufactured with specific lateral or upper clamps for an adaptation to the displacement and for easy recovery of accumulated salts by, for example, salt-collecting trucks adapted for this purpose. Although the desert-houses could be imagined variable in shape and format from small to big, and from circular to rectangular and from a few meters to tens of meters in length by a few or tens of meters width, empirical studies, however, would be required to determine the best dimensions, depth, and shapes that will produce the highest freshwater yield possible with the least possible energy consumed. The depth of seawater houses could also vary from tens of centimeters to a few meters. Anyhow, the dimensions of solar desert-houses should be scalable to larger or smaller magnitudes to supply the best yield possible. It is noteworthy, however, that the larger the solar houses, the greater volumes of freshwater that should be produced.

It was reported that coupling a solar still with a hot water tank generally increases the distilled water output at a scale of 24h, as a result of continuous heating [23]. The increases were higher at night than in the day, since the differences in water and cover temperatures at night are generally higher, resulting in higher production rates [23].

For an automatic recovery of the salts accumulated in the container, the sizes and dimensions of these containers should be adapted to the sizes and capacity of trucks or machines that could be manufactured or adapted to handle them. To increase the seawater evaporation efficiency, solar energy concentrators could also be fixed onto the lateral sides of the solar still to redirect and concentrate the solar insolation directly onto the seawater surface and, hence, increase its conductivity and evaporation.

To facilitate the organization and the management of a desert solar house system in a targeted location, the solar houses could be placed at a given density of rows and columns at regular distances between them, and which should allow a good coverage and freshwater yield collected from each desert-house individually and collectively. It should be taken into account, however, that specific roads between the rows of solar house systems could be traced out to help using maintenance machines or trucks to recover the accumulating salts but also for any other maintenance operation that needs to be performed from time to time (e.g., replacement of roof or damaged parts, the control of the freshwater collecting tubes, the removal of undesirable objects, etc.). Furthermore, to avoid any potential interference with current or future infrastructure on land such as traditional agricultural fields, intercities or intervillage roads, bridges, etc., the seawater-pipeline network could be buried in the soil in a way similar to a sewage or urban water network.

15.3 Floating farms on sea surface

Climate change and water scarcity in wide areas in the world are important challenges for the future of humans and agriculture with substantial consequences on food security and environmental sustainability. The development of alternative clean energy and farming systems are strongly encouraged to sustain environmental resources and to reduce drought consequences and pollution, particularly in dry and arid lands where freshwater is rare. Key elements for such commitments are to generate electrical power from an inexhaustible nonpolluting source, such as the sun, and to produce freshwater from an unlimited source, such as the sea.

Thus, combining solar energy-based seawater desalination technology with a floating agriculture platform in one hybrid innovative system offshore could be a complementary solution to land farming systems to be established in sunny regions to provide a win-win solution to desalinate seawater and produce valuable crops or vegetables on board (floating farms), while offering major environmental and energy advantages. In the next paragraphs, a potential floating photovoltaic-based farm will be discussed as a "seawater floating farm" system or "blue houses" in the sea, by opposition to greenhouses in lands. The potential advantages and shortcomings of the outlined system will be presented.

15.3.1 Solar stills coupled with photovoltaic panels

The sun and sea are, respectively, the largest exploitable resources of renewable energy and water. Sunlight is by far the major exploitable supply of renewable energy providing only 1 h more energy than all of the energy consumed by humans in an entire year [24]. The sun provides about 120,000 terawatt (TW) [25]. With about three quarters (~75%) of the globe as a reserve, the seawater is also an inexhaustible and indirect source of freshwater on our planet. However, due to high concentrations of mineral salts, mainly sodium chloride, and impurities, seawater is unsuitable to be used directly in productive agriculture systems or for human consumption. To be usable in useful life applications, saltwater needs be evaporated or purified from its excess salts to produce freshwater suitable for human drinking, food industry or irrigation of glycophytes (plants that do not grow in saline soils). Solar energy-based seawater desalination is one of the most promising and cheapest desalination approaches that could be used in sunny arid regions, either directly or indirectly to produce water distillate in solar collectors, or by combining other conventional desalination techniques such as multistage flash desalination (MSF), vapor compression (VC), reverse osmosis (RO), membrane distillation (MD), or electrodialysis (ED) [26]. However, under the availability of intensive solar energy most of the time in large areas of dryland, it would be fully possible to scale up the solar still principle and to build large-scale solar-based desalination facilities [27]. In this context, a hybrid system combining solar energy-based seawater desalination systems with "floating farms" platform in one structure would be worthy of investment as a marine "floating farm" or "blue house" by opposition to greenhouses on land [28]. A simple conceptual design of such a system was previously suggested to be built on the sea surface [28]. A floating farm system could be composed of different units: (1) a sunlight capturing unit (or photovoltaics unit) that will capture sunlight to convert it into electric power to heat seawater, (2) thermal desalination unit to desalinate seawater held in floating reservoirs through evaporation with the energy produced in unit 1, and (3) a farming surface with a cultivable surface composed, for example, of sandy, hydroponic, or organic culture support with eventually walk-in growth rooms or corridors, depending on the size, cost, technology being used, and complexity/easiness that will be required in the system [28].

The solar energy will first be gathered by photovoltaic cell panels to be converted into electric power to desalinate seawater thermally (heating and condensation of seawater vapor). Then, the resulting freshwater will be used to irrigate plants growing on a farming surface on board. A photovoltaic floating farm system could indeed be seen as a solar desert-house system built on the sea surface instead of land. To increase the efficiency of the floating system and to augment the overall amount of freshwater produced, other desalination approaches and/or alternative energy methods other than the sun (i.e., wind turbine, wave movement, or seawater currents) could be combined to the photovoltaic-based desalination approach. Although this will increase the overall costs, the yield and efficacy of the system would undoubtedly be increased to make it profitable and economically viable in the long term.

However, to establish optimally a floating farm system on the sea surface, cross-disciplinary studies and joint expertise in different fields would be essential to include, for example, marine engineering, agronomy, system and mechanical engineers, biotechnology, economy, mathematical and statistical modeling, water purification and safety management investigations. Due to uncertain environmental and marine conditions, such as the increase or decrease of seawater level, potential political conflicts, chemical pollutions, etc., floating farms could be ideally designed to be completely mobile or portable and vertically adjustable in function of the seawater level and the surrounding conditions. However, vertically and horizontally adjustable floating farm systems would, again, increase the overall costs but at the same time would offer more resilience and crucial advantages over immobile floating farm systems fixed in the same location as the first option would allow to mitigate unexpected natural adversities, such as any potential pollution around the farming site, or any increase of salt concentrations locally that would accumulate upon the desalination process and which need to be resolved or discharged back into the seawater far from the farming site, or any potential sudden decrease or increase in the seawater levels as a result of uncontrollable climate change.

The international BMT Design & Technology group (Melbourne, Australia) is currently working on floating desalination systems to desalinate seawater with improved capacity up to 150 million liters per day [29]. Some prototypes of semitransparent-bifacial photovoltaic modules for greenhouse roof applications were also recently proposed with potential suitability for high-irradiation regions [30]. Some decades ago, a floating or pelagic solar still assembly was proposed for supporting plant growth and cultivation of sweet water plants in a condensation area of the still [31]. Recently, a concept of "smart floating farms" (SFF) has been suggested as an integrated system combing solar power, hydroponics, and fish farming by a Barcelona-based design company (http://smartfloatingfarms.com). The goal of this system is to produce food on the sea surface without soil (hydroponic culture) such as in vertical, sustainable, and solar-powered farms. The designers estimate that SFF can produce up to 8000 tons of vegetables and 1700 tons of fish annually.[3] Combing seawater desalination approach with farming surfaces on the sea surface can thus be a worthwhile investment, particularly for countries with abundant solar energy and limited sources of freshwater or farming lands, such the Arabic Peninsula, North Africa, and arid lands in Africa, Asia, or Australia.

Depending on the shape, the size, and surface of the floating farm, and the plant species that would be cultured in floating farm platform, a well designed and precise calculations would be required to set up the best proportional rates possible between the farming surfaces and the volumes, thickness, and the physical and chemical characteristics of the system's components and materials to use, which need to be adapted to marine and seawater conditions. For example, what is the photovoltaic panel

[3] https://www.ecowatch.com/giant-solar-floating-farm-could-produce-8-000-tons-of-vegetables-annua-1882045025.html [Accessed 27 June 2017].

surface (potentially combined with other marine energy sources) that should produce a sufficient energy to fulfill the requirements of power and water for the onboard culture surface over the whole plant's life cycle, from sowing to harvesting the plant crop, in addition to all the potential requirements of water and energy before and after the harvesting operations. To reduce manual interventions and to increase the efficiency of the floating farms, maintenance operations and farming processes can also be automated at the maximum possible. To allow such an automation, the freshwater and energy produced in the floating system should be high enough that any excess should be stored in specialized energy concentrators and dedicated water reservoirs to be used later when they are required before and after the plant cycle but also during the growth phases. Any potential excess of energy and/or freshwater produced in floating systems could then be transferred to the neighboring, or inshore, settings for industrial or human consumption.

Floating farms based on solar or photovoltaic panels (or renewable energy more generally) could offer many environmental and agricultural advantages, including, for example, the possibility to avoid or to reduce high temperature and sandy storm damages on desalination stations built on lands in dry regions where such hurdles can harm the desalination system by accumulating sand particles [32]. As for a deserthouse approach on land, the marine salts produced in floating farm systems could also be used as mineral fertilizers for floating farms themselves or sold to land farms to be used in optimal and safe proportions for their crops. In costal countries with arid climate conditions and limited land farming surfaces, floating farms in the sea would introduce new agricultural surfaces that would be limited only by the frontiers of marine international territories and thus would contribute to enhance food production systems and alleviate food insecurity issues if exploited at a grand scale. To make the floating farm systems as profitable as possible, high-valued crops or vegetables or rare medicinally efficient plants could be selected for growing in such floating farms [28]. Floating farms could also be conceptualized to produce crops in the upper floating part and fish or other marine products in the bottom submerged part as a symbiotic or an integrated aquaculture system in which the wastes produced of one part (the upper floating farm) could feed or benefit the population of the other bottom part (e.g., herbivorous fish). In this context, a two-phase polyculture system was proposed combining seaweeds and shrimp farms [33] and salmon [34]. Such integrated facilities would allow exploiting the large amounts of fish or aquaculture wastes (i.e., organic and inorganic phosphorus and nitrogen) as nutrients for seaweeds.

Besides the several potential advantages discussed above, floating farms also present some inconveniences, although they are manageable. An important shortcoming of sea floating farms is the overall budget to invest in, particularly at the first stages where multidisciplinary research and conceptions will be required between different and complementary fields. Floating farms also need to be designed in a robust and adaptable manner to all the potential seawater risks and conditions than their counterparts on land. Although a robust conception of floating farms is costly at the construction stage, it, however, would help saving ulterior maintenance costs invested in the transfer and replacement of materials to and from offshore sea-farm sites. A further downside of a floating photovoltaic-based farm system is that it fits costal sunny countries more than cloudy and cold regions. In cold regions, however, where arable lands are limiting factors and, hence, floating farms in the sea could represent an important extensible farming surface, other combined seawater desalination approaches can be used including, for example, multistage flash distillation, ion exchange, or membrane-based processes such as nanofiltration [35], electordialysis [36], reverse osmosis [37], or freeze desalination using hydraulic refrigerant compressors [38]. In such conditions, the conception of "floating greenhouses" might be more realistic.

Another potential pitfall of a floating farm system is the low-energy conversion rates that could be lower than the optimal thresholds required for full automation and functioning of the entire system. To palliate this challenge, different sources of renewable energy (i.e., solar, wind, and seawater itself) could be combined to increase the overall energy collected from the entire marine system. A combination between photovoltaic and wind energy hybrid system was recently reported to provide electric power [39]. In the same perspective, similar combinations could thus be envisioned for floating farm systems. Increasing the conversion rates and conductivity of photovoltaic panels and minimizing energy loss [40] are other potential venues to take into consideration in the design and experimentation of a floating farm approach. The conversion rates are, indeed, a matter of continual improvement. A table that summarizes the latest findings and conversion rates of solar cell prototypes is regularly updated [41], and it illustrates that the conversion rates are ameliorable over time. The conversion efficiency between the first and last generations of solar cells are clearly noticeable; the efficiency rate of first generations of photovoltaic cells was $<1\%$, but with time it has been significantly improved for different types of cells, including, for example, silicon crystalline cells with up to 25.6% [42], thin film with up to 28%, and close to 40% for multijunction devices [41]. As such, if the conversion efficiency continues to ameliorate at that rate, it could be doubled or triplicated in the future, making the use of solar energy an indispensable source for fixed or floating farm systems in either rich or poor countries. Costal marine regions in sunny and dry regions could then be occupied by large floating farm stations near the shore to feed people on the neighboring coastal lands and beyond.

15.3.2 Floating farm ships

Current large and ambitious projects undertaken in the sea to build, for example, floating cities and floating hotels suggest that building specific ships for farming purposes could also be envisioned [43]. Building farming ships could be indeed cheaper than building full floating cities or hotels in the sea because the construction of farming ships requires much less resources and infrastructure than constructing full floating cities on the sea surface.

An alternative option to the above potentially fixed adjustable floating farms could thus be to build a completely mobile floating farm in the form of a moveable ship as a floating greenhouse on the sea [43]. This is potentially plausible because greenhouse facilities and ship industry are well-advanced technologies in our days, so that it could be relatively easily conceivable to combine knowledge in both fields to build large boats or ships for farming purposes in the sea. Likewise in photovoltaic-based agriculture in the sea [28], the establishment of farming ships would help providing new important farming opportunities for food production and environmental sustainability. Building farming ships would simply need an adaptation of the currently available technologies used in the ship industry with a greenhouse operational system to be assembled in farming ships or farming boats.

Farming ships could offer many important environmental and sustainable advantages, such as to (1) reduce the burden on freshwater by using seawater desalination techniques, or by collecting and storing rainwater from rainy regions to irrigate plants growing on board, (2) introduce new cultivable surfaces particularly in arid and dry regions where arable lands and freshwater are scarce, and (3) potentially to provide complete and self-sufficient farming systems in terms of automated planting, harvesting, processing, and export; imagine a big farming boat of 300 m length by 200 m width and 50 m height and designed to hold 10 vertically tiered-cultivable layers that could provide about 60 ha of cultivable surface in $<0.06\,km^2$ of sea space (1 ha is a surface of $100\,m \times 100\,m$) [43]. This could be a profitable floating field with an economically and nutritionally important crop. The longest ship ever built in the history was the Seawise Giant (also known as Happy Giant, Jahre Viking, Knock Nevis, Oppama, and Mont) with about 460 m length.[4] A smaller and less cumbersome ship for floating farming might thus be conceivable to provide some important agriculture advantages on the sea surface.

Further advantages of farming ships may include that a farming ship could be planted, for example, with a crop species from a hot region, when it is anchored in a hot region with intense insolation, and then just before the crop matures or becomes harvestable or ready for consumption, the farming ship could move slowly to export the onboard crop to its destination timely [43]. A movability aspect of a farming system is indeed an important characteristic of the system (i.e., farming ship) as it would allow it to travel to a new and safer location in the case of danger or any constraining factor such as pollution or conflicts. Farming ships could also reduce the reliance on fossil fuels and cut off the emission of CO_2 significantly, when adopted at large scales with a renewable energy source to operate them (i.e., sun-rotatable photovoltaic panels, wind, and tide-generated energy).

Finally, farming ships could be a complementary solution to other alternative farming systems on land or on sea surface and they could contribute to solving some current and future societal, environmental, and agricultural issues related, for example, to land deterioration, drought, desertification, and sprawling cities [43]. Nonetheless, potential disadvantages may hamper the establishment of farming ship systems which include high costs, a suitability limited to herbaceous and/or small-sized plant species, and high-energy requirements to move giant ships in the sea. However, to palliate some of such inconveniences, farming ships could be anchored in suitable locations (i.e., sunny and safe locations) where it could harvest and store energy and then to move on only when it is necessary (i.e., to exporting the growing crop or to flee dangers). Ship assembly and greenhouse farming technologies are well-developed

[4] http://carnet-maritime.com/marine-marchande/demolition-du-knock-nevis-le-plus-grand-navire-au-monde.html.

conceptions that could be tailored to build or transform ships into a kind of floating fields or floating greenhouses with the least inconveniences and maximum advantages possible.

15.4 Breeding land-sea hybrid plant species irrigable with seawater

In biomedical research, interspecific hybridizations such as human-animal chimeras have been attempted for decades ago to create new hybrid cells with tissues from different species [44]. However, research on human-animal hybrids raises a number of controversial issues and bioethical concerns [45]. In plant research, on the other hand, potential interspecific hybridizations are less ethically controversial and relatively easy to obtain through asexual reproduction methods in plants. Since a long time ago, for example, farmers and agronomists have been using different grafting methods to breed new plant species with agronomical and nutritional characteristics from related species. With the availability of new genetic and biotechnological tools to manipulate the smallest heredity units (genes and nucleus), new interspecific or intergenus species could eventually be produced by using specific and controlled genetic combinations. As such, some land and sea plant species could be used to create new potential land-sea hybrid plant species that combine traits inherited from their land and sea parent species at the same time [46]. The most expected characteristic of such species is to be tolerant to harsh salinity as a trait inheritable from its sea parent, hence potentially irrigable with seawater. If a land plant species growing in dry regions is selected to be the other parent (e.g., tolerant to drought "xerophyte") the new expected land-sea hybrid species could also be potentially tolerant to seawater and drought simultaneously. Theoretically, such species could solve important agricultural and environmental challenges such as to feed animals and cattle in drylands, hence humans as a result, or to be used as a biofuel crop or as a soil-fixing plant to reduce dust and sandy storms in arid regions. The question, however, is how to get such highly valuable species, while avoiding the inconveniences of using seawater in irrigation that would increase the salinity of irrigated lands or underground water. This is what we are going to discuss in the next paragraphs.

Classical breeding approaches may fall short to create land-sea hybrid species, but genetic manipulations may contribute to the establishment of such species with potential interesting characteristics. Three approaches (Fig. 15.2) could be of a particular interest for experimentation toward such purpose; the first is a somatic fusion trial, the second is a nuclear transfer approach, and the third is a binucleation approach [46].

The somatic fusion or hybrid approach has been attempted since years ago to create some new species, including between sexually incompatible species [48]. Nuclear transfer in turn has also been effectively attempted with mixed success. The most famous example of nuclear transfer is the well-known sheep Dolly [49]. The same approach that was used to produce Dolly could thus be attempted between land and sea plant species to create a potential land-sea hybrid species by using nuclear transfer combinations between land and sea species [46]. Along with somatic hybrids

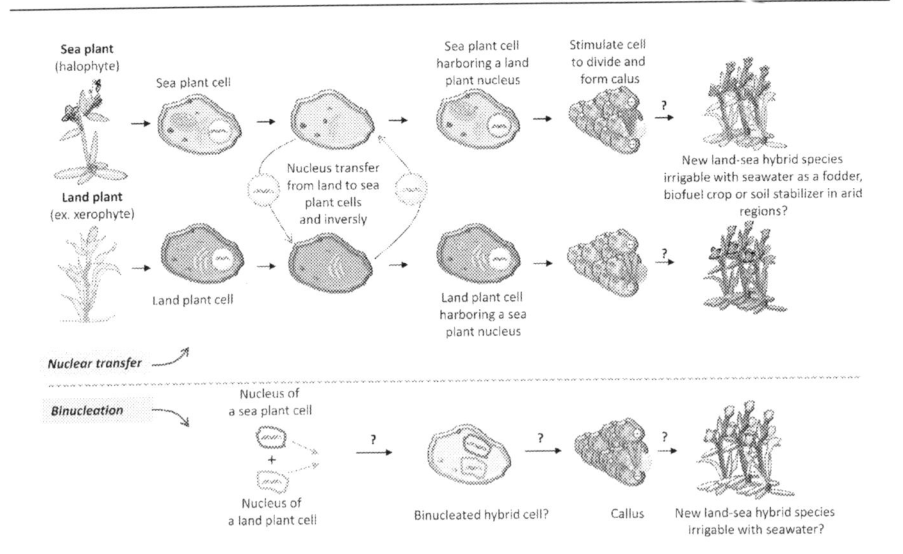

Fig. 15.2 A schema of potential approaches to breed new land-sea hybrid plant species potentially irrigable with seawater. Two plant species, one from land (i.e., xerophyte) and another from sea (halphyte) could be explored for the described approaches. *Top (nuclear transfer)*: the nucleus of a seawater plant (e.g., algae, moss, mangroove, etc.) could be transferred into the cell of a xerophytic plant, and inversely. The resulting cells will then be cultured in vitro to be induced to form calli from which entire plants could potentially be generated. *Bottom (binucleation)*: two nuclei from sea and land plant species could be assembled in one of their respective cells to produce binucleated cells. Then, the binucleated cells could be induced to form callus from which entire plants could be obtained and analyzed for their geno-phenotypes [46,47].
Reproduced with permission from Moustafa K. Binucleation to breed new plant species adaptable to their environments. Plant Signal Behav 2015;10:e1054586; Moustafa K. Toward breeding new land-sea plant hybrid species irrigable with seawater for dry regions. Plant Signal Behav 2015;10:e992744.

and nuclear transfer between sea and land species, a binucleation approach in plants (i.e., producing a plant cell with two nuclei), which has never been attempted before at the best of my knowledge, could be explored as a potential genetic approach to vary plant genomes and eventually produce new allopolyploid genotypes with new characteristics [47]. To generate binucleated cells in vitro, the nucleus of a plant species growing in sea or salt marshes could be attempted to introduce into a cell of a plant species growing in drylands (e.g., xerophyte). Then, the potentially obtained cell could be scrutinized for their genotypes and phenotypes after they are stimulated to divide and form callus [47]. There is a chance that a binucleation approach could produce some interesting genotypes and characteristics, but a failure is also expected although the method deserves to be tested in plants. In humans and rodents, binucleated cells are commonly inducible by different mutagenic and carcinogenic mechanisms [50,51]. In plants, binucleation could also be tried artificially to combine two different nuclei (from land and sea plant species) into one host cell. If successfully obtained, land-sea binucleated hybrid cells could offer some important advantages toward greening the desert using seawater.

Let us assume now that we have successfully obtained a new land-sea hybrid species irrigable with seawater. But the intriguing question is what to do with a seawater salinity of about 30–40 g/L, depending on its source,[5] which would increase the salinization risk of the irrigated lands and underground water if present nearby. To avoid such risks, irrigation with seawater could be restricted to the desert or barren regions that are already excluded from agriculture or farming activities. In such regions, the risk of salinization with seawater would matter less than in arable lands. In fact, large areas of the desert are already deemed unusable for agriculture either because of high salinity concentration or because of water scarcity, so a potential irrigation with seawater would not aggravate the problem more than it is naturally. On the contrary, using seawater to irrigate a plant species tolerating seawater conditions would transform drylands into productive areas. Large areas of the desert could then be transformed into fields looking like rice fields covered with salty water instead of freshwater. To reduce salinization risks, anyway, a more cautious solution might also be considered such as, for example, to isolate the lands intended to be irrigated with seawater with large containers as a kind of large pot ("seawater pot" or "seawater bed") (Fig. 15.3) manufactured of malleable and insoluble containers to hold soil that would be cultured with potential land-sea hybrid plant species irrigable with seawater, similar to indoor plant pots to grow ornamental plants or bonsai but at a large scale for an outdoor seawater farming.

Upon irrigation with seawater, however, salts would accumulate in the soil and on the walls of the "seawater pot" holding the irrigated soil. In this case, the salt could be recovered from the seawater bed by, for example, simple woody or plastic scraping tools and the soil might need specific treatments to remove the accumulating salts when they saturate the soil. The seawater pot could be manufactured or covered with electrochemical salt-concentrating layers that could accumulate salt ions more efficiently on its surface that could then be stripped out later, by the end of the growth season, for example, or when the soil gets saturated with salts.

However, the potential use of seawater pots will require excavation works to remove soil and to install the seawater bed, fill it with soil, and then to return it back before the irrigation with seawater and so on. However, if no underground water is nearby and if a land-sea hybrid plant species tolerating seawater is successfully developed, no specific treatments to remove salts would be required. In such cases, the lands intended to be irrigated with seawater could be seen as a field submerged with seawater or as a shallow salt marsh in which seawater-tolerant species could grow successfully without any particular treatment needed. Lands grown with such species would ultimately be transformed into more environmentally land-living places.

[5] http://www.lenntech.com/composition-seawater.htm.

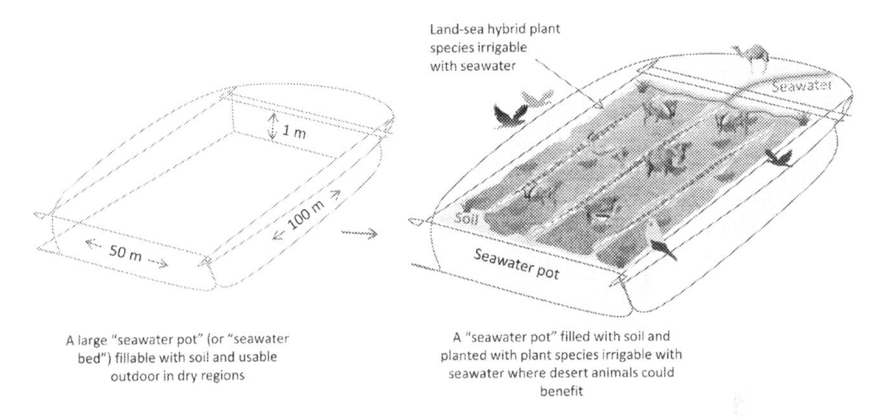

Fig. 15.3 A "seawater pot" or "seawater bed" (a large container) that could be filled with soil and planted with a herbaceous land-sea hybrid plant species potentially irrigable with seawater. The seawater pot could be seen as a type of "garden pot" scaled up to large dimensions to be used outdoor to prevent seawater from leaching out and thus reducing salinization risk to underground water or neighboring soil areas. The seawater pot could be manufactured from supple, impermeable, and insoluble material, for example, of 100m length by 50m width, and 1 m height and buried in the soil at the same level as the soil surface. In seawater irrigation, the pot would retain the seawater salts on its walls, which could then be recovered by specific remediation approaches to remove salts from the soil. The walls of the seawater pot could, for example, be covered with a layer of electrochemical layer that adsorbs and accumulates the salts to be recovered by simple scraping. In such cases, the seawater pot should be discharged and refilled with the soil and replanted in a way similar to a crop cycle approach.

15.5 Conclusion

Freshwater is currently scarce in many parts of the world. Due to the effects of climate change, the worse might come with more stressful and constraining conditions for humans, animal, and plant species. Under recurring severe droughts and low rainfall rates, the provision of freshwater for drinking and farming would be increasingly challenging in wide regions in the world, including Asia, Africa, and the Americas. The Sahara and MENA (Middle East and North Africa) regions are particularly known for their chronic penury of freshwater that might aggravate in the upcoming years under the expansion of urbanism and severity of climate change. Developing alternative freshwater resources and establishing new economic farming systems to sustain agriculture and to satisfy human needs should be the top interests and priorities of scientific programs and public policies. Among the potential solutions to explore in this direction is to reduce food and water wastes, to increase plant productivity, and to breed plant species potentially irrigable with seawater while establishing the economic irrigation systems to save freshwater.

Researchers and stakeholders should explore innovative solutions and set up comprehensive plans to address the problem of sustainability and water availability. The sun and sea are, respectively, inexhaustible sources for energy and water that could be approached smartly, for example, by breeding new land-sea hybrid plant species irrigable with seawater, developing floating farms on the sea surface or using solar energy to desalinate seawater naturally and sustainably to increase biodiversity and to improve life conditions in barren lands. Although such approaches cannot completely replace farming on land for all types of agriculture, they can be considered as complementary solutions to the shrinkage of arable land and/or to produce some valuable vegetables or crops. Large solar stills (seawater desert-houses or artificial seawater ponds) will undoubtedly help restoring biodiversity in dry regions and stabilize their endangered ecosystems. Establishing floating farms in the sea would also bring important environmental and sustainable advantages to face future climate challenges. Such approaches are worthy to explore and to invest in. They are particularly strategic for countries where the agriculture surface is dwindling, urbanism is crawling, and freshwater is rarefying under severe and chronic drought periods. The advantages of such systems could far overweigh their inconveniences at least from a long-term perspective.

References

[1] Despommier D. Farming up the city: the rise of urban vertical farms. Trends Biotechnol 2013;31:388–9.
[2] Moustafa K. Greening drylands with seawater easily and naturally. Trends Biotechnol 2017;35:189–91.
[3] Went FW. Plant growth under controlled conditions. II. Thermoperiodicity in growth and fruiting of the tomato. Am J Bot 1944;31:135–50.
[4] Fath HES. Solar distillation: a promising alternative for water provision with free energy, simple technology and a clean environment. Desalination 1998;116:45–56.
[5] Tzen E, Morris R. Renewable energy sources for desalination. Solar Energy 2003;75:375–9.
[6] Delyannis E. Historic background of desalination and renewable energies. Solar Energy 2003;75:357–66.
[7] Lepaute A. La Chaleur Solaire et ses Applications Industrielles. L'Astronomie 1883;2:197–9.
[8] Mouchot AB. (La) Chaleur solaire et ses applications industrielles. Gauthier-Villars; 1869.
[9] Sumner T. New desalination tech could help quench global thirst. Sci News 2016;190.
[10] Buenaventura A. A short history of desalination; 2016. Available from: http://www.theenergyofchange.com/short-history-of-desalination [Accessed].
[11] Jackson RD, Van Bavel C. Solar distillation of water from soil and plant materials: a simple desert survival technique. Science 1965;149:1377–9.
[12] Murase K, Tobata H, Ishikawa M, Toyama S. Experimental and numerical analysis of a tube-type networked solar still for desert technology. Desalination 2006;190:137–46.
[13] Murase K, Yamagishi Y, Iwashita Y, Sugino K. Development of a tube-type solar still equipped with heat accumulation for irrigation. Energy 2008;33:1711–8.
[14] Ali Samee M, Mirza UK, Majeed T, Ahmad N. Design and performance of a simple single basin solar still. Renew Sustain Energy Rev 2007;11:543–9.
[15] Velmurugan V, Gopalakrishnan M, Raghu R, Srithar K. Single basin solar still with fin for enhancing productivity. Energ Conver Manage 2008;49:2602–8.

[16] Gude VG, Nirmalakhandan N. Desalination at low temperatures and low pressures. Desalination 2009;244:239–47.
[17] Gude VG, Nirmalakhandan N, Deng S. Desalination using solar energy: towards sustainability. Energy 2011;36:78–85.
[18] Campana PE, Leduc S, Kim M, Olsson A, Zhang J, Liu J, et al. Suitable and optimal locations for implementing photovoltaic water pumping systems for grassland irrigation in China. Appl Energy 2017;185:1879–89.
[19] Glasnovic Z, Margeta J. A model for optimal sizing of photovoltaic irrigation water pumping systems. Solar Energy 2007;81:904–16.
[20] Glasnovic Z, Margeta J. Optimization of irrigation with photovoltaic pumping system. Water Resour Manag 2007;21:1277–97.
[21] Constantz J. Distillation irrigation: a low-energy process for coupling water purification and drip irrigation. Agric Water Manag 1989;15:253–64.
[22] Abu-Hijleh BAK. Enhanced solar still performance using water film cooling of the glass cover. Desalination 1996;107:235–44.
[23] Voropoulos K, Mathioulakis E, Belessiotis V. Experimental investigation of a solar still coupled with solar collectors. Desalination 2001;138:103–10.
[24] Lewis NS, Nocera DG. Powering the planet: chemical challenges in solar energy utilization. Proc Natl Acad Sci U S A 2006;103:15729–35.
[25] Lewis NS. Powering the planet. MRS Bull 2007;32:808–20.
[26] Qiblawey HM, Banat F. Solar thermal desalination technologies. Desalination 2008;220:633–44.
[27] Palenzuela P, Alarcón-Padilla D-C, Zaragoza G. Large-scale solar desalination by combination with CSP: techno-economic analysis of different options for the Mediterranean Sea and the Arabian Gulf. Desalination 2015;366:130–8.
[28] Moustafa K. Toward future photovoltaic-based agriculture in sea. Trends Biotechnol 2016;34:257–9.
[29] Group BMT Design and Technology. Floating desalination plant. Available from: http://www.bmtdesigntechnology.com.au/design-solutions/floating-desalination-plant/. Accessed 27 June 2017.
[30] Yano A, Onoe M, Nakata J. Prototype semi-transparent photovoltaic modules for greenhouse roof applications. Biosyst Eng 2014;122:62–73.
[31] Johan B. Pelagic solar still and method for supporting plant growth. Google Patents; 1957.
[32] Reif JH, Alhalabi W. Solar-thermal powered desalination: its significant challenges and potential. Renew Sustain Energy Rev 2015;48:152–65.
[33] Nelson SG, Glenn EP, Conn J, Moore D, Walsh T, Akutagawa M. Cultivation of *Gracilaria parvispora* (Rhodophyta) in shrimp-farm effluent ditches and floating cages in Hawaii: a two-phase polyculture system. Aquaculture 2001;193:239–48.
[34] Troell M, Halling C, Nilsson A, Buschmann AH, Kautsky N, Kautsky L. Integrated marine cultivation of *Gracilaria chilensis* (Gracilariales, Rhodophyta) and salmon cages for reduced environmental impact and increased economic output. Aquaculture 1997;156:45–61.
[35] Shahmansouri A, Bellona C. Nanofiltration technology in water treatment and reuse: applications and costs. Water Sci Technol 2015;71:309–19.
[36] Turek M. Cost effective electrodialytic seawater desalination. Desalination 2003;153:371–6.
[37] Coutinho de Paula E, Amaral MC. Extending the life-cycle of reverse osmosis membranes: a review. Waste Manag Res 2017, https://doi.org/10.1177/0734242X16684383.

[38] Rice W, Chau DSC. Freeze desalination using hydraulic refrigerant compressors. Desalination 1997;109:157–64.
[39] Samrat NH, Ahmad N, Choudhury IA, Taha Z. Technical study of a standalone photovoltaic-wind energy based hybrid power supply systems for island electrification in Malaysia. PLoS ONE 2015;10:e0130678.
[40] Subramani A, Badruzzaman M, Oppenheimer J, Jacangelo JG. Energy minimization strategies and renewable energy utilization for desalination: a review. Water Res 2011;45:1907–20.
[41] Green MA, Emery K, Hishikawa Y, Warta W, Dunlop ED. Solar cell efficiency tables (version 46). Prog Photovol: Res Appl 2015;23:805–12.
[42] Masuko K, Shigematsu M, Hashiguchi T, Fujishima D, Kai M, Yoshimura N, et al. Achievement of more than 25% conversion efficiency with crystalline silicon heterojunction solar cell. IEEE J Photovol 2014;4:1433–5.
[43] Moustafa K. Ships as future floating farm systems? Plant Signal Behav 2016, https://doi.org/10.1080/15592324.2016.1237330.
[44] Behringer RR. Human-animal chimeras in biomedical research. Cell Stem Cell 2007;1:259–62.
[45] Cabrera Trujillo LY, Engel-Glatter S. Human-animal chimera: a neuro driven discussion? Comparison of three leading European research countries. Sci Eng Ethics 2015;21:595–617.
[46] Moustafa K. Toward breeding new land-sea plant hybrid species irrigable with seawater for dry regions. Plant Signal Behav 2015;10:e992744.
[47] Moustafa K. Binucleation to breed new plant species adaptable to their environments. Plant Signal Behav 2015;10:e1054586.
[48] Laiq Ur R, Ahuja PS, Banerjee S. Fertile somatic hybrid between sexually incompatible *Hyoscyamus muticus* and *Hyoscyamus albus*. Plant Cell Rep 1994;13:537–40.
[49] Campbell KH. Nuclear equivalence, nuclear transfer, and the cell cycle. Cloning 1999;1:3–15.
[50] Xia L, Gu W, Zhang M, Chang Y-N, Chen K, Bai X, et al. Endocytosed nanoparticles hold endosomes and stimulate binucleated cells formation. Part Fibre Toxicol 2016;13:63.
[51] Yasui M, Kamoshita N, Nishimura T, Honma M. Mechanism of induction of binucleated cells by multiwalled carbon nanotubes as revealed by live-cell imaging analysis. Genes Environ 2015;37:6.

Index

Note: Page numbers followed by *f* indicate figures, and *t* indicate tables.

D

N

Q

R

S

Y

Z

Printed in the United States
By Bookmasters